System Requirements Analysis

System Requirements Analysis

Jeffrey O. Grady

AMSTERDAM • BOSTON • HEIDELBERG • LONDON
NEW YORK • OXFORD • PARIS • SAN DIEGO
SAN FRANCISCO • SINGAPORE • SYDNEY • TOKYO

Academic Press is an imprint of Elsevier

Academic Press is an imprint of Elsevier
30 Corporate Drive, Suite 400, Burlington, MA 01803, USA
525 B Street, Suite 1900, San Diego, California 92101-4495, USA
84 Theobald's Road, London WC1X 8RR, UK

This book is printed on acid-free paper. ⊚

Library of Congress Cataloging-in-Publication Data

Grady, Jeffrey O.
    System requirements analysis / Jeffrey O. Grady.
        p. cm.
    Includes bibliographical references and index.
    ISBN-13: 978-0-12-088514-5
    ISBN-10: 0-12-088514-X
    1. Systems engineering. 2. System analysis. I. Title.
    TA168.G66 2005
    620'.001'171–dc22
                            2005022277

British Library Cataloguing in Publication Data
A catalogue record for this book is available from the British Library

ISBN 13: 978-0-12-088514-5
ISBN 10: 0-12-088514-X

For all information on all Elsevier Academic Press publications
visit our Web site at www.books.elsevier.com

Printed in the United States of America
    06   07   08   09   10   9   8   7   6   5   4   3   2   1

Table of Contents

List of Illustrations

List of Tables

Preface

This volume was originally conceived in the late 1980s as a way to convert a series of company procedures on a process called system requirements analysis into a textbook. Previously, this subject had been covered extensively only in Department of Defense and aerospace company documents not generally available to the public at large. I had researched many of the representations of this process and found that, while they had in common a rigorous requirements capture methodology, they all omitted a central notion that left the user of the process uninformed about where the process began and where it ended. These several process descriptions also failed to clearly describe how it could be applied for interface, as well as architecture, development in a continuous top-down flow. Commonly, they also failed to show the need for feedback from higher level functional allocation and concept development to lower level functional analysis. Finally, the process coverage was all too often presented inflexibly and failed to show how the using organization could apply it in combination with other requirements analysis strategies across the complete time frame of a system life cycle. Process descriptions I was familiar with were heavy on the proper use of very stylized forms in support of computer data collection systems and weak on the connective tissue describing how the process fits into the overall development process and individual human activity.

At the time I was the Manager of Systems Development at General Dynamics Space Systems Division (1984–1993) and convinced that improvements in requirements analysis and specification development were the most important contribution to his new division's health that the department could mount. Convair had again in 1994 spawned a space division after the original Astronautics that had created the Atlas ICBM had been reabsorbed back into Convair with only a low level of space transport work to support itself. The principal work product of the department was specifications and it seemed to me its weakest link.

The corporation at the time encouraged managers to hire recent college graduates as a way of continuously feeding new knowledge into the organization and working our way out of an employee age hump. Few engineering managers, especially in system engineering, are enthusiastic about hiring recent college graduates. Many system engineering managers believe that other organizations within the company and other companies should act as their farm teams, ripening good system thinkers for the plucking. But, since it was corporate policy that was enforced, I looked for ways to make it work. I found that breaking the department charter down into pieces allowed documenting each of these micro jobs in company procedures. The small jobs had to be cut such that they covered a fulfilling task for the person performing it while accomplishing something important on a program. Each task had to also be small enough that new hires could master the job fairly quickly so as not to pose a budget burden while perfecting their skills. Over time through some training and assignment of these new hires to different tasks, they might become well-rounded system engineers.

This book was adapted from procedures that were also used as text material for the GDSS System Engineering Institute; here engineers were provided instruction in system engineering topics. This was another constraint on department instructions and corresponding task granularity. It had to be possible to describe each in a one-hour lunchtime class session. In 1990, the Director, Dave Clemons, and I, just back from the meeting in Seattle where we had participated in the creation of the National Council On Systems Engineering (then NCOSE, later INCOSE), felt so motivated that we went to the University of California, San Diego Extension, and encouraged the Engineering Director, Dr. John Peak, to start a system engineering certificate program. John accepted the encouragement and I agreed to teach the first course, system requirements analysis.

When so enthusiastically volunteering to teach the course, I had not thought about a textbook. About the time this problem became apparent, a housepainting accident resulted in two fractured arms. Wearing two slings and typing on an Apple IIE, I converted some of the company procedures mixed with new material into a textbook to fill a long-standing void in the area of requirements analysis. This manuscript matured for three years as a course text in the UC San Diego system engineering certificate program before its official publication as *System Requirements Analysis* by McGraw-Hill in 1993.

A subsequent version was crafted by updating the material in five books published in the 1990s into four volumes covering the whole sweep of the system development process from the system engineering perspective. This series is used in a four- and six-course system engineering certificate program offered by JOG System Engineering (see http://www.jogse.com for details).

Years of consulting, lecture work, and experience with many different kinds of companies in the aerospace industry have enabled me to focus on the cause of poor system development performance in four knowledge spaces seldom attacked with synergy: system engineering, functional management, domain engineering (mechanical, electrical of several kinds, software, etc.), and program management. Many companies have tried to improve with unipolar improvement ventures focused on one of these, such as system engineering performance, and all too often have come up short of their goals. The sad story is that they all have to be attacked together and in accordance with a grand plan. The common career competitiveness between the people staffing these activities has to be replaced with a common vision of success with clear knowledge that all of these people are inextricably linked in each other's success and dependent on the contributions of the others for their own success and that of the enterprise as a whole. They are all parts of the enterprise system with prescribed activities and interfaces that must function well for the whole to achieve expectations.

The functional managers must be constantly developing and improving the resources made available to all programs in the form of the best possible people who know what they are doing in the context of a written process description and a set of coordinated tools. The functional managers are the owners of the enterprise knowledge base needed by programs to translate customer needs into practical products satisfying those needs. They are also the owners of pieces of the process under the guidance of, in my view, an enterprise process agent referred to in this book as the Enterprise Integration Team (EIT).

Program managers must have the skill and knowledge to lead teams of specialists toward preplanned program goals but must also have the courage to spend program funds early in the program to accomplish needed system engineering work and the experience, knowledge, and self confidence to demand that this money be spent well achieving specific tasks leading to specific work products within time and budget constraints. If the program manager withholds this funding on the basis of fear of the future, his or her worst expectations will be fulfilled later in the program when realized risks overwhelm it. System engineering is a grand risk abatement process that uncovers problems at the earliest possible date, and it is well known that the earlier a problem is uncovered, the cheaper it is to resolve.

The system engineer must be a quick practitioner as well as a good one. The program manager must exercise a lot of courage to spend program money early and deserves a return on that investment with fast work well done. If this cannot be done then fear will sweep through engineering and program management, encouraging design being started before the problem is fully understood and documented in program specifications.

This book and three companion manuscripts form a grand system development series of books coordinated with the four fundamental sets of knowledge that must be applied to solve complex problems through the exercise of good engineering. We first should define the problem and we do this by writing specifications (this book). Next, the creative juices of the design engineers, aided by a whole host of other specialists, must synthesize those require-

ments into a preferred design solution documented well enough to support manufacture or purchase of the system elements. Finally, the system approach calls for proving that the design solution satisfies the original requirements that defined the problem we were trying to solve. The forth element of this picture is a management infrastructure within which these three activities must reside.

While using the original book, *System Requirements Analysis*, in training courses over a period of several years and talking with many practitioners, I found ways to expand the set of models useful in requirements analysis, including the many software analysis models I had earlier covered by pointing to the many good books in that field. More importantly, I found it possible to put the whole set of models into a framework called a general theory of structured analysis. The framework is admittedly a little thin; the work required to fully describe it is beyond the domain of this book.

For this book, the new notion of RAS-Complete was threaded through the traditional structured analysis process linking together better the analysis work, structured analysis models, specification template structures, and structured analysis work product capture.

This book can be used independently as a textbook for a one-semester/quarter course on requirements analysis in a college system engineering extension program or equivalent company course; used as a basis for a company requirements analysis process description; used as a desk reference by a professional system engineer; or simply be read by a system engineer, system analyst, or design specialist to gain an insight into ways to improve their professional job performance in this very difficult but understandable activity. The book will be used in concert with the other three volumes it was a part of (Grand Systems Management, Grand Systems Synthesis, and Grand Systems Verification) as the basis for four core courses of a system engineering certificate program for delivery through campus courses or in-house in a company.

Acknowledgments

My thanks for the encouragement I have received from the several managers and directors for whom I have worked at Teledyne Ryan Aeronautical and General Dynamics Space Systems Division over the 30 years required to accumulate and assimilate the initial material contained in this book. Many of the ideas and connections included were suggested or stimulated by engineers under my supervision in the process of doing requirements analysis work on different programs in which my employer was involved, and these insights were very valuable. Since much of the material covered by this book has previously been developed only within extensive tomes prepared by military and civilian employees of the Department of Defense, I am also indebted to the many now nameless engineers who contributed to the many SRA descriptions created in the past that have provided a platform on which to build.

Several professional system engineers and educators were kind enough to review the original *System Requirements Analysis* manuscript and, in the process, offered many very valuable suggestions for improvements. These inputs included detection of many typographical errors as well as substantive corrections in several areas and ideas for structural improvements. I am especially indebted to Mr. Barney Morais, President of Synergistic Applications, Inc. and leader of the writing team (as well as a contributing author) for the original Defense System Management College System Engineering Guide prepared by the System Engineering Directorate of the Lockheed Space Division, Mr. Elmer Peterson, President of INTEC, Mr. Wayne Wymore, President of SANDS and responsible for creating the excellent system engineering program at University of Arizona, and Professor Edgar O'Hair from Texas Tech.

Mr. Morais also supplied some material on FRAT that I used in this book and over a period of years has been a great sounding board for ideas potentially useful in explaining how this work may be better accomplished.

Despite the much-appreciated efforts of these gentlemen, some errors and controversial content still, no doubt, survived. These errors are of my own making through stubbornness and oversight and in no way should reflect badly on the reputations of the much appreciated colleagues identified above.

It is an unfortunate reality that one who teaches learns more than those who are taught even when the teacher does the job well. While lecturing in this subject in close to 50 classes, I have benefited from the experiences, knowledge, and questions of hundreds of students and I thank you all. While all of the individual names have not been retained by the author, they participated in classes at General Dynamics Space Systems Division, University of California San Diego, Berkeley, and Irvine campuses, Indiana-Purdue University at Fort Wayne, IN, Loral Aeronutronic, Boeing Defense and Space Division, Newport News Shipyard, Raytheon Systems, McDonnell Douglas C-17 Program, Interstate Electronics Corporation, Sandia National Laboratory, Naval Surface Warfare Center, Naval Undersea Warfare Center, BAE Systems, FAA, Beckman Coulter, Collins, NASA Centers, ITT Automotive, ITT Gilfillan, Lucas Varity, Loral Vought, several offices of the Scitor Corporation, General Motors Electromotive Division, IJ Case Corporation, Computing Devices International, Computing Devices Canada, Caterpillar, Booz, Allan, and Hamilton, and Orbital Science Corporation as well as many public classes for University of California San Diego, University of California Irvine, University of California Berkeley, Indiana-Purdue University, University of Alabama Huntsville, University Consortium for Continuing Education (UCCE), Applied Technology Institute (ATI), Professional Education Institute (PEI), and Technology Training Corporation (TTC) in hotels, and one-day tutorials for INCOSE international symposiums and various INCOSE chapters.

I am still most indebted to my wife Jane for enduring years of mission-oriented behavior on my part while working toward what I hope is accepted as a comprehensive description of this fascinating process called system requirements analysis.

JOG

1

Introduction

Contents

1.1

Introduction to System Requirements Analysis

Contents

1.1.1 The Human Foundation

This book explains a highly organized but flexible method for defining needed man-made systems in an environment of cooperative work by teams of specialized engineers and analysts. The method described is based on three principal axioms. The first axiom is that the human knowledge base is vastly larger than the maximum individual human capacity to make effective use of that knowledge. That is, humans have limitations that force us to specialize and form organized teams to cover the needed knowledge base to master difficult problems. Second, our complex needs for man-made systems involve difficult problems that must be broken into a series of smaller problems (driven by the limitations of individual knowledge and management span) and the solutions to these smaller problems must be stitched into a larger fabric. Finally, because individual specialized design engineers work on only a small piece of the solution to a large and complex problem, they must first understand the requirements for the target of their creative engineering design genius before executing a design solution. Others are depending on them to solve the agreed-upon problem so that their solution to their part of the whole works with synergism in the whole.

I focus on an organized method for identifying what a system should consist of to satisfy a given customer need and appropriate requirements for the things in a system as a prerequisite to design of the system elements. Before beginning our journey into understanding organized system development, we must first dispose of some fundamentals. In this chapter, we define a system, a requirement, and requirements analysis. Then, we look at alternative environments within which to accomplish requirements analysis work, different ways to organize work and people.

Some history of system development will add to your understanding of the need for the organized measures covered in this book. Finally, this chapter summarizes the content of the remaining chapters of this part and the other parts in the book. Information is also included about how you may use the content to advantage as a student or professional practitioner of system engineering.

1.1.2 What Is a System?

Many definitions of the word *system* have been offered, but in the broadest sense, any two or more objects interacting cooperatively to achieve some common goal, function, or purpose constitutes a system. The content of this book applies most appropriately to man-made systems that are organized collections of resources that interact synergistically via the interfaces connecting them to achieve preplanned goals in accordance with a predetermined plan or process. Generally, in these kinds of systems, no subset of the system resources operating independently can totally achieve the same system goal or purpose, because we tend to create them in a least-cost configuration. The key to system existence, and the superior performance of a system over an unorganized collection of independent objects, is the cooperative interaction among the multiple system resources via the interfaces that connect them. So any one system consists of entities interconnected by relationships.

There exist natural systems in our universe, such as the ultimate system, the universe itself, the climatic system on Earth, and the human circulatory system. These systems evolved through natural processes not requiring any human engineering activity. Natural systems can be characterized using techniques described in this book, but we must recognize one fundamental difference between natural and man-made systems. Natural systems are not designed, they simply must be described and understood.

This situation is changing, however. We are manipulating natural systems to an increasingly powerful extent, bordering on redesign on a small scale. As a result, there will likely be an increasing application in the future for organized system engineering methods in natural system fields like biotechnology, agronomy, weather, and aquifers. The requirements impact analysis approach discussed in this book in association with engineering change proposals may apply to these situations more than we would care to think about.

A TRW systems engineer working on the Yucca Mountain nuclear waste storage site told the author that the company first had to identify the degree of isolation provided between the stored material and the local aquifer offered by GFP before determining what man-made features were required. The author asked how government-furnished property (GFP) was involved and the engineer replied, "No, God-furnished property."

A man-made system is developed to achieve a preplanned goal or purpose. These systems require engineering development work to convert the preplanned goal or purpose into a practical solution composed of physical hardware elements that can be manufactured and assembled from available materials. Most often, they also include computer software and, commonly, human operators who interact with the hardware and software to guide the system toward its goal or purpose.

Figure 1.1-1(a) illustrates the ultimate abstraction for a system in the form of a single block that represents the complete system. It interacts with a system environment and itself to achieve the system function, goal, or purpose. The system environment consists of everything that exists, less the system itself, that influences the system. The system goal, stated in a customer need statement, is the requirement that corresponds to the system block in this diagram. It is the ultimate requirement for the system that can be mindlessly brought into existence by the allocation of the customer need to a thing called the *system*. We often speak of decomposition of the system need, and this is where the decomposition process starts, with the ultimate requirement.

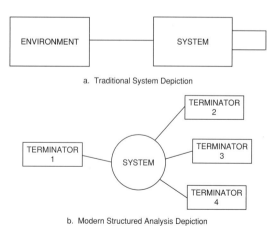

a. Traditional System Depiction

b. Modern Structured Analysis Depiction

Figure 1.1-1   *The ultimate system abstraction*

In a process the author calls *traditional structured analysis*, we progressively decompose or partition the functionality represented by the need into lower-tier functionality as a means of gaining insight into what the system and its parts must accomplish and how well it must perform. The decomposition process stops when we have identified all of the system resources that will yield to detailed design by a single design agent or team in the producing enterprise or can be procured from a single supplier.

In the late 1990s, the author, trying to impress a manager on the International Thermonuclear Experimental Reactor (ITER) program, then based in La Jolla, California, and thus encourage him to purchase a system engineering training program, showed the manager his schematic block diagram of the universe. The author, thinking that the universe included everything there was, illustrated only one block on the diagram containing everything rather than the two block arrangement shown in Figure 1.1-1. The manager looked at it briefly and said, "You may have forgotten a few wormholes."

Figure 1.1-1(b) offers the ultimate system view from the perspective of an adherent of modern structured analysis used for many years, and to this day, by some to develop computer software. The system, whatever it may become during the development process, is shown interacting with several (four in this case) external entities called *terminators*. This is a very useful diagram no matter what modeling approach one might employ. It can also be used to identify all the parties interested in the development effort, often called *stakeholders*.

This book provides methods for understanding the needed relationships between the product systems and the environment in terms of the environmental influences on the product as well as what is called *environmental impact*, which covers the negative influences on the natural environment by our system. The environment is actually much grander than the natural environment. In addition to the natural environment, we need to consider hostile systems effects, cooperative systems interfaces, and noncooperative systems influences. Also, certain aspects of some systems interact with the natural environment to feed back potentially damaging effects called *self-induced environmental effects*.

We must deal with one other kind of system while developing systems, the system that gives birth to the product system, the system of which we humans are members. There was a time when engineers believed that it was sufficient to optimize the product at the system level. We should recognize that we need to optimize at the aggregate system level, including the product and the process. In this book, we consider two components of this creator system: the processes, such as manufacturing, quality assurance, materials, and logistics, and the programmatic aspect, including the organizational arrangement, funding, scheduling, and management of the ongoing process. Some aspects of this process can be generically duplicated from one program to another, and the more of these the better, but others have to be created specifically for the program in question.

The term *grand systems* used in this book was chosen to denote the joint optimization of not only the product and process elements within any one program (no small achievement) but across all programs under development in one enterprise. It is a great truth that, whenever it is necessary to partition a large entity into a series of smaller ones, it is necessary to apply an integrating and optimizing force.

Further, it is necessary to apply this force at every level of work accomplished that focuses on the partitioned entity. This integration and optimization work is often the job of a system engineer.

1.1.3 What Is System Development?

Customers, the ultimate users of systems, or their procurement or acquisition agents, are capable of initially identifying the need for a new or significantly modified system only in the most grand terms. Through observation of their current products and processes or that of their competitors or adversaries, they can see that certain needs are not then being met by the systems available to them, their current needs are not being fulfilled, or there is a risk that in the future they will not be fulfilled. This identified void creates a need that can be characterized in very simple terms by the customer in what we call a *customer need statement*. Some examples of these need statements follow for three different kinds of systems:

- Space transport system. Transport assigned cargo from the Earth's surface to a low Earth orbit.
- Earth moving system. Move 2 acre-feet of naturally deployed soil to a location up to 1 mile away in 1 working day. Do this repeatedly.
- Weapons system. Place a high explosive on a selected target from a safe location and detonate that explosive to destroy the target.

The need statement is a simple paragraph of one or more sentences that clearly defines what the customer expects the system to do in terms of its influence on the natural environment or, more commonly, on other systems operating within and sharing the natural environment. The developer should not expect the customer to be able to initially express its need in precise engineering terms or even to completely understand what it needs. A discovery process must occur and the customer may need help in completing that process.

From the customer's perspective, the process of gaining access to systems that satisfy a need statement is called the *system acquisition process*. This is a management process that creates a contract relationship between the customer, or acquirer, and a system development company and manages the development process in accordance with the contract. The desired result is for the development company, supplier, or contractor to deliver the required product on time and at an agreed price that fully satisfies the need.

System development is the contractor's organized, creative, technical, and management process for translating a customer need into a clear definition of a solution and the production, test, integration of components, and delivery of that solution in a timely way within the cost guidelines defined in the contract to satisfy a customer's desired date of operational capability (called an *initial operating capability*, or IOC, by the U.S. Department of Defense) with a complete operating system that satisfies the need. So, the system development process is the contractor's response to the customer's system acquisition process.

Development of systems for commercial purposes may require a somewhat different vocabulary, in that the customer is commonly not clearly defined in a personal sense or concentrated at the time that the developing company begins work and may not be defined until people make their individual purchase decisions in the marketplace.

Commercial system developers for things like automobiles, toothpaste, and lawn mowers must develop their own understanding of what the customer's needs are and craft a product that, combined with effective advertising and a good reputation, stimulates people to purchase their product.

1.1.4 The Fundamental System Relation

Many attempts have been made to identify the essence of a system. The model offered here is no better than any other, only the one that the author of the book has used as a frame of reference to sort out what is important in a requirements analysis process description. The model uses elementary set theory but is not intended as a serious mathematical system through which systems are defined and designed. It has exerted some influences we probably are not aware of in shaping this requirements analysis process description. Ideally, this influence has been in a positive direction. But, if, in the reader's opinion, this is not the case, this brief explanation of the author's mental model of system development may be helpful in understanding why the author went astray from the reader's expectations.

While a field engineer operating on the opposite side of the world from his company, often alone, the author found it necessary to carry a lot of system information in his head. It was not convenient to carry several volumes while flying operational missions as a technical advisor on unmanned reconnaissance aircraft launch missions staged from a DC-130 during the Vietnam War. While it was not possible to remember all the details of the operations and maintenance processes for any given model, it was possible to remember a framework into which things fit, so that, when questions were posed by the customer about particular operations and maintenance practices, answers came to mind as an extension of his economical system context. Some of this information was also classified and could not be easily carried about in paper form. The model exposed here grew out of that experience as a means to quickly identify important information about new systems and organize this information, minimizing its volume in a model to conform to the author's normal limited memory capacity.

Are there some fundamental entities we can use as a basis for the development of systems? Can we identify some finite list of system descriptors from which all other descriptors can be expanded? This book offers five such descriptors to which every other characteristic of a system can be linked. Systems consist of things and relations among those things, and our definition of a system must absolutely establish whether or not any given object is in the system. If it is not, it is in the system's environment. If it is in the system environment, it will either exert an influence on the system or not. If it does not, it can be disregarded. The things that compose a system can be organized into a hierarchical tree structure. The complete set of things that form a system when organized in this manner are referred to as the *system architecture*. This always begins with the block illustrated in Figure 1.1-1 called the *system*. The system comprises two or more subordinate elements. Each of these is composed of two or more subelements. And so on down through the hierarchical system structure to detailed parts and materials. So, architecture is one of these fundamental system descriptors. We will see momentarily why it is necessary to organize the system architecture in this hierarchical fashion.

The reader should be forewarned that many people use the word *architecture* in a much broader context to include not only what the system is physically composed of but also the scenarios in which this material provides its benefits.

The things, parts, or objects of the system must interact in useful ways for the elements to qualify as a system. The medium for this interaction is called an *interface*. Each interface element is characterized by two terminals and some means of connecting them in the intervening medium. An interface is completed either via an element of the system (a wire harness or fluid plumbing run, for example) or via some characteristic of the system environment (a radio signal from a ground station through the atmosphere and outer space to a satellite, for example). These interfaces are a second fundamental descriptor. Each interface element maps to a pair of terminals, at least one of which is an element in the system architecture. Each interface element is completed between these terminals via some kind of media that is part of the system or its environment.

The architecture elements interact via interfaces with each other and with the system environment, which may include other competing (hostile), independent (noncooperating), or cooperating systems, to achieve the system goals. The environment is a third descriptor, although not a part of the system itself. The environment includes space, time, some subset of the natural environment, a possible hostile environmental element as well as cooperative and noncooperative environmental elements. Environmental influences could be treated as interfaces (in fact, cooperative ones are treated that way) but they are generally singled out as a separate category.

A system is intended to satisfy predefined goals or functions, which form the fourth fundamental descriptor. The highest-level function is the customer need, and this need is also the ultimate system requirement. All other requirements for the system and its parts, in theory, are or should be derived from it. Any requirement not traceable to the need may be a source of unnecessary cost.

Finally, a man-made system should have a prescribed process for operation of the system, and this process is the fifth system descriptor. It includes all the planned steps involved in operation and maintenance of the system.

For an existing system that is actually in operation in its normal environment, we can see the architecture elements interacting with themselves and the system environment via interfaces in accordance with a predefined process to achieve the system goal (satisfy the system need) or function. For new systems we wish to create, we need a model of this vision to make it perfectly clear to everyone working to develop the new system. Figure 1.1-2 illustrates a simplistic set-theoretic approach to understanding the relationship between these entities. If we imagine the architecture, interface, and environmental elements as sets in a mathematical sense, each consisting of many elements, then we can create a mathematical construct for the cross product (x) of the power sets of these entities ($\mathbf{A^* x\ I^* x\ E^*}$). Since a power set (such as $\mathbf{A^*}$) contains every possible subset for a set of elements, including the null and all the elements, we can be certain that at least some of these subsets contain useful combinations of system elements and every useful combination is in the power set. This cross product of the power sets includes every possible combination of system architecture, interface, and environmental elements that could be useful in completing every system process. This set also contains a lot of useless collections.

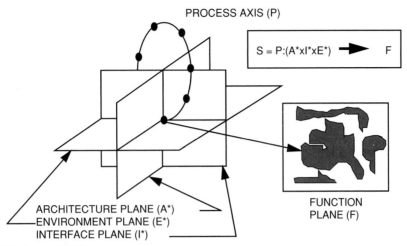

Figure 1.1-2   *The fundamental system relation*

System processes call up the system elements, so let us use the system processes as a relation that maps the cross product of the power sets of architecture, interface, and environment to the system goals or functions. As the system process axis flows through the intersection of the three planes illustrated, it can be thought of as sequentially aligning the three power set planes to pick up the trio of subsets required to achieve a particular system function.

For a new system not yet defined, an unprecedented system, we do not have this full picture, as we do for an existing system, put into the context of Figure 1.1-2. In the most extreme case, we may know only the top-level system function, the need. Our problem in system development is to somehow convert this system need into the full picture of a set of resources that interact among themselves and the environment via specific interfaces in accordance with prescribed processes to achieve the system function. Many systems operate in the cyclical fashion illustrated here but not all. The reason for the cyclical nature of many systems is reuseability of some major element in the system in the interest of economy. In any case, we can say that the system has satisfied its function if, in N cycles of the process axis, the function set is covered (every element of the function set has been satisfied). Thus, we can say that a system is defined as the sequenced process set that maps the cross product of architecture, interface, and environmental power sets to a function set.

With this diagram for reference, it appears that we develop systems in a kind of inverted way. We begin with the ultimate function and work back toward the details of how that function may be satisfied in terms of a set of architecture elements (resources) interacting in a described environment via interfaces in accordance with a planned process. The system development process guides us along this backward road, working from the general toward the specific, dividing up large problems into small ones, solving those small problems, and integrating those small problem solutions to optimize at the system level.

If we make the correct choices during this problem decomposition process, the solutions to these little problems come together in a synergistic way to cover the system's function set and fully satisfy the customer. If we make bad choices, the result almost certainly is some combination of a customer need unfulfilled (function set not covered), unnecessary cost, and late delivery.

1.1.5 What Is System Requirements Analysis?

Requirements are necessary attributes defined for an item prior to efforts to develop a design for the item. System requirements analysis is a structured, or organized, methodology for identifying an appropriate set of resources to satisfy a system need and the requirements for those resources that provide a sound basis for the design or selection of those resources. It acts as a transformation between the customer's system need and the design concept energized by the organized application of engineering talent. The basic process decomposes a statement of customer need through a systematic exposition of what the system must do to satisfy that need. The need is the ultimate system requirement from which all other requirements and the designs flow.

Decomposition is central to the process because we humans and our design organizations have limitations. Many years ago, an engineer, unsupported by specialized engineers, could perform the complete design task for simple systems. To the extent that reliability was a consideration, the design engineer was responsible. Safety features were either present or not in the design as a function of the design engineer's knowledge of what contributes to product safety. The design engineer was a generalist in a broad field, such as mechanical engineering. Systems could be designed that appealed to a fairly narrow range of engineering knowledge, such as mechanical systems or electrical systems, rather than the systems we commonly see now that include a rich mixture of mechanics, electronics, hydraulics and fluid dynamics, aerodynamics, computer software, and many other fields.

Several factors have combined since during the Second World War to change all this, continuing an evolutionary trend to which humankind is no stranger. The amount of knowledge available to us has exploded and our ability to access that expanded knowledge has improved through books and computer networks. Competition has forced us to specialize to take maximum advantage of available knowledge because of our individual limitations. The explosion in available knowledge and competition have made it impossible for large numbers of engineers to take full advantage of large tracts of knowledge individually in an efficient way. We have had to specialize to stay in the game and work toward a competitive advantage.

The mistake that some of us made during the 1960s through 1990s of the last century was to allow walls to be erected between the specialized organizational functions and a serial work pattern to develop. This pattern is often called *transom engineering*, colorfully creating the vision of one specialized engineer throwing the product of his or her work over the transom to the next specialist in line, each isolated from the other by office walls and closed doors. To counter this trend, we need to solve the difficult communications problems that specialization forces upon us.

When a single engineer could effectively apply all the related available engineering knowledge to a problem, there was no communication problem. Competition forces us to find ways to recreate the equivalent of this single all-knowing engineer from a group of specialists who together have mastered a much more extensive knowledge base than the single, more-general engineer of times past. *Concurrent* and *simultaneous engineering* are two terms that came into fashion in the late 1980s to describe a solution for this problem, which never should have been allowed to develop. System engineering is, and always has been, precisely what the advocates of concurrent and simultaneous engineering claimed. The systems approach, unfortunately, has not always been applied with intelligence, leading to some dissatisfaction with the term *systems engineering*.

It is very important that, in reading or using this book, you recognize that the principles described here must be applied by a team of people whose work is coordinated toward common goals by a knowing and decisive management team. Neither the system development/engineering process nor the subordinate system requirements analysis (SRA) process can be applied effectively by isolated engineers or autonomous groups of people. We say more about this throughout the book as a means of emphasizing the importance of team building in applying the techniques described.

Specialization and competition are the principal forces that drive the need to decompose a customer need into a series of smaller problems, each of which yields to a particular kind of engineering expertise. Since we must break up a large problem into many smaller ones, we must find a way to ensure that all these smaller solutions come together to solve the large problem associated with the whole. The method most commonly described for accomplishing this is to determine the necessary characteristics (requirements) for each system element before we begin the design work, then to analyze and test the product of the design and manufacturing processes for compliance with the predetermined requirements.

Complex systems consist of many elements that must be organized into families, each arranged in a hierarchy, because complex systems commonly call for a diverse appeal to technology requiring many different kinds of specialists. These families and elements must be assigned to engineering teams, organizations, or companies specialized in the design of particular kinds of things. Each of these elements must be designed to interplay with the others in the system and the system environment to satisfy the system need.

So, we decompose because we are organized into specialized engineering organizations driven by competition and our limitations and because we are limited in the scope and complexity (breadth and depth) of a problem that any one person can master. Because we are specialized, we must clearly define the requirements for the elements in a public way to ensure that the elements designed by the specialized engineers work together to satisfy the customer need when they are assembled into the system.

The system development process most likely to succeed has evolved from many years of experience. It involves a three-step process accomplished within an infrastructure of sound technical management. The first step in that process is to define the problem as a prerequisite to solving it. Any given detailed design effort should be preceded by the release of a specification that contains all of the essential characteristics and nothing else. The second step involves solving the problem defined in the specification, referred to as *synthesis*. This step is commonly broken into three substeps: creating an engineering design solution, translating that solution into a list of suppliers of materials needed to manufacture and assemble the solution, and the manufacture of the solution. The final step in the process calls for us to prove that the design we select actually does solve the problem defined in the specification. This third step is referred to by the author and EIA 632 as *verification*.

Requirements for a system or an element in a system are captured in a specification. Requirements analysis is the process through which we gain an insight into the proper content of the specification. Requirements analysis is the central component of system requirements analysis, but it is only one component. The other principal components are structured decomposition, architecture synthesis, and specification generation. In Chapter 2, we explore this complete process more fully. This book covers all four components of the requirements process.

The U.S. Department of Defense (DoD) and other customers for large, complex systems, like the National Aeronautics and Space Administration (NASA) and the Federal Aviation Administration (FAA), over the years, evolved very organized system acquisition processes involving formal decision points and progressive iteration and refinement of the system solution. These processes permit a controlled expenditure of funds based on several key decision points, such as an agreement on the system requirements and the system design concept, review and approval of a preliminary and detailed design, a commitment to build sufficient resources to conduct item and system testing, and a commitment to enter full-scale production for delivery of operational systems.

This whole acquisition arrangement is driven by an interest in managing risk. To the extent that a development organization is not concerned about the possibility that unforeseen problems will arise during the development program increasing cost, reducing product capabilities, or extending the period prior to product availability, then it may not be necessary to accomplish all the steps in the full process. The many useful components of this process were worked out over years of good and bad experience in the development of complex systems that can be selectively applied with positive results. The developer should make conscious decisions about the parts of the whole process that will be applied on a particular program based on how much risk he or she can stand.

1.1.6 System Requirements Analysis Timing Considerations

Figure 1.1-3 illustrates an enterprise system development master model showing the front end of the life cycle for any system an enterprise may develop. The system requirements analysis activity must be managed to interact with the design concept development activity early in the development of a system. This interaction takes place within the task illustrated in Figure 1.1-3, which we later refer to as life cycle function F41, Define System. Interaction between these two tracks occurs on what can be called the *system development downstroke*. In later program phases, the principal system engineering thrust is on an upstroke of the development process, where we focus on detailed design, integration, production, assembly, and verification. We focus on system decomposition, requirements analysis, and concept definition on the downstroke and system design, integration, and testing on the upstroke. We decompose and optimize on the downstroke to determine of what the system should be composed and the requirements for system elements. On the upstroke, we synthesize the requirements into designs then manufacture, assemble, and test the pieces to form a physical reality, the system, in a deliverable state.

On the downstroke, the nature of the system requirements analysis activity undergoes a change over these development phases. Initially, it is most concerned with system-level requirements and the decomposition of the system need into component parts that yield to design by particular specialized engineering departments, cross-functional teams, or supplier companies. In this early phase through completion of preliminary design work, the system requirements analysis activity focuses on identification of the technical requirements for the system elements that define what the elements

must do and how well they must do it as a basis for design and item qualification for the application.

Later, the requirements analysis emphasis switches to identification of requirements for the acceptance of the physical product elements by the customer and logistics support of the deployed system. It has long been known that manufacturing and logistics requirements must be embedded in the design as early as possible, because it simply costs less to do it that way and the customer derives better value in the resultant product. We encourage concurrence in these processes with the product design effort, but at the heart of the design effort, there is and always will be a creative product design engineer who must synthesize the many input requirements received from other disciplines into a design concept as a prerequisite to some of the analytical work performed in the manufacturing and logistics support areas.

Unfortunately, the product designer's contribution in the system development process almost becomes lost in the discussion of an organized development process, but we must never forget that we do all this organized work to provide the specialized designer or integrated design team a safe territory within which to apply their creative genius in search of an acceptable design concept or solution to fragments of the global system development problem. At the very hub of all of this structure is a corps of talented design engineers who translate today's science and technology into practical solutions to problems characterized by sets of requirements known to be consistent with solving the customer's need for the system.

On the downstroke, we decompose the customer's need into families of things that satisfy the need. In this book, one of the methods covered emphasizes the use of an activity called *functional analysis* to decompose needed functionality into the elements that constitute the system. The functions become primitive performance requirements, telling what the system must do, and these requirements are allocated to the things. The development of concepts for the items identified through this process should follow behind

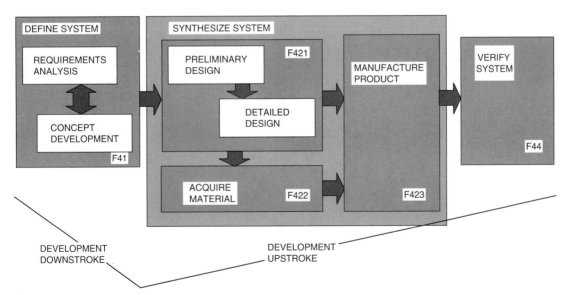

Figure 1.1-3 *System development strokes*

in time, but not too far behind, the identification of performance requirements and allocation of them to system elements. The desired result is that these two activities proceed interactively with the functional analysis leading.

When the timing between these activities is well coordinated by the system engineering function, lower-tier functional analysis takes advantage of higher-level performance requirements allocations and resultant architecture concepts. This results in identification of functions that are progressively more clearly related to the preferred system concept as the analysis progresses. A functional flow diagram, graphically depicting how the functions are sequentially related, begins as a means of gaining insight into what the system must do, matures into a system process diagram focused on the specific resources to which the functionality was allocated. What begins as a functional analysis should gradually evolve into a process analysis fueling the need to identify logistics support requirements.

If the functional analysis progresses too far ahead of the functional allocation, then there is a danger that the lower-tier functions will be difficult to relate to the evolving architecture and concept. If the concept development activity gets too far ahead of the functional analysis activity or is accomplished without having accomplished this organizing activity, there is a danger the specialized design engineers will apply available point designs that may not satisfy all system needs or take advantage of the latest technology.

The allocation of performance requirements to architecture elements is the triggering step in the concept synthesis process, and through this organized methodology, the fundamental twin system development tracks of requirements analysis and concept development interact systematically. Subsequent synthesis steps assemble the allocated performance requirements into system elements depicted in a system architecture block diagram. Concept development responsibilities are then identified for each element in the architecture.

Responsible persons in the design community initiate action to develop alternative concepts for assigned system architecture elements, which satisfy the requirements listed for that element. These alternative concepts are subjected to a trade study, where no clear preferred alternative is easily identifiable or predefined by the customer or current product line. The concepts, approved in an engineering decision-making forum, are included within a baseline concept definition documentation keyed to the corresponding system element architecture identification. During the period when engineering and manufacturing development occurs, these concepts are more fully defined in engineering drawings prepared in preliminary and detailed design activities.

This cyclical process is repeated, working downward through the expanding system functional definition allocating the functions to system elements in the architecture (elements that already exist in the architecture or elements that must be added). The functional analysis process provides a systematic way of understanding the system and subelement functions and their sequential relationships. The architecture conclusions that result are linked into family structures in the architecture synthesis that provides a framework for the expanding definition of the system synthesis.

Often when engineers discuss the top-down structured development approach, they conclude that a chicken-and-egg problem is buried somewhere within it. Management personnel responsible for cost and schedule performance can easily arrive at the conclusion that the process has no beginning and no end, only an endless, expensive middle. Unfortunately, most written descriptions of the process do not supply the reader a clear understanding of a beginning and an end. In the previous paragraphs, we provided the beginning and end definition. Let us now attempt to make it more clear. Do not leave this chapter before you understand the next paragraph in the context of ideas developed to this point.

The process begins with the customer need, the ultimate requirement that applies to the top element in the system, the system itself, generally phrased by an ultimate user. The need is allocated to something called the *system* without any thinking required. The process ends when the decomposition activity has identified needed system functionality and allocated it to a series of elements along the lower fringe of the system architecture, all of which will surrender to preliminary and detail design by a particular specialized engineering group or are to be purchased from some outside supplier in accordance with an approved set of requirements. We now need to understand what happens in between and how to make it happen within the cost and schedule constraints of a program. Most of the remainder of the book focuses on those goals.

1.1.7 Development Approaches

We may select from several very organized and well-recognized structured development approaches. Alternatively, it is conceivable that an unstructured approach could be used where every design agent operates autonomously. The dangers of the autonomous, or ad hoc, approach are too obvious to belabor, but it is not all that uncommon for engineering organizations apply this approach. This book focuses on a top-down structured approach applied throughout the development process and system hierarchy. In this approach, the customer need is decomposed, working from the general to the specific, from the system level down toward the parts, materials, and processes level, from the simple toward the complex.

It is possible to develop a system from the bottom up, with a heavy emphasis on integration of many component solutions that may or may not be completely compatible. It is also possible, and not all that uncommon, to employ a middle-out approach, where a particular collection of principal elements are organized to satisfy the customer need. This is followed by organized work upward to integrate these major elements into a system concept and downward to expand the design concept for each major element. Outside-in is also possible, focusing on external influences as a beginning.

While this book focuses on the pure top-down approach, in reality we commonly are forced to apply a combination of all four approaches noted. The customer has some resources it wishes to apply from the residium of older systems that no longer serve its needs in their entirety. The customer, in this case, specifies that the new system must use these resources (government- or customer-furnished property), if at all possible. The developing contractor, interested in applying a top-down approach, must do so within an environment where some architecture elements are predetermined. It may be necessary to work from the middle out to integrate these items into the new system elements.

It is also possible that some elements of a program encourage a bottom-up approach, because no one has a clue how to satisfy an implied need. A rapid prototyping approach, working on several different fragments, some of which may work together, can sometimes lurch to an acceptable answer faster than the highly organized top-down approach. This requires very creative, very lucky people. When it works, it is a lot of fun for the participants. It commonly does not work, and there is seldom any humor surrounding the aftermath.

The structured approach offered in this book was developed many years ago and has evolved primarily in the context of large military weapons systems programs. Many people have concluded that this process is too cumbersome and expensive to apply at all, much less to apply to the development of small-scale systems. Alternative approaches have been developed featuring small variations on the themes expressed here. Other solutions have migrated back from software methods that originally evolved from general system technology. The author maintains that the traditional structured development approach with the few refinements offered in this book remains a very powerful tool that can be profitably applied to the development of systems over a wide range of size and complexity. This process can be applied intelligently, and it can be applied badly. Ideally, this book will help increase the number of people who can apply it intelligently.

It will be interesting to see in coming years how expert, parallel, and super computer systems as well, as a clearer understanding of what is now called *chaos*, will influence our view of the optimum system development process. There may be better models used by nature for the evolution of systems—ones we currently do not understand. We may gain insight into new approaches from research into mathematical chaos. Would it not be ironic if the science of chaos provides the underpinnings for an improved general system technology, since many people have for years suffered the chaos of their company's attempts to apply the traditional system approach. Whatever the future may bring, for now, the traditional structured development approach is still the most powerful mechanism man has created for transforming complex needs into cost-effective, timely solutions.

1.1.8 Degree of Precedence Alternatives

This book focuses primarily on the solution of an unprecedented problem, but this is only one system development situation to be faced by the developer. The problem posed by the customer could be one of improving or modifying some existing system, which is in operation and has some history of use. This book follows the notion attributed to the architect Louis Sullivan, that form follows function, suggesting we should first understand what the system must do then determine how to compose it. This approach works very well for unprecedented problems but is not necessarily the most-efficient way to approach heavily precedented problems characterized by an existing system operating observably in the real world. In Chapter 1.3, an alternative approach for structured analysis is offered for this situation.

1.1.9 Organizational Alternatives

It has been implied that teamwork is necessary to accomplish system requirements analysis with good results, and this is very true. Some companies have system engineering organizations that perform this task. Some companies distribute the task throughout the organization, require all engineering activities to employ a sound system engineering process, and require all organizations to cooperate in a more or less common system requirements analysis process. Some companies do this job well however they are organized but many do not.

We might ask ourselves, "Is there a best way to organize to attain and maintain a world class system requirements analysis capability?" Traditionally, American industry has organized in one of three ways: functionally, projectized, or matrixed. In functionally organized companies, employees work collocated by function, such as engineering, manufacturing, and finance, on one or more company programs. A department supervisor assigns work from the programs to his or her workers to meet all program schedules. In a projectized company, the employees work collocated in a program environment. The program does the hiring and releasing of employees, which tends to happen quite often.

The matrix organization was conceived as a way to blend the strengths of functional and projectized management. In a matrix, all the personnel are members of functional departments from which they are temporarily assigned to programs based on the changing business situation. Day-to-day supervision is performed by project management. When a program no longer needs the services of an engineer, he or she returns to the functional organization for reassignment. Each functional organization maintains common standards for its charter tasks across all programs.

There are advantages and disadvantages to each of these methods of organization, depending on many factors. The reader, no doubt, is familiar with these from other studies or experience. The question is whether one of these or some other organizational pattern is superior in encouraging excellent performance of the requirements analysis process or the complete system development process? The author's opinion is that it really does not matter how your company is organized, so long as the people who perform the development tasks on a project can be effectively organized into physically collocated concurrent engineering teams oriented about the way the product is organized.

In this book, we encourage and assume that the generic company uses some form of matrix management structure to provide employees with an administrative and technical home, a basis for continuous process improvement of standard methods, an efficient transition between project assignments, and project-oriented day-to-day management. The project organization may be structured in any reasonable way, so long as, at one or more levels in the product system organization, project personnel may cooperate to concurrently develop the requirements, design the product, and develop the production process and deployed system operational and support concepts.

Figure 1.1-4 illustrates a typical matrix structure for a large aerospace company or division that includes a functional system engineering organization that embraces specialty engineering, system analysis, and the core system development body (requirements analysis, system integration, and interface development). Engineering also includes a functional director for each of the major design disciplines shown down the left margin. Program managers working with a program chief engineer, the program integration team (PIT) leader in Figure 1.1-4, manage the development

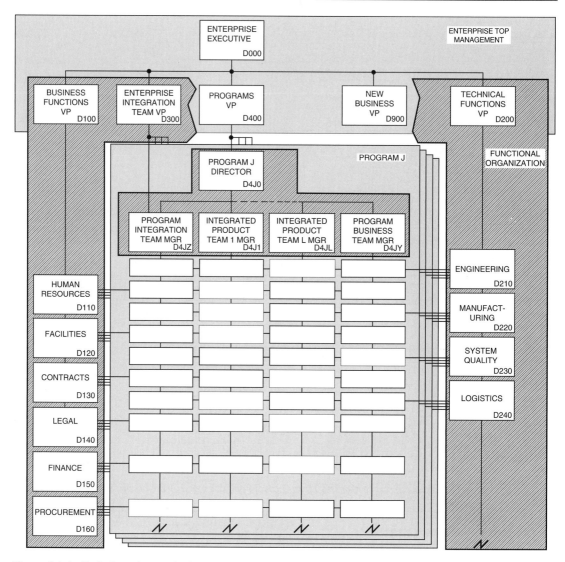

Figure 1.1-4 *Typical matrix organization*

work of the product teams assigned in accordance with a program plan. The understanding is that all the program personnel in each team case are physically collocated by program. This is the model we use throughout this book.

We accept that the program chief engineer is allowed to organize assigned personnel from the many functional departments according to how the product system architecture matures, that he has the authority to apply these specialized personnel in teams to the development of the evolving architecture. He assigns system engineers as the principal engineers to lead the development work for high-level architecture items, above the level where real design work occurs. He assigns good engineers with leadership qualities from wherever he can get them to lead concurrent development teams for major system items that require a primary appeal by one principal discipline. He

assigns a design specialist as the principal engineer for lower-tier items.

In each of these cases across the depth of the system architecture, the assigned team leaders must collect a team of program personnel expert in their field and appropriate to the object of their work and mold them into an effective development team. In the previous paragraph, the chief engineer was characterized by masculine terms out of convenience. The fact is that the system development process should be an equal opportunity situation. There is no evidence to support the notion that people of any particular race, color, creed, religion, or sex offer any advantage in performing or managing this process. What is needed is experience and knowledge and the ability to use them effectively, creativity, energy, an acceptance of a need for discipline, and an ability to work effectively with other people.

Concurrent engineering is a term that struck a resonance in industry in the late 1980s and early 1990s. The concept is characterized by physically collocated cross-functional teams of engineers from many disciplines working together to develop the requirements, product and production process concepts, product and production process designs, and logistics solutions. Most experienced system engineers and managers find very little new in the concept, since it is, in reality, simply a rebirth of the system engineering approach practiced rather poorly at many companies at the time, perhaps. Some never left it. Unfortunately many did or never found it in the first place. In some engineering organizations, the work is still accomplished in a serial fashion within engineering departments. Concurrent engineering, like the general system engineering process, seeks to address the disintegrating effects of specialization brought on by competition and exploding knowledge. Many years ago a single engineer, with an ability to effectively use a high percentage of a relatively small pool of engineering knowledge across many fields, was able to perform useful design work in a fairly autonomous way. Today, it is more common to find, applied to a problem, a group of highly specialized engineers, each very narrowly knowledgeable about one facet of a tremendously expanded pool of knowledge.

The central problem in system engineering today is to blend these many specialists into the equivalent of one very effective engineer in command of this much expanded pool of knowledge and be able to apply it to the solution of difficult engineering problems. Concurrent engineering is not a new specialty engineering discipline, rather an umbrella beneath which many specialists can effectively become one temporarily to achieve a common purpose.

Throughout the discussion of the system requirements analysis process, we find references to a need for teams of engineers to work together to solve difficult problems. It is possible for these teams to consist of functionally organized engineers intermittently working together toward a common objective. But, it is much more effective for these teams to respond to the leadership of one team member while working in close physical proximity for the duration of the team effort.

1.1.10 Data Environment Alternatives

The principal output of the system requirements analysis process is a stream of specifications covering the necessary attributes of the system and its elements prior to the onset of detailed design or procurement work on these elements. As the predecessor of this book was originally being written in the late 1980s, it was common practice to publish these specifications in paper form after creating them using computer word processor or desktop publishing application programs. Some companies had converted to a computer database system from which the specifications could be printed and within which traceability and verification connections could be maintained. As it was being rewritten in 2004 and 2005, a more common case was that programs were using effective computer requirements applications but still publishing the specifications in paper form. Later in this book, we explore the future of this work.

In this book, we create a requirements analysis process that can be used in either of these environments. As we will see in Part 7, however, a world-class system requirements analysis capability must employ a computer database environment. It is inconceivable that anyone would use a

typewriter to prepare a specification today but a word processor is not really that far ahead of a typewriter and many people used word processors in 2005 to prepare specifications.

1.1.11 Some History and References

The U.S. government military establishment and its contractors have been the principal architects of an organized method of system development and requirements analysis. The reason for this is that the worldwide confrontation between the USSR and the United States over the period 1946 through 1991 stimulated each to master the other's military capability. Each development made necessary a counterdevelopment involving more difficult problems and unproven technology. Development experience showed that the life-cycle cost of these weapons programs became extremely high where an unstructured development approach was used, due to false starts and the need to modify the systems subsequent to fielding.

Many of these systems also required the safe use of very dangerous products like nuclear weapons. System features had to be developed before systems were deployed to assure that the systems would not be more dangerous to those they were designed to protect than to the intended enemy. So, in addition to increasingly complex needs driven by competition and limitations on the amount of available money to fund development and operation of systems, an organized system development process was encouraged by a great potential for risk and danger from bad systems.

The system requirements analysis process grew out of the lessons learned by the U.S. Air Force on several major weapons system acquisitions, the U.S. Navy on the Polaris submarine program, and the U.S. Army on various weapons programs. The Air Force made an attempt to codify this process in a series of Air Force systems command manuals prepared in the early 1960s. AFSCM 375-5, System Engineering Management Procedures, a 301-page tome, held that "No two systems are ever alike in their developmental requirements. However, there is a uniform and identifiable process for logically arriving at system decisions regardless of system purpose, size, or complexity." The author subscribes to that premise and it is one of the underlying principles of this book.

AFSCM 375-5 provided detailed procedures for performing the system development process. These procedures required engineers and analysts to complete a mass of forms, Forms B among others. Some people argued that government contractors should be allowed to develop their own internal procedures to manage their system engineering activities and that the government should publish only a standard providing guidance and a set of minimum characteristics. In July 1969, the Department of Defense issued MIL-STD-499 titled "Engineering Management." This was a 25-page document with no detailed procedures covering how to do the job, only the minimum requirements the government expects of a contractor. The original MIL-STD-499 required validation of a contractor's system engineering process in much the same way that DoD certifies contractor cost and schedule control systems. The concept was so radical that DoD permitted a test application on only three specific programs. The test results indicated that validation was not cost effective. A contractor's procedures for system engineering were only as good as the people implementing them. Accordingly, MIL-STD-499A was published as a guide only. It required, when applied to a program, the contractor to prepare a System Engineering Management Plan, subject to customer approval, that defined how the contractor would organize for and implement an effective system engineering process.

The Space and Missile Systems Organization (SAMSO of the U.S. Air Force systems command) prepared SAMSO Exhibit 68-62 in August 1968 to cover the system requirements analysis program for the Minuteman ICBM program. In November 1977, SAMSO published SAMSO STD 77-6 to guide the application of the system requirements analysis process on the MX weapon system. The Ballistic Missile Office of the U.S. Air Force systems command published BMO-STD-77-6A for the same purpose to cover the small ICBM program in July 1986.

Other than MIL-STD-499, which did not go into the same degree of detail, all these documents and an ill-fated attempt to publish an Air Force–encouraged military standard on the subject in 1985, applied the same general rigorous system requirements analysis procedures contained in AFSCM 375-5 to the several programs noted. All these documents were researched in writing this book. All of them contain gems of wisdom buried in an elaborate appeal to the use of forms to capture a mass of detailed information about the system being developed in a very organized way. This process was accomplished in a manual way with typewriters and paper initially, but it soon became evident that computers could be applied such that the forms were computer data collection devices that were converted into 80-column cards by key punch operators for main frame data entry using punched card input/output (IO). The process description documents just identified are actually requirements documents for the computer systems supporting the requirements analysis process and provide relatively little information about how the humans should perform the work.

Two other early references to the requirements analysis process are noteworthy. The U.S. Army published a system engineering field manual, FM 770-78, in April 1979. This document does not make the intense appeal to forms characteristic of the U.S. Air Force documents, but it covers an organized approach using what are called *Requirements Analysis Sheets* (RAS). In October 1983 the Defense Systems Management College (DSMC) at Fort Belvoir, Virginia, published its *System Engineering Management Guide* (based on an earlier Lockheed Space and Missiles document developed under the leadership of Bernard Morais) for use in its excellent system engineering management training program. A second edition came out in December 1986, and the manual continues to be updated, some say by the lowest bidder. In 2001, the latest version, a very good one, was available on its website. AFSCM 375-5 and FM 770-78 are available on the International Council on Systems Engineering (INCOSE) website, http://www.incose.org.

Only a few large aerospace contractors ever become proficient in the performance of this process, as defined by the Air Force in the several documents just referenced. TRW, Lockheed, General Dynamics, McDonnell Douglas, Boeing, and Martin Marietta are six contractors from a short list that had extensive experience. Many others avoided becoming expert in the Air Force description of the SRA process because (1) they think it demands a degree of top-down direction, teamwork, and conformity alien to their organizational style; (2) it is too complex; (3) none of their customers showed an interest or identified a requirement for it on their product line; or (4) they never heard of it.

Today, increasing competition in world markets is forcing many companies to look critically at how they are organized and the degree to which they are capable of working in an environment characterized by intense teamwork among members of a group of specialized engineers and analysts. Many companies are finding that they have become fragmented into a series of autonomous organizations, working in a serial fashion on problems that demand teamwork in parallel or concurrently. It is time we all take another look at this very organized process called *system requirements analysis* and try to find ways to adapt the best parts of it to the concurrent engineering environment evolving in American industry.

1.1.12 Overview of the Book

1.1.12.1 How it came to be

This book was originally written to satisfy the needs for a textbook the author needed for a requirements analysis course he started teaching at the University of California, San Diego, in 1990; and at the time it was rewritten in 2005, he was still teaching the course. The original book was a collection of practices that the author had previously written for General Dynamics' Space Systems Division. In 1993, this book was published by McGraw-Hill under the title, *System Requirements Analysis*. When McGraw-Hill ceased publishing the book, the author rewrote it as the second volume in a series of four manuscripts he used in teaching a system engineering certificate program that his company, JOG System Engineering, offers to companies and universities. This four-volume set, titled *Grand Systems Development*, addresses the four fundamental elements of the system development process. These four volumes focus on the following areas of the overall process:

1. Grand systems management. The overall process for controlling the development work in accordance with a plan.
2. Grand systems requirements. Defining the problem as a prerequisite to solving it. The results are captured in specifications with which the designs must comply.
3. Grand systems synthesis. The creative design process through which we solve the problem defined in the specification. Integration and optimization work ensures that the small problem solutions are mutually consistent with system-level concerns.
4. Grand systems verification. A process of proof that the design is a good solution to the problem defined in the specification.

The book was rewritten in 2004 and 2005 for publication by Elsevier Publishing to remove earlier references and update the content for several significant new insights into system development. The author continues to use the other three volumes in training programs his company offers but uses this book as volume 2 in the series for the requirements course.

1.1.12.2 The remainder of this part

This introduction is intended to provide you some background in system requirements analysis and a beginning common vocabulary that we can build upon toward an elegant process description. The second chapter continues building this vocabulary with an overview of the larger process within which system requirements analysis is located. An effort has been made to systematically decompose the aggregate system development process into smaller pieces, in which humans may cooperate to accomplish system development goals. The author has tried to apply the systems approach to the system that we are members of—the enterprise and program. The content of this book is a product of decomposing the complex task of system development into a series of smaller interrelated problems that yield to solutions by specialists so as to contribute to the solution of the large problem. This is exactly what we wish to do in our development work on product systems.

We, as members of the development organization, in fact, are elements of the system developing the product system. The complex act of developing a system involves the interaction between the product system being developed and the system of the developing organization. If the developing agent has no organized approach to understanding the product system needs, applies an autonomous collection of functional organizations with which to develop the product, and has no appreciation for the interaction between these two systems, it should be no surprise when such an enterprise produces a product that fails to provide the customer with good value. Chapter1.2 expands on this introduction to provide a more-detailed description of an organized development process. The part, chapter, and paragraph numbering scheme employed throughout this book is in the form X.Y.Z, where X is the Part (1 in this case) and Y is the Chapter (2 in this reference). This may be followed by Z, a period-delimited subsection number within that chapter.

1.1.12.3 The other parts of this book

Engineers would generally much rather think the creative thoughts that result in designs than agonize over the requirements for the object of their creative work. All too often, we saddle the creative design engineer with dogmatism, format, and style that confuses the true, intended content of requirements statements. Part 2 begins the requirements analysis odyssey with an extremely simple requirements writing scenario. Let us make sure we all understand what a requirement is in its most fundamental form before focusing on proper specification format. This simple requirements writing scenario can be used to great advantage early in the development of a system to capture requirements that otherwise may be lost. If you have not already found from experience in industry that it is hard work to write requirements, you will eventually. There is, however, a simple way to master good requirements writing and it is described in Chapter 2.1.

Ideally, all requirements should be quantitatively defined, but qualitative requirements may be appropriate in uncommon circumstances. The first chapter in Part 2 also covers several ways to establish appropriate values for quantitatively stated requirements. The combination of the progressive scenario, focused on simple attributes that must be controlled, and value assignment methods presented provide a basis for all later performance requirements, design constraints, and programmatic requirements analysis work in later parts.

Chapter 2.2, Requirements Traceability Relationships, explores the lineage between the requirements for two different system elements. It was noted earlier that the ultimate requirement for a system was the customer's simple need statement. Every requirement for the system and its elements should evolve from this single requirement, with the exception of some that relate to interfaces between the system and other systems, perhaps. One test of a good requirement is whether it is traceable to a parent requirement. If not, it may result in overdesign and added unnecessary cost or a solution that is not responsive to the need. This chapter also explores a need to provide traceability between the requirements for an item and the analytical processes through which they were derived or the sources from which they came. If a distributed requirements analysis responsibility pattern is used by the developing organization, the system engineering community must audit the requirements developed for lower-tier system elements to ensure that the requirements analysis job is done well at the lower levels. Some customers require insight into the results of such a study, a traceability study.

A common failing among system requirements analysis devotees is a belief that the only way to achieve success is within the narrow confines of a set of very complicated paper forms or computer screens that must be filled in within prescribed time frames. This becomes a religious experience and the whole reason for existence with some people. While many very good systems have been developed this way, what is important is that the requirements be written down and made public on the program in a timely way with respect to the design and procurement processes. In Chapter 2.3, four requirements analysis strategies are offered within which program personnel may systematically develop an insight into appropriate requirements for the elements of the system.

On a given program, only one of these strategies may be imposed across the breadth of the program, or we may choose to use each independently at different phases in the program. It is possible to use combinations of these strategies or even all four simultaneously without fear of being struck dead on the spot for breach of system engineering discipline. Part 2 covers only two of these strategies in some detail and provides an introduction to the structured analysis strategy covered in detail in Part 3 and emphasized throughout this book.

Part 3 focuses on a very organized process, traditional structured analysis, to decompose the customer need and lower-tier functionality into an architecture for the system and identify primitive performance requirements. The allocations of performance requirements are fed to an architecture-building process. We then map the architecture to a specification tree, configuration item identification, and the organizations responsible for design. These allocations also become the primitive performance requirements for the items to which they are allocated.

This book partitions all requirements analysis work into two major domains, that related to systems and hardware entities and that related to problems solved through development of computer software. Within the systems and hardware domain, the book applies two major analytical subdomains: (1) product performance requirements analysis and (2) product design constraints analysis. Structured models are offered for all the kinds of requirements the analyst will ever need to identify for any specification. A general theory of structured analysis is offered (Chapter 3.2) into which several recognized models are inserted. Models that have proven useful for hardware and software are included (the former in Part 3 and the latter in Part 4).

Chapter 3.1 covers the beginning of a new system development with the application of structured analysis, first in combination with mission and operations analysis to define system level requirements, then to decompose system functionality as a precursor of performance requirements analysis for lower-tier items. Two alternative techniques for performance requirements analysis driven by functional analysis are offered. The analyst may allocate raw function statements to architecture items and subsequently convert them into complete performance requirements with values. Alternatively, the analyst may complete the conversion of a function statement into a performance requirement prior to allocating it to an item. There are advantages and disadvantages to each technique as well as fierce defenders among practicing system engineers in each case. Chapter 3.3 describes the heart of the traditional structured analysis process, functional analysis; and Chapter 3.4 extends this to performance requirements analysis. Chapter 3.5 discusses the identification of system architecture based on the allocation of performance requirements to architecture entities.

Chapters 3.6 through 3.8 focus on the development of design constraints. Performance requirements tell what a system must do and how well it must do it. Constraints provide boundary conditions within which the design engineer must remain while satisfying the performance requirements. Three kinds of constraints are recognized: interface (Chapter 3.6), specialty engineering (Chapter 3.7), and environmental (Chapter 3.8). Interestingly, performance requirements can be defined on the functional side of the problem analysis but design constraints must be identified after you know what the things are, that is, on the physical side of the analysis. Chapter 3.9 offers alternative functional analysis methods, including hierarchical functional analysis, IDEF-0, process analysis, and state diagramming. Chapter 3.10 offers an extension of the requirements analysis sheet ideas exposed in Chapter 3.3 to cover every kind of requirement using several plane matrices to show the relationships between them in what is termed a *RAS-complete analysis*. Chapter 3.11 shows how to connect the models appropriate for gaining insight into the several kinds of requirements to the templates used for specifications and a way to capture the results of the structured analysis in such a fashion that a program can maintain lateral traceability between the requirements in specifications and the models through which they are developed in a configuration-controlled way.

A system engineer working today had better realize that it is not enough to be a good hardware engineer prior to becoming a system engineer. You are going to have to be able to communicate with software engineers. Today, all the hard functionality flows into software; and the hardware is composed of structure, power, sensors, and loads. Do not expect to be able to do justice to the job of being a system engineer without enough knowledge of software development to allow you to talk to software engineers. Part 4 includes some valuable pieces of the needed vocabulary in the form of models software people use to understand complexity.

Part 4 covers the early application of flowcharts to software requirements analysis during a time when system, hardware, and software engineers used essentially the same model, flowcharts. It then moves to the other early process-oriented models including modern structured analysis and the Hatley-Pirbhai extension for real-time systems. Data-oriented methods are covered in Chapter 4.3, useful if the product being developed is a database. Chapter 4.4 covers the several early object-oriented analysis (OOA) methods concluding with the most recent software modeling language, unified modeling language (UML). The author believes that UML contains modeling artifacts that can be applied to problem spaces expressed from the perspective of the three principal facets that system and hardware models relate to: functionality, architecture, and behavior. The author therefore, believes that there will be success at the end of the tunnel now being entered by the Object Modeling Group and International Council on System Engineering with a new modeling approach, SysUML, that applies UML to system analysis. Chapter 4.5 discusses a method applied by Department of Defense in the development of large, complex information systems that tie together sensory systems with the weapon systems meant to target the objects sensed. This process is often referred to as *sensor-to-shooter engineering*, but the correct name at the time of this printing was the DoD architecture framework evolving from an earlier term, C4ISR, meaning command, control, communications, computers, intelligence, surveillance, and reconnaissance.

At one time, both hardware and software people used the same model, flowcharting. Software people evolved a long sequence of models more effective for analyzing problems that are solved with software while hardware and systems people have continued to use flowcharting, referred to in this book *as functional flow diagramming*. UML includes within its structure a flowchart that can handle the functional analysis construct of traditional structured analysis and other modeling artifacts that can be used for the other two facets. It remains a possibility that requirements analysis for hardware and system could be accomplished within UML, leading to a closure of methods between hardware and software people, probably increasing the communications between these far-too-distant disciplines and the closing of the chasm between them. Chapter 4.6 considers this eventuality.

There are some great advantages to specification style, format, and content standards that many of us would prefer did not exist. But, a common specification format and style is important. Part 5 reinforces this notion and addresses specification trees, applicable documents analysis, specification libraries, and other related topics.

Part 5 provides several chapters on specification structures, formats, and standards (MIL-STD-490A and MIL-STD-961D/E). Chapter 5.4 covers applicable documents. Chapter 5.5 covers the release of specifications, revisions, and changes and configuration control. Most of the content of Part 3, and the whole book, relates to the development of Chapter 1 specifications but Chapter 5.5 gives guidance on the development of Chapter 2 specifications. Part 6 deals with management of the overall process. The author believes that a top-down flow of requirements analysis is preferred (there are other approaches) and shows how this directionality can be sustained and paced with respect to the concept development and design work. Technical performance measurement is covered in Chapter 6.2, providing guidance on the selection of TPM (technical performance measurement) parameters, tracking, predicting, and reporting the evolving requirement values and risk status and the use of action plans to overcome perceived problems with specific parameters. Part 6 also deals with validation as part of the risk management process, requirements value management, and configuration control of the many product engineering representations in addition to drawings.

Chapter 6.6 offers an introduction to requirements verification and customer audits, where the customer is shown evidence of the extent to which the requirements have been satisfied by the design and the extent to which the manufactured product complies with the design and manufacturing instructions. Part 6 closes out in Chapter 6.7 with ways to prepare a program to perform requirements analysis well and enable low-risk management of the process.

Allocation and traceability are two reasons why computer databases have become popular in requirements analysis and specification generation work. Part 7 explores how computers can be useful in the system requirements analysis and management activity. They can also be an important part of simultaneously improving productivity (decreasing cost) and improving the product. On the other hand, it is possible to pour a lot of money down a rat hole while trying to acquire a computer database capability for your requirements analysis work. Chapter 8 concludes the book.

1.1.13 How to Get the Most Out of the Book

This book was written with two users in mind. It is intended to be used by professional engineers who need a desk reference on requirements analysis that provides useful answers to problems they encounter in their job. The second application is as a textbook in a university extension system engineering certificate program attended by professional management, design, specialty, analysis, test, and system engineers interested in improving their knowledge and skills. Compromises were knowingly made to match book content to the best possible balance in satisfying these two needs that are not entirely coincident. Finally, this book with related presentation and supplementary material provides a set of student manuals employed in the system development certificate program and tutorial series offered by the author's company.

The student user will be constrained by the class structure to read and understand chapters in a particular order.

Ideally, the lecturer will follow the chapter order, since the sequence was specifically selected for that purpose. The student can best learn to apply this material in a class that requires a team project such as that described in the companion workshop manual used in classes by the author. For the student teams to get their project underway early in the class sequence, some things have to be covered early and may be presented out of sequence. For example, preparation of part of a specification is a common assignment, so Chapter 5 has to be covered fairly early in a course.

Students are encouraged to approach this material in a teamwork fashion, meeting away from the class as their schedule permits to discuss their insights, which will certainly differ. Remaining uncertainties or dazzling conclusions can then be opened up in class for broader conversations. Quizzes may be used to stimulate team discussions if the process does not ignite spontaneously.

Students who are practicing system engineers will immediately recognize the utility of experiencing the course as a member of a small team. Others may not appreciate this until later. System engineering, and the requirements analysis subset, is a team activity, dependent as much on interpersonal skills as technical know-how. Practicing these skills in a class project will enforce many points the book makes along these lines in a much less dramatic way.

It is unlikely that any practicing professional system engineer, or anyone else for that matter, will find this book entertaining reading to the extent that they will want to read the book from cover to cover in the order offered. Until a professional system engineer is familiar with the book's contents, the table of contents and the index are the best route to locate a reference that relates to an immediate problem.

1.2

System Development Process Overview

Contents

1.2.1 The Ultimate Process Step—The Enterprise Vision

When we develop unprecedented product systems, we begin work from the grandest vision of the system description, the customer need. The reason for this is that it is the way users of systems become aware that they are not currently capable of satisfying some unfulfilled need. We do not think of the details then synthesize them into a whole. No, we start with a grand vision and expand that toward the details, working from the known toward the unknown, from the simple to the complex. The customer need is a simple statement telling what must be accomplished by some system yet to be fully characterized. Through a structured analysis process, we can expose increasingly greater detail about the system to be, reaching an understanding of the problem to be solved captured in specifications.

This process can also be applied to systems that already exist but must be modified to accomplish a new or significantly changed need. In this case, the physical system already exists, so the architecture is known. It is still helpful to understand the need and expand that into a functional view of the details. Just as in the case of an unprecedented system, we can allocate the exposed functionality to things in the existing architecture, gaining insight into things that the existing system cannot accomplish. As a result, we make decisions about what elements in the existing architecture are no longer useful, what are okay as is, what need to be modified in particular ways. In re-engineering a system, we still work from the needed functionality in accordance

with Sullivan's ideas but we need to recognize the realities of the existing architecture and look for the least disruptive and costly way of achieving the new functionality.

This product system development logic also works for process systems, which we use to create product systems. The ultimate functionality is a little different, however. Product systems are created by enterprises that function for various reasons. The management of an enterprise may perceive its purpose is to achieve a profit, provide products of value to customers, or capture market share, for example. We refer to this ultimate enterprise process system functionality as the enterprise vision statement, akin to the product system customer need statement. This statement must be generic in nature, in that, at any one time, the single enterprise may be involved in the development of seven product systems for seven different customers.

Commonly, multiple-program enterprises permit each program to invent its own new process with little interference, leading to little chance of progressive improvement over time based on lessons learned. It is imperative that a multiple-program enterprise have a generic process that can be adjusted to satisfy program-peculiar needs. This generic process should be derived from the vision statement, just as we derive product system functionality from the customer need. Figure 1.2-1 illustrates how the customer need and the enterprise vision statement apply to a generic process captured in a life-cycle flow diagram for the enterprise. This process system applies for every product system that the enterprise will ever create. The enterprise vision is the

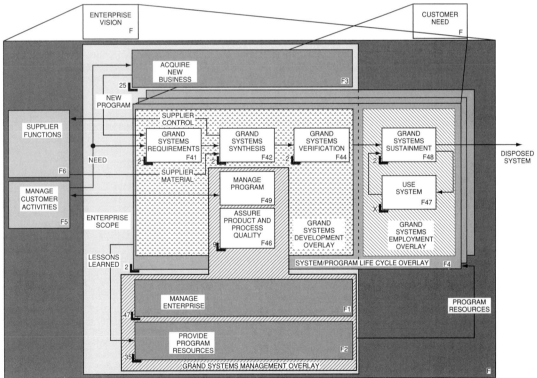

X: REFER TO PROGRAM SYSTEM DEFINITION DOCUMENT FOR EXPANSION

Figure 1.2-1 *Vision and need statement relationships*

ultimate process functionality expanded to identify nine process system functions, each of which must be further decomposed and defined as suggested in the figure.

These two ultimate functionalities must coexist in the enterprise. While the enterprise attempts to satisfy multiple customer needs through several programs in different stages, it must also seek to satisfy its own vision statement, and these goals may not be mutually consistent. This is why this book encourages an enterprise integration team—to adjudicate between all of these different expectations and encourage the best approach in the aggregate, which is called in this book *grand systems development.*

1.2.2 Product-Line Effects

The reader probably can identify with the development experience in the context of the life-cycle flow diagram of Figure 1.2-1, no matter what product line he or she has been involved in. However, the product line of choice for the enterprise has an effect on the details of the process selected for development of product systems. Different kinds of systems demand an appeal to different technologies and related analysis techniques, models, and simulations.

A customer need that is satisfied through use of some kind of seaborne vessel generally does not require an extensive aerodynamics analysis, unless very high speed is needed, although a similar process must be applied beneath the surface with a noncompressible fluid medium. Likewise, aerodynamics is not a needed discipline in spacecraft development. So, there is an obvious correlation between the many different engineering domains and needs as a function of the product concept and related technologies. But, that is not the only product-line effect in one's development process.

Product system implementation in software versus hardware determines a radically different manufacturing process component. Instead of metalworking machines on the factory floor and circuit-board manufacture on a production line, we find an office facility staffed by many software analysts and programmers operating engineering workstations. Even electronic hardware design and manufacture is falling under this same pattern. Application-specific integrated circuits (ASIC) are designed using desktop design software operating on engineering workstations. The resulting lines of code are translated into a format directly compatible with the machinery capable of creating integrated circuit chips. The engineering steps involved in building electronic circuitry in this fashion are very different from those involving old style "black boxes" full of wiring and large electrical components and you cannot do the job with a T square and drafting board.

1.2.3 Customer-Base Effects

The Department of Defense evolved a rigorous acquisition process over many years of trial-and-error procurement. While doing so, it built up a tremendous storehouse of standards covering everything imaginable. Every step in the multi-phased process was prescribed in these standards referenced in contractual documents. In this environment, aerospace contractors became adept at compliance with external standards and paid scant attention to the definition of their own internal process. In the early 1990s, the DoD became aware that it was not being well served by military standards and began a course of action to adjust to commercial standards wherever possible.

The logic was inescapable. The upkeep of 30,000 standards was a drain on funding for development. The government was having trouble creating and maintaining standards in some areas where commercial practices were changing rapidly, like software and electronics, increasingly the key to successful systems. Industry driven by a commercial marketplace full of demand for innovative electronics products was evolving parts that were as good as those called for in military specifications but not specifically identified as acceptable for procurement. At one time, the DoD was responsible for purchase of most of the integrated circuit production and could steer the production base at will. In a world where children buy more electronics than Uncle Sam, the DoD lost a lot of its leverage. Thus, a move to commercial off-the-shelf designs integrated into military systems and reliance on commercial standards for manufacturing and management as well as other fields appeared consistent with a new reality for military and space system development. As this book was entering its final publishing processes in mid-2005, the DoD and NASA were beginning to have second thoughts about the depth to which they had plunged into this new reality because of the demonstrated system losses in space systems, in particular. Therefore, one may expect some return to product development oversight and emphasis on system engineering. The mature reader, of course, will have experienced this oscillation before. System development is as complex as anything humankind tries to accomplish, and the process of moving to a condition of perfection is fraught with false starts and accompanied by many sad stories.

In this new world, aerospace enterprises are expected to have a written process description and actually follow it. This is a significant change from the 1980s, when a contractor had to write plans for implementation on a contract on which it was bidding but would not be held strictly accountable for following them under contract. This was especially true in the system engineering domain. Every room in the Pentagon and all of its annexes could likely be wallpapered with the pages of system engineering management plans never followed.

A customer may contractually interfere with an enterprise's definition of its internal processes through contractual imposition of external standards or an enterprise may be held accountable for following its own internal process definition determined through customer process audits. Alternatively, a customer may not express any interest in what happens internally as long as the delivered product satisfies its requirements.

This book covers a process definition approach that permits an enterprise to follow its own internal process no matter which of these kinds of customers it encounters. The problem is that the enterprise must have a written process description. Here, too, this book helps in the creation of one that any customer will find acceptable. Customers that concern themselves about an enterprise's internal process are reflecting past concerns from programs that yielded bad products late at an overrun cost. It takes few of these experiences to develop a great interest in a contractor's process in an attempt to foresee risks and avoid them. A contractor who has a sound process and a way of applying it to programs will survive any audit a customer wishes to apply.

Also over many years, the DoD evolved a multi-phased contracting approach that continues to have a great deal of merit. This is a risk-mitigating structure that recognizes that, early in a program, we know very little, certainly not enough upon which to based production decisions. In a phased approach, some work is accomplished to define the product and process, followed by a major customer review. Based on the results at that review, decisions are made about the next phase, which might include canceling the complete project. It is well known that the biggest life-cycle cost in any system occurs in operation and maintenance, termed the life-cycle model's *use system* function in this book. The next biggest cost commonly is production.

Development cost, while it can be large by individual standards, is generally small relative to these later cost elements. So, when a customer lets a contract for $5 million to study a new program potential, that $5 million spent early may very well save $2 billion in system life-cycle cost. Figure 1.2-2 illustrates a multi-phased structure with DoD and NASA phasing structures overlaid as they were defined in the early 1990s.

When a customer applies a multi-phase acquisition approach to a contract, the enterprise's process description must be able to handle it. The process description encouraged in this book is configured so as to be compatible with any phasing concept. The steps in the process included can be applied cyclically but generally reflect a three-step design phasing arrangement: concept development, preliminary design, and detailed design. Where the risks permit, one could shorten this to concept and detailed design, but in general, this three-step process is a good idea. Figure 1.2-3 illustrates how the author sees the generic process being applied to a multi-phase acquisition process prescribed by a customer.

In early program phases, some of the early work in the generic process is accomplished. In the next phase, some of this work must be redone, because in the process of doing the work in the first phase, the program team learns enough to know that the solution could be improved. This is called *iteration* of the evolving solution. In addition, the program would have applied requirements analysis at the system level in the first phase and have to apply this same process in the next phase for the next lower tier of the system. This process continues until every item in the evolving system architecture has been defined in a specification. The generic requirements analysis process may have to be applied 50 times throughout the whole program.

This process continues from phase to phase until the product system is delivered, at which time the complete generic process has been consumed by the program. What is achieved through program phasing is the progressive reduction in program risk while avoiding obligation of customer funding until program risks are thoroughly understood and minimized.

1.2.4 Structured Process Analysis and Process Definition Expansion

The vision statement, a very few sentences, obviously is not expressed in sufficient detail to fully express the generic enterprise process. We need more detail. A structured analysis, or decomposition, process can be applied to provide this detail. Several structured analysis models are offered in Part 3 but we select one of them, process or functional flow diagramming, for this application. An enterprise might very well select a different structured analysis model from those in Part 3 with no loss of validity. Later in the book, process and functional flow analysis are differentiated, but for our purposes in this chapter, they are treated synonymously.

This decomposition begins, from the contractor's perspective, with the vision statement, which should give us some ideas for a list of n things that must be done for us to claim that we satisfied the vision statement. Then, we should try to connect these process steps in some kind of flow network, showing which of them must be accomplished as precursors to other process steps. Some of these

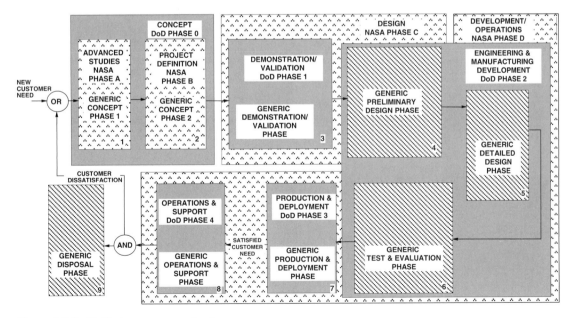

Figure 1.2-2   *Multi-phase program structures*

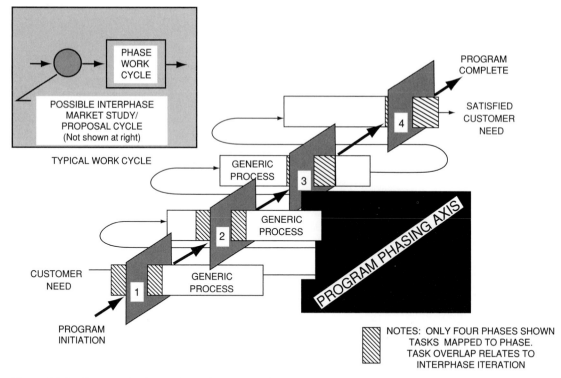

Figure 1.2-3 *Program phasing and generic process relationships*

activities may have to be accomplished concurrently, together, or in parallel.

This top-level enterprise functionality could be listed in a lot of different ways, but for purposes of further discussion, we must agree on one. Any enterprise must accomplish a few fundamental things or pass from the scene. It must have an existence even when there are no programs in house and must retain knowledge determined by the product line and customer base selected in keeping with the vision statement. Also, the enterprise must be led and managed. All this functionality can be allocated to enterprise top management and a functional organization structure.

The enterprise must constantly seek new business, for the business it has at any one time is fleeting. If it is successful in completing a program, the income from that customer ceases and the lifeblood of commerce flows from another program. This function can be satisfied by an organizational entity that seeks new business.

Finally, the enterprise must run the programs it has in house to develop, manufacture, and deliver products of value to customers. At any one time, several programs may be in the house, all at different points in their life cycle. Each of these programs must be managed by a program manager to whom the functional organizations have donated appropriate resources in keeping with the funding provided by the program. Figure 1.2-1 illustrates a generic process at a very high level that would satisfy this very high level implementation of the enterprise vision statement.

The new business acquisition process may be augmented by customer participation in the management of the acquisition, where the customer is highly organized as in the case

of the Department of Defense. Where the customer base is not organized, as in commercial sales, the enterprise may have to satisfy this function internally.

We can further decompose the product and process system life-cycle, as illustrated in Figure 1.2-4, applying the fundamental business functions shown in the Figure 1.2-1 program and system life-cycle process component. Clearly, we must create, or develop, product systems on programs, these products must be modified during their life, and someone must use the product system over its life. At the end of its useful life, the system must be disposed of.

The aggregate functionality may be associated with the enterprise and the customer in a number of different ways. It is conceivable that all the functionality could be associated with either but not a common situation in either case. The case illustrated in Figure 1.2-1 offers a common partitioning of aggregate functionality between customer and enterprise with the customer involved in the overall management of the acquisition and the contractor responsible for the primary development and manufacturing functions. Commonly, the customer is the operations and maintenance agent responsible for system sustainment.

The enterprise vision (illustrated here with no clear linkage but linked in an umbrella fashion as shown in Figure 1.2-1) is identified as process system function F and subfunction IDs have been assigned to illustrated subordinate functions. As shown in Figure 1.2-1, we can continue to decompose F4 to fully characterize the enterprise's generic process for development of new product and process systems. Functions F1, F2, and F3 can similarly be decomposed.

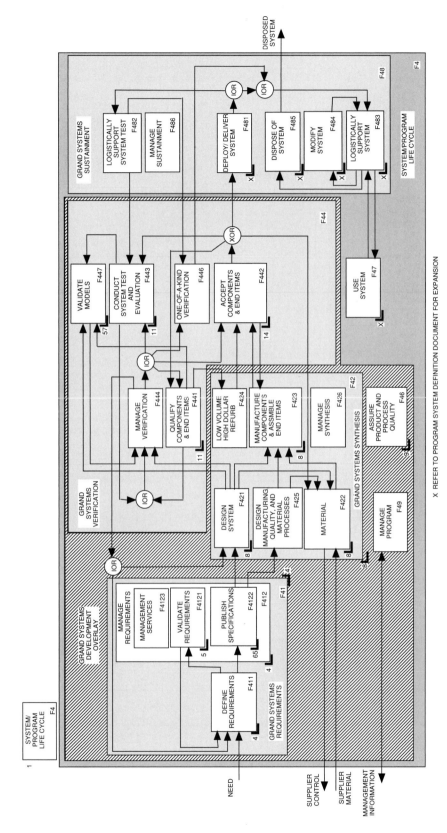

Figure 1.2-4 *Product and process system life cycle*

X REFER TO PROGRAM SYSTEM DEFINITION DOCUMENT FOR EXPANSION

Functions F1, F2, F3, and F4 can and should be captured in the generic process definition in the context of repeatable modules that can be applied to each item in the product system. As noted earlier, the requirements analysis process must initially be applied to the system. Thereafter, this process must be applied to all the things immediately subordinate to the system. This process flows down to lower levels, resulting in many applications of the requirements analysis process module that can be generically characterized. Function could be decomposed in different ways, but the author has chosen the following:

F41	Grand systems requirements	F48	Grand systems sustainment
F42	Grand systems synthesis	F481	Deploy/deliver system
F421	Design product and process system	F482	Logistically support system testing
F422	Material and procurement	F483	Logistically support system
F423	Manufacture system	F484	Modify system
F44	Grand systems verification	F485	Dispose of system
F46	Assure product and process quality	F486	Manage sustainment
F47	Use system	F49	Manage program

Functions F41 through F46 can be largely expanded in a generic fashion to define an enterprise-common process to be applied on all programs. Functions F47 through F49 must be expanded in the context of the customer need statement for the particular program and system. Function F484 is conditional on reaching a point in the life cycle where it is necessary to significantly modify the previously developed system. Function F5 is the business of the customer. Chapter 3 focuses on the expansion of Function F47 during the define system function (F41) to give insight into the necessary characteristics of the product system, including the product system architecture to which the generic process modules are linked to form program process planning. It looks like a bootstrap process is at work, but it can function effectively in the context of early program work. Function F41 is accomplished at the system level, identifying the major things in the system architecture. Later program process applications of function F41 can be based on earlier planning oriented around the product architecture. In other words, early work defines higher-tier architecture while identifying lower-tier entities in the architecture. Subsequently, the same process defines the next-lower-tier entities while identifying the next lower tier.)

Function F41 must also develop the program-peculiar characteristics of the logistics support (F483) and use system (F47) functions. This includes logistics support (training, technical data, spare-parts processing, maintenance, and maintenance levels and locations) and operational techniques.

Figure 1.2-5 illustrates this simple yet confusing collection of ideas. The principal confusion element is that the program process must create both the program product and program process, and part of the work in developing the program product definition is used to organize the program process. High-level product definition is fed back to early program process definition to combine with generic process definition to form lower-level program process definition.

1.2.5 Documentation Media

This section continues the process expansion beyond that shown in Figure 1.2-4, producing a complete generic process diagram contained in a document called an *enterprise definition document* (EDD), an example of which is available on the publisher's website. This document is a definition of the process system that the enterprise implements. It is similar to a product system definition document (SDD), also available from the publisher's website in the form of an exhibit from a student manual used in a corresponding course taught by the author's company. It contains the basis for the enterprise organization and clearly defines all the work the enterprise must perform on programs and in support of its own continued existence. The publisher's website also includes a set of other document templates that are referred to as *exhibits*. These documents include templates for system, item performance, item detail, and software specifications as well as a template for a technical performance measurement (TPM) plan and report.

The EDD is essentially a process specification and should be a fundamental input to the program planning work for any program, in that it supplies the generic process definition as well as the map of the functional organizations to this process, yielding the resource strings to pull into the program along with the processes and related resources.

1.2.6 Lower-Tier Development Functionality

To further expose the needed functionality, we have but to study each of the functions listed in Figure 1.2-4 and consider in more detail what things have to happen to encourage success in the function we are expanding. This can continue down to some level of granularity best determined in the organizational plane. When we are confident we understand the full functionality of all of the organizational entities we have created and we have created all of them that are necessary to accomplish the vision statement, we may conclude that we have exhaustively examined the needed functionality. Grand Systems Management/Division II/Exhibit H (EDD)/Appendix A (available on the publisher's website) expands on the functionality expected from our enterprise in some detail. A map is included in Table A-2 of that document between the exposed tasks and responsibilities coordinated with the functional organizational diagram shown in Exhibit H (EDD)/Figure C-1. The typical program organization diagram is also illustrated in Figure C-1. Some of the lower-tier functionality is examined in some detail in subordinate sections. There may be disagreement about the best way to define the process, but the author hopes the example included in this book offers encouragement for crafting a detailed definition that satisfies the enterprise's process needs as a basis for its functional organizational needs and generic program process.

1.2.6.1 Grand systems requirements, F41

Figure 1.2-6 expands on the define requirements (function F411) that is part of the grand systems requirements function (F41) of Figure 1.2-1 for generic program X. Subordinate sections discuss the exposed functionality. This set of processes may very well be going on in parallel at different

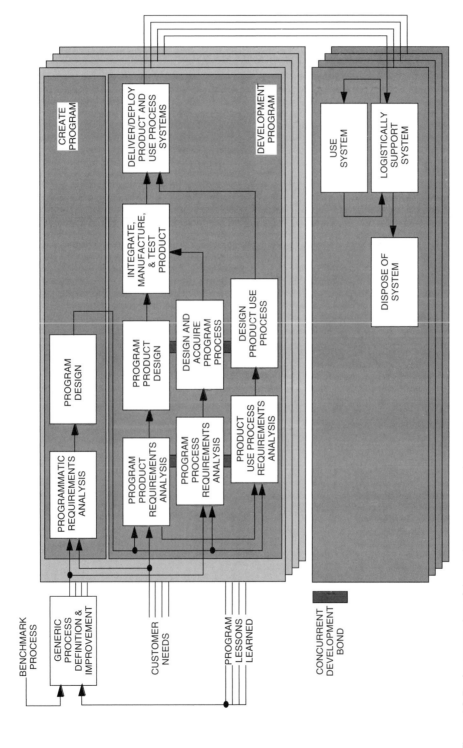

Figure 1.2-5 *Aggregate view of the product, process, and program*

points in the life cycle in several programs being run at the same time in the enterprise. The author maintains that an enterprise should possess a common process model that it applies to all programs (possibly with some tailoring). The employees applying this common process become more and more qualified over time to do this work through repetition of the common process. Given a common process, it is also possible for the enterprise to develop or purchase the most-effective tools to accomplish the work and evolve the tool set over time in an economical manner, while encouraging employee familiarity with this tool set in the most cost-effective way.

Function F411 includes two fundamental tracks not entirely clear from the diagrams used in this section. These two tracks focus on the two kinds of teams needed to create product systems. The grand system (combined product and process) is the responsibility of the program integration team (the system team) that assigns development responsibility for specific system items to integrated product and process teams (IPPT). Some of the tasks in our process diagram must be accomplished by the PIT at the system level and subsequently accomplished by the several IPPT for lower-tier elements. We first go through the cycle for the PIT, the integration and optimization agent for the program, then come back for the IPPT cycle. For example, the PIT should accomplish all the functions in Figure 1.2-6 for top-level specifications, including the system specification and the top-level item specifications for the architecture elements to be assigned to the top-level IPPTs. The IPPTs formed by the PIT (in function F4112) accomplish functions F4113 and/or F4114 for the items they are assigned, while the PIT continues to accomplish the other functions in Figure 1.2-6.

1.2.6.1.1 Program Integration Responsibility for the grand systems requirements function, F41, is assigned to a combination of the PIT and IPPTs in a program setting. Every program activity must have a group in the functional organization that is responsible for providing all programs with personnel, tools, and practices corresponding to that responsibility. Therefore, these activities are allocated to a system engineering functional department. On a program, the PIT is initially responsible for system definition (mission analysis, technical program plan development, and system specification preparation), applying one cycle of function F4111, F4112, F4113, and F4114. Depending on the kind of problem faced by the program, the teams identified by the PIT for lower-tier elements have to apply some combination of functions F4113 and F4114 to develop lower-tier specifications. Engineers assigned by many functional departments must cooperate to accomplish all the PIT work, as is the case for all of the IPPTs established on a program. But, to see this multiplicity of functional department responsibilities clearly, one must look one or two levels deeper than the level at which work will actually be accomplished.

1.2.6.1.1.1 Initial System Analysis A new program may entail solving an unprecedented or a precedented problem (steps buried in F4111). In the former case, the problem has no current solution and the solution probably will involve new development and design. In the latter case, an existing reality has been providing service for some period of time. The program is likely to require some form of modification of the existing system. The work in F4111 may be accomplished by the customer (user or acquisition agent), a contractor during a proposal effort, or a firm contracted to do a study.

Mission analysis is an important subprocess of the unprecedented case. It is a process for translating a customer need into a critical mass of information about a preferred system solution that will satisfy the intended mission. To the extent that this work is accomplished by the proposing or developing contractor, the principal responsible functional department is system engineering working with very experienced people from several analytical and design departments. This process may begin with only a need conveyed by a potential customer through a request for a proposal or deduced by the enterprise, based on marketing analysis. The need acts as the ultimate function for the intended life cycle for that particular system and program and is allocated instantly to the top architecture block, system. The first expansion of the need exposes a life-cycle model and, in particular, the use system function (F47) that covers operational application of the system toward satisfying that need. Continued work identifies lower-tier functionality, which, in turn, is allocated to lower-tier elements in the architecture, thus forming the physical view of the system.

Also, in mission analysis, we must seek to understand the ways the customer will use the system to be developed, and the results of this analysis feeds the development of the functional view of the system, helping to visualize design concepts for the elements of the system that must satisfy higher-tier functionality. Part of the customer interaction process is a question and answer activity that attempts to bring forth from customer representatives answers about the intended application. A key part of the mission analysis process is the development of alternative solutions and trade studies focused on identifying the preferred solution among those considered viable. Figure 1.2-7 illustrates this process.

The need is expressed by one or more life-cycle master flow diagrams, and the difference between them is expanded as necessary to differentiate between them. We then allocate the top-level functionality to things that accomplish that functionality, forming one or more architecture block diagrams (ABD) for each functional view. For each architecture diagram, we identify the corresponding top-level interfaces (SBD), environmental factors (E), a logistics concepts (L, possible basing differences between the concepts, for example), a brief scenario (S) communicated pictorially or in the form of an event list telling simply what happens in the employment of the system, and a set of descriptive data (D) upon which the selection decision will be made.

Through requirements analysis we uncover a set of requirements that all candidate solutions must satisfy and a set of selection criteria used as the basis for selection between alternatives. A trade matrix is used to succinctly capture the intersections between alternative characteristics, derived from the candidate descriptive data, and selection criteria. A decision maker selects the preferred system from the candidate systems studied, based on their relative ranking in the trade matrix in accordance with a previously developed value system formed by the selection criteria used in the study, weighting factors telling the relative importance of the selection criteria, and a means of normalizing the selection criteria. To easily combine the effects of the different selection criteria, we use what are called *utility curves* to translate the units of the selection criteria into positively sloped unitless ranges of numbers that can be added to give an aggregate effect for each candidate.

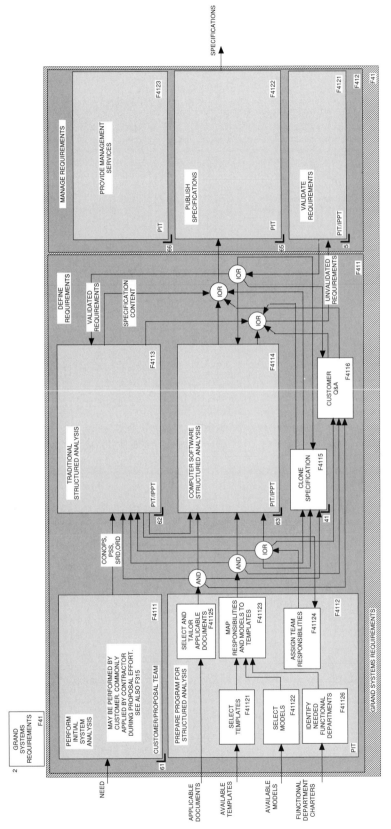

Figure 1.2-6 Grand systems requirements

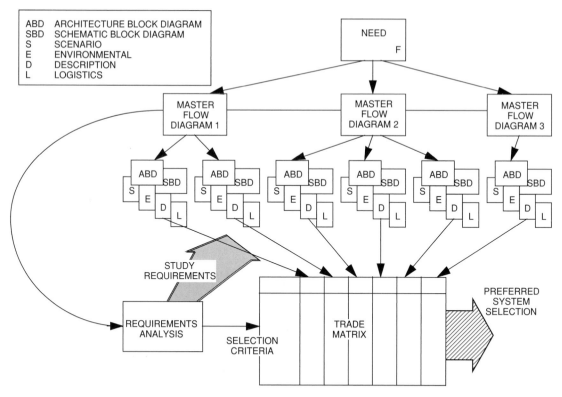

Figure 1.2-7 *Structured selection of the preferred concept*

In the case of a precedented development, diagrammed in Figure 1.2-8, an existing architecture may have been providing satisfactory service to the customer for some time. The customer has found it wanting in some fashion, generally because some factor in the environment of the system has changed. An analysis of the physical processes the system applies is generally helpful, followed by a comparison with those that will be necessary to satisfy the new functionality expected of the new system. We can then study the existing architecture and determine which of several actions are appropriate: remove the item from the architecture, modify the item to satisfy new functionality, or retain the item as it is currently designed. In addition, we may have to add new items to satisfy the new functionality and possibly that of deleted items.

1.2.6.1.1.2 Publish Specifications, F4122 All the work illustrated in Figure 1.2-6 is initially accomplished by the PIT for the system and possibly top-tier items. Ideally, the PIT published the specifications for the items for which the top-tier IPPTs are responsible, so that, when each top-level IPPT is formed, the team leader has the specification defining the technical problem the team must solve and the program planning data organized by WBS (work breakdown structure), which should also be the basis for the product architecture, teams, and the specification tree. Functions F4113 and F4114 are repeated by the assigned teams for the items for which they are responsible, however many times are necessary to completely define the corresponding problems. All the IPPT requirements work is coordinated and

integrated at higher system levels by the PIT. This may include team reviews chaired by the PIT or audits conducted on team results in process. The PIT should maintain traceability data initially established by the team responsible for the specification. The PIT must also manage the margin and budget accounts, risk, and traceability issues as explained later in this book.

The PIT reviews all product specifications, process specifications and plans, and product concepts. During the development of these documents and concepts, the PIT must coordinate between and across the responsible IPPTs to ensure that needed information is exchanged between the teams. This is encouraged by making it clear to the IPPTs what interfaces exist in the product that touch the product items for which it is responsible and pairing up the teams to attack those interfaces that join product entities for which they are jointly responsible. Much of this work is accomplished through cross attendance in IPPT meetings.

1.2.6.1.1.3 Traditional Structured Analysis, F4113 Each IPPT continues to expand the functional model begun by the PIT to gain insight into item performance requirements and the physical product entities at the lower tiers. Constraints analysis completes the content of lower-tier specifications in terms of interface, environmental, and specialty engineering requirements. This process may have to be performed repeatedly for each of the items for which the team is responsible. Each specification must be reviewed and approved by the PIT or the program manager prior to the beginning of any detailed design work.

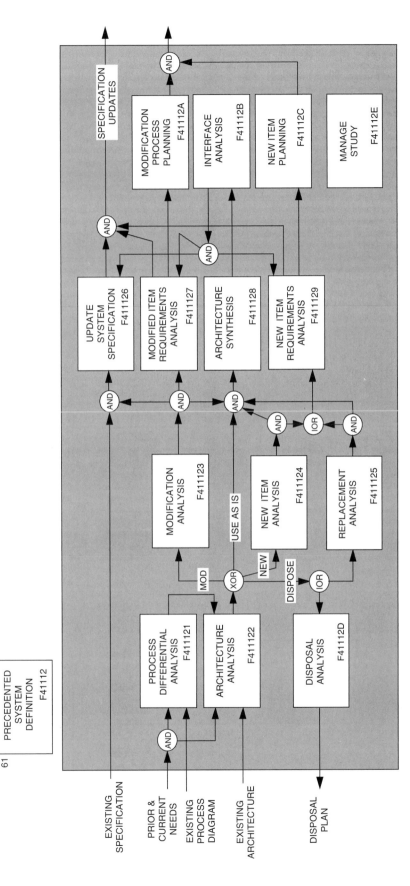

Figure 1.2-8 *Precedented development work*

Specifications are developed for all things in the systems architecture that is also determined in this process. This task also covers development and maintenance of requirements traceability, requirements management, and other related tasks. The understanding is that the PIT applies this process initially (in function F4113) to develop the system specification and possibly the next-tier specifications. Then, the IPPTs apply this process to develop all the lower-tier specifications for which they are responsible.

Allocation of functionality results in identification of things that have to be fitted into the physical model of the system, the architecture. The principal outputs include

1. *Requirements analysis sheet.* The RAS, in paper or computer screen format, captures the relationships between the functions, the derived entity capability (performance) requirements, and the product architecture entities that accomplish the functionality.
2. *System description document.* Captures the system diagrams defining system composition. It includes (a) a functional flow diagram, illustrating needed system functionality; (b) the aggregate RAS (in full or by reference to a computer database tool containing it), capturing the allocations of functionality to architecture; (c) an architecture block diagram, a hierarchical block diagram, defining the things in the system; (d) a schematic block diagram, defining the relationships between all the things shown on the architecture block diagram; and (e) any other documentation products of the system definition process.
3. *Drawing breakdown structure.* The engineering drawing overlay of the product architecture, telling what engineering drawings will be produced.
4. *Specification tree.* The specification overlay of the product architecture, telling what specifications will be prepared and what format they will follow.
5. *Manufacturing breakdown structure.* The manufacturing overlay of the product architecture, defining the groups of things moving from one major production area to another.
6. *Configuration/end item list.* Identification of the things through which the program will be managed.
7. *IPPT responsibilities.* Team responsibility boundaries relative to the architecture.
8. *Work breakdown structure.* The product component of the WBS is a finance overlay of the product architecture, upon which the whole program plan is based.

The top-level specifications are approved by team and program management as a prerequisite to development of lower-tier specifications and top-level preliminary designs. Throughout this work, the PIT must integrate and optimize the evolving requirements pool and cause the modeling work accomplished by several teams to appear as one analysis. This can be accomplished by the PIT being responsible for a system definition document contributed to by all of the teams. As specifications are reviewed for release, their structured analysis basis must also be reviewed and, if approved, the latter is accepted into the SDD as the specification is released. Figure 1.2-9 illustrates this process, showing how the several requirements analysis activities feed appendices of the SDD and, in particular, are captured in Appendix G, the RAS, where the requirements are allocated to product entities.

In the lower-tier details, to be exposed in Chapter 3, several different structured analysis models are discussed that could be applied selectively, depending on the nature of the problem to be solved and the skills and experience of the team doing the work. The approach encouraged in this book for systems and hardware analysis is called *traditional structured analysis*, using functional flow diagramming. If the design is going to be implemented in computer software, one of the software modes is preferred, such as modern structured analysis or unified modeling language, in function F4114. It is a problem that no one structured analysis model is universally accepted as appropriate for every kind of problem and implementation intent (hardware, software, and personnel accomplishing procedures). In this book, the author explains all these models and offers some hope that a variation on UML will evolve into a universal model in the near future. For now, the process offered in this book encourages the use of traditional structured analysis for systems, hardware, and procedural activity and UML for computer software entities.

Every requirement identified in every specification is screened for ways to prove the design is compliant. This process starts with each requirement, which is extended into a verification requirement linked to verification tasks for which plans are developed, implemented, and results of accomplishing those plans documented in reports. The reports provide the evidence of compliance. This process begins with PIT development of system verification requirements and planning data followed by IPPT implementation of the same process for their areas of responsibility under the coordination, integration, and review by the PIT.

During preliminary design to follow the requirements, verification focuses on item qualification planning; and during detailed design, the verification planning focuses on acceptance planning. All the design and analysis documentation developed for an item is completed as a coordinated package and moved through a formal release process that ensures that all enterprise documentation requirements have been satisfied in the package and the content is mutually consistent within the context of the item and the system at large.

Concurrently with program product development, we must also develop the program processes so that they are mutually consistent. The teams work to specify the process problem in the form of process requirements expressed in process specifications and plans.

1.2.6.1.2 Computer Software Structured Analysis, F4114
Traditional structured analysis works where the object in question is the whole system or the design to be implemented in hardware. It is not effective where the product is to be implemented in computer software. Function F4114 includes several models that are effective for software: modern structured analysis, database modeling, early object-oriented analysis, unified modeling language, and Department of Defense architecture framework (DoD AF). These are discussed in Chapter 4.

1.2.6.1.3 Validate Requirements, F4121 Requirements validation is a subset of the program risk activity, focusing on the product performance aspect of the program. All the requirements in top-level specifications are passed through a risk filter to determine the degree of difficulty anticipated in complying with them. Where a concern is uncovered, tests, analyses, demonstrations, and examinations are accomplished to improve confidence that those requirements

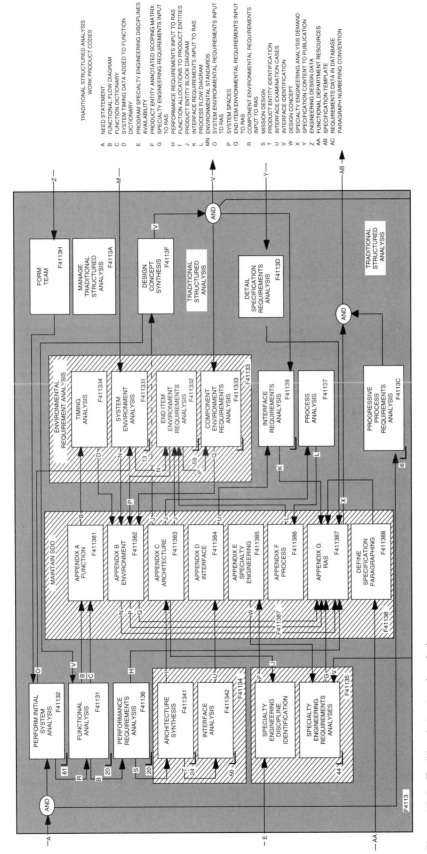

Figure 1.2-9 Traditional structured analysis

can be complied with. If this is going to require some period of time, entails fairly involved processes to validate a requirement, and the requirement is a significant driver, the concern should be considered for selection as a technical performance measurement parameter, which can be tracked over time and coordinated with specific action plans to compliance to the requirements value by current demonstrated capability. All this work should be complete prior to holding the preliminary design review.

1.2.7 Grand Systems Synthesis, F42

The second step in the system development process is to solve the problem, and this commonly occurs in three transformations. First, engineering-dominated design teams transform the requirements into design solutions using the selected technologies and domain knowledge. Second, the design is transformed into a list of sources for the needed parts and materials through an effective procurement process. The received materials and parts must be processed and, possibly, temporarily stored on their way to the production process. Finally, the design is implemented using the procured materials and parts in a manufacturing process.

1.2.7.1 Design grand system, F421
Design work is accomplished in preliminary (F4211) and detailed sequences (F4212). The design covers the product and its related processes. The design work of all the teams is integrated and optimized by the PIT. Item design and related process designs are reviewed and approved as a prerequisite to detailed design go-ahead. The principal purpose of preliminary design is the mitigation of risk.

1.2.7.1.1 Item Team Preliminary Design, F4211   Figure 1.2-10 offers a detailed view of this process. A trade study process is encouraged when the preliminary design selection will not yield easily to sound engineering judgment. At the end of function F42111L5, the actual preliminary design is accomplished in the task identified as document preliminary design, F42111L6. All the work in this function (F4211) is contributed to by people from many specialized functional departments and must be managed in the context of a cross-functional IPPT.

The PIT does no design work in the conventional sense. In Figure 1.2-10, the IPPTs contribute to the process design but the PIT is in overall control of that evolving design. Also, the PIT accomplishes all the integration and optimization across the teams for product and process design at the system level. Each team is the system agent for any subordinate teams.

The preliminary design system integration function, F42114, is expanded in Figure 1.2-11 and is a PIT responsibility. The PIT is initially responsible for system definition (mission analysis, technical program plan development, and system specification preparation), applying one cycle of function F411, and subsequently for overall system development and integration once it has assigned integrated product team responsibilities to system elements and initially staffed the teams. People from many functional departments must cooperate to accomplish all the PIT work like all teams established on a program. The program teams should be led as program entities staffed from the functional departments.

1.2.7.1.2 Item Team Detailed Design, F4212   The approved preliminary design is expanded to completely describe the preferred solution and submitted to a detailed

design review as a prerequisite to manufacture and shipment to the customer. Figure 1.2-12 offers a view of this process. As in preliminary design, these tasks are accomplished by specialists from many functional departments coordinated by the IPPT management within the context of the team and across the teams by the PIT, accomplishing integration and optimization work.

The PIT accomplishes activities similar to those in the preliminary design phase with emphasis on compliance of the detailed design feature with requirements, evolving physical interface compatibility, manufacturing, quality, logistics, and verification (qualification and acceptance) detailed planning.

Generally, two levels of design reviews are held in a development effort, a system design review and item reviews. The former is held to expose the system design concept and determine if it satisfies system requirements within the context of controllable risk. Item reviews are generally held in two sets, preliminary design review and critical design review. These two reviews are closeout actions for functions F4211 and F4212, respectively. The former demonstrates that the performance requirements can be satisfied within cost and schedule constraints and commits to the development of a particular detail design. The latter approves the detailed design and commits to manufacturing the resources needed for test and evaluation. In a simple system, one design stage and one review may be sufficient, but in a complex system, it will be necessary to use a two-step process to encourage sound risk control. In either case, the PIT should hold item reviews presented by the IPPT followed by a system review presented by the PIT for the program manager that closes out all the item reviews.

1.2.7.2 Material operations, F422
All product materials identified in program engineering documentation are acquired from approved sources through contractual and purchase relationships with other firms, associate contractors, and the customer. As shown in Figure 1.2-4, these materials are made available in accordance with schedule needs for manufacturing purposes in task F423, verification tasks in function F44, and logistics support purposes in function F48. Some materials may also have to be acquired and expended in association with the engineering work accomplished to gain confidence in satisfying challenging requirements.

1.2.7.3 Manufacture system, F423
The product is produced in a physical sense. Materials acquired in task F422 are processed in accordance with engineering and manufacturing planning crafted in tasks F421 and F425. The manufacturing functional department is responsible for this activity for physical product manufacturing. Software manufacture is commonly an engineering responsibility.

1.2.8 Grand Systems Verification, F44

The system engineering process involves three fundamental steps: define the problem (in specifications), solve the problem (through design), and prove that the design solved the problem (verification through test and analysis). Verification planning accomplished in tasks under functions F41 and F42 are implemented to develop evidence of compliance,

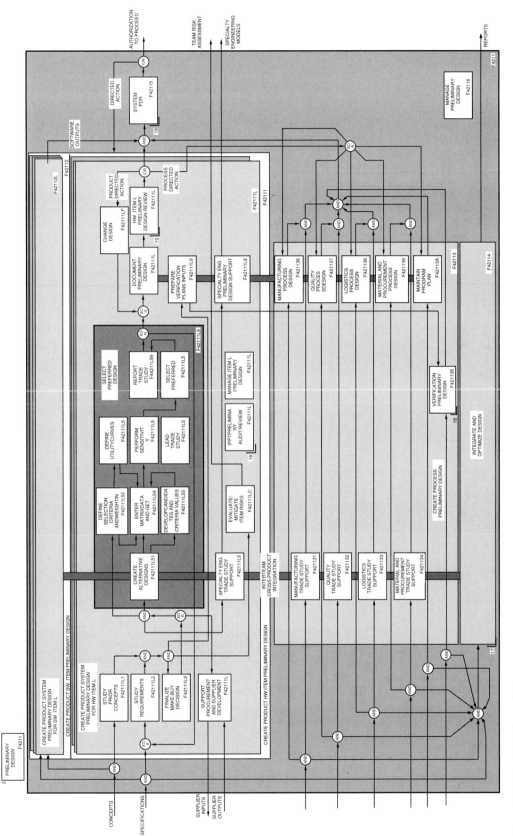

Figure 1.2-10 *Preliminary design*

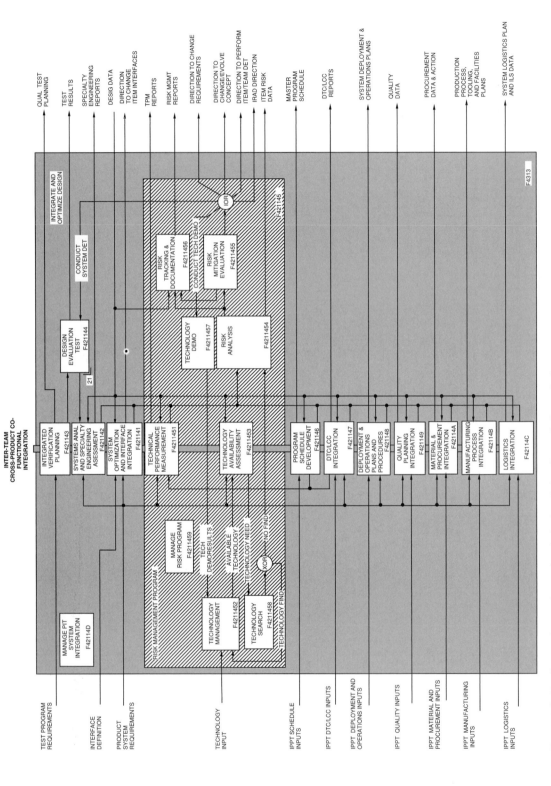

Figure 1.2-11 *System integration and optimization during the preliminary design*

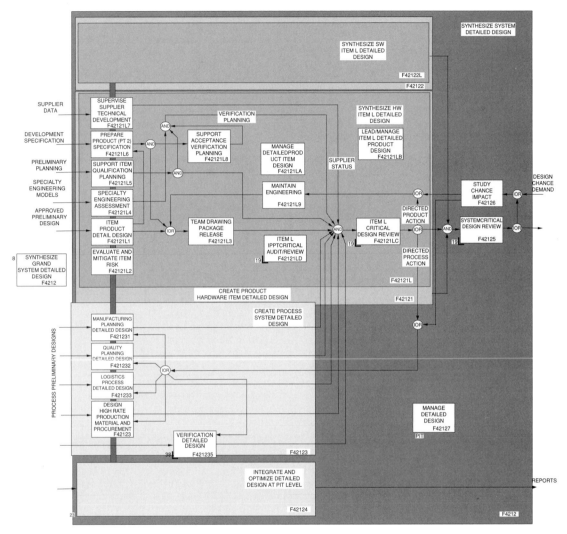

Figure 1.2-12 *Detailed design*

which is accumulated, correlated with planning data, and developed in status reports. The verification process consists of three cycles: item qualification to prove that the product design is adequate, item acceptance to prove that the manufacturing process satisfies product detail requirements, and system test and evaluation to prove that the whole system satisfies system goals. Engineering is commonly the principal responsible functional organization for all formal verification, although manufacturing can be made responsible for item acceptance.

Management skill is brought to bear to implement the content of verification plans and procedures, track the status of the activity, report the status, and take action to prevent problems from occurring and correct problems that do arise. The qualification work is brought to a close through a functional configuration audit (FCA), which is held to monitor the evidence of qualification compliance in the form of test and analysis reports. The conclusions are captured in meeting minutes, directing any additional work that must be accomplished and identifying any residual actions that must be completed prior to closure of the FCA. Generally, FCA closure is necessary to permit system-level testing involving the item in question. One or more FCAs may be held for each IPPT, concluding with a system-level FCA to close out any remaining open items. Alternatively, in a smaller system, a single FCA at the system level may be adequate.

The first article through the manufacturing process should be subjected to a physical configuration audit (PCA) to verify that the manufactured article complies with all manufacturing, engineering, and quality requirements. Subsequent production articles are simply subjected to acceptance and shipped.

The PIT must manage the overall work flow during qualification verification. Some tasks may be actually accomplished by the PIT but most of them will likely be done by an IPPT. In all cases, this work should be accomplished in accordance with the careful planning accomplished as a prerequisite.

1.2.9 Grand Systems Sustainment, F48

Throughout the development period, logistics support resources are defined as well as product resources. These resources from the manufacturing (F423) and material acquisition (F422) functions are applied to support field use of the product system through all or some portion of the product life cycle. At some point, the logistics support responsibility may shift from the producing contractor to the customer, which requires the resources to continue the support through product life. Logistics is the principal responsible functional organization.

1.2.9.1 Logistical support system, F482 and F483

When the system enters system test and evaluation verification, it requires logistic support, since things fail in testing as well as normal use. Therefore, during the logistics planning accomplished earlier, provisions must be made for spare parts, special support equipment, technical data, and training relative to the system testing crew and process. In the interest of cost and lack of an abundance of resources so early in the program, compromises may have to be made, such as providing factory support at the depot, use of some factory test equipment rather than the final field support equipment planned, direct use of logistics analysis results rather than polished technical data, and cannibalization of parts to satisfy immediate needs. It is generally unwise to provide lavishly for resources in this situation, because the results of system testing and evaluation commonly require changes in them.

When the system completes system testing and evaluation verification and any needed design changes have been completed, the logistics resources must be ready to flow to outfit the user's initial operating capability and support any user testing that must be accomplished. Thereafter, the system must be supported over its lifetime. Commonly, at some point, logistics support becomes normalized, with the user taking over this responsibility. Prior to that point, contractor personnel may be employed to maintain the system or only to provide consultation through field engineering services.

1.2.9.2 Deploy/deliver product system, F481

The manufactured system resources must be moved from the point of manufacture to the point of operation and placed in initial operation. This commonly involves transportation. Deployment is accomplished in accordance with plans made during function F41. Logistics is the principal responsible functional organization. The military word *deploy* is used, but this process is every bit as important in commercial systems, where we have to decide how we intend to deliver the product into the hands of the customer. Shall we sell directly to the public with FedEx delivery, apply some wholesale-retail formula, or sell through our own stores, for example? To what extent is warehousing needed and how shall inventory be managed throughout the distribution system?

1.2.9.3 Modify product system, F484

The system is modified to meet customer needs. This commonly involves coordinated hardware, software, and procedural changes. This function includes the whole process of defining the change, implementing it in design and manufacturing, and verifying compliance. This is essentially a mini program focused on a specific change, which may involve a minor change or a substantial update to the system.

1.2.9.4 Dispose of the system, F485

During the development of the system, thoughts of disposal may not be very popular with the program manager, who is spending all his or her energy to bring the program into being. It is necessary to develop the disposal process as part of task F41, however, since it may very well imply requirements for the product and supporting resource base. The system is analyzed for hazardous materials, as such materials call for development of plans for disposal in a safe manner. Ideally, these materials should have been identified and minimized in function F41. Some system resources may have residual capabilities that can be applied in new systems, either as-is or after modification or update. Other system resources may be scraped. Logistics is the principal responsible functional organization.

1.2.10 Use Product System, F47

The system is placed in operation as planned in function F41 and verified in function F44. Over the life of the system, it will need to be modified for improved performance or to correct problems observed in operation and maintenance. Logistics is the principal responsible functional organization within the contractor, although it commonly requires help from system engineering for operational details.

1.2.11 Manage Program, F49

The overall program is managed toward the program goals defined in program planning. The principal functional department is Programs. The program manager works with an initial staff to build the program definition, and staffing during the proposal effort or early program studies commonly is either customer or enterprise funded. This book tells how the program manager may clearly define the program based on cataloged enterprise resources mapped to specific program planning identifiers. In so doing, the enterprise functional resources are clearly identified, and to the extent that the program manager can pay for them through program workhours, the program acquires the resources needed to implement the program from functional departments in accordance with its plan and schedule.

The program plan, developed during preprogram work (proposal development or commercial program development) must be maintained, accounting for changes driven by improved knowledge about the program's needs. The PIT and program office must cooperate to ensure that the technical work demanded is coordinated with resources available to accomplish that work.

All program documentation other than memoranda should be formally released, controlled, and correlated to particular design baselines defined by specific document sets or computer database versions. The configuration of design representations (models, simulations, mockups, etc.) also must be controlled. During the work accomplished in function F41, there is normally no physical product, so this work is documentation oriented.

1.2.12 Assure Product and Process Quality, F46

The traditional function of quality assurance is to inspect a produced product relative to some standard and determine if the product complies with that standard. To be effective, this quality function must be independent of the manufacturing agent. The practices effective in performing this function are well described in Juran's *Quality Handbook*. If we should apply these practices relative to the product, does it not make sense to apply them to the process as well?

Individual and team efforts to accomplish program work cannot always be depended upon to follow program plans and enterprise practices. Commonly, no force other than program management monitors and reacts to this problem, and it is not all that uncommon that program management is encouraging the deviation from the standard enterprise practices defined in functional department manuals. The author believes that the effective performance of system engineering on programs and the continuous improvement of performance in doing that work demands adherence to written functional department practices in order to encourage common processes, employee work repetition, and all of the benefits derived from common processes.

Therefore, the enterprise quality assurance function should be applied across both product and process entities. Process implementation should be checked in accordance with some plan that is both effective and affordable. In this book, quality assurance is assumed to be responsible for a capability maturity model (CMM) application or other assessment of enterprise processes and their implementation. In this regard, it is important to verify that enterprise processes compare favorably to world-class practices defined in recognized standards for particular practices and the programs implemented within those enterprises follow enterprise practices faithfully. Combined with continuous process improvement, this moves the enterprise closer to a condition of perfection.

1.3

Process
Variations

Contents

1.3.1 The Situation

1.3.1.1 The central model

The content of this book was originally built primarily for one business situation, that being an enterprise solving unprecedented problems for large or concentrated (collectively procuring) customers. The reason for this is that the systems approach evolved in this environment to satisfy primarily military customer needs for effective systems during World War II and the Cold War that followed until approximately the year 1990. Especially since the end of the Cold War, a new commercial focus has evolved worldwide, driving commercial customer interests, commercial enterprise competition and economic interests, and commercially available technology to the forefront. Even the American Department of Defense has recognized the ascendancy of the commercial sector in its interest in commercial off-the-shelf (COTS) equipment and software.

True, many commercial companies developing new products generally look with some disdain on the seemingly complex and costly process applied by the DoD in the development of new products, which is the basis for the discussion in Chapter 1.2. Some of this concern is well founded and some an overreaction of an organization in denial now applying an ad hoc approach without having seriously considered alternatives.

1.3.1.2 DoD process rationale

One reason that DoD applies its very well-defined and rigorous process is that it must, by law, not favor particular suppliers but select suppliers based on a fair evaluation of alternative offers. Commercial firms have no such obligations

and can make these decisions strictly from a business case. Also, military systems carry with them some risks with which commercial products do not commonly have to deal. Kipling's poem about the loss of a kingdom driven from want of a nail is not that farfetched in the use of military systems. The results of failure of a military system in the dynamics of warfare or a NASA system to transport humans to the surface of Mars can have dire consequences very different from failure of a commercial system used to farm a field. Finally, no organization has been involved in the development of more-complex systems to satisfy such difficult needs than the American Department of Defense. Over many years and many unsatisfactory development efforts, the DoD evolved a very rigorous process that is as effective as humankind has devised to solve very complex problems. Figure 1.3-1 illustrates the DoD life cycle in 2004.

Over the years, in development of military and space systems, one sees an evolution toward more- and more-complex systems. This is driven by the reality that we always view our needs from an advancing plateau. The systems we use today give us insight into new possibilities, where those systems are but subsystems of grander systems composed of existing and new systems. This process continues at each step in the evolution, and no organization has had an opportunity to pass through more layers of this process than the DoD. Separate gun and radar systems become connected into a weapons system capable of tracking the target and directing the gun. The U.S. Navy, in its procurement of the DD X, the 21st century combatant ship, drastically reduced the number of people needed to run the ship through automation applications. This requires that the whole vessel

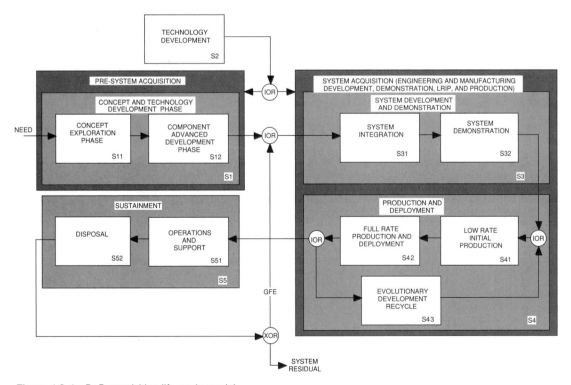

Figure 1.3-1 DoD acquisition life-cycle model

be treated as a system. Even that image is insufficient, in that the ship sails into action as one node in a network of ships, manned and unmanned aircraft, personnel on the ground, vehicles ashore, and satellites interacting over a network of computers via radio-completed links. This kind of complexity is uncommon in commercial systems to date.

1.3.1.3 Other U.S. government life-cycle models

The National Aeronautics and Space Administration (NASA) independently evolved its own acquisition life-cycle process, illustrated in Figure 1.3-2. The Federal Aviation Administration acquisition life-cycle model is shown in Figure 1.3-3. The requirements analysis process covered in this book can be coordinated with any of these acquisition processes, generally with the earlier stages of the process.

1.3.1.4 Commercial firm future

As commercial firms follow this same sequence that the DoD has been following for decades, evolving more- and more-complex systems, they too will come under the influence of similar development approaches. Some commercial companies appealing to the author's company for assistance in improving their system engineering capabilities have indicated that the need to solve the complex problems they find themselves dealing with drives them to the use of electronics and software implementations and ad hoc approaches that have been found wanting. Some of these firms moved toward the phased development approach interspersed with major reviews common to military development programs and a more serious effort to define requirements for items as a prerequisite to seeking design solutions for those items.

A small commercial firm developing an in-home medical instrument found that the Food and Drug Administration (FDA) required that it be able to show proof that it satisfied FDA requirements. Therefore, the company needed a database system for capturing the traceability between specification content and evidence from test activity. The FDA required an effective requirements verification process that the firm had not anticipated.

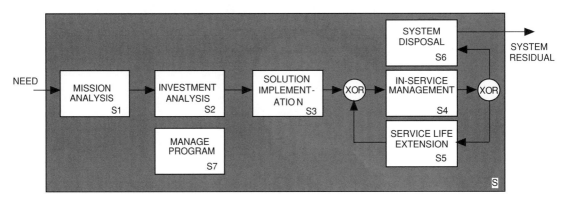

Figure 1.3-2 *NASA acquisition life-cycle model*

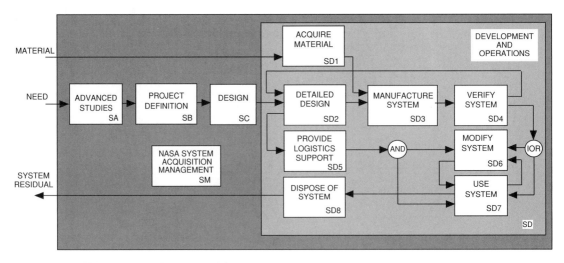

Figure 1.3-3 *FAA acquisition life-cycle model*

A medical laboratory apparatus producer concerned about its ability to prepare good specifications included a software manager who thought it was a waste of time to apply structured analysis as a means to identify appropriate content for specifications. He thought it was appropriate to proceed with the design and work through any problems observed. This approach has been fairly conclusively shown to be fraught with cost, schedule, and performance risk, whether the developer is producing military or commercial products.

A company developing farming equipment realized that it was possible to collect yield data while harvesting the fields coordinated with field GPS coordinates. This tremendously valuable information could then be coordinated with field survey data relative to soil moisture, nutrient concentration, and seed density to encourage micro rather than macro farming techniques, potentially yielding more product at less cost and environmental impact from excess fertilizer runoff. By providing a farming office complete with desktop computer and software, the farmer might determine appropriate settings for planter equipment in the next planting season with finer area granularity. This is a farming system connecting together tractors and planters, harvesters, and an office system to achieve a common purpose that encourages increasingly efficient farming and potentially more profit.

A construction equipment manufacturer found that its current multiple prototype product development process, where complete machines would be built and subjected to evaluation in several cycles, had to be changed in the interest of reduced cost and schedule. The solution it approached was an increased use of simulation. A commercial developer of heavy equipment found that it could more economically develop the electronics systems for its equipment if it could modularize a family of electronics units that could be selectively employed in a family of machines. This means a more complex interface and design problem than

dedicated packages for each machine. A farming equipment maker wanted all its products to have a common feel no matter the size of the machine. The author drove three sizes of these machines and found they had succeeded admirably, having applied some very effective system engineering and human factors engineering in the process. Not only were the controls in the same place physically, but the gains had been crafted to compensate for the different mass of the machines.

A developer of amusement park rides discovered that the risk of rider injury and the corresponding risk of litigation was not being adequately addressed in its design work and elected to improve the verification of safety requirements.

1.3.1.5 The JOG system engineering prescription for specifications

The author's company, in its grand systems development certificate program, recommends the process sketched in the common process diagram, some charts of which were explained in Chapter 1.2. The steps in the process, however, are sufficiently strung out on the common process diagram set that it may not be perfectly clear how simple this recommended process is. Therefore, it has been summarized in Figure 1.3-4 and explained in this and subordinate sections as a brief prescription involving eight steps. Steps 1 through 4 must be accomplished in a generic fashion, while steps 5 through 8 are accomplished for the specific program. These steps are suggested in the context that the enterprise commonly has multiple programs in work at any one time, each involving significant system development complexity; that they are organized as a matrix with the functional departments providing programs, with good resources constantly being improved and programs responsible for accomplishing and managing the work thorough physically collocated cross-functional teams oriented about the product architecture; and that the program teams in each case include a program integration team that is the system agent

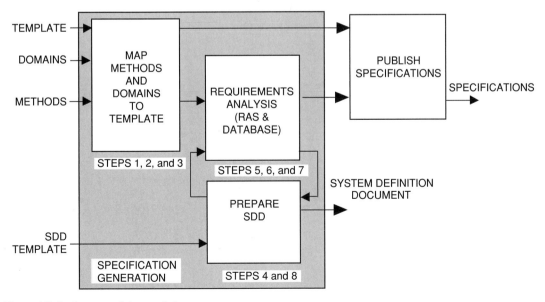

Figure 1.3-4 *A successful prescription*

and two or more IPPTs responsible for development of specific product architecture items.

1.3.1.5.1 Template Preparation
The enterprise should pick a set of specification templates, one for each kind of specifications that the organization has to develop. This includes system, hardware, software, parts, materials, and processes as well as interface control documents, of course. But it may also include special formats for support equipment or other product types. Ideally, the enterprise selects templates that its customer base prefers, but if the enterprise has multiple customers, they may not agree on a single set. In this case, it may be necessary to select alternative sets or tailor some sets for equivalency. The author's preference is MIL-STD-961E for systems and hardware and MIL-STD-498/EIA-J-STD-016 for software. Chapter 5 discusses specification formats and content controls.

1.3.1.5.2 Map Templates to Functional Departments
Map each template paragraph to the functional department from which analysts will come to do the related program requirements analysis. Commonly, the analysis work corresponding to a specification paragraph 3.1.3, covering required states and modes complying with MIL-STD-961E, commonly is done by engineers from a system engineering department while a reliability department would supply people to perform the analysis for MIL-STD-961E paragraph 3.1.5, titled reliability. Specification standards are discussed in Part 5.

1.3.1.5.3 Map Templates to Structured Analysis Models
Map each template paragraph to a structured analysis model that will be used to by the analyst to gain insight into that requirement. Requirements are identified through structured analysis modeling and defined through an appropriate analysis applied by the assigned analyst.

1.3.1.5.4 Provide for Configuration Management of the Model Base
Establish a system definition document that will be used to capture and publish the work products produced in the structured analysis to identify requirements. A generic template should be available for use at the proposal or program start. Chapter 3.11 offers a model for this kind of document.

1.3.1.5.5 Perform Structured Analysis on Programs
Apply the agreed-upon models to the program problem space to identify requirements linked to structured analysis model artifacts. Derive requirements from these models and capture the requirements mapped to their model in a RAS-complete implemented in a database application. The enterprise should use models for all system and hardware specification development and also for software specification development. So, every requirement for every specification should be traceable to a model from which it was derived. The RAS-complete is described in detail in Chapter 3.10.

1.3.1.5.6 Allocate All Requirements to Product Architecture
All requirements appear in the RAS-complete. Use the RAS-complete to allocate requirements derived from model entities to architecture. This is simply a matter of annotating each line or record in the RAS-complete in the assigned product entity column.

1.3.1.5.7 Coordinate the RAS-Complete with the Template Structure
Associate each item in the architecture with a particular specification template and format the RAS-complete content relative to that item using the template paragraph structure. The template can be loaded into the RAS-complete for each architecture item providing paragraph headers, and in most cases, the requirements derived from the models are subparagraphs within that structure. If the RAS-complete is in a database application, we simply set the filter for the architecture item of interest, order the content by paragraph number, and print the specification to printer or screen.

1.3.1.5.8 Capture Modeling Work Products in SDD
Capture modeling work products (diagrams and tables, generally) in the SDD and configuration manage the model base through document revisions. The SDD could be a published as a paper document with several appendices, each capturing a particular type of work product, or retained in a computer database and accessed online. Refer to Chapter 3.11.

1.3.2 Alternative Sequence Models

The treatment in Chapter 1.2 suggests what is called a *waterfall development model*. Some people have concluded that the waterfall model is no longer appropriate, but this model is the basis for all known models except chaos. The V model simply bends the process in the middle and thus encourages a connection between requirements definition and verification process, which is not so obvious in the waterfall model. The spiral model accomplishes the waterfall at a higher repetition rate than once per program. Thus, the process covered in the prior chapter can be made to work in any of these situations.

1.3.3 Concentrated Versus Distributed Customer Base

People in commercial companies sometimes argue that the DoD approach is not appropriate for their situation because, while the DoD provides their competitors with a single customer for their products, they have to deal with a widely distributed and diverse customer base. Therefore, they cannot focus on a single need but have to broaden the design features to appeal to the widest number of potential buyers.

Regardless of the degree of organization of the customer base, the supplier or contractor must thoroughly understand the customer needs. True, on a military development, the DoD does its best to characterize a single product to be developed, making it somewhat easier for the supplier to understand, one would think. The reality is that the customer in this situation seldom can fully describe its need with an adequate degree of completeness, requiring a lot of work by the contractor to fully characterize that need. Successful DoD contractors must maintain an aggressive marketing effort to stay tuned to evolving customer needs. Commercial suppliers must also maintain a marketing effort to ascertain the needs of their chosen customer base. There is essentially no difference between these situations.

1.3.4 Precedented Versus Unprecedented Systems

During the Cold War and since, American military power has been based on superior technology because America could not field as large a force as anticipated adversaries.

The intent has been to fight more efficiently than adversaries, and this requires continual development to stay ahead of potential adversaries. The result is that the DoD had to develop a lot of systems that had no precedent. Their creation involved developing new technology needed to satisfy very demanding requirements. In many cases, these systems solved problems never before solved, that is, unprecedented problems. In these development programs, the principal DoD priority was on product performance with relatively little interest in the cost necessary to achieve that performance. As a result, the methods developed by the DoD were heavily influenced by a need to develop unprecedented systems, often referred to as *clean sheet of paper systems*.

Most commercial systems evolve from other earlier systems. The 2001 model Chevrolet was a variation on the 2000 model to a very large extent. This process might be called a *spiral development process* today. Each variation of the system is heavily precedented by the previous model. Each model advances the state of the art in small steps. As covered in Chapter 3.1, the development approach that seems to work best on an unprecedented problem is Sullivan's notion of form follows function. Part 3 offers several models that apply this approach, where we first identify needed functionality then allocate it to a physical architecture.

When developing the next generation of a precedented system, you already have a physical architecture of the prior system solution and, ideally, the structured analysis data upon which that physical reality was based. It is often more efficient to base the next system on the current architecture, rather than repeat the approach useful on unprecedented problems, and determine the architecture changes to satisfy the functionality differential between the old and new versions.

Since the demise of the Soviet Union, the DoD has not had the funding to apply its earlier demand for unprecedented systems leaping as far ahead of the competition as possible with each new system. As a result, military system development now follows a more precedented sequence than earlier and is more in step with commercial development.

1.3.5 The Three Gross Models

The Wright Flier was purchased by the U.S. Army in accordance with a contract that covered, perhaps, seven pages, including performance requirements for the airplane. Subsequently, as aircraft and other articles of war become more complex, the Army and Navy found that they had to be more specific about their needs in contracting for them. Specifications became popular as a means of conveying those requirements to suppliers. These specifications commonly were developed using what would now be considered an ad hoc approach and published using typewriters and various kinds of copy and printing technologies. This document driven development approach still is in use by many developers.

Some developers, however, have switched to databases for capturing the results of the requirements analysis. Often these database systems are simply used to print paper documents, which is only an extension of the document driven approach. But, when the developer uses the content of the database directly without publishing a paper document for its own purposes, this could be considered database driven development.

Figure 1.3-5 illustrates these two development models as well as a third, which is just now beginning to creep into use, model driven development. In model driven development, databases are used but they are connected so that the relationships between them need not be completed by sneaker nets, paper reports, or extensive human communication. Team members enter data into the databases for which they are responsible and anyone on the program can call up a view of the combined information of value in performing their work, in the process creating new information, which is added to the aggregate information set. Documents in the traditional sense need not ever be published in paper form.

All three of the development models are in use today and will continue to be for a long time. Figure 1.3-5 suggests that structured analysis eventually will rise to 100% utilization on development programs, but that is probably a bit too optimistic.

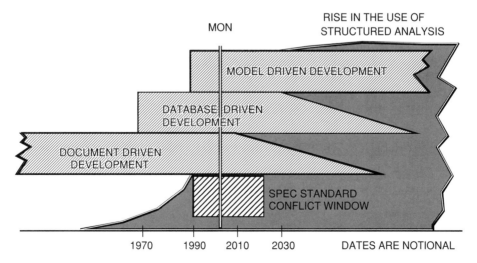

Figure 1.3-5 *Development models*

The crosshatched area on the figure notes a temporary condition where current software specification templates, such as those provided in EIA J-STD-016, are out of sync with the artifacts in recent structured analysis models such as OOA and UML. This condition will end when we conclude that it is no longer necessary to publish paper specifications and find that we can develop good systems by direct appeal to the data in the model driven databases. It is shocking to some system engineers that MIL-STD-961E currently permits a specification to be developed as a model rather than a paper document.

1.3.6 The Lowest Common Denominator

A commercial or aerospace developer can shed management and technical processes that evolved through experience on many DoD programs at its own risk in the interest of reducing development cost and schedule demands up to a point. Three fundamentals should not be eliminated nor should the sequence be tampered with. At the very foundation of the systems approach appears the insistence that one should first define the problem that must be solved and the content of specifications provides that definition. Second, the development organization should synthesize the requirements contained in those specifications using physically collocated cross-functional teams staffed with the right talent and knowledge and implementing an effective concurrent engineering process. This synthesis process must result in three views: (1) the product design, (2) the procurement pattern, and (3) manufacturing process definition. Finally, the product design resulting from the synthesis of the specification content should be subjected to planned specification-driven verification inspections yielding objective evidence of design compliance with the requirements. All this work must take place within an infrastructure of sound technical management. These fundamentals should not be tailored away from the comprehensive process that has evolved on the DoD program at no small cost. Yes, it did cost a great deal of money to develop some tremendous systems using this process while it was being perfected. It will generally cost less to employ the evolved process than to attempt to execute a development using an ad hoc approach.

2

Requirements Foundation

Contents

2.1 Requirements Fundamentals

Contents

2.1.1 Primitive Requirements Statement

Requirements analysis and requirements writing are necessary but difficult, sometimes tedious, tasks. In this chapter we will explore four strategies—one or more of which, hopefully, will reduce the requirements analysis difficulty by providing mechanisms for gaining insight into the attributes that should be controlled by written requirements statements. The fact is that it is not hard to write requirements. It is hard to know what to write them about and to determine appropriate numerical values to include within them.

In this chapter we will develop a progressive scenario that will be useful in writing simple requirements statements regardless which requirements analysis strategy you choose. Given an organized method for identifying the requirements to write and a simple, progressive method to write them, the task remains to assign appropriate values to our requirements and we close out this chapter with that subject.

So, first we need to understand the very essence of a requirement statement. If we strip away all of the style and format common to a paragraph in a specification, we are left with a residue that we will call a primitive requirements statement. This primitive statement is very easy to phrase. Any engineer without special requirements analysis training can do it. Any engineer who hates English grammar can do it. You can combine several of these primitive statements to form a very simple requirements document for an item. This simple document can be useful in easily and inexpensively documenting the requirements for the item very early in the development process before formal specifications are needed. As the need for formality increases to support detailed design or procurement, this primitive requirements list may be converted into a specification of the appropriate type using conversion logic covered in this chapter.

We will call this simple requirements document a Concept Requirements List (CRL). It is intended for use in studies, the early phases of development contract work, and as an initial uncluttered way to capture the requirements for any system element at any phase of development. The understanding is that the engineer is using a structured analysis strategy in this work as discussed in Parts 3 and 4 but it is possible to apply one of the other three strategies, just with a lower chance of success.

2.1.1.1 The essence of a requirement

A requirement is an essential attribute or characteristic for a system or an element of a system. The attribute is coupled with value and units information for the attribute by a rela-

tion statement. Following is an example: "Weight less than or equal to 2400 pounds." In this case, "weight" is the attribute, "less than or equal to" is the relation, and "2400 pounds" is the value and units data. In the case of non-numerical requirements, the relation statement may be words like "is," "shall comply with," or similar text.

A normal requirement statement is a written statement of a requirement in one or more complete sentences in a language familiar to the customer (normally English in the U.S.) using the idiom of the particular business sector (aerospace for example). Good sense and common specification standards require that the content of a specification include complete sentences organized in a particular way. Each requirement statement in a specification must satisfy several characteristics including: (1) proper grammar, (2) appropriate use of shall, will, and other key words, and (3) rigid compliance with a format. But the need, during early contract phases, to immediately prepare complete requirements statements satisfying a specification style guide can act as a barrier to the early, timely identification of needed technical requirements as a prerequisite to early design concept development and procurement work.

In an effort to unburden early concept development work of unnecessary rigor, reduce the cost of such work, and improve the early requirements identification capability on a project, a Concept Requirements List (CRL) is recommended. This document is composed of nothing but a cover and a numbered list of primitive requirements statements. Figure 2.1-1 illustrates the construction of a primitive requirement statement. The primitive statement requires no punctuation and no knowledge of or concern for sentence structure. It requires no knowledge of specification style. It is a simple, ordered string of the four entities noted in Figure 2.1-1. The list of relations is not exhaustive in the figure.

2.1.1.2 Document style and format

The structure of a CRL is very simple. In paper form it consists of a cover page and as many pages of requirements as needed. Each page contains primitive requirements statements. These statements are numbered from 1 through "n". No effort is made to format the primitive statements by category (performance or constraints) or to follow a special prescribed sequence. The statements may simply be listed in the order conceived by the element principal engineer.

The document may be published either in book form or presentation form. In the former, the document is published with the printed lines running parallel with the 8 1/2 inch side of 8 1/2 by 11 inch paper. In the latter, the lines run

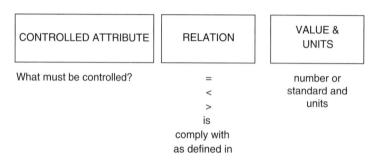

Figure 2.1-1 *Primitive requirements statement structure*

parallel with the 11 inch side of the page. The presentation format is intended for inclusion in briefing or presentation materials.

The cover should contain an identification of the element by item title and architecture number, WBS identification, or part number. The cover must also include a date and document number. As a minimum, the engineer responsible for preparing the document (normally the principal engineer for the element) and the project, program, proposal, or study Chief Engineer or lead engineering person should sign the cover page to indicate it is an element of the concept baseline. The principal engineer, his or her supervision, team leader, or the Chief Engineer (or equivalent) may require other document approvals.

The body of the document is composed of one or more pages of numbered primitive requirements statements. Figure 2.1-2 illustrates a typical page from a CRL. Simply number the primitive statements from 1 to "n." If, after the initial list is published, it is necessary to delete a statement embedded within the list, it and its number should be deleted without changing the other statement numbers. If the document is prepared on word-processing equipment rather than from a computer database, the number of the deleted statement may be retained followed by the word "Deleted." If desired, the CRL may include any of the simple block diagrams defined in subordinate paragraphs. When included they should be referenced in a primitive requirements statement.

a. *Architecture Block Diagram (ABD).* An architecture block diagram is a hierarchical diagram consisting of simple blocks depicting the elements of a system illustrating the family structure of the system. If an ABD is included in the CRL, it should consist of one block for the subject block and one block for each of the immediately subordinate elements arranged below it and interconnected by a series of lines denoting hierarchy. The primitive statement would be, "Element architecture as defined in Figure X."

b. *Schematic Block Diagram (SBD).* A schematic block diagram illustrates the interfaces required in a system or element thereof by connecting blocks from the architecture block diagram with lines indicating an interface requirement between a pair of elements. A schematic block diagram illustrates the interfaces between the element that is the subject of the document and all other system elements. This diagram consists of a block titled with the name of the element and one block for

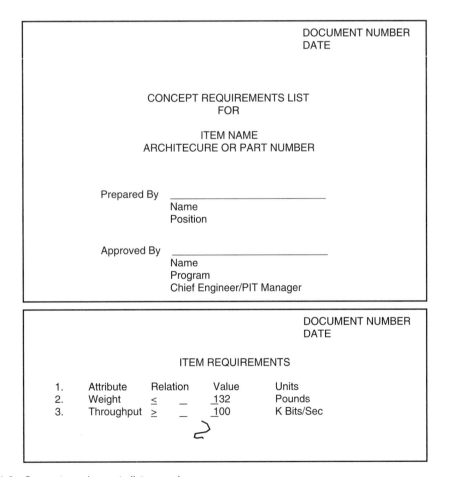

Figure 2.1-2 *Concept requirements list example*

each unique element the subject element interfaces with. One or more lines are drawn between the subject block and each interfacing block. Alternatively, an n-square diagram could be used. The primitive requirement statement referencing the diagram would be, "Element interfaces as defined in Figure X."

c. *Functional Flow Diagram (FFD).* A FFD may be included to illustrate the identification and sequence of functions the element must satisfy as a precursor of performance requirements captured in other primitive requirement statements. The referencing primitive requirement statement would be, "Element functions as defined in Figure X."

d. *Process Flow Diagram (PFD).* As an alternative to a FFD where the elements of the system are already well established, and functional analysis is not necessary as a structured system decomposition tool, a process flow diagram may be included. The PFD is similar to the FFD but it is created as an analog of the planned real world operating system process. A functional flow diagram is frequently prepared when the exact composition of the system is not yet known. The primitive statement would be similar to that shown for the FFD.

e. *Timeline Diagram (TLD).* A timeline may be included that defines critical timing of signals or events that the element design must respect. It should consist of a Gantt-type chart where the bars are defined on a vertical axis and their length indicates time duration against a horizontal time scale. If a FFD is also included, the FFD blocks and the timeline blocks should be coded with corresponding numbering. The referencing primitive requirement statement would be, "Element timing as defined in Figure X," where _____.

2.1.1.3 Primitive requirement statement conversion

At some point the CRL should transition into a specification with the normal, complete requirements document structure. At that time, the principal engineer may convert the CRL content very easily into complete English sentences as described in the paragraphs below. At the same time it will be necessary to organize the requirements into the prescribed paragraphing structure. Depending on how the organization is structured and how the work is divided between departments, the design engineers may be tasked with creating the CRL content and translating it into specification language. Alternatively, the design engineers could be held responsible for the CRL and a specifications group could translate the technical primitives into specification style and format.

By way of example, assume an original primitive requirements statement for a liquid hydrogen valve of, "Closure time less than or equal to 0.5 seconds." This primitive statement may be converted into fully compliant specification text as follows.

a. *Add a Subject.* Add a subject to the front end of the statement. In the case of our example, the subject is "valve." So the statement would now read, "Valve closure time less than or equal to 0.5 seconds." Often the word item is used generically, as in "Item valve closure time . . ."

b. *Add a Shall Verb.* The statement now will read, "Valve closure time shall be less than or equal to 0.5 seconds."

c. *Provide a Sentence Ending.* In this case the sentence can be ended by simply adding a period as fol-

lows: "Valve closure time shall be less than or equal to 0.5 seconds."

d. *Provide a Paragraph Title and Number.* Identify the kind of requirement and assign the appropriate paragraph number. This can be done by consulting the appropriate requirements document standard or customer-provided data item description. In this case our requirement is a performance requirement that would normally be assigned a number within the paragraphing structure 3.2 reserved for performance (entity capability) requirements in most requirements document structures. Also assign a paragraph title, normally linked to the controlled attribute. Assuming that we had already assigned paragraphs 3.2.1 through 3.2.3, the complete conversion would be: "3.2.4 Valve Closure Time. Valve closure time shall be less than or equal to 0.5 seconds."

A compromise between the two requirements numbering extremes (full compliance with a customer standard and simple unique numbering) might entail numbering requirements in four basic sets as follows: performance, interface, environmental, and specialty engineering.

e. *Additional Data.* It may be necessary to add defining, clarifying, or explanatory information to ensure understanding and to avoid ambiguity. Add this kind of information at the end of the converted statement as in this example: "3.2.4 Valve Closure Time. Valve closure time shall be less than or equal to 0.5 seconds. Valve closure time is measured from the time the controlling signal reaches 50.0% of its nominal value until flow through the valve drops to zero."

2.1.1.4 Total effect of changes

The difference between the primitive and complete requirement statement in this example can be seen in the bold characters of the following statement:

3.2.4 *Valve Closure Time.* **Valve closure time** shall be **less than or equal to 0.5 seconds**. Valve closure time is measured from the time the controlling signal reaches 50.0% of its nominal value until flow through the valve drops to zero.

2.1.1.5 Variations

The process of creating a CRL has been kept as simple as possible in this chapter. As stated, this activity may be accomplished in the context of any of the three requirements analysis strategies covered in Chapter 3 of this part. In applying the progressive scenario within the structured strategy environment, one would also be interested in capturing other information beyond the attribute, relation, value, and units. If the primitive requirements are retained in a networked computer database, this is greatly facilitated. Of particular importance, the source of the requirement should be captured. It will later be important to know whether the customer provided this requirement, it was driven out by internal trade studies, or was consciously included for future growth capability of the product line.

We will find that the structured strategy encouraged in this volume provides two basic mechanisms (structured analysis and constraints analysis) to gain insight into requirements. In each case, these mechanisms drive out primitive characteristic statements that then need to be translated into full requirements statements. How this progressive concept can be applied very easily in a freestyle environment will be intuitively obvious in Chapter 3. We will also clarify how it can be applied in the context of a cloning strategy in that chapter. Parts 3 and 4 develop the structured requirements analysis strategy.

2.1.1.6 Document example

Figure 2.1-2 illustrates the CRL document concept. In each case, a cover page with signatures is followed by a simple list of requirements each stated in terms of an attribute, a relation, a value, and units. The book style appears with the data included in portrait orientation and the presentation style places the information in landscape orientation on the page. In the presentation style, the data should be placed on the page in accordance with the program template for presentation materials. The cover page may be used as the presentation introduction page for element requirements or left off where these materials are a part of the Principal's presentation that also includes concept materials.

2.1.2 Requirements Value Definition Methods

Some requirements will be of a qualitative kind, invariably, existential requirements being an example. But the norm should be that requirements contain numbers making related magnitudes clear.

2.1.2.1 Why is quantification important?

It is important that requirements be quantified. A requirement states a necessary attribute of an item yet to be designed. How can the designer possibly design an item for an attribute without the attribute magnitude being characterized? In the absence of quantification, there is an excellent chance for failure in the form of: (1) exceeding the minimum necessary cost due to overdesign or (2) failing to account for needed capability that reduces item value to the customer. Quantitative requirements are also necessary in order to adequately verify the produced item satisfies its requirements. Test planning activities focused on requirements verification must define pass/fail criteria and this is very hard to do for qualitatively stated requirements.

In organizations where designers are held accountable for definition of lower-tier item requirements as well as design, it is not always easy to convince the designers of the importance of clearly defining the requirements and values for them prior to developing a design solution. Many design engineers feel that this is an unnecessary impediment that constrains their creativity without useful benefit. Others feel that this is an essential and inseparable part of the design process and that such data will have to await the development of design solutions.

Depending on the intensity of feelings, these attitudes may be difficult to change, but they are clearly wrong-headed. Properly defined quantified requirements provide a maximum permissible solution space within which designers may creatively pursue solutions. They provide a safe framework within which creativity can be allowed full reign. Development of systems too large for one person to master the complete design task requires cooperative effort between many designers and analysts who all must be aware of the needed attributes. We have to date found no better alternative than to document the needed attributes and make them available as widely as possible in the form of specifications.

The principal impediment to group participation in requirements analysis is that individual engineers fail to understand that in order for them to have access to the information they need to support the development of their item, other people have to provide information they can share. Likewise, others need information for which they are responsible. Everyone has to contribute to the greater good. In our definition of a system earlier, we said that a system is greater than the sum of its parts. This is true of teams performing system engineering work as well, provided everyone on those teams is doing their part. Team members may be motivated by their own enlightened self-interest, a selfless interest to serve the needs of their fellow engineers, or only in response to knowledge that their pay, position, and employment longevity may be dependent on their requirements analysis performance.

In Part 1 we discussed various organizational alternatives within which the requirements analysis process may have to function. The responsibility approach encouraged in this book involves a distributed process where designers have to perform a requirements analysis for their lower-tier item prior to switching to a design solution search and design engineering work. We should all recognize that this approach puts a great deal of stress on the designer. Design engineers are creative people. Design work is creative work. Structured requirements analysis work involves organized thought processes and very little creativity. Creativity and structure seldom reside together in any one person. It is probably true that the more creative the individual engineer, the more difficulty he or she will have adapting to an organized requirements analysis methodology.

Some companies have avoided this problem by centralizing their requirements work in a system engineering organization populated by people who like to work in a structured environment. The specifications are given to the designers complete and the designer's task is to synthesize those requirements into an effective design solution. The author accepts the position that, while this approach will work and avoid designer stress encountered in a distributed approach in the process, we will realize a better mix of requirements quality and synthesis excellence by involving the design engineer in the requirements process.

In order to apply the distributed approach successfully we must ensure that everyone understands the stressful situation in which we place the design engineer and provide an environment within which the designer can be successful. The primitive requirements writing scenario outlined earlier was motivated by exactly these attitudes. This book encourages the availability of several models that are relatively easy to apply to uncover an appropriate requirements set for a single item and a simple way to initially write these requirements. If an organization chooses to do so, a centralized specification group may then massage this intermediate product into a polished specification.

2.1.2.2 Value definition methods

The assignment of a value to a requirement can be done through an application of good judgment based on information acquired through some rational process including: (1) a market survey to find out what industry is doing or what customers prefer, (2) experience with the product line and customer needs, (3) an appeal to authority such as the customer or program engineering management, or (4) reference to industry or customer standards for values proven in past practice or testing. These approaches can provide a quick answer of merit proportional to the credibility of the source.

The simplest analytical approach to defining an appropriate value for an individual requirement is to find a way to express the value mathematically in terms of other variables for which appropriate values are known or can be assigned to study their effect on the value of interest. The mathematical relationships are solved for one or more sets of independent variable values therefore defining the dependent variable value.

As a simple example, given that we are interested in specifying the volume and form factor for a servo amplifier unit for a procurement specification, we might first accept that the box will be of a rectangular form. The box volume, V, is therefore the simple product of height, length, and width. If our company prefers standard single channel amplifier circuit boards 6 inches wide and 4 inches high, we need only now determine how many channels our amplifier needs, power supply size, and box interior space needs. We might allow 1/2 inch on each side of the circuit cards for mounting hardware and case and 2 inches for circuit card electrical connector. The simple box volume equation has helped us determine that this three channel amplifier box must be $6 \times 7 \times 8$ inches in size or 336 cubic inches. Our space requirements and interest in standard circuit board size could be stated for potential suppliers as:

> 3.2.2.5 *Unit volume.* Servo Amplifier Unit volume shall be less than or equal to 336 cubic inches.

> 3.2.2.6 *Unit packaging.* The item shall use 4 by 6 inch circuit cards.

The maxima and minima methods of differential calculus can be applied to mathematical relationships to define optimum rates of change. Integral calculus provides a means to conclude appropriate distances, areas, and volumes and their analogs in various fields. Set theory, combinatorics, probability, and statistical theory can be used to quantify desired product features like reliability.

A suitable value for a requirement can be selected based on an evaluation of a wide range of possible values keyed to other requirements using parametric analysis. This approach develops curves relating two or more quantities in graphical form and influenced by one or more other variables called a parameters. A simple example of such a relationship is the linear equation $y = mx + b$. The quantities x and y are the variables that can be applied to two requirements categories while the slope and intercept points, m and b respectively, are the parameters. Using this relationship, it is possible to characterize every possible linear relationship between two variables x and y.

We can form similar parametric relationships between variables influenced by a quadratic relationship using the conics. Even more complex relationships can be formed by higher-order equations, trigonometric and exponential relationships, and application of integral and differential calculus. It is also possible to create parametric relationships from experimental data acquired from tests where a parameter is varied in some organized way to observe how it influences the relationship between the variables.

Analysis of the relationships will disclose boundary conditions driven by, for example, the laws of physics, realistic values based on the current state of related technology, or time that limit the solution space. If we can develop a pair of parametric relationships involving a common variable and parameters where one variable is directly related to the common variable and the second is inversely related to the common variable, we may be able to develop a convenient "bathtub" relationship illustrated in Figure 2.1-3 that gives the best combined values for the pair of requirements represented by the two different variables.

The operations analysis or research field has developed, or extended the application of, many useful analytical tools for defining requirements values. Linear programming is useful in analyzing relationships between entities that can be expressed in terms of straight lines or planes. It is an organized sequence of steps (thus the word programming) for developing a series of algebraic equations leading to a conclusion about the best values for two or more quantities related linearly. This method seeks to allocate resources in an optimum way. The resources can be anything we are smart enough to apply the method to. Lawrence L. Lapin's *Quantitative Methods for Business Decisions*, Harcourt, Brace Jovanovich, 1981, offers an excellent detailed discussion and many examples of this and other techniques identified in this section. This concept can be extended to quadratic and stochastic models as well.

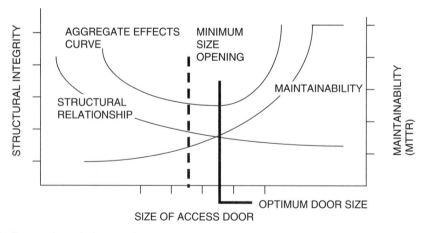

Figure 2.1-3 *Parametric analysis example*

Dynamic programming, like linear programming, seeks to maximize positive attributes (profit, payload capacity, range) while minimizing undesired attributes (cost, schedule time). It applies the techniques of mathematical programming to the more complex situation where time is involved.

Queuing theory can be brought to bear on requirements that can be related to a situation where one or more servers are providing a service to people or things arriving at unpredictable times. This technique is very useful in maintenance requirements work as well as in computer and telephone system service expectations.

All of the techniques discussed above can be carried out with pencil and paper computation. They can be more effectively employed in the form of a computer model that can be manipulated rapidly for given values and the values changed to observe effects of the changes. Moshe F. Rubinstein in *Patterns of Problem Solving*, Prentice Hall, 1975, defines a model as "an abstract description of the real world; it is a simple representation of more complex forms, processes, and functions of physical phenomena or ideas."

Reliability and mass properties models allow assignment of values to system elements while maintaining system level values based on all of the constituent values. They do this by performing the mathematics of the model on the constituent values to calculate the system value. In the case of mass properties, the mathematics is simple addition. The reliability mathematics is a little more complicated but the same process is involved. Models can be used to define requirements values and observe the system effects or as a basis for allocating a scarce available resource, like weight.

In determining appropriate values for requirements like propulsion system thrust, guidance unit accuracy, and allowable airframe asymmetry, we commonly have to appeal to more complex computer models properly called simulations. A simulation commonly involves a model defined in mathematical or logical terms and a series of step-by-step computations allowing the model to imitate the manner in which the real system might perform in real time. The simulation will be run with a particular set of detailed requirements values to produce an output condition possibly in terms of system cost or the values for a predetermined set of measures of effectiveness. Other runs with other combinations of input variables will produce different outputs. We can choose the input combinations that produce the best aggregate output values.

Monte Carlo simulations offer a relatively inexpensive way to define a reasonable reality for time requirement values where uncertainties make it difficult to select one. This approach works by defining a mathematical relationship between input variables, a probabilistic definition of each input variable, and a very large number of simulation runs each with a randomly selected set of inputs. The outputs are samples for one or more probabilistic variables. This approach can have a high price tag, but is less expensive than actually building equipment and performing a large number of experiments. Since the quality of the output is driven by the number of runs, it can require a lot of computer time if the problem is complex and computer speed is slow.

One of the most effective requirements valuation methods is allocation by apportionment where the value for an attribute is determined by subdividing a parent value for that attribute into component parts in accordance with some mathematical rule and a rationale for relative values across the subordinate items. Allocation and budgeting, two

views of this methods, are covered in detail in Chapter 2.3. Weight, design-to-cost, and reliability are easy examples of attributes that will yield to this approach.

As explained in Chapter 2.3, allocation involves breaking a single attribute value into parts and assigning those component values to subordinate items. Budgeting commonly involves one or more attributes across several layers of system architecture that can be related in some way. For example, in determining how we will control system characteristics that contribute to guidance accuracy for an aircraft we may budget a maximum error for the guidance platform, a mounting error for the platform to the airframe, errors induced or allowed in testing the guidance system, and initialization or update errors based on the degree of precision in our knowledge of where we are at the time of initialization or update. These become requirements for a number of things in the system architecture all related to system guidance accuracy.

These approaches are driven by the economic reality that there is never enough of critical commodities, especially when you are pushing the state of the art, and that the available value for a commodity must be shared in the best aggregate way by the subordinate items. The need to respect the parent value leads to healthy debate between the principal engineers for the subordinate elements about why they need particular values, the total of which is more than what is available. Hard choices are made, hopefully based on factual input from the lower-tier principal engineers and the analysis engineer responsible for the parameter across the system.

Each attribute approached through allocation or budgeting requires some kind of mathematical model, like those described above, as the engine for systematically producing coordinated values. The model may be as simple as one or more equations or as complex as a computer simulation running on a super computer. In mass properties requirements definition, a simple weights statement is sufficient. The mass properties engineer decomposes a top-level weight figure into component parts. In each branch of the architecture the sum of the item weights must be less than or equal to the parent item weight. Since budgets commonly involve several attributes measured in different units, they may require a more complex relationship than a simple additive or multiplicative model.

2.1.3 Requirements Derivation

Many design engineers like to separate all of the requirements that pertain to the item for which they are responsible into two sets: (1) source or customer requirements, and (2) derived requirements. This may be a useful distinction for an engineer working on what he or she perceives as an isolated development task. The system engineer should realize that in the development of an unprecedented systems, every requirement but one, the system need, is derived from the need.

In the overall requirements analysis process, the derived distinction tends to have little special significance since all requirements except the ultimate customer need are derived. Particular engineers focused closely on their own problems, may attach a great deal of significance to the term derived. One motivation for this feeling is that externally identified requirements cannot be changed while derived requirements are under the control of the engineer and are often enforced by the design decisions made by the engineer.

Another basis for this attitude is that some engineers feel that derived requirements are inextricably intertwined with the design solution; they are characteristics of the design solution. These requirements are valid because of the way the product is designed. This attitude may lead to release of the item requirements document subsequent to design development rather than before. This attitude is, of course, not in keeping with the principles covered in this book. Development requirements are needed attributes defined as a prerequisite to design solution development. If they are developed in any other way, time spent doing so is money wasted.

It is true that the design solution at one level can and should influence the requirements for lower-tier elements. Some engineers, responsible for development of multiple layers in the system architecture, can become confused about the difference between tiered requirements and tiered solution characteristics. The whole set of characteristics, whether properly called requirements or solution characteristics, becomes simply identified as derived requirements. This is where a specification tree and discipline in the requirements development sequence relative to requirements synthesis efforts can be very helpful.

It is not necessary to engage designers with these attitudes in heated debates about terms. It is important to ensure that they understand they must respect the requirements identified for items they are responsible for designing. Where a distributed requirements analysis responsibility pattern is used, these designers also must understand their responsibility to define the requirements at each level prior to synthesizing them into a design solution. Whether particular requirements for an item are properly described as derived or not is of little consequence, since all requirements are derived at some level.

The author prefers a particular use of the word derived. It is applied to the derivation of requirements from driving models. A requirement is identified by deriving it from a model and then defined through the application of a relevant analytical discipline. For example, the fact that item A132 must have a reliability requirement is identified by the reliability math model including item A132. The required failure rate applied to that item is determined through analysis of the item, its application in the system, how parent failure rate is distributed among the items, and other factors. A performance requirement is derived from a function included in a functional flow diagram and defined as a result of analysis of just how well the system must accomplish that functionality. This book will encourage the use of a requirements analysis sheet (RAS), ideally captured in a requirements database in a computer application, as the means by which we relate all requirements and the models from which they were derived and the product entities to which they are allocated.

2.1.4 Kinds of Requirements

There are two kinds of requirements for systems and hardware entities that must be developed. These are performance requirements and design constraints. It is convenient to partition all requirements into these two sets because they must be identified through fundamentally different methods as will be seen in Part 3.

2.1.4.1 Performance requirements
Performance requirements define what the system or item must do and how well it must be done. Several modeling techniques have been developed to encourage a structured approach to finding these performance requirements and a model of some kind should be used because they tend to be exhaustive. Structured analysis performed with models encourages full coverage of the problem space and avoids identification of extraneous requirements.

2.1.4.2 Design constraints
2.1.4.2.1 What Is a Design Constraint?
A design constraint, or simply a constraint, is a boundary condition within which the designer must remain while satisfying the aggregate of the performance requirements for the item. Some engineers assign a more general meaning to the word constraint as anything that bounds their design prerogatives. Note we are using the word more precisely. All requirements constrain the solution space but in this book design constraints are those requirements not related to performance of the product.

We wish to impose no more boundary conditions than necessary on the designer because they reduce the solution space within which the designer may seek an answer to the problem defined by the performance requirements. At the same time, it has been proven through years of experience in DoD procurement of countless systems that the total cost of a system over its life time (Life Cycle Cost) can be significantly reduced over what it might otherwise have been by imposition of certain constraint conditions in specialty engineering areas dealing with logistics support.

Most engineers who have experienced a flawed development cycle, resulting in a large number of design changes during the integration period occurring in the design and production phases, would agree that large sums of money can be saved in clearly defining as early as possible the interfaces between items and the environments within which the items must perform their function. The fundamental economic choice again comes to the fore; we can spend the customer's money in an organized requirements analysis process or we can do so on design changes later. It is surprising how many development teams choose to do the latter, often by default, even though it has long been known to be more expensive than solving these problems on the front end.

Three major categories form the body of constraints we will discuss in this chapter: (1) interface, (2) environment, and (3) specialty engineering. Each item in the system architecture should have developed for it, during a requirements analysis period, a set of requirements embodying both appropriate performance requirements and design constraints. In each of these three categories we will seek in this book to uncover an organized approach to the identification of a list of primitive requirements attributes, which can be expanded into full requirements statements using our progressive requirements analysis scenario tuned to a particular methodology found effective for the analysis category.

Integration and accountability are very important topics relative to constraints analysis because, as we will find, it takes a large number of specialists to perform constraints analysis and the requirements product produced by these people can easily include contradictions resulting, if unattended, in added cost. Since we engage in constraints analysis specifically to control Life Cycle Cost, it would be very disheartening for it to result in added cost.

2.1.4.2.2 Design Constraints Analysis Timing
Design constraints analysis for a given item should be accomplished concurrently with functional analyses and the resulting hardware and software allocations. That is,

it should be performed in parallel with the performance requirements analysis recognizing a significant difference. We can identify performance requirements on the functional plane before knowing what the architecture item is to which the requirement will be allocated. That really should be considered the norm. Design constraints have to be identified in context with the architecture items they are associated with. They must be identified on the physical plane. The system engineering purist may feel that the complete requirements analysis process, including constraints analysis, should be finished before the design concept is selected. Unfortunately, that is not always possible. Some design constraints can only be properly defined in quantitative terms after the design concept or even preliminary design has been selected.

The development team or principal engineer must be watchful that compatibility is maintained between the evolving constraints and the approved design concept. Where new, needed constraints are identified that conflict with existing design conclusions, they must be re-visited and conscious decisions reached about how to handle the conflict. We should seek to minimize cases where the design must be changed to reflect the late development of requirements, of course. If carried to the extreme, this can result in what we are trying to prevent through careful requirements analysis, the addition of unnecessary cost.

2.1.4.2.3 Major Design Constraint Categories

In Part 3 we will thoroughly investigate each of the three major classes of constraints. As a way of clarifying the field of play in these discussions, let us see if we agree on a brief description of each: interface (covered in Chapter 3.6), specialty engineering (covered in Chapter 3.7), and environmental (covered in Chapter 3.8).

Interface requirements analysis identifies physical and functional relationships between system elements and between system elements and the system environment. Contractors and development teams are individually responsible for interfaces where both terminals of the interface are within their design responsibility or that of one of their subcontractors and jointly responsible when one terminal is their responsibility and the other terminal the responsibility of an associate, teaming contractor, or other development agent. In all of these cases, one contractor or team member should be assigned principal responsibility.

In Chapter 3.6 we discuss interface analysis at some length as a means to determine how the elements of a system must inter-relate. Interface analysis is to interface requirements identification as functional analysis is to performance requirements identification. It provides an organized environment by identifying attributes that can be expanded into quantified requirements statements. As in performance requirements analysis, we seek to apply a structured technique involving a graphical abstract to focus our thinking on interface needs and their corresponding requirements. In this case it is the lines on schematic block diagrams or the marks in the intersections of an N-Square diagram.

With the exception of interface requirements, requirements are collected about the architecture elements. Interface requirements are captured about pairs of architecture elements and the interface shared by the two architecture elements. We can choose to place the interface requirements in an interface control document (ICD) that bridges between two specifications. Alternatively, we can capture the interface requirements in the terminal specification pair. If we choose the former, we should not also include the requirements in the specification pair because you will not be able to keep the content of the two specifications and the ICD in sync. The two terminal specifications should treat the ICD as an applicable document referencing it in Section 2.

Environmental requirements analysis identifies conditions that are expected to be encountered during storage, shipment, and operation from natural, hostile, non-cooperative and self-induced sources (e.g., thermal, vibration, g-loading, acoustics, humidity, etc.). Environmental requirements will require the most complex model of all of the models applied in traditional structured analysis entailing three different models.

One of the principal differences between the engineering craft now and several decades ago is the emergence of many specialty engineering disciplines. This has come about through explosion in available knowledge forcing us to specialize as we have discussed before. The specialties include reliability, maintainability, producibility, mass properties, electro-magnetic interference/electro-magnetic compatibility, and many others that will be listed later. In each case, the specialty engineer works to identify a minimum set of characteristics that each system element must have to satisfy program requirements for that specialty engineering discipline.

2.1.5 Requirements in Time

It is a common attitude that requirements should always be defined before any design work is accomplished. This attitude is not always expressed in actual work but, even among designers who race to a point design solution before any attempt to understand the problem, it is commonly felt that requirements should come first. It turns out that this is not always the case, actually. When developing a new item, it is not always possible to thoroughly understand the problem before developing some alternative solutions. Performance requirements can be defined prior to even knowing what the item is but design constraints must be identified after knowing what the item is and some of its features exposed.

Where it is necessary to develop a new product, it is necessary to define the requirements for the item to be developed. These requirements should be design independent and not predetermined a specific point design solution. Structured analysis is a very useful method for understanding what requirements are appropriate for the item. The specification within which these are captured has been called a Part I, development, or currently a performance specification by Department of Defense. This specification is the basis for design development and item qualification. These specifications should be complete and approved by PDR and before any program funds are spent on detailed design work for the item covered by the specification.

Given a particular design solution, an organization that uses two-part specifications will develop a Part II, product, or detail specification that is the basis for product acceptance. Note that this specification must be developed knowing the design solution and the requirements are focused on identifying how we may determine if a particular product article is acceptable for delivery to the customer.

Organizations that prepare one-part specifications can evolve into a set of bad requirements analysis habits based on the above logic. The way that a one-part specification should be developed is to first develop the proper development content devoid of design solution bias and release that specification as what DoD would now call a performance specification. Subsequently, this specification should be edited, based on the design, for acceptance characteristics and re-released as a detail specification. A faulted way of doing this work is to recognize that the one-part specification cannot be completed until the design is complete. The responsible engineer, not wanting to cause unnecessary cost, may decide to wait to release the specification until it is complete. Over time, the fact that some of the requirements are developed after the design is complete may subconsciously encourage engineers to accept that it is okay to prepare the complete specification after the design is complete. Nothing could be further from the trail toward engineering excellence.

2.1.6 The Remaining Road

In this chapter we have constructed a simple way to write requirements statements and reviewed several ways to define appropriate values for requirements. We need to apply these techniques to particular requirements, but what requirements? How do we know what the right attributes to control through requirements are? That is the subject of Parts 3 and 4 of this book. Prior to tackling this problem, we will first study the relationships between requirements in a system, margins and budgets, and then take up four strategies for identifying characteristics that should be controlled.

2.2

Requirements Traceability Relationships

Contents

2.2.1 Requirements Are Not Islands

Individual requirements statements contain information about essential characteristics or attributes a system or system element must possess. In Chapter 2.3 we will discuss margin and budget relationships between individual requirements. In Chapter 6.4 we will discuss the problem of integration of these statements for a given element to ensure that none of the requirements in an element set are in mutual conflict. In addition to these relationships, there are five other relationships of note between requirements and between requirements and other sets of data.

This chapter covers linking requirements up to the source from which they sprang, a rationale for their identification, and requirements for other elements with which they are related. This is called vertical or parent-child traceability. In addition, as shown in Figure 2.2-1, we may also be interested in longitudinal traceability that establishes linkage between a requirement, its design features, and verification information. The third kind of traceability of value is traceability to the modeling source from which the requirement was derived, or lateral traceability.

Traceability information is seldom included in a specification, but it is very useful during the development and modification processes to have access to this information. This information can be developed after the fact, but it is much more difficult to accomplish than it is during the time the requirements are being defined. The cost and ease or difficulty of doing the work and editing the traceability data over time is influenced in a huge way by the methods we use to capture it. It can be developed in real time with the requirements analysis process at little added expense if an effective computer database tool is in use.

This information is valuable because it helps to answer questions about why particular requirements have been identified some time after they have become a part of the system requirements. As the development process moves forward there will be many occasions where you will need to know just how important a particular requirement is. Changes in program personnel, the passage of time, and the tremendous volume of information involved in requirements analysis work for a large system will make it very difficult for you to answer customer questions along these lines based solely on human memory. This information needs to be documented somewhere. All of this traceability work does cost time and money and one might very well conclude that it is not an effective use of those resources. Often the benefits from traceability appear to be derived later rather than sooner but in the near term it encourages better product

requirements and it tends to reduce program risk. The question, therefore, becomes, "How much risk can you stand?"

2.2.2 Vertical Traceability

2.2.2.1 Requirements source traceability

A requirements source statement identifies where the requirement came from. At the system level this could be from the customer need directly, as a result of allocation of master functional flow diagram functions to the system level, a system mission analysis report, a conversation with a customer representative, or many other sources.

The most common source for lower-tier requirements should be as a result of the requirements analysis process performed for the item in question. A reliability requirement for a parent item will end up being allocated to all of the child elements as will the weight of the parent item. Similar flow down processes performed by specialty engineering and system analysis disciplines captured in analysis reports are a normal source of lower-tier requirements.

MIL-STD-490A, the Department of Defense (DoD) specification practices standard for many years, now replaced by MIL-STD-961E, cautions against including extraneous information in a specification. Source information would be considered extraneous if it were included directly mixed with the requirements. The DoD specification standard allows for a Section 6 called "Notes" that can be used for capturing all kinds of information including source information. Ideally, requirements would not be included in Section 6, however.

A way to do this is to include a paragraph, say 6.2, in the specification Section 6 as follows (only a fragment of the referenced table included):

"6.2 *Requirements sources.* Table 6-1 identifies the source for each requirement in Sections 3, 5, and appendices except for titles-only paragraphs.

Table 6-1 *Requirements sources table*

Paragraph number	Source
3.2.1.2	Contracts letter 34-567 dated 11-10-90
3.2.1.3	Customer System Requirements Review minutes
3.2.1.4	Derived from customer need as a function of our solution using a chemical rocket
3.2.1.5	Chief Engineer decision at ERB-1302
3.2.1.6	System Analysis memo ANAL-90-153
3.2.1.7	Telecon 09-15-90 John Foster with Capt Bill Stanley in SPO system engineering

The table has a line item for each paragraph in the specification that includes a real requirement followed by the source definition. Note that in this example meeting minutes, reports, memos, and even telephone conversations (telecon) are referenced as sources. These are all valid sources as are many other possibilities. If the analyst maintains this kind of table in Section 6 as he or she builds the document content, it is possible to inexpensively capture the sourcing data for future use but the best way to capture all of this traceability data is in a computer database. If the

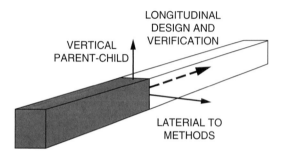

Figure 2.2-1 *Three-Dimensional traceability*

database of choice does not happen to have appropriate fields and relationships to permit source traceability capture most of the commercially available tools apply a variable schema that can be changed to support it.

2.2.2.2 Requirements rationale traceability
The source information covered above does not tell why a requirement was included and this information can be very useful. The model built for source information can be easily expanded to account for rationale information. Simply add another column to the table titled "RATIONALE." In this column you will include reasons why the requirement is included and/or the basis for the value identified.

We can alter the earlier Section 6 paragraph and table as shown below for a hypothetical specification. In both of these tables it would be an improvement to include the paragraph title to convey more clearly what the requirement concerns without leafing back and forth through the document. The paragraph title was not included here in the interest of space. Anyone who has tried to maintain information like this using a word processor (or, heaven forbid, a typewriter) knows it can be a little maddening. We will see in Chapter 7.1 that it can be made less so through a computer database approach. For now just think about the utility of this information and not how hard or easy it is to maintain.

6.3 *Requirements source and rationale.* Table 6-1 identifies the source and rationale for each requirement in Sections 3, 5, and appendices except for titles-only paragraphs.

Table 6-1 *Requirements sources and rationale table*

Paragraph number	Source	Rationale
3.2.1.2	Contracts letter 34-567 dated 11-10-90	Customer furnished equipment driven
3.2.1.3	Customer System Requirements Review minutes	The kill radius was changed to 30 meters
3.2.1.4	Derived from customer need as a function of our solution using a chemical rocket	Reaction time selected to account for best possible capability
3.2.1.5	Chief Engineer decision at ERB-1302	Conflict between range & payload capacity resolved
3.2.1.6	System Analysis memo ANAL-90-153	Worst case impact angle is 13 degrees
3.2.1.7	Telecon 09-15-90 John Foster with Capt Bill Stanley in SPO system engineering	The Capt feels that lateral coverage is more important than in-line coverage by factor of 2

Requirements source and rationale capture are examples of a phenomenon understood by many parents called

delayed gratification. The value of the action taken today does not appear to match the near term cost but the value is greatly appreciated at some time in the future. The value of the money spent doing this work produces little immediate utility but many months or a few years later it can be very valuable to be able to recall the details of work accomplished earlier. As a result, it is sometimes difficult to justify to a program manager the expenditure of program funds to accomplish this work. Once a program manager experiences a serious program problem that has been solved by reference to this information, he or she will become a devotee of capturing it. For example, Col. Shaftly asks at a critical design review who the person was that came up with the idiotic idea to exclude a redundancy requirement. It can be helpful to your cause if not your popularity if you are able to reply, "I regret to inform you, Sir, that it was one Col. Shaftly, on June 13th two years ago at the preliminary design review."

2.2.2.3 Requirements traceability and allocation/flow down
In functional analysis we use the term allocation to mean the identification of an architecture element which will be responsible for satisfying a performance requirement. This term is also used by some to name the process of deriving a sub-tier requirement value from a parent architecture element requirement value also often referred to as flow down. The author generally uses these two words interchangeably as either is a good term for the general process of working the requirements analysis job from the top down. Weight, cost, and reliability are common examples of requirements that easily yield to this technique. This methodology provides a systematic way to determine an appropriate value for a requirement when you know the parent value for the same requirement. It is useful in performance requirements analysis sometimes but most often in the design constraints disciplines. Flow down clearly establishes a traceability condition that can be easily captured while it is being practiced. Actually, traceability should be recognized as bidirectional phenomena.

We will identify three allocation or flow down techniques called apportionment, equivalence, and synthesis. Each must be used in combination with an accompanying analytical methodology to determine the appropriate sub-tier requirement values. In the case of weights flow down, the accompanying methodology is the organized partitioning of a weight figure into smaller values, the sum of which is the parent weight figure. Some constraints disciplines, such as reliability, deal with probability figures where the methodology entails the same basic process used in weights but a different rule of mathematical combination.

Value apportionment is effective where the requirement category is common to several levels of architecture, is numerically quantifiable, the value can be partitioned in accordance with some mathematical rule or relation into two or more parts that will be the requirement values for the parent element's subordinate elements, and the sum (or some other mathematical relation) of the parts yields the parent requirement value.

For example, given that the Design To Cost (DTC) requirement for element A135 has been assigned a figure of $5800, and the element A135 has three sub-elements A1351, A1352, and A1353, we would allocate this $5800 to the sub-elements by determining how this cost should be divided between them based on their relative complexity and other factors. We might allocate this $5800 DTC figure as follows:

Element	DTC($)	Rationale
A1351	1000	Change one of four circuit board designs.
A1352	2000	Major redesign, no current supplier.
A1353	2000	New design for widget; some technical risk.
A135M	800	Margin account.
TOTAL	5800	

Note that something called margin has been included. This is a part of the total that is not apportioned to any of the lower-tier items. It is withheld to enable risk management during the design process. If we find it very difficult to satisfy an item requirement value, managers can dole out this margin if necessary.

Another flow down rule that can be applied is that of equivalence. In this case, every sub-element is simply assigned exactly the same value as the parent. Clearly, this rule cannot be applied to attributes that follow an apportionment pattern as it would violate the unforgiving rules of mathematics. Some qualitatively stated examples where it can be applied follow:

a.   Item color shall be olive drab.
b.   The item shall have no sharp edges or points that could cause injury to personnel as a result of casual contact in normal use.
c.   No flammable materials shall be used in construction.

These are all attributes that could be applied to parts of the item in question as well as to the item itself. These kinds of requirements are excellent candidates for inclusion in boilerplate material since their phraseology does not change from one item to another.

The third flow down rule involves complex situations, such as guidance accuracy and required engine thrust levels, where two or more requirements must be combined, or synthesized, in some complex way to derive the lower-tier requirement generally revealed through some complex simulation or other process. This topic was expanded upon in the previous chapter.

2.2.2.4 Parent-child requirements traceability

Vertical requirements traceability, for a given requirement of a given system element, is a condition of clear knowledge of the ancestry of that requirement in terms of the parent requirements that make it necessary for the design of the element to respect that requirement. Every requirement for every system element should theoretically be traceable to the customer need as illustrated in Figure 2.2-2.

Allocation is the process of flowing down requirements from parent elements to child elements in the architecture. Traceability is the connectivity that results between the requirements for one element and the requirements for parent elements. Traceability is generally thought of as having an upward-looking perspective whereas allocation has a downward-looking perspective. Clearly, if allocation (flow down) is well done and some minimal records kept, then traceability is well established in the process. Even in this environment, however, it may not be known how well the flow down process may have worked. A traceability analysis may be performed as an audit of selected requirements traceability paths to assure the engineering team or customer that requirements possess a condition of traceability.

2.2.2.4.1 Why Traceability?

Customers and a contractor's own management have the right to know if the current system concept satisfies the customer need and whether or not there are any excesses in the concept. We should be striving for the minimum adequate

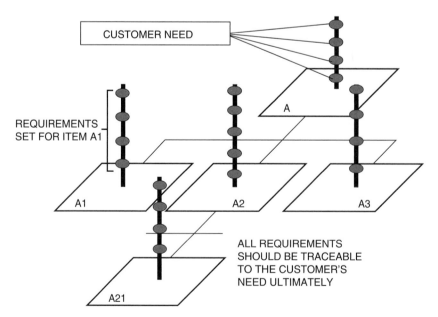

Figure 2.2-2   *Requirements traceability*

solution that respects the customer need and the amount of money they are willing to pay. A record of the traceability between requirements for system elements affords customer and company management an opportunity to monitor development team progress by providing insights into potential gold plating. This data can stimulate the right questions that help to maintain team focus on efforts to satisfy the customer's need.

Well maintained traceability data also can be very useful later in program development when changes are being evaluated. The traceability data can be used to trace higher-order requirements changes throughout the design to help determine what must be changed in the lower tiers to satisfy a higher-level requirements change.

2.2.2.4.2 Traceability Mechanism

To establish a condition of traceability, we need some way to hook up requirements for one element to requirements for other elements. Downward traceability is almost always a one-to-many relationship between a requirement and those it traces to but it most often is a one-to-one relationship in the upward direction. A simplistic approach that will work in a word processor environment is to include a traceability matrix in Section 6 of the specification such as that illustrated in Figure 2.2-3 depicting a Table 6-3 from a development specification.

Let us assume that the example item is a performance specification for a launch vehicle segment core vehicle, architecture element A11. In this example the requirements in the "ITEM REQUIREMENT PARA NBR" column are from that performance specification. Paragraph 3.2.3 has been assigned a requirements ID of AG4G for traceability coding in a computer database and has a title "Reliability."

In the four right columns are listed the architecture ID for the parent item (A1, the launch vehicle), requirement paragraph number, requirement ID that it traces to in the requirements set for architecture element A1 and the title. If a requirement traces to two or more parent requirements, there will be two or more line items for that requirement with different entries in the right hand columns.

Sometimes it is convenient to report requirements traceability for a series of items or documents. This would be useful, for example, included in a parent item specification showing the downward traceability to the several subordinate paragraphs. Figure 2.2-4 illustrates one approach. The parent paragraph column may correspond to paragraph

numbers in a segment specification. The DOC 1 through 6 columns would contain the paragraph numbers from a series of next lower tier item specifications.

In a computerized database application, the traceability information is included in the database. This can be done as a simple file of two pairs of fields for Architecture ID and Requirement ID. The Requirement ID could be the paragraph number or a cleverly conceived or arbitrary code. In the author's opinion, the best requirement ID approach is to use a simple, unique, arbitrarily assigned code. You can very quickly drive yourself crazy with cleverly devised requirements ID codes that contain more information than a simple unique requirements ID. Most commercial requirements tools automatically assign a unique requirements ID that can remain in the background. The human deals with paragraph numbers but the traceability data is actually linked through the IDs. The reason this makes sense is that during development changes in requirements and their traceability can add a lot of parasitic traceability work just driven by paragraph number changes. This work is simply avoided when using requirement IDs. Working traceability with a word processor is a very painful process especially where paragraph numbers are used for traceability directly. Good database tools simply show the operator a pair of windows, one for each specification, and the analyst highlights one requirement title in each window and clicks a link button.

The current DoD specification standard, MIL-STD-961E, does include a paragraph in Section 3 for requirements traceability and the tables discussed above could be placed in that paragraph rather than in Section 6. The Software specifications covered in MIL-STD-498 and EIA J-STD-016, which was created from MIL-STD-498, set aside Section 5 for traceability.

2.2.2.4.3 Traceability Across Interfaces

The traceability we have been discussing up to this point can be referred to as vertical traceability. It involves traceability of requirements heritage up through the architecture vertically. The traceability situation is, unfortunately, not that simple. Many requirements are driven by interfaces with other items in a lateral fashion within the architecture. We have said that systems exist on the basis of multiple items interacting via interfaces. The more richly or intensely the items interact, the more heavily their functionality is interdependent and the more intensely their requirements

Table 6-3 Requirements Traceability

ITEM REQUIREMENT				ITEM REQUIREMENT			
ARCH ID NBR	PARA NBR	RQMT ID	TITLE	ARCH ID NBR	PARA NBR	RQMT ID	TITLE
A11	3.2.3	AG4G	Reliability	A1	3.2.1.5	KJU8	Reliability

Figure 2.2-3   *Single tier traceability matrix*

PARENT PARAGRAPH	DOC 1	DOC 2	DOC 3	DOC 4	DOC 5	DOC 6
3.2.2.1	3.2.2.1	3.2.1.1	3.2.2.1	3.2.2.1	3.2.2.1	3.2.2.1
3.2.2.2	3.2.2.2	3.2.1.2	3.2.2.2	3.2.2.2	3.2.2.2	3.2.2.2
3.2.2.3	3.2.2.3	3.2.1.3	3.2.2.3	3.2.2.3	3.2.2.3	3.2.2.3
3.2.2.4	3.2.2.4	3.2.1.4	3.2.2.4	3.2.2.4	3.2.2.4	3.2.2.4
3.2.2.5	3.2.2.5	3.2.1.5	3.2.2.5	3.2.4	3.2.2.5	3.2.2.5
3.2.3	3.2.3	3.2.3	3.2.3	3.2.3	3.2.3	3.2.3
3.2.4	3.2.4	3.2.5	3.2.4	3.2.5	3.2.4	3.2.4
3.2.5	3.2.5	3.2.6	3.2.5	3.2.7	3.2.5	3.2.5
3.2.6	3.2.6	3.2.4	3.2.6	3.2.6	3.2.6	3.2.6
3.2.7.1	3.2.7.1	3.2.7.1	3.2.7.1	3.2.8.1	3.2.7.1	3.2.7.1
3.2.7.2	3.2.7.2	3.2.7.2	3.2.7.2	3.2.8.2	3.2.7.2	3.2.7.2
3.2.7.3	3.2.7.3	3.2.7.3	3.2.7.3	3.2.8.3	3.2.7.3	3.2.7.3
3.3	3.3	3.3	3.3	3.3	3.3	3.3

Figure 2.2-4 *Multiple document traceability matrix*

will have to be interdependent. The traceability across interfaces can be handled exactly as covered earlier with tables relating Architecture ID-Requirement ID pairs, but the analyst must consciously know to look for these connections.

Two examples of interface traceability will help. First, a simple one then a more subtle one. A requirement for architecture item A1567 defines a bolt pattern for mating to architecture item A23. These two items are in entirely different branches of the architecture yet these two requirements are clearly bi-directionally driving each other. Our traceability matrix should show this relationship.

Architecture item A1214, an on-board digital computer, shall be designed to tolerate an acoustic noise level of 145 db across a frequency spectrum of 10 hertz to 800 hertz. This requirement is traceable from an engine requirement covering the allowable noise spectrum and a compartment sound insulation effectiveness requirement.

2.2.2.4.4 Multiple Traceability Paths

We may have implied that requirements traceability is a one-to-one proposition up to this point. Let us dispel that illusion. It is true that upward traceability for many requirements is one-to-one. Many specialty engineering requirements like reliability fall into this pattern. But, let us consider some that are not. The one-to-one pattern breaks down commonly for the requirements that are also the most complex and for which developing values is the most difficult.

A requirement for an aircraft heading accuracy is properly traceable from the accuracy of the gyro compass, the flux gate airframe mounting accuracy, the amount of structural asymmetry tolerated in the airframe and the aircraft's control authority over that asymmetry, Doppler drift angle

solution accuracy, and the magnitude of computational errors in a central computer. The requirement for an aircraft maximum airspeed of 580 knots is traceable from an engine thrust requirement, an airframe surface roughness (smoothness) requirement, an aircraft weight requirement, and an airframe drag index. You can see from these two examples that the requirements that present a multiple traceability path pattern are commonly among the same requirements that attract a budgeting interest discussed in Chapter 2.3.

2.2.3 Longitudinal Traceability

Program requirements traceability runs in three directions when properly and completely performed. Two of these dimensions are illustrated in Figure 2.2-5. Traceability should run vertically from the original user system documentation through to the formal system specification and from there down through all of the program specifications. In addition, there should be good traceability from the content of the specifications through the design and verification information stream as also depicted in Figure 2.2-5.

Section 3 of the system specification defines the necessary characteristics of the system and each of these requirements should be traceable to one or more system test and evaluation process requirements contained in Section 4 of that same specification. These requirements should drive the content of the verification plans and procedures as it relates to the system level and through the actual verification work performed (not illustrated) to the report upon the test and evaluation work accomplished. For each item Part I or performance specification, the Section 3 content should be design independent and traceable to the content of

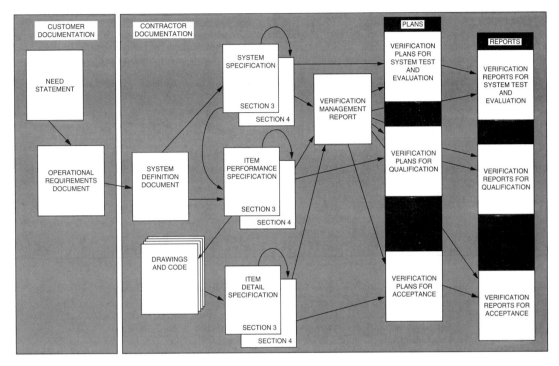

Figure 2.2-5 *Program-wide requirements traceability*

Section 4 that provides requirements for the item qualification process. The engineering drawings and software code reflect the design of the product driven by Section 3 of the corresponding item performance specification. The requirements content of Section 3 of the item Part II or detail specification should be design dependent, driven by the product design. This content includes a selection of the most significant design features that should be used as the basis for the article acceptance process at the end of the manufacturing process. The content of Section 3 of these specifications should be the basis for the content of Section 4 of these specifications which provides the requirements for acceptance verification. This matter will be further explored under requirements verification in Chapter 6.6.

Design traceability relative to the requirements is generally poorly documented where the product is implemented in hardware. About the only way it is ever covered is at the major design reviews provided the design presenters have been encouraged to present: (1) the requirements as you understand them, (2) the design features, (3) the correlation between the requirements and the design features, and (4) identify remaining risks. The traceability is therefore captured in the design review presentations. In the case of software, where software tools are employed which can automatically generate code, good design traceability can exist within the tool though it may not be easily reportable.

2.2.4 Requirements Traceability to Process

In section 2.2.2 we concerned ourselves with the pure traceability between requirements for one item to requirements for another. Requirements source and rationale

tracking covered in the beginning of this chapter were also part of the traceability activity. Next we discussed traceability to the verification events that prove they have been satisfied. The picture is completed by a method for linking up the requirements with the requirements analysis process through which the requirement was originally identified.

We should understand by now that requirements analysis is not accomplished well in today's work environment by individualists, but rather by a team of specialists cooperating toward a common goal. Each of these specialists is performing requirements analysis for a specific discipline across the whole system or in the context of a particular architectural scope. Each of these specialists should have some means of independently collecting and reporting the results of their analysis. Some common reporting format examples are: Reliability and Maintainability AAA Report, Weights Statement or Mass Properties Report, various LSA Reports, and Life Cycle Cost Analysis Report.

In addition, many other analyses take place that do not form a component of a larger, organized analytical process. These analyses can, however, be reported upon through a letter to a supervisor or team leader. The combination of formal internal reports, formal reports that are contract data deliverables, and internal analysis memo reports provides a basis for capturing the results of analyses having a bearing on requirements definition. Ideally, each requirements analysis event (one requirement for one item) should correspond to some analysis report and be referenced to that report. All requirements types will not fit this pattern, but those that do should be traceable to their analytical source.

Earlier in the chapter we discussed source and rationale recording and that provides a means to capture the traceability of requirements to the requirements analysis process. For each requirement we can refer to the analysis report that identified and established the value for the requirement.

It is also useful to know who was responsible for identifying requirements in the system hierarchy. This can be done by adding the analyst's name to the tabular data covered in paragraph 2.2.2.1 above. When a requirement is later questioned, it is then possible to call in the named engineer and quickly obtain an answer to the question. Otherwise, in the extreme, it may be necessary to repeat or conduct another costly analysis to confirm a requirement's need or value.

2.2.4.1 Single sheet traceability to process

While requirements analysis process traceability information can best and most economically be captured in a computer database approach, it is possible to do it manually. Figure 2.2-6 illustrates a manually implemented requirements analysis sheet (RAS) that could be used for each requirement derived using the functional analysis process.

Note that this sheet will capture only a single allocation of a function to an architecture element and the definition of only a single performance requirement from that allocation. If a single functional allocation resulted in three allocations to architecture and each of those resulted in three performance requirements for the corresponding architecture items, it would require a total of nine sheets to

REQUIREMENTS ANALYSIS SHEET XYZ SYSTEM			
FUNCTION ID			
FUNCTION NAME			
DESCRIPTION			
INITIAL EVENT			
TERMINAL EVENT			
TIME DURATION			
MEAN			
VARIANCE			
REQUIREMENT ATTRIBUTE			
VALUE/UNITS	TARGET		UNITS
	MARGIN		
	ACHIEVED		
REQUIREMENT STATEMENT			
PARAGRAPH TITLE			
PARAGRAPH NUMBER		REQUIREMENT NUMBER	
ARCHITECTURE ITEM ALLOCATION			
	NAME	SIGNATURE	DATE
ANALYST			
PRINCIPAL ENG			

Figure 2.2-6 *Manual requirements analysis sheet*

capture the lot and much of the information on the sheet would have to be repeated.

You could overcome this problem by making it possible to record all of the requirements derived from a single function on the same sheet. The problem with this approach is that it will then not be possible to physically pile up the sheets that relate to specific architecture items together unless you make copies and deal out a copy to each architecture element where an allocation has been made. By agonizing through this thought process, that many have experienced, you begin to see why a computer might be useful as a requirements capture environment. The computer can be applied to this problem to establish and maintain the many-to-many relationships between functions and architecture. You can allocate the functions to architecture and expand the simple function statements into performance requirements with only a single strike of the keys for each unique piece of information. The computer can be programmed to print the specification directly from the database and within the database traceability and verification records can be efficiently maintained.

Another reason that it makes little sense to use these sheets in a manual environment is that it becomes necessary to write down the requirements twice, once for the RAS and once for the text of the specification. Instead of saving money, you may spend more money without matching benefit. You may also validate the incorrect notion for your company that the SRA process is expensive and does not add value.

The manual single-sheet RAS was used prior to computer use. As computers were applied to requirements analysis the manual RAS was used as a computer data collection and entry form in an environment where the engineer filled out the form with a pen or pencil and a key punch operator transferred the approved data to an 80 column card which was then loaded into a main frame computer. Main frame computers were applied to requirements analysis work before the computer hardware and software technology was ready doing more damage than adding value. The analyst was separated from the data through forms, keypunch input/output, and computers locked away in computer rooms. Current applications have returned ownership of the data to the analyst as was the case when analysts filled out manual paper forms before computers were used.

Today, the analyst could use a manual RAS and have computer operators enter the data into the computer database. But, this approach can be cost-effectively short-circuited by engineers entering the data directly into a window on a computer screen themselves based on the results from on-going analyses. In any case, the data that is entered into the computer is the result of human thinking, not wonderful and mysterious actions taken by the computing machine and its software.

2.2.4.2 Specification template traceability

The enterprise should evolve a set of templates, one for each kind of specification they will be called upon to develop. The content of each template should be mapped to a preferred method for defining the content for each major paragraph heading, the functional department from which a program would obtain the specialist needed to perform the related work (DEPT), and the appendix (APP) of the program system definition document (SDD) where the model data is captured. Table 2.2-1 illustrates a template trace. The left columns are simply the paragraph numbers and headings from MIL-STD-961E for a system specification. The

METHOD column identifies the preferred method. The SDD is covered in Chapter 3.11.

The appendix references in Table 2.2-1 relate to the appendices in the SDD as illustrated in Figure 2.2-7 assuming that modern structured analysis is being applied as the modeling method. The SDD provides a means to capture all of the requirements analysis modeling artifacts in a way that the model basis for the content of the specifications can be captured and even baseline managed.

Functional analysis produces functional flow diagrams that are captured in Appendix A that are used as the basis for defining performance requirements that are allocated to particular product entities defined in Appendix C. The performance requirements are captured and allocated in Appendix G, the requirements analysis sheet. Environmental requirements are defined in four different requirements analysis activities and he results captured in the RAS in Appendix G. The interface requirements flow into the RAS. Finally, specialty engineering requirements analysis work can also be captured in the RAS.

The RAS then becomes a transform device intervening between the analysis work and the flow of requirements into the specifications. Table 2.2-2 shows a few requirements of each of the four kinds that have to be defined for systems and hardware items: (1) performance requirements derived from functional analysis using a model ID beginning with the letter "F," (2) specialty engineering requirements using a model ID beginning with a letter "H," (3) interface requirements using a model ID beginning with a letter "I," and environmental requirements of four kinds (natural (QN), hostile (QH), self-induced (QI), and non-cooperative (QX)) using the lead model ID letter "Q."

The table maps requirements derived from particular models identified by model IDs (MID) to the requirements derived from those functions identified by requirements ID (RID) assigned within the database used to capture the requirements and the product entities to which those requirements have been allocated in which specification they will appear identified by a product ID (PID). (See Figure 2.2-7.)

2.2.5 Grand System Traceability

The U.S. Air Force Systems Command developed a powerful program planning and management discipline called the integrated management system. In this system, you develop a system specification from the need and additional information and then develop a Statement of Work (SOW) that expands on the work breakdown structure (WBS) telling what work must be done to create the product system described in the system specification. There was interest at one point in requiring contractors to show traceability from the SOW to the system specification because all of the contract work should be driven by the product requirements. This is not a common requirement but it did open the possibility of expanding requirements database tools use for both product requirements and process requirements.

The integrated management system concept expands this traceability downward from the SOW to an integrated master plan (IMP) that tells how the contractor will manage and accomplish the work defined in the SOW. Work tasks are defined in both narrative form and a list of significant accomplishments each linked to completion criteria. All of these tasks are related to major milestone events and organized in time into an integrated master schedule (IMS).

Table 2.2-1 *Specification traceability to responsibility and method*

Paragraph	Title	Dept	Method	App
3	REQUIREMENTS	D216-2	–	–
3.1	Functional and Performance Requirements	D216-2	Traditional Structured Anal	A
3.1.1	Missions	D216-2	Mission Analysis	A
3.1.2	Threat	D216-2	Threat Analysis	B
3.1.3	Required States and Modes	D216-2	Functional Flow Diagram	A
3.1.4	Entity Capability Requirements	D216-2	Performance Requirements Analysis	A
3.1.5	Reliability	D216-4	Reliability Math Model	E
3.1.6	Maintainability	D216-4	Maintainability Math Model	E
3.1.7	Deployability		Flow Charting	E
3.1.8	Availability	D216-4	Availability Math Model	E
3.1.9	Environmental Conditions	D216-2	Standards Assessment	B
3.1.10	Transportability	D231-1	Logistics Analysis	F
3.1.11	Materials and Processes	D216-7	M&P Analysis	E
3.1.12	Electromagnetic Radiation	D213-3	EMI/EMC Analysis	E
3.1.13	Nameplates and Product Markings	D211-3	Boilerplate	E
3.1.14	Producibility	D224	Manufacturing Requirements Analysis	E
3.1.15	Interchangeability	D231-1	Logistics Analysis	E
3.1.16	Safety	D216-5	System Safety Analysis	E
3.1.17	Human Factors Engineering	D216-5	System Safety Analysis	E
3.1.18	Security and Privacy	D216-6	Threat Analysis	E
3.1.19	Computer Resource Requirements	D213-2		E
3.1.20	Logistics	D230	Logistics Analysis	E
3.1.20.1	Maintenance	D231-1	Logistics Analysis	E
3.1.20.2	Supply	D231-2	Logistics Analysis	E
3.1.20.3	Facilities and Facility Equipment	D231-1	Logistics Analysis	E
3.1.21	Personnel and Training	D231-4	Logistics Analysis	E
3.1.21.1	Personnel	D231-4	Logistics Analysis	E
3.1.21.2	Training	D231-4	Logistics Analysis	E
3.1.22	Requirements Traceability	D216-2	–	–
3.2	Interface Requirements	D216-2	Schematic Block Diagrams	D
3.2.1	Government-Furnished Property (GFP) Interfaces	D216-2	Schematic Block Diagrams	D
3.2.2	External interfaces	D216-2	Schematic Block Diagrams	D
3.2.3	Internal interfaces	D216-2	Schematic Block Diagrams	D
3.3	Design and Construction	D210	Boilerplate	E
3.4	Precedence and Criticality Requirements	D216-2	Boilerplate	E

The complete development process can be expanded in a top-down structured way from this skeleton with traceability of each element all the way to the ultimate requirement, the customer need. Traceability between the system specification and the SOW can be captured in a database containing both. This traceability can be continued down the product requirements path through specifications to procurement specifications and from there to procurement statements of work also linked to the program SOW. The author has developed a variation on the USAF integrated management system called a grand systems planning process and described it in the student manual used in a Grand Systems Management course. The modified approach provides for traceability throughout the enterprise work planning and program planning processes.

2.2.6 Traceability Reporting

It is very hard to justify accomplishing traceability work using a manual approach or implemented in a word processor. It results in a considerable cost of questionable compa-

rable benefit. On one program, the author can recall, the customer demanded to review the traceability data. The contractor program manager explained that traceability had not been covered in the contract. After a long argument, the customer provided additional funding to accomplish the traceability work. The program specifications had all been crafted in a Wang word processor system by a small army of typists. The traceability data was built using tables also published from the Wang word processor. When complete, the customer reviewed the data and found it to reflect good traceability. Two years later, the customer asked to review the updated traceability data and was told by the contractor program manager that they had only funded the initial traceability analysis and not the continued maintenance of the data. After a shorter argument the additional funds were made available and applied to updating the data which had changed significantly over the intervening period due to content changes, yes, but also due to paragraph number changes. The computer application applied (Wang word processor) did not maintain traceability via requirements IDs. The only means available was paragraph numbers.

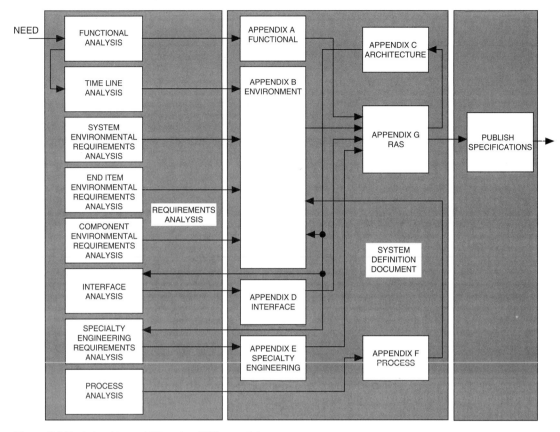

Figure 2.2-7   *Lateral traceability using RAS complete*

Table 2.2-2   *Sample requirements in a comprehensive RAS*

Model Entity MID	Model Name	Requirement Entity RID	Requirement Statement	Product Entity PID	Product Name
F4735	Maintain V/H Ratio	G6YE	$100 < AGL \leq 5000$ Feet	A133	Radar Altimeter
H11	Reliability	DF6T	MTBF > 2000 Hours	A131	On-Board Processor
H11	Reliability	HY7R	MTBF > 2500 Hours	A132	Guidance Set
H11	Reliability	J918	MTBF > 2400 Hours	A133	Altimeter
I127	Radar Altitude	G74H	$0 \leq$ Altitude ≤ 5000 feet	A133	Radar Altimeter
I127	Radar Altitude	QJU7	$0 \leq$ Altitude ≤ 5000 feet	A131	On-Board Processor
QH7	Explosive Force	U83G	TBD-9	A1	Air Vehicle
QI3	Vibration	5TYU	1.5 g, 5.5 to 200 hz	A1	Air Vehicle
QN1	Temperature	FR5Y	$-40° \leq$ Temperature $\leq 120°$	A1	Air Vehicle
QX5	Electromagnetic Interference	K957	TBD-3	A1	Air Vehicle

In order to accomplish traceability work affordably, you have to apply a good computer database system of which there are several on the market. There are three specific traceability reports we should require from such tools. Some customers like to receive a total system traceability report but it appears to the author to be a report of questionable utility. We should be able to print out (screen or paper) a traceability report for all of the requirements in a given specification which would produce a pair of columns (paragraph number and title) for the specification under study and the corresponding parent numbers and titles. This report is useful when a specification comes up for review and approval permitting the briefer to claim a particular degree of traceability.

Given any requirement for any item, we should be able to see the trace upward as far as it goes, to the system specification or until a condition of non-traceability is reached. Given a failure to trace, we would then inquire why that condition exists.

A third report that is useful is a trace downward but it should be an operator aided trace where the computer lists all of the lower-tier requirements that trace from a particular entered requirement. The operator then selects the requirement of interest and the computer gives the next tier of requirements. This process can repeat as long as desired until the requirements listed are in a lowest-tier specification.

2.2.7 Traceability Audits

We perform traceability analyses generally to prepare for traceability audits performed by the customer in association with major program reviews. A good engineering organization will, however, also be interested in following the evolving condition of traceability as the development process proceeds as one way of protecting against bad or unnecessarily complex and costly design solutions.

In the organizational model used throughout this book, a program integration team is responsible for traceability and should therefore perform a traceability audit, or insist that one be run by the responsible team, as a prerequisite to specification approval. We should be able to depend on system engineers establishing and maintaining sound requirements traceability throughout the team hierarchy. However, the program integration team (PIT) must conduct traceability audits on requirements sets developed by principal engineers or development teams below the system level. Depending on available budgets and priorities, there may not be enough resources to perform a 100% audit of lower tier traceability. In that case, the PIT may have to prioritize which items are in the greatest need of a traceability audit.

These audits may be accomplished on a selected list of requirements across the architecture, on every requirement for selected architecture items, or in accordance with some other rule. Generally, there will not be sufficient budget to perform a 100% traceability audit on every requirements set below prime item level. It is true that all of the principal engineers can be tasked with performing a traceability audit on their own requirements (and a good idea this is), but to keep the fox out of the hen house, it is necessary that at least some of these traceability paths be audited by an outside agent. The results of the audit should be communicated to the responsible principal engineers with direction for correction of any deficiencies noted and a due date. This condition is then maintained as an open action until satisfactorily resolved.

When selective development, or plunging, is used, it results in the temporary creation of islands of requirements not connected with parent architecture requirements. Eventually the structured requirements analysis process will work its way down to these islands. At that time it will be necessary to conduct a requirements traceability audit on the selectively developed requirements with respect to those flowed down in the structured process. The purpose here is to identify the parent requirement for each selectively developed requirement, thus insuring that the selectively developed requirements reflect an optimum set that will lead to the development of system elements that will interact so as to fully satisfy the customer need.

We would assume in this traceability analysis that the parent requirements, developed through structured requirements analysis, were correct in any case where the traceability audit identifies a mismatch. Generally this will be the proper conclusion, and the proper response is to change the selectively developed requirements for good traceability, including any affects on lower-tier requirements (if more than one level was selectively developed).

If, on the other hand, the analyst concludes that the parent requirements must change and engineering management concurs, then that change must be rippled up through the parent requirements and possibly laterally and back down some of the other branches of the architecture tree. Before your organization commits to plunging, it had best have an effective means for tracing requirements impacts provided by an effective computerized traceability matrix.

Each of the major customer design reviews is, among other things, an opportunity for the customer to audit the requirements set. The contractor must be prepared to conduct selected traceability audits at, or in preparation for, any of these reviews. If the system is computerized and the customer has a terminal allowing routine access to the database, it is best to prepare for these reviews by coordinating with the customer persons interested in requirements traceability and encouraging them to make their own traceability audits prior to the review and to discuss with you their results. Then the review may only cover the areas of concern that they have exposed or provide a status review of requirements development.

2.3

Requirements Allocation, Margins, and Budgets

Contents

2.3.1 Requirement Value Determination

The numerical values we select for requirements statements should be achievable. They must be determined based on the needed performance of the system and what the technology of the times will support. Otherwise, the program will endure considerable risk that it cannot achieve those requirement values. Three techniques are available to help programs establish achievable requirements thus supporting sound requirements and program management. These techniques are called allocation, margin management, and budget management.

2.3.2 Requirements Allocation

The word allocate means to set apart, assign, or allot for a particular purpose. As the word applies to requirements analysis work, the word means to partition a value assigned to a parent requirement into parts that are assigned to child item specifications. For example, how shall the weight defined in a parent item weight requirement of 2320 pounds be allocated to five child items? If we take no action then the engineers responsible for the five lower-tier items may identify their weight requirements independently without any conscious effort to make sure that the total of those independently determined values adds up to something less than 2320 pounds. Allocation is an organized process for determining child values in allocable requirements categories. Requirements that are numerically defined which stretch across several items in a hierarchy related in accordance with a mathematical formula can be allocated.

For a given parent requirement we have three principal ways to allocate the value to child requirement values. The work can be done from the top down by a system engineer. The principal engineer for the parent item could take on this responsibility. A third, and probably better, approach is for the allocation work to be accomplished by a specialty engineer from that discipline to do the allocation work. The allocating engineer could simply assign numbers that add up to the right total but should try to consider to what extent available technology supports particular values for the subordinate items. The justification for this approach is that if values are assigned that are unachievable by one or more lower-tier engineers, then the numbers can be simply re-allocated. The alternative approach is for the allocating engineer to use a bottom up approach permitting each child principal engineer to identify his or her preferred values. The allocating engineer must then add up the preferences and compare the result with the parent value. If the sum is greater than the parent value, then some integration work will be required of the allocating engineer to negotiate a mutually agreed set of values with all of the lower-tier principal engineers.

Often this is difficult work because all of the engineers want the best chance of succeeding in their development work and that best chance is coincident with the least demanding set of requirements. In the end, it may be necessary for the allocating engineer to make an arbitrary decision that balances the degree of difficulty as best he or she can determine it. During the design effort, if it develops that the value cannot be satisfied for a particular item, re-allocation is always available.

In the discussion above, the implication has been made that the rule of combination is addition. This is not necessarily the case. In the general allocation case, we deal with a particular parameter (such as weight, reliability, or life-cycle cost) and a particular mathematical rule of combination. Where the parameter is weight, the rule is clearly addition. Where the parameter is reliability it could be addition if we used failure rate as the way of measuring the reliability parameter or it could be a probabilistic relationship if using mean time between failure (MTBF) or reliability number.

For all allocable requirements, there will generally be an engineering discipline responsible for the values. So, the allocation process should involve the parent and child principal engineers and a collection of discipline engineers for the several kinds of requirements. No matter how the values are allocated, the parameter principal engineer must capture the agreed upon values in their mathematical models and corresponding specifications. This work also requires assertive integration work by the principal engineers and system engineers.

2.3.3 Margin Management

Engineers are vitally interested in implementing designs that possess margins in specific areas relative to features of their design. Engineers more often refer to these as design margins than requirements margins, perhaps, but we are discussing the same artifact. They are interested because it is often uncertain what the actually achieved capability will be for their design until testing establishes actual demonstrated capability. Engineers call the difference between the required value and the demonstrated capability a margin. We will use the word margin in this chapter for essentially the same meaning but with a little difference. From a management perspective margin is not necessarily good in that it has the effect of providing a product that performs better than needed. This extra capability carries with it commonly added cost or schedule impact. Margins that engineers individually retain private knowledge of are particularly disadvantageous because they cannot be managed in an open or public fashion. Some engineers strive for a margin because past experience has included one or more design efforts that failed to satisfy the required capability and the engineer does not want his or her design efforts to be found wanting on this new program.

Ideally, our means of documenting requirements would capture the required value, any margin established, and the current demonstrated capability. With this information, it would be possible for a requirements database system to be told to find some management space when confronted by a problem in meeting a particular requirement value.

2.3.3.1 What are formal margins?

A formal design margin is an unallocated portion of a quantifiable, allocable requirement value that is used by design engineers and management to avoid and resolve problems that arise in system development. The best and simplest example of a margin is found in weight requirements. Weight growth is a common problem on every program, so it is frequently chosen as one parameter to manage under a margin approach. At each level in the system architecture the required weight value includes an unallocated portion. Figure 2.3-1 illustrates the general approach to margin accounts and will be used here to explain the fundamentals.

Architecture element A135 will have a certain weight figure, say 2320 pounds, applied to it as a maximum acceptable weight. The principal engineer for that element must participate with the principal engineers for the subordinate elements (A1351 through A1355) and the mass properties engineer to allocate that weight to sub-elements. They will do so by cutting out a portion as a margin in accordance with a general weight margin plan, which may require a 10 percent margin figure for weight at that point in the program.

This 10 percent margin (232 pounds) will not be allocated to lower-tier elements but will be reserved to insure that element A135 does meet its maximum weight requirement target of no more than 2320 pounds. The design engineer for element A135 will strive to meet a target weight of 2088 pounds (2320 minus the 232 held in reserve). In general, REQUIRED VALUE = (TARGET VALUE + DESIGN MARGIN VALUE). Figure 2.3-1a shows margin accounts at each level for one particular branch down through four layers. Margin figures would also be applied in all of the other branches as well. Figure 2.3-1b illustrates how margin figures for these four layers of the system might be allocated.

If the area of Figure 2.3-1b is proportional to the weight of 3320 pounds, then A135 margin area would correspond to 232 pounds in this example. The remaining 2088 pounds will be allocated to the five lower-tier elements A1531 through A1535 based on the perceived need for their weight budgets. Likewise margin figures would be assigned at lower tiers as illustrated for A1351 and A15311 margins. Generally, these margin figures will be progressively smaller because there is a progressively smaller pie to partition but the amount of the margin should be proportional to the risk potential.

The result will be that we have distributed throughout the architecture a way to absorb weight growth that we know will occur. If the design group for element A135112 finds that it is impossible to design that element within the allocated weight, and convinces his or her team leader of that

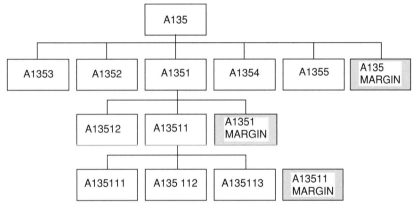

a. Margins in an Architecture Context.

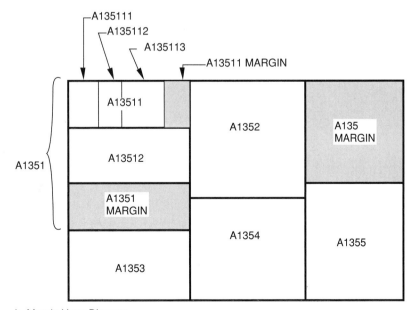

b. Margin Venn Diagram

Figure 2.3-1   *Requirements design margin accounts*

position, additional weight may be allocated from margin account A13511, or a higher-level margin account by the parent team or program integration team.

Figure 2.3-1 actually suggests a cascading of margin that is not a good idea. If 232 pounds has been set aside for margin at the A153 level, there are five elements subordinate to A153, and they all have fairly equal weight risk, the sum of lower-tier margin should be 232 pounds. That is, at each level we should distribute the parent margin. Otherwise you will have much more than 232 pounds of margin at level A153.

Ideally, when a system is mature, all of the design margin accounts (this does not include safety margins discussed below) will have been consumed in accounting for development problems. Margins are in reality a way of encouraging planned overdesign early in a program and may contribute to providing capability that is in excess of that needed to satisfy the system need. Residual margins do contribute to system robustness but may also represent added cost. In general, however, a well managed design margin approach will result in minimizing system cost even if there are residual design margins at completion of the design process. Most complex system development efforts are pressing the technology in so many directions that there will be adequate risks to flush out all of the early margins.

A customer may require that the margin program be structured to conclude with a fixed residual in one or more requirements categories specifically to ensure design robustness in the planned operational environment. This can be done by identifying a robustness component and a management component in the margins accounts where robustness is required.

2.3.3.2 Selection and maintenance of design margin parameters

Requirements that have a history of trouble during development, such as weight, should be selected for inclusion in the margin program. Margin parameters must also be quantifiable and have a clear rule of apportionment associated with them. Other common margin candidates are: cost (life cycle and/or non-recurring), reliability, maintainability, gain, and electrical load.

Early in a program, preferably during proposal activity, the engineering community must select the initial margin account parameters. Candidates for this list should be developed by a review by the team managers and the Chief Engineer at an engineering review meeting. Subsequent changes to this approved list should be subjected to continuing review. Often the customer will require margin management for a specific minimum list of parameters identified in the contract, a contracts letter, or formal review minutes. It may occur that program management will wish to manage a larger set of margins than required by contract to provide assurance of compliance. In this case, the formal margins that must be maintained by the customer will be a subset of the compete set being followed by the program.

The parameters that cry out for selection for margin accounts are those parameters that your company historically finds it difficult to satisfy. Weight is a universally applicable margin candidate and Design-To-Cost is another. Your engineering department should have a standard list of parameters like this and be continuously working to develop methods that will, on future programs, allow the development team to use margin accounts on a monotonically decreasing list of parameters. Margin accounts cost money to operate, but they can allow the team to avoid costly mistakes where the company has historically been unable to control parameter growth. It makes sense to try to find the root cause in each case and attempt to change the process to avoid the need for margin account management where possible. Here you see one example of how your company can integrate program performance with its continuous process improvement efforts, an example of lessons learned.

Each margin account must be assigned to a specific engineer who must assertively manage the account across the system architecture. This margin account principal engineer will generally be someone from the specialty discipline responsible for the kind of requirement involved. For example, it would make sense to appoint someone from the weights group to manage the weights margin account and someone from reliability to manage a reliability margin account. This same person or persons from the same group in each case would be involved in allocating the corresponding quality to each element of architecture.

The margin principal must also require the engineers responsible for requirements and design development for each element within the margin umbrella to help identify appropriate margins and to respect the margins established. As the element concept and design matures, the margin principal must assess periodically how well the element principals are doing in satisfying their challenges in that parameter. Where the principal engineer comes forward with a problem meeting a requirement value protected by a margin or an independent analysis or test reveals growth in that parameter for an item, the margin principal must take action to bring about a rebalancing of the account.

Rebalancing may occur as a result of direction to the element principal that the original requirement must be met. Alternatively, where the growth is unavoidable, margin accounts can be accessed to provide the engineer with the relief needed. It is also possible to rebalance accounts through inter-account transfers without actually consuming margin.

Someone on the program integration team must be identified by name to manage the complete margin program. This person will interact with each margin principal engineer periodically to assure that margins are being respected and to help management spot potential problems before they become very difficult to solve.

2.3.3.3 Safety margins

The margin accounts addressed above are called design or requirements margins (depending on your perspective) and their purpose is to provide management with a built-in capacity to solve problems in the evolving design. They provide a reserve of requirements values that may be drawn upon. In certain fields another kind of margin is applied to insure that designs are more than adequate for their purpose by multiplying an adequate requirement value by a factor greater than 1 and using the result as the requirement. These margins are called safety margins or safety factors.

The utility of safety margins is that it is possible to avoid some development testing if safety margins are properly applied in accordance with a customer specified applicable document referenced in the contract or system specification. The theory is that if a structure, for example, is designed for twice the strength required, then it will surely withstand the forces applied in its normal experience and need not be subjected to the severe testing that would have to be applied if it was designed without a safety margin.

As in the case of design margins, the requirements for which safety margins will be applied and the safety factors for each should be determined early in a program and approved by the program manager. Subsequent changes in the list should also be subjected to formal review.

Analysts responsible for requirements that have a safety margin applied must account for the safety margins in determining appropriate values for architecture elements as well as the design margin values. In general, on parameters where more is better, the following relation holds: DESIGN TARGET VALUE = SAFETY MARGIN VALUE (REQUIRED VALUE + DESIGN MARGIN VALUE). On parameters that do not have a safety margin requirement, the safety margin simply has a value of 1. In either case, the design engineer must strive to satisfy the target value in the design effort. On parameters where less is better (such as cost and weight), the design margin is subtracted from the required value.

2.3.3.4 Inclusion of margin accounts in requirements data

A computerized requirements system should be configured to accept design and safety margin figures for requirements at all levels of indenture. The system should provide the capability to analyze the current distribution of margins in selected requirements categories and print out these distributions. It should be possible to easily redistribute margin values. One approach in both computerized and manual implementations is to enter the margin number in brackets after the requirement value. For example:

3.2.2.1 *Weight.* Element weight shall be less than or equal to 2320 pounds [Design Margin = 232 pounds].

In the manual application, this figure would remain on the requirements analysis sheet or specification. In the computerized application, the margin figure would be entered in the margin field. Throughout the design period, the designer and management personnel would be focusing on responding to the target weight figure of $2320 - 232 = 2088$ pounds and this is the figure that the designer would compare the current value against at reviews.

2.3.3.5 Design margin account transfers

The whole purpose of establishing margins is to allow their consumption to solve problems during the design phase and still meet required performance. The engineering team must, therefore, have some organized way of transferring margin account values into target values. The control mechanism for this process is the Engineering Review Board. When it can be shown that a target value cannot be satisfied, the principal engineer must come forward to seek relief from any remaining margin. The principal should request a formal review, present the case for a margin transfer (with the recommended source identified, if possible), and respond appropriately to the decision-maker's direction.

There are essentially three ways to transfer or redistribute design margin values. First, a margin for a parent element may be redistributed vertically in whole or in part within the same requirements category to a sub-element.

Second, a margin figure may be redistributed laterally within the same requirements category to one or more elements in different branches of the architecture. Finally, a margin figure may be transferred from one category to another vertically or laterally in the architecture. As an example of the latter, let us say that element A153 has a Design to Cost (DTC) margin of $80,000 and the designer for element A162 has difficulty meeting the reliability target figure of 0.998 (current estimate is 0.988) and he is right at his DTC target. The A162 principal finds that it would be possible to satisfy his reliability target figure if the element could cost $5000 more because he would be able to use a more reliable valve. One solution to this problem would be to redistribute $5000 from the A153 DTC margin account to the A126 DTC requirement target value.

In the more complex redistribution situations suggested in the previous paragraph, it may be necessary to determine transfer relationships between accounts in order to help decide which of two or more transfer possibilities is the more favorable. This is done by comparing the transfer ratios. In the previous case, it would have been possible to satisfy a reliability figure of 0.988 without a cost increase. So, the transfer ratio in this case would be $5000/(0.998-0.988)$, or $5000 per hundredth in reliability. This figure of merit may be compared with similar figures for other transfer possibilities.

2.3.4 Budget Management

Budget management is commonly a more complex issue than margin management because a budget is often spread across more than one engineering discipline whereas a margin account commonly applies to a single discipline, such as weight. For example, we may establish a guidance error budget that has components in gyro drift, gyro mounting accuracy, Loran unit errors, local Earth density and conductivity anomalies, airframe asymmetry, allowable airframe asymmetry, and aileron setting accuracy.

A budget principal engineer must allocate a budget from the total allowed error to the principal engineer for each element of the system involved in the budget. Each allocation must be converted into units corresponding to the item in question. The budget principal periodically gets updates from those with budget responsibilities and determines how the design process is maturing. If the total requirement is at risk, the budget principal identifies the cause, takes action to abate the risk, and monitors progress on the abatement action plan. The objective is to allocate a specific parameter value to the several specifications involved in the budget case so that when the items are developed or procured and come together, they will function as required.

Budgeting is essentially the same as allocation of a parent value to child items. One could put a fine point on a difference by saying that allocation is a process for partitioning a parent value to determine child values while budgeting is a management approach for controlling potentially risky requirements values. But, the same process is involved. We are partitioning a parent value into a set of lower-tier values. In allocation, the relationship is between a parent item and the subordinate child items. In budgeting, the parent value and all of the lower-tier elements need not necessarily be within the same architectural entity.

2.4

Requirements Analysis Strategies

Contents

2.4.1 The Four Strategies

All requirements analysis work can be collected under one of four fundamental strategies referred to in this book as: (1) structured analysis, (2) cloning, (3) freestyle, and (4) question and answer. A program engineering team can select one of these strategies for use throughout a program in all phases, use some single combination of them throughout the development process, or move from one strategy, or set of strategies, to another as the program and system mature.

As an example of the latter and by way of introducing the strategies, let us see how we could vary the requirements analysis strategy across the sweep of the program life cycle. In the concept program phase the only specification required may be a system requirements document or system specification. This system level document could be quickly created by a very experienced system engineer using the freestyle strategy where the engineer creates the specification from scratch on a word processor. The content is driven by a clear understanding of the customer's need attained through discussions (question and answer) between the analyst and customer personnel, the results of a mission analysis and other studies, and focused on a general solution through glasses tinted by the company's product line history.

In reality, this experienced engineer using freestyle is very likely relying on a system specification format drilled into his or her subconscious through years of working this same problem. But, since a formal specification standard is not used, the engineer's approach could be called an example of the freestyle strategy. In the freestyle strategy the requirements analyst does not rely on an organized external format, mechanism, or process to gain an insight into requirements appropriate for the item. A negative term applied to this approach by many is ad hoc. This approach is not recommended but we have to recognize the reality that many specifications are created this way.

During the development phases, the engineering staff could switch to a structured approach where the team performs a very organized, top-down analysis of the system need. Through this process, the team progressively exposes a preferred system architecture. For each item in the architecture a set of requirements and a responsive concept description is developed. Designers translate these data into engineering drawings and manufacturing plans the production of articles defined by the drawings.

As the system moves into the production phase, a need for new customer capabilities, or a demonstrated failure to satisfy current requirements, may inspire engineering change proposals. These proposals may expose a need for other system elements not covered in the current specification and drawing libraries. In preparing requirements documentation for these new elements, the engineering team could elect to employ the cloning strategy involving the use of specification standards or existing specifications as a means of gaining insight into an appropriate set of requirements for these added items. For example, we may find it necessary to procure a valve that is different from all other valves in use. We can create a new valve procurement specification by copying the specification for an existing valve into a new computer file for the new valve. The principal engineer simply edits the document for the few differences and changes item nomenclature.

Another document that may be used as a standard is a generic "boilerplate" document that contains all of the common requirements for a particular type of item. To use this boilerplate, the principal gets a copy and simply edits it with the results of the associated requirements analysis work. As a final example of cloning, the principal could make a copy of the parent item specification and edit it based on the results of a flowdown analysis. These strategies can be very effective on a mature program where the program requirements are well established and the team has evolved a clear definition of the generic program requirements.

2.4.2 Freestyle Strategy

It is possible for an experienced system engineer, familiar with the product line appropriate to the customer need, to craft a system requirements document for the system based on a customer need statement and minimal additional information entirely from scratch on a word processor. We call this a freestyle approach since the engineer makes no obvious appeal to any kind of structure as a way of gaining insight into the appropriate set of requirements. In reality, the engineer who does this job is probably relying on years of experience with system requirements identification within the context of a particular product line and a memorized format. But, since it is all being done internally within the engineer without recourse to external mechanics, we call it freestyle.

As suggested in Figure 2.4-1, the outcome will not always be cheerful. Freestyle carries with it the danger of possible incompleteness due to lack of rigor in the analysis process. It requires a very experienced analyst to have any hope that this approach will succeed. Regardless who does the work, the results should be scrutinized carefully by project personnel most familiar with the customer's needs. It is a great truth that we cannot see our own mistakes, however egregious, but that any one else will see them clearly. So, we should always take advantage of the low cost of criticism through peer review.

When this strategy is followed as the only requirements analysis approach on a large program, it can be characterized as chaos. There is a likelihood that the principal engineers responsible for the many lower-tier specifications will not effectively communicate with each other in any organized way and that the requirements for lower-tier items will not be determined from the system requirements. As a result, the system created from the integration of the lower-tier items designed and produced from those requirements will probably not satisfy the need to the customer's full satisfaction.

While freestyle can be useful when accomplished under certain conditions by a skilled practitioner, this book does not encourage nor further develop the strategy. The reader who is committed to this technique is encouraged to migrate to a more organized process using at least some of the methods explained in this book.

2.4.3 Cloning Strategy

A program that is already established with a large library of specifications, however they were created, can have a very successful requirements analysis experience without applying the structured analysis approach preferred in this book. Such a program has a tremendous storehouse of requirements already in existence in the form of the many specifications on hand. While one may question the quality of that storehouse as a function of how it was obtained in the first

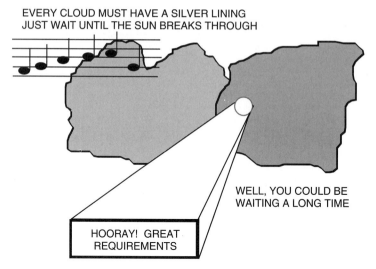

Figure 2.4-1   *Freestyle is for experts*

place, it has probably withstood the test of time. These documents can be used as an inspiration for others through a process called cloning, which is a scheme for using an existing document as the basis for another. Figure 2.4-2 illustrates three approaches using this technique that have been applied in industry. Each is explored briefly.

2.4.3.1 Specification standards

A specification standard is a document that contains a set of standard requirements applicable to a range of system items. When an engineer must prepare a particular specification, he or she can select a copy of the closest standard and edit it, in accordance with a set of instructions, into the new specification. In the process of editing, the results of the requirements analysis process are included in the document in place of generic place-holders. A more commonly used term for this kind of standard is the term "boilerplate."

A boilerplate can be prepared for a whole class of specifications, such as a procurement specification boilerplate, or be more finely cut as in the case of a valve procurement

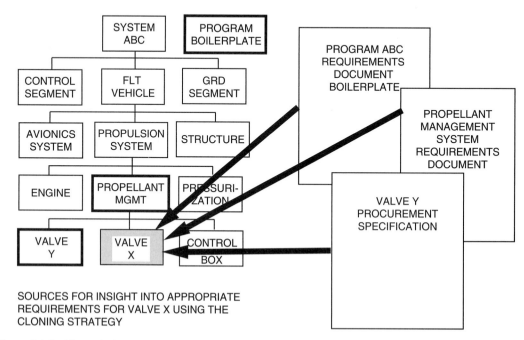

SOURCES FOR INSIGHT INTO APPROPRIATE
REQUIREMENTS FOR VALVE X USING THE
CLONING STRATEGY

Figure 2.4-2   *Three cloning methodologies*

specification boilerplate or an aerospace ground equipment procurement specification boilerplate. This thinking carried to its extreme begins to blend into the notion of a general specification that is augmented with a series of detail specifications that give item differences.

Essential elements of a boilerplate include standard structure, content definition, and formatting, an approved applicable documents list coordinated with contract requirements, and standard environmental requirements that account for planned test program needs. We want to avoid being too specific in the boilerplate in other areas or our boilerplate conversion task will include deletions of extraneous material rather than addition of item-specific material. We can never be sure that we will have deleted all of the extraneous material and this can lead to unnecessary cost.

A principal engineer uses the boilerplate to gain insight into at least some of the requirements attributes that should be considered and places the results of the requirements analysis process implemented for the item into the context of a copy of the standard to arrive at a complete program-peculiar specification. This technique is very effective for Sections 1, 2 (to the extent that the principal appeals to a standard list of applicable documents, not all of which are used), and 5, the constraints portion of Section 3, and the text and verification matrix format of Section 4. It cannot be effective in the performance requirements because these vary so widely as a function of the specific kind of item. A standard is also of little value in interface requirements definition.

A first class boilerplate uses a facing page scheme. The left hand page provides instructions for completing the specification data on the right hand page. If a company commonly uses specification standards, or boilerplates, on two or more programs and anticipates using them in the future, they should maintain a generic set of specification standards, or boilerplates, that apply to their product line and customer base. They are created and maintained using computer word processors commonly used on programs to create specifications and retained in a ready status in electronic media for use on programs. A database could include these standards as well, of course.

Each program, upon initiation, translates or converts this series of generic company specification standards into a set of program standards while getting ready for the preparation of individual requirements documents. The difference between these two kinds of standards centers on detail. The generic documents may be as simple as an outline. The program standards must include program-peculiar applicable documents with tailoring (not all of which will be necessary in each specification), any special program/contract-peculiar language required by the customer, the results of a system level environmental requirements analysis and prime item zoning information, if applicable, and standard program verification data requirements.

The program specification standards should be made available to personnel assigned principal engineer roles in a media compatible with program word processing capability. Where a program uses a computer database tool for requirements capture, the boilerplates must be in a form that is compatible with the database capability to load the database from the standards or otherwise introduce standard content. In a word processing environment, the standards should be located on a computer network server in read-only mode such that a principal engineer may copy the standard to his or her workstation and begin the process of creating a specific document.

The program standards must be prepared as quickly as possible to ensure that principal engineers have available to them the tools they need to immediately begin identifying and capturing requirements. Generally, the task that will delay program standard development most acutely is the preparation of program environmental standards. With the exception of the environmental standards, the bulk of the program standards should be available within a period of no more than two weeks from program go-ahead. The environmental standards should be available no more than 30 days from program go-ahead and made available to each principal engineer then in work on a specification to cut and paste into their document.

The program standards availability date can be improved, where there is no possibility of an ethical dilemma, by using the time available during any proposal periods of inefficient personnel utilization such as the waiting periods between proposal submittal, questions and answers, best and final, and contract award. The proposal manager should be informed of the intent to use available time for this purpose in order to ensure that proposal team priorities are fully coordinated and any time charging and ethical concerns are resolved honorably.

2.4.3.2 Like item approach

The like item approach entails the use of an existing specification covering a similar item as a basis for a new specification. A sound approach here is to obtain an electronic copy of the existing specification and edit it into the new one by changing a few appropriate details. If, for example, a program must add a new motor operated valve different from all other valves in a fluid system, the new specification could be created very quickly by selecting the specification for another motor operated valve in the system as the basis. The results of the requirements analysis task performed for the valve provide the differences to be introduced into the new document. A title change and a few other cosmetic changes and the new specification can be made available very quickly. The task is not made a great deal more difficult when the source document covers a valve operated by a different stimulus (solenoid operated, for example).

2.4.3.3 Parent item, flow down, or allocation approach

The third cloning approach involves using the parent item specification as the basis for the new specification rather than a like item specification. In this case the requirements analysis approach that will be most effective is a flow down methodology. The requirements called out in the parent item are allocated to the subordinate item and form the basis for the differences introduced through editing a copy of the parent specification. This topic extends to a discussion of margins and budgets, of course.

As we saw in the discussion of value apportionment in the previous chapter, margins are reserved accounts of requirements values that will be used to insure that requirements are satisfied in the evolving design and to provide management with problem-solving potential in association with technical risk abatement. Unfortunately you will not be able to afford to run margin accounts on every requirement in a system. It does cost more to do this work and the work would be wasted on most requirements. We should be selective in applying this technique and run margins only on requirements that involve some design risk.

Requirements budgeting is an organized way to apportion the responsibility for satisfying a requirement among many other requirements for one or more architecture elements. A budget should be managed by a principal engineer, who must determine the effect of the current and projected values of these subordinate requirements on the requirement being managed. In many cases this budgeting process only affects a single requirements category in a particular range of architecture elements and in these cases it is equivalent to allocation or flowdown. In other cases, like guidance accuracy, the budgeting process may extend across several branches of the architecture and involve many different requirements categories. One might budget the guidance accuracy in terms of weight, thrust, aerodynamic drag asymmetries, thrust misalignment, guidance unit mounting accuracy, and other very different requirements categories.

2.4.3.4 Flow down scope limitation

The parent item flow down technique is very effective for certain kinds of requirements but should not be used as the sole strategy for requirements analysis on a program. It is effective when applied to particular partitionable requirements such as weight, reliability, maintainability, and life cycle cost where the corresponding practitioner uses a mathematical model and a partitioning mathematical rule such as the parent value is equal to the sum of the child values.

Where this approach is applied program wide for all specifications and all requirements, one must determine all child requirements for all specifications by considering how a particular requirement in a parent specification should influence the requirements set for each child item. Some requirements, frankly, do not flow down very cleanly or easily.

As an example of this approach, let us assume that the program is organized into teams and each team must develop all of its own specifications. The team would then focus first on deriving all of the requirements for its top level architecture item from the parent item specification which may be the system specification or some subordinate specification developed by a parent team. Other teams may be at work independently deriving their top specification requirements in an identical fashion. Even if the teams coordinate on the flow down of each requirement, there will be a concern for completeness of the child item requirements. Each child item specification is presumably limited to only the requirements named in the parent item specification or those derivable from them.

Some of these relationships will be fairly obtuse. For example, the aircraft specification may require a cruise speed of 620 knots. While developing the requirements for the propulsion system, we may be alert enough to understand that the speed requirement imposes a need for a certain amount of engine thrust or horsepower, that the structure specification will have to call for a certain smoothness condition and drag maximum, and that the structural integrity of the airframe can sustain the stresses related to this speed. There may be other lower-tier effects that should be identified as well and what is the whole set at the immediately subordinate level and beyond. We may never be confident that we exhaustively identified all of the appropriate lower-tier requirements for the aircraft speed requirements. In addition, some of these lower-tier requirements may be properly subordinate to other aircraft level requirements.

Flow down is not a structured analysis process in that it does not include any built-in mechanism to encourage identification of requirements unless the child item require-

ments have the same title as the parent requirements. That is, because we respect reliability as a parent requirement we are encouraged to respect reliability as a child item requirement. A structured analysis process commonly uses simple graphical sketches with particular symbols or parts that tend to stimulate the mind to think of particular kinds of requirements.

2.4.4 Question and Answer Strategy

The third strategy involves the developing agent asking the customer questions about their needs and formatting the answers into requirements for inclusion in top-level specifications. This process should be part of every requirements analysis activity. It is not effective as a sole requirements identification technique because it is not, of itself, structured or organized. The key to this process providing any useful information is that good questions are asked of people who know the answers. So, those who would ask the questions need some kind of machinery to encourage good questions and all of them of interest. Also, these questions must be asked of people who have knowledge.

Q&A combined with structured analysis provides the best combination since it uses simple pictures (models) to stimulate a conversation between one party with questions and another who may have answers. These models communicate with the minds of the parties using the most efficient way of getting information into the mind of man, that is through vision, and therefore encourage discussion and refinement of the meaning. They encourage common understanding and completeness as well.

The Q&A process can also be improved by asking the right kind of questions. The contractor party should ask open ended questions that stimulate the customer party to talk rather than yes-no answer questions that can be answered too quickly. Many readers will be able to remember having been interviewed by a hiring manager who spent the whole interview talking about himself or herself. A hiring manager should, of course, be looking for information about the applicant and gets access to it by encouraging applicants to talk about themselves. The same motives work when trying to gain information about a customer agent's knowledge about their need. Many books covering quality function deployment (QFD) contain good information that will help engineers improve their customer interviewing skills. Refer to chapters dealing with the voice of the customer. The book *Exploring Requirements* by D. C. Gause and G. M. Weinberg is also a good source on this matter.

2.4.5 Structured Analysis Strategy

The final of the four requirements analysis strategies we will discuss is the structured approach. Structured analysis is synonymous with the term system requirements analysis as applied in aerospace industry. Parts 3 and 4 provide a detailed explanation of this organized strategy and the remainder of the book is primarily focused on this strategy. The structured analysis strategy provides an organized, systematic environment within which to decompose a large problem into a series of smaller ones such that solution of all of the smaller, more specialized problems results in a solution to the larger one.

While one might wish that there were only one structured analysis process, the fact is that as of this year no one had devised a single model that had been shown to be the best structured analysis approach for all situations. They are divided in particular between problems that will surrender to hardware solutions and those that will surrender to software solutions. The author is convinced that the arrival of unified modeling language (UML) on the field does offer a potentially universal approach for hardware and software where traditional structured analysis (TSA), useful for grand systems and hardware is but a subset of UML. This is further explored in Chapter 4.6.

3

Traditional Structured Analysis

Contents

3.1

System Beginnings

Contents

3.1.1 What's in a Name?

For the initial development work, the author chose the term system definition over many others that the reader may prefer such as mission analysis or concept development. It is intended to generally cover the front end work on a program that moves the customer and contractor from a condition of relative ignorance about the need toward a clearly defined set of system requirements and a preferred system concept supportive of those requirements. It begins with the customer need statement and terminates with the system design review (SDR). Structured analysis techniques, as covered in this part, are valuable tools for uncovering the right questions to ask about the problem space proscribed by the customer's need statement but to be most effective they must be accomplished within a context richer than a simple user-oriented need statement.

In a large customer procurement situation, this process may be accomplished by the user, an acquisition agent for the user, a contractor working under a study contract, or some combination of these possibilities. The period of time may be fairly brief or it may stretch over a period of several years implemented through multiple study contracts involving trade studies, technology development, and even a competitive development and testing of alternative systems as in the case of the DoD F-22 or JFX Programs, for example.

In a commercial situation the term system development may be considered an overstatement but a similar process must be accomplished that results in the enterprise uncovering a product opportunity that has potential customer appeal where a product can be designed, manufactured, and distributed to turn a profit for the enterprise.

3.1.2 In the Beginning

New, man-made systems come into being initially through the application of what this book refers to as system development under the control of a developing agent. That development agent accomplishes the related work under a contract with the acquisition agent in the case of new systems for the U.S. Department of Defense and for other large organized customers, as well. Where the development agent is a commercial enterprise, they commonly act on their own business motives to develop a product that they believe can be manufactured and sold for a reasonable profit. While very fine products can be conceived through unstructured creative thought, success in developing systems to solve very complex problems is more often going to be realized by the developer applying an organized process especially where time and money constraints apply.

Very complex problems commonly require no small amount of work to properly bound and characterize the system level problem. This process is actually a program in miniature calling for requirements analysis, synthesis of concepts and accomplishment of trade studies to select preferred concepts, and some analytical work, simulation, and even some testing work to validate the requirements (gaining confidence in one's ability to satisfy the requirements). There are two general approaches suggested in Figure 3.1-1. The problem the enterprise is trying to solve is either an unprecedented one, meaning that the enterprise has not in the past solved this problem and they know of no other solution that they could apply, or a precedented one, meaning the enterprise has previously solved this kind of problem. There are other variations possible covered in Chapter

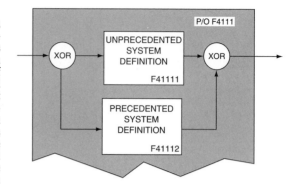

Figure 3.1-1   *System definition alternatives*

1.3 but this beginning process will be discussed in this simpler context. The remaining chapters in this part expand upon the introduction offered in this chapter.

3.1.3 The Meaning of the Term

The term system definition is used in this book to denote the initial transform between the content of a customer need statement or commercial equivalent and a formal go ahead to implement a system development process for a specific system thus defined. The goal is to establish a critical mass of knowledge about the application that can be built upon through the system development process described in this book. The key differences between task entry and exit are the degree of specificity available on the system and the degree of knowledge about risks that still remain. Commonly, the entry condition is an expression of customer need and the exit corresponds to a fully formed set of requirements stated by the user as a minimum or an at least nearly contractually sound system specification, ideally. In the best case, the requirements concerns would be sufficiently mitigated that the program could be undertaken with some confidence within the context of a particular amount of available funding and development time. At the heart of this process resides a contractor person or team and a customer and/or stakeholder person or team. Several techniques have been successfully applied to acquire knowledge of the customer's need and many of them are explored in this section.

3.1.4 Unprecedented System Definition

Figure 3.1-2 illustrates the unprecedented process beginning with the need statement. Three principal activities are involved: requirements work to clearly define the problem, concept and program design work to craft a solution that is achievable, and management work. This work will often be performed as a study rather than a full-fledged program. It is often accomplished by the using customer or by a contractor to transform the user requirements into a form that can be used as a basis for a development contract such as a preliminary system specification. Often initial user requirements are not properly quantified for use as the basis for a development contract so engineering work is necessary to translate what appears clear to the user into precise contractual terms.

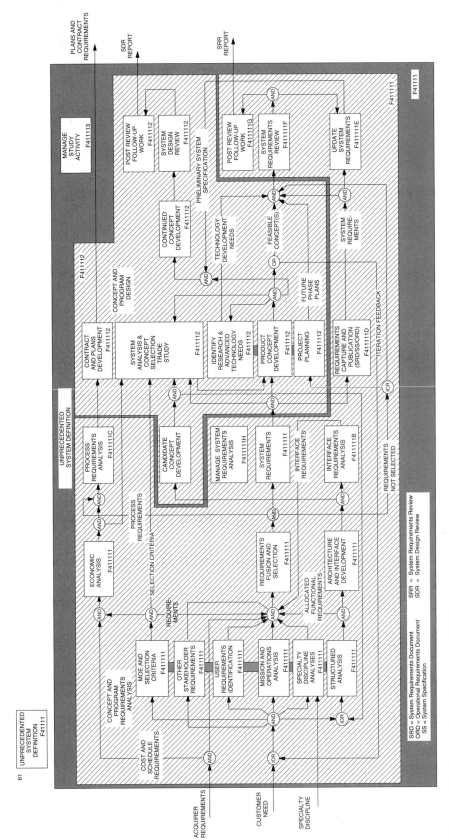

Figure 3.1-2 *Unprecedented system definition*

3.1.4.1 Customer interaction

While a structured analysis process, such as those covered in the chapters of this part, is very important in understanding a customer need, this effort will not be successful without more information about the customer and their expectations. It is essential, to the maximum extent possible, to engage the customer in a discussion of their operational and logistic needs.

In this process, the goal is full disclosure of the information the customer has about their need but one should recognize that this may not be very much. It is easy for contractor engineers to develop a bad attitude about what they perceive as customer ignorance or arrogance at this point in a program. The reality is that the customer may very well be ignorant in the finest sense of that word (lacking in knowledge). One often hears contractor system engineers speak disparagingly about customers and their lack of knowledge helpful in properly characterizing the system. These same engineers may fail to account for the fact that the customer is paying their enterprise to help them solve the problem not yet fully characterized. These negative thoughts should be discouraged. Contractor engineers should be consumed with a positive attitude about the customer helping them to understand what is possible while trying to understand their need.

This phenomena of a progressively improving understanding of the need is at the root of a problem that will influence every program where there is a substantial element of new development. As the contractor begins to evolve a system concept with validated requirements giving the customer some glimpse of the possible with the current state of the art, the customer begins to get new ideas about their need. The result is often called requirements creep. It may be impossible to stop this from happening and it may not be in the interest of either party to do so. It is important, however, that the contractor be protected against requirements growth without compensation through a well structured contract that establishes a technical baseline as early as possible and works cost, schedule, and product performance issues related to improvements driven by new insights

as trades yielding more resources to the contract for more capability included in the system.

The extent to which a program will be saddled with requirements growth is inversely a function of the degree of skill that is applied to the process of talking to the customer about their need. This is not a process that can be easily diagrammed because there are so many possibilities. But there are some guidelines that should prove helpful in maximizing the flow of information.

First, there are some reasons why the customer may not be as forthcoming as a contractor engineer thinks they should be. If an Air Force Captain has been working on a program for two years prior to the time a contract is let, the Captain may feel that everyone knows the information that he has been dealing with for the past two years and not want to bore you with details that you must already know. So, it is helpful to inform your counterpart that nothing is too simple for your ears in these conversations. Second, with large customers like DoD, during the competitive phase of an acquisition, the customer representatives may be prohibited from discussing the system with any one contractor. This problem can be overcome to some extent through formal questions submitted by the contractor. The down side is that all competitors get the answers. Therefore, this avenue is both a way to get real answers to questions as well as way to ghost your competition with questions you want them to think are important to you. Finally, the customer just may not know the answer.

Ideally, both customer and contractor would actively and cooperatively work to seek out knowledge of the customer's need with perhaps the first step being to identify all of the stakeholders for the system. These are people and organizations that have some interest in the system to be developed including those that might consider the system to be disadvantageous. All of these stakeholders will have ideas about the system that can be translated into requirements. For simplicity, we can classify all of the stakeholders into user, acquirer (acquisition agent), and others as suggested in Figure 3.1-3. Figure 3.1-2 tasks F411111 (Mission and Operations

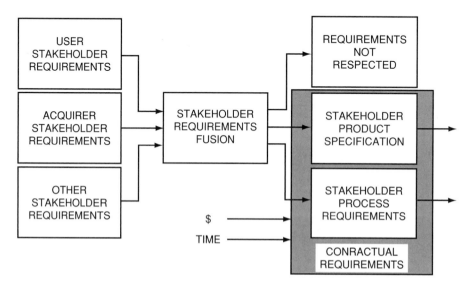

Figure 3.1-3   *Requirements fusion and partition*

Analysis), F411112 (User Requirements Identification), F11113 (Other Stakeholder Requirements), and F411115 (Structured Analysis) all define requirements of possible merit and the results of each activity must be reviewed for possible selection for inclusion in process or product specifications.

Some one or team should collect all of these user requirements as noted in task F4111117 and subject them to a fusion process where they are sorted into three sets as follows:

a. *Stakeholder Process Requirements* Requirements that should influence the development process. These requirements should flow into the contract, statement of work, and program planning.
b. *Stakeholder Product Specification* Requirements that should flow into the system specification.
c. *Requirements Not Respected* Requirements expressed by some stakeholder that have been specifically rejected for inclusion in the contract.

The latter category can be a great source of program difficulty if not handled openly and with discipline. Many of these requirements will commonly have been identified by the user community based on their understanding of their mission. The acquirer may conclude that some of these requirements cannot be satisfied with available money in the near term. In these decisions, the user and acquirer should come to a clear understanding about how those rejected requirements will be treated. The technical ones could be included in the initial system specification as goals or included in future plans for a later product block, or made the subject of further research efforts outside the contract being discussed thus giving the user community some hope of their eventual inclusion in the product. If they are simply rejected or not consciously dealt with, the user community may be motivated to continue to consciously or subconsciously respect them leading to problems later in the program when the system is tested to determine if it meets user requirements (OT&E on DoD programs).

At each major review the user community may seek to focus attention on failures of the system briefed to comply with these user requirements that were not included within the contract. If the contractor and acquirer are not alert to detect the out of scope nature of the problem raised, requirements creep can set in exposing the program to cost and schedule risks.

Hard feeling from the user community may survive several years of system development and come back to haunt the program in test and evaluation. Given that the system is proven to satisfy system specification requirements in a development test and evaluation (DT&E) conducted by the contractor, the user may subsequently subject the system to test plans for the operational test and evaluation (OT&E) program conducted by the user that are based on a pre-contract system operational requirements document that includes performance requirements rejected in developing the formal system specification because of their cost impact. The fallout from resulting reports of failure of the system to satisfy test requirements can linger long after these claims have been debunked as having been based on testing that was out of scope.

The standard for development of systems prepared by the Electronic Industries Alliance, EIA 632, identifies a process named product validation which seeks to prove that the system satisfies its user requirements. The author believes that there should be no conflict between the user requirements and the contract requirements such that when we succeed in proving that the system and all of its elements have satisfied their corresponding requirements through an effective verification process entailing item qualification, item acceptance, and system test and evaluation, that we have proven that the system satisfies its user requirements. The approved user requirements must be in a proper system specification.

All of these requirements issues must be dealt with at the system requirements review (SRR), often accomplished within a few weeks of contract award, with clear direction that the resultant requirements defined in the contract and system specification are the system requirements and no others will be considered unless they are formally added to the contract through the normal configuration control board (CCB) process for reviewing engineering and contract changes.

The astute contractor will, however, maintain a list of unrespected system requirements throughout the contract run, report upon them at each major review, and take every opportunity to move requirements on this list into the system specification. For example, when an engineering change proposal is being considered, the contractor may offer the acquisition office an opportunity to move one or more unrespected requirements into the specification for a relatively small cost. During the development effort it may be found that parts of the system have been unintentionally over-designed and that the system as a result can meet a higher performance goal than covered in the specification bordering on the values originally required by the user. While the contractor may not be able to get paid for that added capability, it can offer the capability to the acquisition program manager and expect some form of consideration relative to other pending issues yet to be resolved.

3.1.4.2 Mission and operations analysis

Mission analysis and operations analysis provide a window into a deeper understanding of the customer's need. This is a creative process depending on engineers and analysts with experience in the product line and familiar with customer and company history in solving similar or precursor needs. The ultimate customer requirement is expressed in their need statement. This need statement should first be expanded into a master life-cycle diagram in task F411115, Structured Analysis. One or more of these functions will relate to accomplishing some mission leading to successfully satisfying the whole need or some part of it.

Often, satisfying a system need must necessarily involve a certain dynamic to move some object from one place to another in time and space while surviving possible threats from hostile forces. The analyst must conceive one or more ways in which the need could be satisfied in terms of a basing concept, a mission scenario, and a logistics support concept. Mission models are developed that allow the analyst to manipulate some parameters while observing effects on others. Scenarios are often useful in helping to understand mission opportunities. These may be created in the form of an event list, icon flow diagram, functional flow diagram, or state transition diagram. In any case we must identify some key parameters to evaluate alternative concepts.

3.1.4.3 MOE and selection criteria development

We must understand what our customer thinks is important in numerical terms. This can be stated in a limited list of Measures of Effectiveness (MOE) which are numerically stated capabilities that the system must have in a few critical areas like range, availability, and payload weight capability. Other measures of effectiveness may also be used as a basis for selection of alternative concepts. In these cases, the values may be bounded but alternatives scored based on the value they offer in these parameters. For example, alternative 1 may have a life-cycle cost (LCC) of 2.4 billion dollars while alternative 2 has a LCC of 1.9 billion dollars. Given that both alternative 1 and 2 satisfy all of the requirements (expressed in other MOE), we would be moved to select alternative 2. A selection is seldom that simple since there will be several MOE upon which we must make a judgment and the final answer will have to be arrived at in terms of a best mix of features or a least onerous combination.

3.1.4.4 Requirements work

Several requirements analysis steps identify user and other stakeholder requirements, linked together by a vertical bar on Figure 3.1-2 signifying intense cross task coordination. It is important to consider both the user and acquirer requirements and to look for any user requirements that have not been included in the acquirer requirements that will be the basis for any contract that covers the development process. On a DoD program, it is not uncommon for the acquirer to shave off some user requirements in the interest of matching program cost to available funding.

An unprecedented system should be defined based on the results of a structured analysis (F41115) to determine appropriate performance requirements and physical architecture. This work should be coupled to the search for measures of effectiveness (F41114) and selection criteria for trade studies that will probably be necessary to choose the best available candidates. Refer to the several chapters in this Part for models appropriate for structured analysis. Part 4 covers models appropriate where the product will be implemented in computer software.

Task F41117 is an integration task to sort out any conflicts between the several classes of stakeholder requirements and to consider possible omissions. We should take special note of any differences between acquirer and user requirements where user requirements will not be satisfied as these cases can become serious problems during user system testing and evaluation (referred to as OT&E on DoD programs) at the tail end of the development process.

Architecture identification based on structured analysis in some cases may require a trade study to ensure careful selection of the all around best candidate for the needed functionality. Also, a study of the interfaces between the top-level system items may give insight into architectural complexities that suggest alternative architectures.

The requirements identified from all sources should be brought into a common context by fashioning a system requirements document (SRD), Operational Requirements Document (ORD), CONOPS, or preliminary system specification. The content of this document, validated through concept definition and risk analysis, should be subjected to a system requirements review (F4111F).

Structured analysis is an organized method, requiring a broad appeal to knowledge, for partitioning a complex problem into a series of smaller problems each of which, if possible, will yield to solution by a narrower span of knowledge or technology and of a granularity (depth and breadth) better matched to human and team proportions.

Many discussions of the system requirements analysis process appeal only to functional flow diagramming as the means to gain insight into needed system functionality. That is one approach but not necessarily the only fruitful one. It is possible to use more than one method on one development effort. In fact, it is downright foolish to force the use of only functional flow analysis. The reason is that it is not as effective for some purposes as other methods, computer software requirements identification for one. Also, we will find that it makes a lot of sense to encourage the structured analysis process to gradually blend into a process analysis as the top-down development work passes through the level in the architecture corresponding to clear responsibility assignment to specialized design organizations.

If your engineering organization employs a central system engineering function to define requirements throughout the architecture, it may be a good idea for that organization to select a single particular methodology and train all of their system engineers on that common method. It should be remarked that if your organization's product line includes computer software as well as hardware and systems composed of both hardware and software, you should consider embracing a pair of approaches as a minimum because a model that works well on hardware commonly does not work well for software and vice versa. Adherents of every structured approach, however, do claim that their preferred method will serve both purposes better than any other single model or combination of them but the author remains unconvinced.

A sound structured analysis process will not only yield useful information about needed system functionality but encourage questions of merit to be posed to the customer about their need. Thus the several activities shown in Figure 3.2-1 complement each other when all performed concurrently. Tasks F11115, F411118, F41111A, and F41111B of Figure 3.1-1 are described in subsequent chapters of this part of this book and are applied in the same way here except that we are at the top of the system functionality and system concept and our knowledge about the problem and solution is minimum at this point.

3.1.4.5 System environmental definition

Every system takes the ultimate form of a single block marked system interacting with its environment. We need to define with clarity what that system environment is, how it relates to the product system, and precisely where the boundary between them is. The first step in this process is to identify the spaces within which the product system will have to function. This might be as simple as America's roads. A military system on the other hand may have to operate on roads but also over unimproved surfaces. An aircraft must operate on the ground between flights but also fly in Earth's atmosphere. A space vehicle must exist for some time on Earth subjected to gravity and natural environmental stresses to the extent that it is not protected from these stresses in clean rooms. But, this craft is also going to have to survive the ride to space and function for years in space exposed to the rigors common in space. So, a system may be exposed to a relatively simple set of environmental stresses or a combination of several very different sets as a function of the number of different environmental spaces within which it must operate or endure.

The natural environmental stresses for each different space can be readily determined from the appropriate environmental standard defining environmental parameters with precision. It is not enough to simply reference these standards, however, because they commonly offer a definition that covers many possibilities. One should consider every parameter listed in the standard and evaluate the ranges and conditions included in the standard and how these compare to a realistic assessment of the stresses appropriate for the system in question. Failure to tailor environmental standards for reality is one key way that excess cost enters into a system development effort. At the same time, failure to consider all of the appropriate environmental factors is also a source of great program risk to be discovered too late for economical correction.

The natural environment is not the only problem. We must also seek out the characteristics for four other environmental subsets: self-induced, cooperative, non-cooperative, and hostile. Self-induced environmental stresses are driven by some feature of the product itself that interacts with the natural environment to create stresses that are applied to the product so long as that product feature is active. For example, a chemical rocket powered launch vehicle rises off the launch pad because of the tremendous power of its rocket engines. These engines interact with the atmosphere to generate powerful acoustic noise forces that impinge upon the space transport vehicle as it rises off the pad. These stresses could be eliminated, of course, by simply turning off the engines but that would be counterproductive to our principal function of transporting a payload to space. So, the launch vehicle must be designed to tolerate these stresses that are self-induced. Commonly, the drivers for these stresses are energy sources within the product but other sources are possible.

Cooperative environmental connections involve other systems with which the system in question will have to interact. These influences are more often handled as external interface relationships rather than environmental relationships because there is another party with whom one can cooperate. Software people model this environmental element using what they call a context diagram illustrated in Figure 3.1-4. With clear definition of these interfaces/cooperative environmental elements, this establishes the boundary between the system and its interfaces with other systems with which it will have to cooperate.

Unfortunately all other systems will not be cooperative with this new system. Some systems will be uncooperative. This does not mean they are purposely trying to interfere with your new system operation, just that they are sharing some of the same space with yours and there are unplanned interactions. A good example of this effect is electromagnetic interference. All systems at all involved with electronic signals or electrical power generate some form of electromagnetic signature. If another system happens to be sensitive to these stresses, interference may occur. It is a service

to all other systems that our system should minimize these emanations and some customers require certain maximum levels in accordance with levels prescribed in a standard.

Some other systems may carry the notion of non-cooperation to the extreme qualifying as hostile systems that purposely try to defeat your system. These environmental effects have to be studied from two perspectives: (a) in what ways could another system interfere with our planned operations and (b) what threats exist now or may become operational during the life cycle of our system. These questions require a threat analysis that lists the possible threats and the nature of their potential intrusions. The hostile threat could entail physical security, computer security, destructive effects delivered at distance, and any number of other hostile acts. Given certain defined threats, the design may have to be changed to reduce the probability of interference or severity of the impact.

3.1.4.6 Specialty discipline analyses

Many of the specialty engineering disciplines will have to be consulted during the early program work to define system and next tier requirements and support an understanding of what they imply about the system to be selected. This may include reliability and maintainability, which may be combined to form an availability requirement that, in turn, may be combined with life-cycle cost to form a system effectiveness parameter. Some of the system level specialty parameters may be selected as MOE.

One of the key specialty problems involves defining the preferred system basing and logistics concept. At what geographical locations will the system be based? Over what geographical span will the system function? How many levels of maintenance shall there be? How intense in time shall operations be relative to the time available for repair? Different alternative solutions may very well encourage different basing and logistics concepts that impact life-cycle cost and other parameters useful in the selection process.

3.1.4.7 Concept and program design

In addition to defining appropriate requirements for the system to be, we need to come up with an acceptable design concept. We wish to exit the system definition process with a sound system specification and matching concept. The reason for this, rather than simply a system specification, is that the concept definition process validates the requirements as well as giving us insights into additional appropriate requirements for the lower tiers if not the system level.

This concept definition process requires a stable architecture from which to proceed and is completely enmeshed with a trade study process for selecting the best alternatives. One could describe this whole process as one big trade study or collection of trades. Figure 3.1-5 illustrates this process for an unprecedented development. The need statement is the ultimate requirement that can be expanded into a life-cycle model that will be essentially the same for every system developed by the enterprise as discussed in Part 1.

The use system function of the life-cycle model may be decomposed in one or more ways to satisfy the operational need as shown in Figure 3.1-5. Each of these alternatives may be associated with one or more architectures combined with a schematic block diagram illustrating top level interfaces, an environmental definition (that may have many elements in common across the candidates), a logistics concept giving the distribution or deployment scheme, basing concept, and maintenance plan, a simple scenario describing

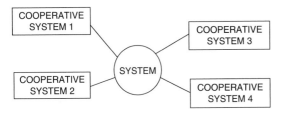

Figure 3.1-4 *System context diagram*

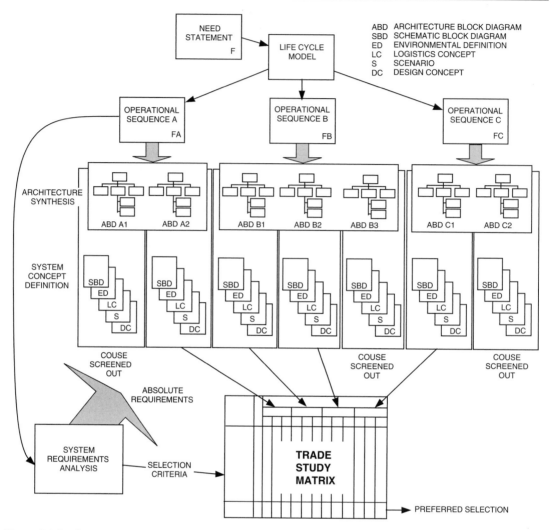

Figure 3.1-5   *General preferred concept selection*

how the system will behave, and a conceptual design. The pre-design work accomplished must determine the capabilities of each candidate.

Prior to identifying candidates we should determine a value system based on a selection criteria, relative weighting of those criteria, and a set of utility curves transforming a raw value range for each criterion into a unit-less monotonically increasing function. Pre-design work determines raw values for each candidate in each criterion and these are entered into a trade matrix in the column for the right candidate. Raw values are translated by the utility curves and acted upon by the weighting values to form a trade value that can be added for each candidate.

The candidate with the highest score is probably the best overall candidate, but we know relatively little at the time we do this work. A sensitivity analysis of the trade matrix will tell you how sensitive the data is to small changes in weighting, candidate values, and utility curve shape. If the preferred candidate changes with small changes of these values, we say the decision is very sensitive to small changes and characterized by more risk than if the preferred candidate is stable under small changes.

3.1.4.8 Manage the study
Clearly the system definition study requires good management, as does any development activity. On a small program this may take the form of working supervision where the team leader is also a worker. On a large study this work may form a complete development program phase with a specialized manager.

3.1.4.9 Program funding profile requirements
Economic analysis (F4111119) should be applied to the problem to help determine the affordability of alternative solutions and calculate acquisition and life-cycle cost figures. After the demise of a worldwide military threat to

American dominance, economics took on a much more important role in the DoD acquisition process recognized in the initiative of the late 1990s called cost as an independent variable (CAIV). At the height of the cold war, cost was a dependent variable, dependent on what ever it took to achieve desired performance goals. MIL-STD-490A on specifications specifically stated that cost should not appear in a specification since it was a programmatic matter. As this book is being re-written, cost commonly appears mixed in with the technical performance parameters in specifications in terms of a unit cost, life-cycle cost, or acquisition cost. The current military standard on specifications, MIL-STD-961E, does not include a specific paragraph for its inclusion but does not explicitly prohibit the inclusion of cost in a specification.

3.1.5 Trade Studies

Early in the development period, it is necessary to make very difficult decisions before a program has collected enough knowledge to make those decisions based on clear knowledge. A trade study process encourages good decisions based on a clear understanding of a well-defined value system. One important application of the trade study process is to select the best all around design concept for a particular architecture entity.

3.1.5.1 Trade study mechanics

The trade study approach calls for identification of two kinds of requirements as the basis for evaluation of alternatives. There are some real requirements that all candidates must satisfy just to be accepted as a viable candidate. Often these will be performance-oriented parameters. We need to identify a second set of parameters that will be allowed to vary and these are the selection criteria. Cost is a common selection parameter but it could involve performance characteristics. Now, each candidate solution will have to be developed sufficiently well that we can assess the corresponding selection criteria values. This process has a significant cost so to the extent that candidates which have no chance of being selected can be deleted based on sound engineering judgment, there is an opportunity to avoid spending scarce resources. The remaining candidates should be evaluated in the trade matrix.

Figure 3.1-6 shows a typical trade matrix including columns for raw score (RS), utility value (UV), and trade value (TV) for each of six possible candidates. The raw score is a real world value derived from evaluating the candidate for the criteria in question. That RS is plugged into the criteria utility curve, introduced below, to derive a utility value. That utility value is then multiplied by the weighting (WT) value that reflects the priority the decision maker places on that criteria relative to the others. The result is the

SELECTION CRITERIA	WT	CANDIDATES																	
		A			B			C			D			E			F		
		RS	UV	TV	RS	UV	TV	RS	UV	TV	RS	UV	TV	RS	UV	TV	RS	UV	TV
TOTALS																			

Figure 3.1-6 Typical trade matrix

trade value. The trade values will always be positive numbers in a range dependent upon the selection of the utility curve vertical axis numerical value range and the value of WT for that criteria. If we selected a maximum WT value of 10 and all of the utility curve max values were 10 then the maximum trade value would be 100, for example. We can select the utility curve shapes to always provide trade values based on an increasing positive value being a higher score by using indirect relationships (negative slope) in the utility curves for real world values that have an increasing value with decreasing goodness.

Finally, we sum the criteria TV figures for each candidate and enter the result in the aggregate effect row. This score can now be compared with the other candidate scores to suggest a preferred choice. One problem remains. At the time that a trade study is performed, the amount of supporting knowledge is very limited and error is entirely possible. Therefore, the analyst should apply a technique called sensitivity analysis to the results. We vary the numbers over a small range, say 5 percent, and observe whether the selection would be changed as a result. If the selection is stable under variation then we say it is insensitive to variation of the value system within a certain range and we can feel comfortable that even though we may not have identified the values with perfect precision, the errors contained in the trade study are insufficient to effect the selection result.

A utility curve translates raw values into a positive valued, positively sloped function. You need a utility curve for each selection criteria. Figure 3.1-7 illustrates several possible utility curves. Curve A shows a direct linear relationship. Curve B shows an inverse linear relationship. Curve C illustrate a non-linear relationship favoring mid-range values. If, for example, the curve is for vehicle range and the real world values are in hundreds of miles, for a candidate raw value of 10 (1000 miles), the candidate would be awarded a utility value of 8.8 which is fairly good in a 0 to 10 utility

value range. The same utility value range should be used in all utility curves so that the only prioritizing is accomplished by a single factor, the WT factor. When you distribute priority into several places, it is hard to keep track of your value system as changes are made and changes will be made as the trade study moves forward.

One might ask, "Whose value system should we be respecting in this process?" The decision-maker's value system is the immediate choice but the decision-maker should determine if that should be his or her own personal/professional one or if he or she should be respecting the customer's values and, if the later, whether it should be the acquisition or operating customer's system. The analyst who accomplishes or leads the trade study should sit down with the immediate decision-maker and determine answers to these questions before the trade study has consumed a lot of resources.

Another question of merit for the decision-maker involves the information the decision-maker wants to use in making the selection decision. Some decision-makers would prefer to apply an analytical approach using numerical values and the techniques shown above would be perfect for this person. Another decision-maker might prefer to make the decision in a more intuitive fashion where the candidate scores were illustrated in colors such as blue for outstanding, green for good, orange for marginal, and red for bad. Alternatively, the candidates could be simply ranked in each criteria.

If the numerical approach given above is fully explained to even the intuitive decision-maker, he or she will often see the wisdom of the numerical approach. The thing that a decision-maker fears is the hidden decisions in which the decision-maker was not involved that the numbers represent. If you can show the decision-maker how all of those numbers were obtained through utility curve transformation and WT factor multiplication and that the value system included in all of these conversions reflects the value

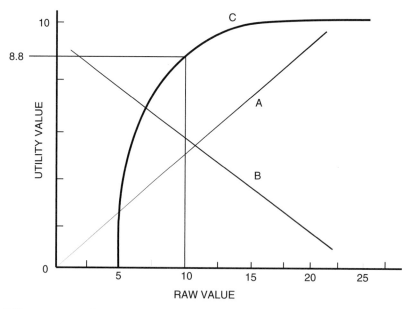

Figure 3.1-7   *Utility curve examples*

system agreed upon prior to the study, a lot of the resistance to a numerical approach will be set to rest. The big advantage of the numerical approach is that the sensitivity analysis results make more sense than in the case where colors or vague terms are applied.

3.1.5.2 Post selection tasks

Once the preferred selection has been made in task F411122 of Figure 3.1-2, it must be further developed in task F411124 and coordinated project and contract planning work accomplished. The approved technical requirements must be integrated into a particular specification (or alternative) format in preparation for the system requirements review (SRR).

The SRR should present the requirements defined for the system, why they are appropriate or necessary, and to what extent they represent program cost, schedule, and technical risk. If there were user requirements that specifically were not respected in finalizing the specification, they should be listed and agreement on all sides reached to either continue to exclude those requirements or fold some of them into the specification. Subsequent to the review, all action items must be dealt with including any agreed upon changes to the specifications. This is the ideal time to submit the system specification for customer approval but many acquirers are reluctant to sign the specification at this time because that triggers formal baseline management involving engineering change proposals (ECPs) and other costly mechanisms. A delay in approving the system specification should be coordinated with the kind of contract that the contractor will be asked to accept and the nature of the anticipated development risk. Ideally, the contractor and customer should share the risk burden in some fair way.

Following SRR, the contractor should continue to refine the top-level requirements and more fully develop the system concept in terms of architecture depth and technical detail in order to expose any residual risks. Throughout this work more knowledge will commonly expose concern for prior decisions resulting in iteration that ripples through that prior work. System design review (SDR) should close out the conceptual work exposing a sound, low risk system concept that can be shown to be compliant with the approved system requirements.

3.1.6 Rigor Versus Creativity

The work accomplished in system definition (F4111) is very creative in nature and it requires a degree of freedom from rigor to be successful. It is possible for management to suppress the creative energy of the team by applying an excessively structured approach more appropriate for the next phase of work. At the same time, the team performing this work must publish and protect the baseline that they have evolved. At the end of this work the team should have created and brought under some form of configuration control and library security all of the following documents as a minimum:

a. *System Specification:* Top-level technical requirements for the program. Ideally, this specification would be approved by customer and contractor management and a functional baseline established based on its content. Alternatively, a preliminary system specification, final operational requirements document, or system requirements document might be agreed upon.

b. *SSR Report:* The minutes of the SRR supplemented with action item responses. Appended should be the final presentation materials.

c. *SDR Report:* The minutes of the SDR supplemented with action item responses. Appended should be the final presentation materials.

d. *System Definition Document:* This preliminary release should include the results of the structured analysis work, generally consisting of diagrams and tabular data. All too often this material is lost and it is too important for that to happen.

e. *Design Concept Document:* A compilation of the approved design documentation keyed to the architecture and interface definitions in the system definition report.

f. *Analysis Reports:* All of the letter reports covering various technical matters. This should include specialty engineering, system analysis, and programmatic reports that may be of use in later program phases for historical purposes showing why certain features and requirements were selected. This also includes the trade study reports, and briefing materials.

3.1.7 Precedented System Definition

When developing a precedented system we would be well served by the prior program if that program had created and preserved a system definition document containing a full set of system diagrams. We also would have a complete set of engineering, manufacturing, and verification data available as well as a real system operating for its user. While the author prizes Sullivan's idea of form follows function that should be operative in developing an unprecedented system, it is probably more practical on a precedented system to enter the system definition work from an architectural perspective as shown in Figure 3.1-8.

If we do not have a process diagram for the existing system, we should build one. This is possible by simply observing the existing system in operation. At this point in the program, the diagram does not have to be expansive in depth. The new system will presumably have to do something different or more wonderfully than the existing system and we should study that functionality differential and mark up the process diagram accordingly to represent the new system.

Next, map the existing architecture to the process diagram block at some level of indenture. In each case, consider if the item will fully satisfy the functionality implied by the process block. If so, it can be a part of the new architecture without modification. Another alternative is that the item in question could be modified in some way to satisfy the new need. Another alternative is that the item in question just will not work in the new system and it would not be cost effective to modify it to work. Some of these may have to be replaced with new items or new items may have to be added to supplement the existing system assets.

Any existing specifications surviving into the new system should be updated based on a requirements analysis process selected from the techniques offered in Parts 3 and 4. New specifications may also have to be developed and this work could be accomplished here, delayed until the system requirements analysis process following this activity, or both.

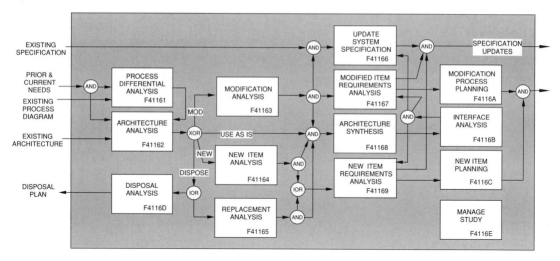

Figure 3.1-8   *Precedented system definition*

3.1.8 Concluding Reviews

The final principal work products of the early program development work should be a system specification adequate to support a development and procurement contract coordinated with a validating concept. Department of Defense refers to these reviews as a system requirements review (SRR) and system design review (SDR) respectively. The SRR should expose any disagreements between devel-

oper and acquirer regarding the content of the system specification resulting in agreements on content and meaning or changes to that specification and the SDR should expose design concepts that can be believably depended upon to successfully implement the requirements in the specification. MIL-STD-1521B provides a good basis for these reviews even through it is no longer maintained by DoD.

3.2

A General Theory of Structured Analysis

Contents

3.2.1 What Is Structured Analysis?

Structured analysis applies a general model to the solution of a specific problem such that the model encourages us to think of all of the factors of importance in arriving at a clear definition of the problem while allowing us to remain undistracted by unrelated factors. Structured analysis provides an organized, systematic way of thinking about a complex problem. This process provides a model within which we can think about the important characteristics of a problem space. Structured analysis is not a problem-solving domain, rather a problem definition domain. Some of the models intended for software development link to structured design methods but the analysis component is focused on defining the problem to be solved.

All structured models known to the author feature one or more simple diagrams the features of which relate to elements of the problem space. These diagrams communicate powerfully through the sense of vision to the human mind triggering thoughts about the features depicted. In order to make the visual message as powerful as possible, the model must be visually simple, often consisting of rectangles or bubbles connected by directed lines. If we use the model to capture all of the related details that we are interested in, it will obscure the visual power of the diagram. As a result, these models often include one or more dictionaries linked to the features of the diagrams in which we can include the details.

This chapter attempts to present insights into the general modeling processes and encourage an overall understanding into which several particular structured analysis models will knowingly fit. There is no one universally accepted method for structured analysis as of the time this was being written. Many different practitioners, analysts, and authors have offered many different methods and models. Some of them have been widely respected and adopted by large numbers of people and organizations. Others are applied in isolation by people who find them useful. There are five that have achieved some renown and these five include many of the model components that comprise others not covered. All five, listed below, are covered in this part and Part 4. In some cases, variations have been developed that extend the utility, simplify use, or improve the model in some other way and many of these are covered as well. Derek Hatley and Imtiaz Pirbhai, for example, extended the utility of modern structured analysis to better model real-time systems by adding several constructs.

a. Traditional Structured Analysis (TSA), developed in association with military weapons and space systems programs. It uses functional flow diagramming as a main element of its model. There are several variations on the use of using functional flow diagramming in this model: IDEF-0, behavioral diagramming, or enhanced functional flow diagramming. The author believes that FRAT encouraged by Professor Mar and Mr. Bernard Morias is a variation on the same theme.

b. Modern Structured Analysis (MSA), used in the development of computer software, features dataflow diagrams and other constructs with extensions for real-time system development.

c. Database Modeling Analysis (DMA) applies table normalization or IDEF 1X to develop relational databases.

d. Early Object-Oriented Analysis (OOA) offers a bridge between computer processing-oriented analysis methods like dataflow diagramming and data-oriented methods like entity-relationship diagrams.

e. Unified Modeling Language (UML) improved upon the early OOA models in an attempt to gain agreement on a common or unified approach.

3.2.2 Structured Analysis Goals

The goal of structured analysis is completeness and avoidance of the unnecessary in terms of identifying characteristics that the product must have as a prerequisite to designing that product. This idea strikes many design engineers as being just plain backwards. They would commonly uncover the characteristics as a function of the design solution. The structured approach, in this interpretation, is only necessary when the problem being solved is very difficult, it requires that knowledge from several different specialized disciplines be brought to bear, and there are resource limits applied to developing the solution. In other situations not fitting this profile, we may be able to tolerate some inefficiency and live with the small risks that might present themselves when the design community seeks to solve the problem with a design without benefit of a carefully developed specification. But, most of the simple problems were solved a long time ago. Today's problems are increasingly complex often involved in linking two or more systems previously operated independently. The problems are so complex that we need many people involved in their solution for the reasons earlier developed and once we get many people into the act, we need some instrument that captures our agreements to ensure that we are all working on the same plan and toward the same goals. Specifications satisfy this need by providing a definition of the problem to be solved.

3.2.3 Where Does It Appear in the Process?

Figure 3.2-1 illustrates the overall development process used throughout this book. Structured analysis resides early in the development period, of course, because we must identify the problem before we attempt to solve it.

Ideally, these structured methods would be applied to the general problem-solving process preferred by an enterprise and there are few systems more complex than the one that creates product systems. An enterprise's common process should be developed through structured analysis and torqued into a program plan for the specific product development. In this book structured methods are applied to process system development as well as product system development. Structured analysis as it applies to the product system occurs within function F4111, Perform Initial System Analysis, F4113, Traditional Structured Analysis, and F4114, Computer Software Structured Analysis as shown in Figure 3.2-2.

Within function F4113 we will find several models that may be selectively applied to gain insight into the problem we must understand and solve. All of the structured methods discussed earlier appear here. Traditional structured analysis including several variations is covered in this part. Part 4 covers structured analysis models appropriate for product that is going to be implemented in computer software. It is not suggested that the individual, team, program, or company select a single method for all applications, rather that all of these methods are useful in some situations and product lines. Therefore, a system engineer should be versed in applying any of them or participating on a

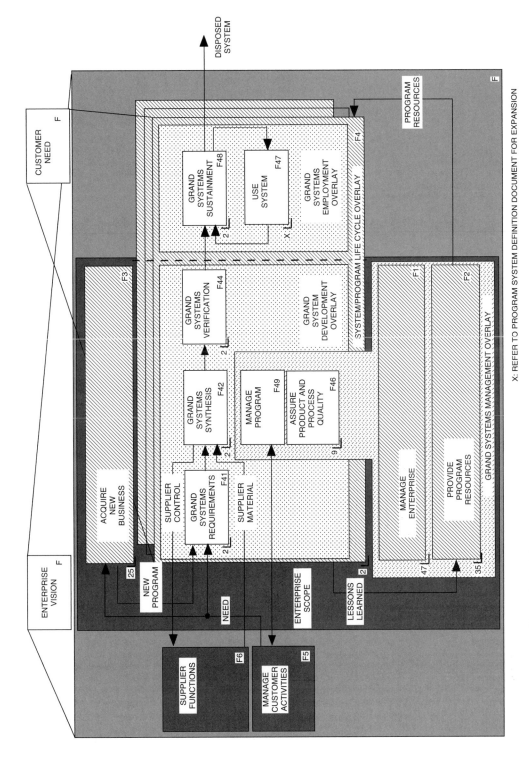

X: REFER TO PROGRAM SYSTEM DEFINITION DOCUMENT FOR EXPANSION

Figure 3.2-1 *System life-cycle model*

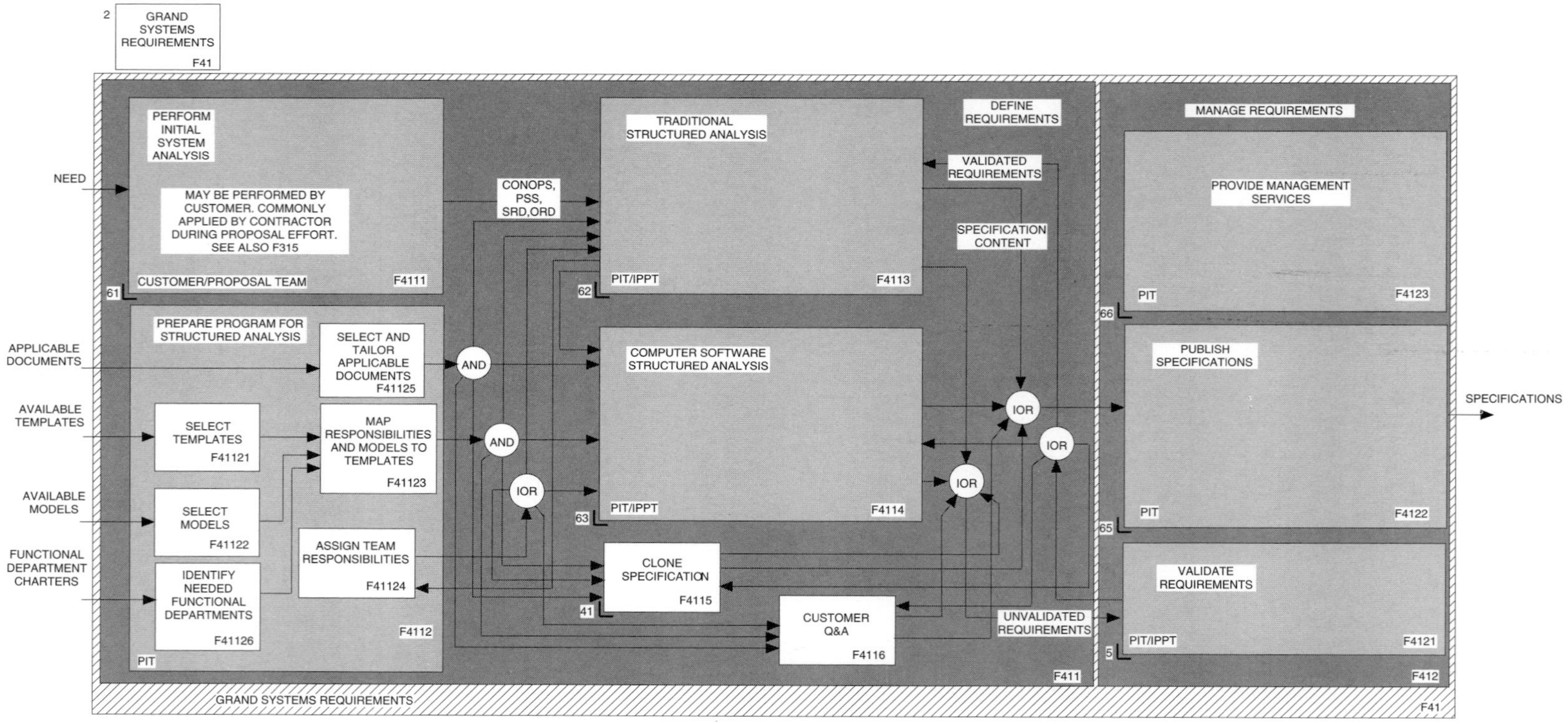

Figure 3.2-2 Grand system requirements analysis process function

team applying any of them. Some of these methods work best for hardware or software and regardless what the background of the system engineer he or she should master approaches that are effective in each. An engineer who became a system engineer through the hardware route needs to understand that systems increasingly are created with the difficult functionality allocated to software so they must adapt or pass from the scene as a system engineer.

3.2.4 Comparative Overview of Approaches

The first approach covered starting in the next chapter, traditional structured analysis, is most useful for grand systems of yet to be determined character and will work well below the system level for hardware entities that exhibit a space time dynamic with big things moving in space and time. The author has had trouble applying it in applications that are relatively static from a physical perspective. This approach is function driven following Sullivan's form follows function encouragement. Problems that will reveal themselves while the analyst is focusing on what the system must do (functionality) will yield to this approach. Problems that are most easily analyzed from a perspective of how the system must work (behavior) often will not. This is not a good approach for computer software development, though this is where the software process started in what was called flowcharting.

The modern structured analysis approach will work well for computer software systems where computer processing of data is very intense and interest in the data is important but incidental. Adherents of the Hatley-Pirbhai variant of modern structured analysis claim that HP is appropriate for all systems no mater their eventual implementation domain but especially where control under real-time conditions is important. Database modeling approaches are useful in developing computer database systems where the processing of that data is incidental to the warehousing of and access to the data.

Mr. Mack Alford, as the chief scientist for Ascent Logic, was the man behind RDD-100 development. It was an application of an earlier method called input process output (IPO) developed as a merge between computer flowcharting and data analysis. Mr. Alford renamed the technique behavioral diagramming and claimed that it provided a single model for simultaneously analyzing both what a system must do (functionality) and the data it must act upon in accomplishing that functionality though the lateral flow could be some commodity rather than data. By combining these two facets of the problem, one could say that we therefore understand how the system must behave. Other methods have been devised to inquire into a systems behavior and these will be covered in the chapters on traditional structured analysis, modern structured analysis, and object-oriented analysis in the form of state diagrams. RDD is not included on Figure 3.1-2 but is discussed in this part under the heading behavioral diagramming.

The object-oriented approach makes an overt three-dimensional attack on the problem space simultaneously probing it for needed objects or things (what the solution must consist of), functionality (what the solution must do), and behavior (how the solution must work). It was conceived as a means of merging both computer processing and data-oriented approaches into a single model but it also exposes a three-faceted structure that all structured analysis models should address in some way. Most authors in this field encourage entry into this modeling method via the objects which causes some cynical system engineers to claim that it stands Sullivan's ideas on their head encouraging function follows form and it may be a reason that many authors express difficulty in identifying objects.

Some readers will wonder why quality function deployment does not appear on the author's list of structured methods. QFD does not fit the pattern offered by the approaches listed above. The basis is a set of matrices that permit joint analysis of two sets of data arranged on the two axes. The number of matrices used is different for different supporting schools and authors, from 6 to 18. Generally, the analyst puts information in the left side of the matrix and establishes relationships with information listed on the top. In the next matrix, the top information from the previous matrix becomes the left side data for the next cycle and so on. This model simply establishes relationships between information pairs and does not provide graphically expressed devices in models that encourage thought about specific aspects of the problem space. The author's conclusion is that it is a requirements listing tool linked to a design implementation tool and most useful for incremental improvement situations involving hardware production. It is addressed under Chapter 9 of this part along with several other methods that are useful in specific situations.

3.2.5 Polyfaceted View of Problem Spaces

All problem spaces can be viewed in the context of an n-faceted construct where n is the number of simultaneous facets through which we stare at the problem space. There are several three-faceted approaches which focus on functionality, behavior, and static physical elements possessing the functionality and behavior needed to solve the problem. These are generally the kinds of modeling approaches we wish to explore in detail in this part and Part 4.

Picture the problem space illustrated in Figure 3.2-3. We have many options for trying to collect information about this problem enroute to a solution. If the problem is sufficiently simple, a single engineer can apply his or her native skills and knowledge to understand it by brute force and guile without plan from within the engineer's domain of knowledge. This approach corresponds to $n=0$ and is often called an ad hoc approach. There is no plan for understanding the problem, we simply attack the problem and ideas come to the fore. Often this approach involves trying to match known solutions from the past to the problem space followed by a leap to a point design solution based on past successes. While this may be fun and could result in a good solution in simple cases, it is not an effective approach where complex problems are being dealt with and serious money is at stake. Many people must contribute to understanding a complex problem and solution so there must be a way for all of these people to communicate the parts of the problem that they are responsible for to everyone else in a public way open to critical comment. Once again, this is what specifications do.

A three-faceted structured approach encourages us to stare at the problem space through one or more organizing facets. As we do so, the problem becomes organized in context with these facets and the unknown aspects of the problem space begin to disappear. In the process, the physical objects of which the system must consist fall out onto the

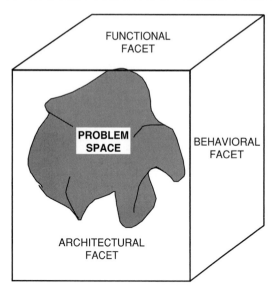

Figure 3.2-3 *Three-faceted problem space*

physical, object, or architecture facet. At the same time, using a structured approach exposes performance requirements that the design of those things must respect. In order to expose these performance requirements, we examine the problem space from the perspective of what the system will have to do (functionality) and how it must work (behavior) in order to solve the problem. A truly universal model might also include a relational facet completed through interface definition in hardware systems, relational references between key fields in relational database development, or object associations in OOA. The relational aspect is buried in the behavioral facet in this explanation. In software systems it would also be useful to recognize a data and processing facet but the author wished to limit the number of facets to three in this discussion.

Table 3.2-1 lists four three-faceted approaches and identifies the model structures used in each for the three facets shown in Figure 3.2-3. These approaches will be discussed in chapters in this part and Part 4 in the order listed and the model structures covered in detail in the first application listed. The modeling structures named in bold are the normal entry facets for the indicated approach. These four

three-faceted models share the same three facets but do not all apply the same entry facet. In the general case, then, a structured analysis process may have n facets where n is a positive integer (zero excluded as it represents a non-structured process), one or more entry facets, and a particular preferred sequence of moving between the facets.

3.2.6 Entry Facet Differences

Many system engineers have been steeped in the logic of Sullivan who held that form should follow function. Clearly, Sullivan followed a two-faceted approach to development of new building designs: functionality and physical. The traditional structured analysis process involving functional analysis, implements Sullivan's ideas. The principal artifacts employed in the model are illustrated in Figure 3.2-4. It is based on the notion of expansion of the need, the ultimate statement of functionality, into more detailed information about what the solution must do to solve the problem. This functionality, once exposed, is allocated to physical things forming the physical product architecture. One then attempts to understand behavior by identifying interfaces between the things driven by the way you allocated functionality to the things in the system and timing behavior across these interfaces.

As we stare at the problem space through these three facets, we assemble these three views until there are no more unknowns in the problem space, We have completely characterized the problem space in terms of appropriate requirements for the solution.

The analyst could apply as an alternative to functional flow diagramming, in the functional facet, hierarchical functional analysis, process analysis, or IDEF-0 (borrowed from SADT) to the same end. Alternatively, *n*-square diagramming could be used rather than the schematic block diagram suggested in Figure 3.2-4.

Early software modeling efforts were done with a single axis model implemented using flowcharts that one could argue are functional flow diagrams. Input-Process-Output (IPO) charts expanded this single facet to include dataflow offering a two-axis approach, three when combined with identification of the software entities that would implement the indicated functionality and behavior. Mr. Mack Alford when developing the Ascent Logic requirements tool requirements driven design (RDD) applied a model called behavioral diagramming that was essentially the old IPO developed earlier for software modeling purposes.

When we apply the modern structured analysis process popularized by Yourdon, DeMarco, Constantine, and others,

Table 3.2-1 *Four three-faceted modeling approaches*

Structured approach	Modeling structures		
	Functional facet	*Physical facet*	*Behavioral facet*
Traditional Structured Analysis	**Functional Flow Diagram**	Architecture Block Diagram	Schematic Block Diagram
Modern Structured Analysis	**Data Flow Diagram**	Software Product Hierarchy	P Spec and State Diagram
Early Object-Oriented Analysis	Data Flow Diagram	**Object/Class Diagram**	State Diagram
Unified Modeling Language	**Use Case Diagram** **Activity Diagram**	Object/Class Diagram Component Diagram Deployment Diagram	Sequence Diagram Statechart Collaboration Diagram

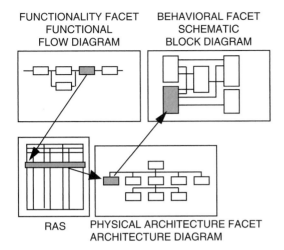

FUNCTIONALITY FACET
FUNCTIONAL
FLOW DIAGRAM

BEHAVIORAL FACET
SCHEMATIC
BLOCK DIAGRAM

RAS   PHYSICAL ARCHITECTURE FACET
ARCHITECTURE DIAGRAM

Figure 3.2-4   *Traditional entry facet and sequence*

we seek first to identify the processing, or functionality, that must occur in the computer and how the data must flow between these processes using a dataflow diagram. Given a particular processing bubble, we then try to uncover the needed behavior, perhaps using a state diagram or input-output matrix. We associate processing bubbles with particular architectural entities, computer programs, or subordinate units of software. So, here too the functionality facet seems to take the lead in exposing the problem space to view.

The system engineer, familiar with one of these techniques, is startled to find that all modeling approaches do not encourage Sullivan's sequence. Some analysts apply state diagramming as a single facet directly to a problem space and from the states and sequences exposed deduce needed functionality and necessary product characteristics.

The OOA approach of Rumbaugh et al., in *Object-Oriented Modeling and Design*, encouraged identification of the objects, akin to physical architecture, as a precursor to understanding functionality using dataflow diagrams and behavior using state diagrams. Booch in his *Object-Oriented Analysis and Design* offered the same encouragement as did Coad and Yourdon in *Object Oriented Analysis*. This is opposite to what Sullivan suggested—function follows form—but many of the early adherents of OOA may have drawn support from another architect named Frank Lloyd Wright who felt that an architectural design had both form and function and both were important. This is not to say that early OOA is therefore suspect. Rather, one who would try to understand complex problems as a precursor to solving them (an important aspect of the systems approach) should realize that there is more than one way to probe a complex problem space and that some of these techniques have particular value for particular problem and product types. A system engineer should be familiar with all of these techniques to the point that he or she can understand what is expressed in the models and documentation common to the method and interact with those who have built that documentation to express the results of their analysis.

It is possible to master multiple methods after one recognizes that each new method that comes onto the field recycles some of our old friends. This is not a reasonable cause

to fight over who has the biggest slide rule, best airplane, or finest modeling method. Rather, it is an opportunity to see how the elements of this new method play in the context of an *n*-faceted problem space viewer and thus allow us to fit it into our personal inventory of methods. We should recognize all of these techniques as three-faceted problem space viewers that will enable the communication process between different specialists encouraging the sharing of information about a problem space as we seek to understand that space that is so difficult that none of us can master it by ourselves.

3.2.7 An Entry Continuum

While the author is an adherent of Sullivan's ideas of form follows function, there are situations where it makes sense to enter the problem space primarily from the architecture perspective. Given that you must develop a new system that is heavily precedented by an existing system leading to a minor modification rather than a new build, it probably makes more sense to begin with the architecture of the existing system than to begin a new structured analysis effort. One might ask, "Well, what are all of the problem space entry possibilities?" Figure 3.2-5 offers one view of this continuum. The vertical axis corresponds to the degree of precedent in the development over the range 0% (unprecedented) to 100% (fully precedented). The former relates to a new development situation while the latter to a minor modification situation. In between are major modifications over a wide range of possibilities. The relationship between the two axes illustrated in Figure 3.2-5 follows some shape between the qualitative limits noted.

For a new, unprecedented development one should fully apply Sullivan's perspective. In a development approaching no change (fully precedented) an architecture entry would likely be the most cost-effective. In the architecture entry, one studies the existing architecture in terms of appropriateness for the new functionality and sorts it into those items that are adequate as they are, those that are inadequate and must be deleted, those that can be modified for acceptability, and those new items not then existing.

3.2.8 Model Documentation

It is not uncommon for the results of the structured analysis to be lost after its initial benefit has been derived. It is uncommon for programs to have a way of documenting and controlling the models used early in program work. This is important information and should be captured while doing it and formally released in some fashion. The author uses a document referred to a system definition document (SDD) to do so. Figure 3.2-6 illustrates one way of doing this where traditional structured analysis is used for system level entities as a minimum.

The analyses on the far left are system requirements analysis tasks each of which produces work products captured in appendices of the SDD. The appendix data is then used by the analysts responsible for lower-tier requirements analysis. The results of this work, requirements, flow into a requirements analysis sheet (RAS) where requirements are allocated to a particular product entity and the specification template assigned to that entity. The responsibility for continuing the structured analysis can be retained by the PIT for all levels or can be passed down level to level with each layer being responsible for accomplishing such analyses as

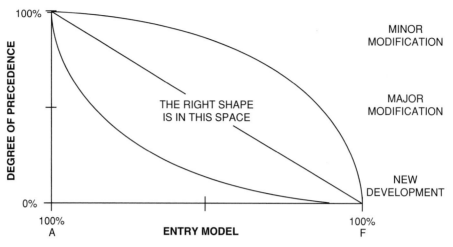

Figure 3.2-5   *The problem space entry perspective continuum*

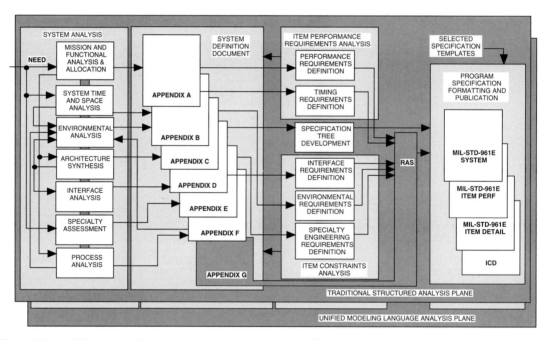

Figure 3.2-6   *SDD coordination with program processes and specification templates*

needed to prepare all of the specifications between their own level (prepared by a parent team) and the next lower team level.

3.2.9 Completeness and Avoiding Model Madness

While structured models can yield helpful results when wielded by a skilled practitioner, they can lead to analysis paralysis when the process employing them is not well managed. It is, frankly, fun to use these models and easy to succumb to a mission-oriented attitude toward expressing

perfection through them. It will always be possible to change them for the better but there is a point of diminishing returns where one should simply conclude that they have extracted all the value they can out of the model for the money they have available to spend on them.

The use of models should be seen as a means to an end not an end in itself. It is necessary to keep one's mind on the goal and be ready to declare victory when that end has been achieved. The analyst using the model may not be the best party to make that decision about completeness because of the seductiveness of the modeling process. Team and program management must monitor the on-going results of the modeling work against clear completion criteria defined prior to the start of the work like every other engineering job in the system development process. When that criterion has been satisfied, it is time to call an end to it and move on to other tasks.

So, what is that end goal that we can use to determine when we have completed the modeling and thus the requirements analysis? The dictionary definition of the word *requirement* that the author likes is, "an essential characteristic." A specification for an item (system or component) is complete, therefore, when you have identified all of the essential characteristics and excluded all other content. This condition is not knowable in a numerical or logical sense for every kind of requirement but it is for some kinds of requirements.

As previously noted all system and hardware essential characteristics can be partitioned into performance requirements and design constraints. The latter can be further partitioned into specialty engineering, interface, and environmental requirements. You can be sure that you have identified all of the interface requirements if you have determined in what ways the item must interface with all other elements of the system and that you have identified appropriate interface requirements for each of them as discussed in Chapter 3.6. Where interface requirements are captured in specification pairs rather than an interface control document, one should go one step further here and make sure that the specification on the other end of each of those interfaces has defined the same interfaces in a compatible fashion.

We can be sure that we have identified all of the specialty engineering requirements by making a list of all of the specialty engineering disciplines that must be considered for system items and that we have identified requirements for each of those disciplines identified for the item in question. The constraints scoping matrix described under specialty engineering requirements in Chapter 3.7 offers one approach to organizing this view.

Environmental requirements completeness is a little more complicated in that it depends on the specification level. System environmental requirements have been comprehensively defined when you have identified all of the spaces within which the system will have to function, you have identified standards that cover each of these spaces, you have selected appropriate requirements from each of these standards, and you have tailored each of the selected requirements appropriately for the application as discussed in Chapter 3.8. You have identified all of the end item environmental requirements when you have completed the environmental use profile work for the item as discussed in Chapter 3.8. Finally, you have identified the component environmental requirements when you have completed the end item zoning and mapping of components to zones.

That leaves only the performance requirements that tell what the system must do and how well it must do those things. Completeness here cannot be ascertained with the same degree of precision perhaps as in the case of design constraints. We have to make an engineering judgment of the quality of the functional model and the goodness of the allocation work accomplished.

When you train your analysts make sure they are made aware of the tremendous utility of problem space modeling but also the need for quickness in their application.

Table 3.2-2 Model coverage references

Model	Useful on	Text ref
Traditional Structured Analysis	Systems, Hardware	3
Functional Flow Diagramming	Systems, Hardware	3.3
Performance Requirements Analysis	Systems, Hardware	3.4
Architecture Synthesis	Systems, Hardware	3.5
Interface Requirements Analysis	Systems, Hardware	3.6
Specialty Engineering Requirements	Systems, Hardware	3.7
Environmental Requirements Analysis	Systems, Hardware	3.8
Hierarchical Functional Analysis	Systems, Hardware	3.9.2.1
Enhanced Functional Flow	Systems, Hardware	3.9.2.2
Behavioral Diagramming	Systems, Hardware	3.9.2.3
IDEF-0	Systems, Hardware	3.9.2.4
State and Event Analysis	Systems, Hardware, Software	3.9.3
Mathematical Methods	Systems, Hardware, Software	3.9.4
Scenarios, Strings, and Threads	Systems, Hardware, Software	3.9.5
Process Flow Diagramming	Systems, Hardware	3.9.6
FRAT	Systems, Hardware	3.9.2.5
Quality Function Deployment	Systems, Hardware	3.9.7
Software Models	Software	4
Flowcharting	Systems, Hardware, Software	4.2
IPO	Systems, Hardware, Software	4.2
Modern Structured Analysis	Software	4.2
Hatley-Pirbhia	Software	4.2
IDEF 1X	Software	4.3
Early Object-Oriented Analysis	Software	4.4
Unified Modeling Language	Software	4.4
DoD Architecture Framework	Software	4.5

These models are simply wrenches to be removed from the tool box, used on a tough nut, and replaced in the tool box. They are working people's precision tools from a tool box not children's toys from the toy box. We owe it to the program manager or team leader to do this work well but also quickly.

3.2.10 Detailed Coverage of Models

This book covers many of the past and present models used to accomplish structured analysis in the development of systems. Table 3.2-2 lists these models and tells where in the text (Part. Chapter. Paragraph Number) they are covered in some detail.

The reader may ask why one should be interested in so many different models. What is wrong with picking one and living with it? At the time this book was written there was no single model that could be depended upon to expose every needed requirement for every specification that must be developed for a program. The author has met system engineers who were able to apply a particular model across the HW/SW divide but it is not a common talent.

It is unusual today when a system is developed that does not entail some software and that software must run in hardware. Therefore, every system will involve some combination of hardware and software and there is no current model that can be employed for both. The author is convinced that that will change during the decade starting with the year 2000 but until the universal model is available, we will have to use multiple models.

A hardware dominated system engineer could argue, of course, that he or she does not understand software and would derive no benefit from learning anything about software models. Any system engineer with that history and future outlook will have a very hard time maintaining employment in this profession because all of the hard functionality is being accomplished in software and there is not enough left of the system about which to wrap a profession. System engineers have to be able to talk to software engineers and an excellent way to develop the ability to do so, besides talking to them, is to learn how they model the problem space.

The author has found that even within the system modeling work, there are some problems that will model very well in functional flow diagramming while other problems do not. It helps to have alternative models that you can bring to bare on a problem space when your preferred approach is not yielding useful results. You will not know what models work for you unless you try different models on actual problems encountered on programs.

3.3 Functional Analysis

3.3.1 The Heritage of Structured Analysis

There are currently many models used to accomplish structured analysis of problem spaces. During the development of many military systems over the period 1950 through 1980 government acquisition offices and contractors developed and applied an organized approach commonly called functional analysis. The author has chosen to refer to that technique, with other models linked to it to make it complete, as traditional structured analysis. Figure 3.3-1 illustrates this process that will be explained in Chapters 3.3 through 3.11 of this Part. It is an expansion of block F4113 of Figure 1.2-6. After the process is explained using a particular kind of functional analysis in this chapter, other variations will be covered in Chapter 3.9 that sometimes work better than this first method offered. The person who would understand problem spaces should seek to perfect his or her skills in a range of models because none of them seem to work equally well in all situations. There is an art form as well as sound logic in the use of models and some of us find some of them useful and others not so as a function of the problem space, the nature of the conceptual solution, and, possibly, the way we are wired. Individual system engineers will not know what works for them unless they try different models in their normal work.

This problem space modeling technique works for the author very well on very large problems the solutions of which will be characterized by a large space-time dynamic. That is, there are going to be a lot of physical things moving around in space and time, such as a space launch vehicle moving from the Earth's surface to deliver a payload into Earth's orbit. The author has found it very difficult to apply this model in less dynamic situations like automobile traffic control where state diagramming seems to work very well for him. This model has a lot in common with the methods used by industrial engineers to define processes and the flowcharting technique that was used in the early days of computer software development. Software people today would be well advised to use something other than this model or flowcharting for software development but please note that in the earliest days of structured analysis both hardware and software people used essentially the same model. System engineers and hardware people drew the diagrams horizontally probably driven by the engineering drawing logic of left to right and top to bottom while the software people drew them on a vertical axis probably because of the ease of printing them on line printers using ASCII characters. We will return to this point later with the observation that in the future we will very likely return to a single model used by hardware, software, and system engineers. That model will include flowcharting renamed. Traditional structured analysis is perfect today for the initial work on an unprecedented system application where it is not yet clear in what way hardware, software, and people through procedures will interact in the solution.

3.3.2 Form Follows Function

An architect named Sullivan concluded from his study of natural systems in the later half of the 19th century that his designs for buildings should be based on the functionality they were intended to satisfy, that is, that form should follow function. Under this theory, we first seek to understand what the system must do and then we create a form or design that satisfies that functionality. This thought process is not necessarily a natural one for engineers many of whom are very solution oriented and find it difficult to consciously take what many of them would consider an extra unnecessary step enroute to a solution to the problem. Many engineers apply their creative ability directly to the problem as they see it in what a system engineer would refer to as an ad hoc approach. This ad hoc approach poses no real disadvantages so long as the problem is small and relatively simple such that it will surrender to a single engineer or a single engineering discipline like mechanical engineering.

When the problem is large and complex, it is very risky to apply the ad hoc approach because the problem must be solved by many people all specialized in different disciplines working together toward a common goal. They need a common vision of the problem (supplied by a specification) and the intended solution path to avoid developing conflicting parts of the overall solution. That common vision must advance from the general to the specific, from the unknown to the known. This common vision will be most effective if it is captured in a visual form as in a picture because of the tremendous power in the human mind to comprehend visually presented information versus the difficulty in comprehending the meaning of 1000 words. Thus, if we are to follow Sullivan's prescription, we need a pictorial presentation of functionality as a beginning. This pictorial view should be composed of simple artifacts that relate to specific elements of the problem space to which we can connect other more complicated ideas.

Also, a problem characterized by both considerable depth and breadth is very difficult to resolve by individuals or groups and must be decomposed into a set of related smaller problems of a scale that fits human capabilities. Our method of functional depiction should support a mechanism of logical expansion such that large and complex ideas can be partitioned into a set of related simpler ideas.

Finally, while we seek to understand the problem space through examining the necessary functionality of the physical objects that will have to accomplish that functionality, we must make the transition from that functionality to those physical entities. It is the physical entities that will have to be designed and as a precursor we should first understand the problem the design must satisfy. There should, therefore, be some kind of linkage between the functionality and the things in the system as well as the functionality and the performance characteristics of the system. Functions define what a system or item in a system must do. Performance requirements capture those functions in the language of choice telling what the system or item must do and how well it must do it forming performance requirements applying to the things that will satisfy the needed functionality.

The identification of needed functionality as a prerequisite to design and the association of that functionality with physical things of which the system shall be composed, and identification of the performance characteristics corresponding to those things based on the functionality is called functional analysis. There are several methods or models people have found useful to this end. The first one we will discuss is called functional flow diagramming. Such a diagram is composed of blocks representing the functionality connected by directed line segments that show sequence. Logical AND and OR symbols are used to connect the blocks into serial and parallel flow patterns as necessary.

This chapter deals with functional analysis (F41131 of Figure 3.3-1), through which we build the functional flow

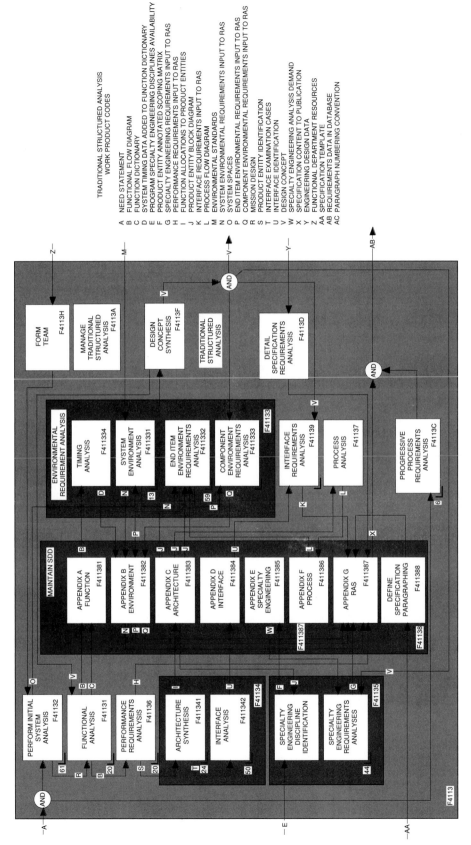

TRADITIONAL STRUCTURED ANALYSIS
WORK PRODUCT CODES

A NEED STATEMENT
B FUNCTIONAL FLOW DIAGRAM
C FUNCTION DICTIONARY
D SYSTEM TIMING DATA ADDED TO FUNCTION DICTIONARY
E PROGRAM SPECIALTY ENGINEERING DISCIPLINES AVAILABILITY
F PRODUCT ENTITY ANNOTATED SCOPING MATRIX
G SPECIALTY ENGINEERING REQUIREMENTS INPUT TO RAS
H PERFORMANCE REQUIREMENTS INPUT TO RAS
I FUNCTION ALLOCATIONS TO PRODUCT ENTITIES
J PRODUCT ENTITY BLOCK DIAGRAM
K INTERFACE REQUIREMENTS INPUT TO RAS
L PROCESS FLOW DIAGRAM
M ENVIRONMENTAL STANDARDS
N SYSTEM ENVIRONMENTAL REQUIREMENTS INPUT TO RAS
O SYSTEM SPACES
P END ITEM ENVIRONMENTAL REQUIREMENTS INPUT TO RAS
Q COMPONENT ENVIRONMENTAL REQUIREMENTS INPUT TO RAS
R MISSION DESIGN
S PRODUCT ENTITY IDENTIFICATION
T INTERFACE EXAMINATION CASES
U INTERFACE IDENTIFICATION
V DESIGN CONCEPT
W SPECIALTY ENGINEERING ANALYSIS DEMAND
X SPECIFICATION CONTENT TO PUBLICATION
Y ENGINEERING DESIGN DATA
Z FUNCTIONAL DEPARTMENT RESOURCES
AA SPECIFICATION TEMPLATE
AB REQUIREMENTS DATA IN DATABASE
AC PARAGRAPH NUMBERING CONVENTION

Figure 3.3-1 *Traditional structured analysis*

diagram. In Chapter 3.4 we will discuss the derivation of performance requirements (F41136) from the functions identified in the functional analysis and the allocation of those performance requirements to architecture (F411341). In Chapters 3.6, 3.7, and 3.8 the design constraints, that complete the requirements analysis process using traditional structured analysis, will be covered. Chapter 3.6 covers interface identification (F411342) and requirements analysis (F41139). Chapter 3.7 covers specialty engineering discipline identification (F411351) and specialty engineering requirements analysis (F411352). Chapter 3.8 covers environmental requirements analysis (F41133) that requires three different models plus timeline analysis for the timing element. The work in building timelines is very closely connected to functional flow diagramming but the author has chosen to associate that work with the environment.

The student should notice that in the case of each kind of requirement found in a system or hardware specification the content can best be built through a two-step process. First we use a model to identify the need to control a particular characteristic and then we apply a particular form of analysis to define the identified requirement.

Chapter 3.9 discusses several different alternative models to functional flow diagramming. Chapter 3.10 coordinates all of this work with a set theoretic display of the process and an insight into a better way to connect interface identification with functional analysis by coordinating the N-square diagram with the requirements analysis sheet (RAS). Chapter 3.11 provides an environment within which the analyst can affordably capture the results of the traditional structured analysis process in a document that can be configuration controlled.

3.3.3 Functional Flow Analysis

There are three fundamental parts of an effective functional flow analysis: creating functional flow diagrams and selectively creating companion timeline diagrams (Chapter 3.8), transforming the function names into performance require-

ments prior to allocation if desired, and allocating the functions depicted on the flow diagrams to architecture. The first two parts are accomplished on the functional or problem plane of the process and the latter provides the transition from the functional plane to the physical plane as shown in Figure 3.3-2. The physical plane is composed of the system architecture or product entity structure, a hierarchical depiction of the physical things that compose the system.

Performance requirements can be extracted or derived from the function names before allocation or after but the best approach generally is to do so before allocation. The rationale for this is that allocation of function names to architecture often will take the form of a one to many relationship in the higher tiers of the analysis whereas allocation of performance requirements derived from functions will most often follow a one-to-one relationship.

Figure 3.3-3 offers a process diagram for doing functional flow diagramming and will be used while discussing the first of the three basic activities that collectively form the functional analysis approach encouraged in this book. This figure is an expansion of block F41131 of Figure 3.3-1, Traditional Structured Analysis.

3.3.3.1 Function identification and sequence

The first step in functional flow diagramming is identification of the ultimate function, the need, which is a user statement of some capability that does not at the time exist. The initial work, including identification of the need, may be accomplished by the user, by the acquisition agent, by a firm contracted to do the early concept studies, or in a commercial firm by the organization with the development responsibility. The input received by the contractor may be the customer need, possibly expanded into a mission analysis report, concept of operations (ConOps), or operational requirements document (ORD) developed through earlier marketing and program development work, hopefully, also accomplished in an organized fashion. As the process progresses, the inputs to continuing work will be a set of higher-

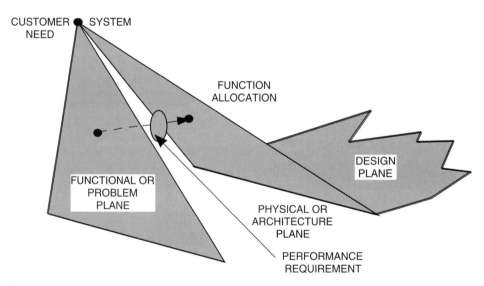

Figure 3.3-2 *Development planes*

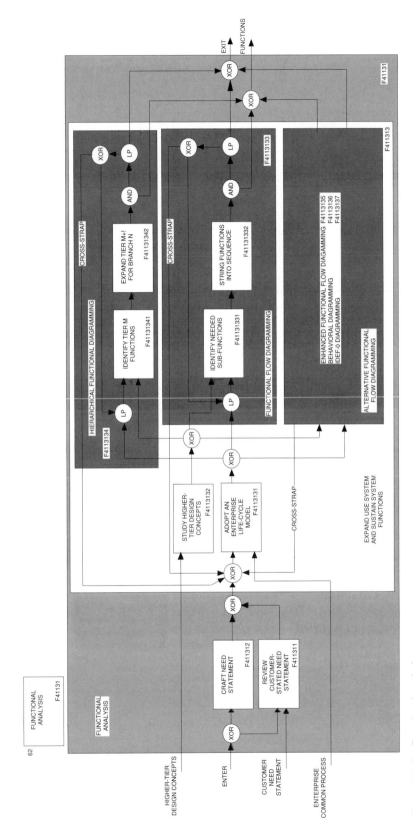

Figure 3.3-3 *Functional analysis*

tier functional flow diagrams, feedback on how the higher-tier functions and derived performance requirements were allocated to architecture, and higher-level design concepts that are responsive to the requirements identified for these architecture items.

The need should be expanded into a life-cycle model, an example of which is provided in Figure 3.1-1. The work can then focus initially on decomposing functions F47, Use System, and F48, System Sustainment, using the process diagrammed in Figure 3.3-3. Most of the performance requirements for the system will be derived from these functions and their subordinate functions as discussed in Chapter 3.4.

The analyst should at this point select a preferred modeling method: (a) functional flow diagramming, (b) IDEF-0, (c) enhanced functional flow diagramming, (d) behavioral diagramming, or (e) hierarchical functional analysis. This chapter will proceed with functional flow diagramming. Refer to Chapter 3.9 for the alternative approaches and how they are accomplished.

Assuming we have selected traditional structured analysis, after identifying the need that is represented in functional flow terms by a single block identified as function F and containing the need statement, possibly paraphrased, within it. The first expansion of the need should be a life-cycle model and for a given company it is recommended that it be essentially the same model for all of the company's programs. The functions that must be further analyzed are what the author refers to as the Use System (F47) and Sustain System (F48) functions. The process starts with accomplishing function F41131521 of Figure 3.3-3 to expand function F47 of the life-cycle model for this particular system and string the resultant functions into a sequence. Then, assuming that this does not end the analysis, we must loop (LP) back to expand the functions just previously identified into a next lower-tier set of functions. The loop logic symbol is not commonly used in simple functional flow block diagramming, rather it is from enhanced functional flow block diagramming that is explained in Chapter 3.9. It simply provides a special case of the OR logic symbol calling for iteration until a particular condition has been realized and that condition in this case is analysis complete.

This work should be accomplished by a cross-functional team because no one person has the knowledge and experience of a group of specialists from different domains. It is especially important that software as well as hardware people are involved so that there will be careful consideration of alternative allocations of the functionality. The software people should not be the last to hear that they will be responsible for implementing functionality that might better be done in hardware. For some time to come most system engineers we encounter will have come from a hardware background and they should be accompanied by someone with software skills in doing this work. This is changing, of course, and it will continue to because it is increasingly the case that the hard functionality is migrating to software with hardware providing structural framework, electrical power, sensors, and loads at the periphery. So, system engineers beware, behold software and learn something about it at least the ability to talk to software people. Part 4 of this book offers an initial opportunity to do so. It is also true, of course, that people only familiar with software development should not be selected to do this work but include some people with hardware skills.

We create functional flow diagrams to gain insight in an organized way into what the system (or element) must do and in what sequence it must do it. Ultimately, this flow diagram will be used as a basis for gaining insight into needed system architecture and primitive performance requirements statements corresponding to those architecture elements. As an alternative, the analyst could create a function list, outline, or hierarchy, but this format will not provide sequence information or opportunities for gaining insight into possible needs for functional simultaneity or alternative action. We will return to this question in Chapter 3.9.

The principal work products that can be used to capture the results of the functional analysis are:

a. A functional flow diagram (FFD) that defines the functions that the system must perform to satisfy the system needs statement or system goal in accordance with the content of the mission analysis report (or equivalent statement of user needs). This is a block diagram using blocks, lines, and logic symbols to define the sequential connectivity between the functions depicted by the blocks.

b. A function dictionary is a tabular listing of system functions illustrated on the FFD (one tabular line item for each block on the FFD) containing the function ID code, function name, initial and terminal event definition, and a description of the function. A tabular timeline can be included here with columns for mean and variance figures.

c. Initial entries in a requirements analysis sheet (RAS) involving the functions identified. These will be supplemented with corresponding entries for derived performance requirements (Chapter 3.4) and product entities (Chapter 3.5) to which they are allocated to complete the requirements analysis story for performance requirements.

d. The author considers time as an environmental parameter and is therefore covered in Chapter 3.8 but the development of a system timeline is very closely connected to the functional flow diagram.

Functional flow diagramming applies a common pattern of using simple models for complex ideas. We want a simple diagram that encourages good visual communication with the human mind but we also want a lot of details. These goals are incompatible. A simple diagram will satisfy the first part and a coordinated dictionary expanding on the detail will satisfy the second. If we put all of the details contained in a dictionary on the face of the model, the connection of the eye to brain and understanding will not work as well. The reader is encouraged to note as we move through the several models that system and software engineers have found to be useful in analyzing problem space that some models are visually complex providing a rich expression of the reality we are trying to understand but that these model do not couple into the mind easily through vision. Other models are visually simple and do connect to human observer/analyst understanding easily. But in so doing do not carry a very rich story with the passage. Truthfully, functional flow diagramming is near the latter extreme in this continuum so it is a good model with which to begin our study.

Functional analysis is a top-down, structured method for identifying and analyzing performance requirements and partitioning them into discrete tasks or activities that can be performed by specified system elements. A function is a characteristic action to be accomplished by one of the system elements composed of hardware and facilities, software, personnel (procedures), or any combination thereof. This process calls for the identification of these characteristic actions and assembly of them into series-parallel strings, sequences, or flows called a functional flow diagram (FFD).

The functional flow diagram is perceived as a model of actual system operation where a function is performed by a set of system resources perhaps not yet fully identified. Note the significant difference here from the process flow diagrams commonly used in logistics support analysis (LSA) work. While using LSA process flow diagrams, the analyst is fully aware of the system resources that have been previously identified through a functional decomposition process. The objective in this case is to develop a process to efficiently use the resources and to describe how the human and support resources are involved in the process.

At the time the FFD is first created, the analyst generally does not know specifically what resources will be identified to satisfy system functions, though the normal contractor product line may tend to color the purity of method. The creation of the FFD is intended to help answer these questions of what are the system functions and what resources would best satisfy these functions. Functional flow analysis provides a systematic methodology for identifying functional requirements that must be allocated to architectural elements through trade studies and an appeal to sound engineering judgment acquired through experience developing similar systems.

3.3.3.2 The top function
This is a top-down structured process, so it is necessary to begin at the top. The abstract system definition gets no higher than the definition of its top-level function which is the need statement for the system or some paraphrasing of it that is acceptable to the customer. Different large customers like DoD have different names for this top-level function and sometimes change their name for it but the author has decided to remain with the need as the top-level function. Indeed, many system engineers fail to make the connection that the need has anything to do with the functional analysis process. In this chapter we will discuss the sequence of a user identified need being expanded into a contractor's life-cycle model, one function of which is "Use System," that we can then continue to analyze and decompose to understand the customer's need in more detail.

This need statement should be prepared and approved before the mission analysis is begun but may end up being modified in the process of performing that analysis. It is the principal input for the mission analysis. The need can usually be phrased in a simple sentence or paragraph. Some examples of need statements fashioned from the perspective of the user follow:

a. Move 1000 cubic yards of naturally deployed soil 5 miles in 12 hours.
b. Transport 50,000 pounds of payload from Earth surface to low Earth orbit.
c. Safely transport 1 to 5 people on road networks in comfort.

Whether the program is a commercial venture or a contracted effort for a large customer, the user need is always supplemented by a business case from the developer's perspective that should involve some expectation of profit. If the development is a contract activity, an acquisition agent's concern for availability of funding relative to the user's statement of need is appropriate. In the case of Department of Defense (DoD), there was a time when cost was treated as a dependent variable, meaning that the cost was dependent on whatever it cost to satisfy the performance goals. DoD in the early 21st century has fully converted to development applying cost as an impendent variable where cost, performance, and schedule all play an independent variable role in the trade study process to define appropriate requirements and select a preferred solution.

Therefore, the top-level FFD will always be a single block bearing a title that is the system needs statement (or some paraphrased restatement of it). This is system function "F," sometimes called the level zero function. It is not necessary to prepare a FFD sheet illustrating this single block (seems a little silly even thinking about it), but the function should be entered into the function descriptions. This function is also the very top-level functional requirement. Every other system function and functional requirement (operational requirement) should properly flow from this one function. There should exist logical continuity between this function and every other decision made on a program as they are all driven by an interest in satisfying this function.

The customer should have supplied a system need statement in some form, although it is not uncommon that this statement is buried in a statement of work, briefing materials, requirements document, or verbal information communicated to the contractor Proposal/Program Manager. Sometimes the need statement was derived in an earlier study phase, the customer has rephrased it in current program documentation, and it has become lost even to the customer's consciousness. In these cases, it is important to work closely with the customer to reconstruct a mutually agreeable need statement. In the event that the program is in a competitive proposal phase where the customer will not respond to your questions, the analyst should work with his or her own program office toward phrasing a company version of the need statement that can later, after the program win, be updated to reflect customer insights.

If the program is totally company funded, as in commercial enterprises, it may be that the analyst must create a need statement from scratch. In this case it is necessary to completely understand who the eventual customer is and what the objectives of the program office are. It is a difficult task to fabricate a system need statement, and the Program Manager may not support the task because he or she thinks it is either already very obvious or that it is too early. It is important to write down the need statement formally because it is almost certain that everyone on the program does not have the same understanding about it that the Program Manager has. If there are several alternative system need possibilities (because all of the alternatives have not yet been fully explored), they should be written down, and these possibilities explored for the most appropriate statement of the customer's need.

The Program Manager is the proper decision-maker about the final form of the need statement within the company structure. The need statement should be released in some form that permits control of its status. This could be in the form of a memo, email message, or program database content with a control number and date where the control number has a revision letter tacked on for each change. The method should permit rapid dissemination of changes because the chief arguments against writing down the need statement are that: (1) "It will change too often" and (2) "We will not be able to keep everyone informed about changes." These are actually pretty good reasons to write it down and communicate it well but they are seldom perceived as such at the time it has to be done. The need statement, ideally, would be recorded in the program requirements database as function F.

The need statement, paraphrased if necessary to form the top system function title, should begin with an action verb like all function titles. For example, if the system purpose is to carry payloads to orbit, the need statement might be phrased, "Transport Assigned Payload from Earth to Orbit." The key word here is "Transport." A weapons system for destroying hostile tanks might have a need statement, "Identify and Destroy Hostile Tank Vehicles." These examples are intended to communicate the message that the need statement should be simple, brief, and very much focused on the essential operational purpose of the system from the user's perspective. This need may be instantly allocated to the ultimate element in the product architecture, the system, by definition and this is the beginning of the process. The first functional flow diagram is, therefore, a single block corresponding to the need and the product entity that will accomplish that need is the system.

3.3.3.3 Life-cycle master flow diagram

The next level should be a life-cycle flow diagram composed of several function blocks that identify major life-cycle functions from creation of the system to disposal of it at the end of its life. Subsequently, these life-cycle functions may be further decomposed as necessary. For most enterprises, one life-cycle flow diagram will be adequate for every system they will ever create. It is only in the details of development, manufacture, system use, and logistic support that the uniqueness of particular systems comes to the surface. Therefore, one can craft a single flow diagram as the basis for an enterprise common process and that is what has been done in the series of books on grand system development that the author uses in his company's system engineering certificate training program. The author considers this book as Volume Two of that series. Figure 3.3-1 is generic in nature and an expansion of block F3113 on Figure 3.2-2 in a string beginning with a one-block diagram titled Enterprise Vision that is the ultimate function for the developing enterprise and from which the enterprise life-cycle common process model is expanded.

Figure 3.3-4 offers a simplified view of a typical life-cycle master flow diagram. The symbols used on functional flow diagrams are defined in Figure 3.3-5. The blocks represent activities the system must be capable of accomplishing. The directed lines that join them denote the flow of activities from one major function to another in relative time. Arrows on lines define the sequential direction of movement through the diagram. Two kinds of logic symbols, AND and OR (not used on Figure 3.3-4), provide for separating and joining the flow where necessary. The enterprise vision statement can be expanded to form the generic life cycle for the enterprise. The life-cycle diagram is the first expansion of the need statement for a given enterprise program as well as the expansion in a generic sense for the enterprise ultimate function, their vision statement.

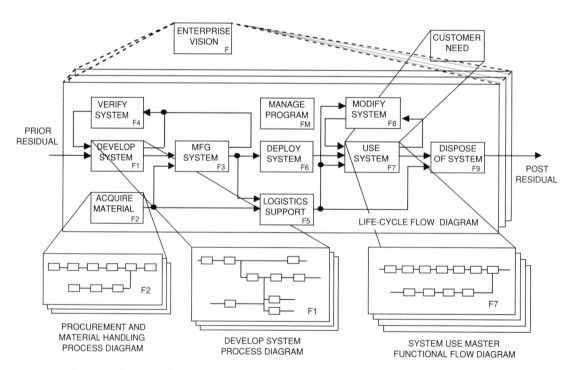

Figure 3.3-4   *Life-cycle master flow diagram*

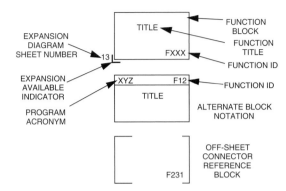

Figure 3.3-5 *Functional flow diagram style sheet, blocks*

3.3.3.4 Flow diagramming details

Figure 3.3-5 offers a style guide for the symbols used on flow diagrams. Blocks that are expanded in greater detail on another sheet should be signaled with some mechanism. The author uses a corner at the lower left of the block annotated with the sheet number where you will find that block expanded or decomposed. An alternative block depiction might use a divider with program acronym and function ID above and the function title below. Some practitioners use an incomplete block like that shown in Figure 3.3-5 to reference connections from one sheet to another but the author does not prefer that structure.

Each function must be named with a short title that communicates the essential meaning of the function. Function titles should begin with an action verb in all cases where the function in question is under the full control of the system. In cases where the function covers actions that are totally controlled by natural forces or the other system environmental elements, the author sometimes permits himself to use titles that are passively stated but this practice is not especially recommended.

Each function block should be uniquely numbered. The author uses a base 60 ID system (0-9 integers, A-Z less O, and a-z less l) to avoid delimiting periods needed in a base 10 system where there are more than nine functions on a branch at any one level. The function ID is a key field in any database that might be used to capture the results of the functional analysis and the periods do not sort properly in inexpensive databases because of ASCII character priorities. Period delimitation of base 10 function IDs may be preferred when commercial requirements analysis tools are used that support proper sorting and indexing of these strings.

If the ultimate function, the need, was assigned the function ID F and there were six functions on the first expansion of this function, the next tier functions would be identified F1, F2, F3, F4, F5, and F6 using the author's system. If F3 expansion entailed five subordinate functions they would be identified F31, F32, F33, F34, and F35 in this notation.

The author prefers to apply a unified flow diagram combining the enterprise vision expansion into a common process system flow diagram and a customer need expansion into a product system life-cycle flow diagram which is a subset of the former. Both of these may be identified with the function ID "F" depending on whether one is thinking in terms of the generic enterprise life-cycle master flow diagram or a specific program life-cycle master flow diagram.

As shown on Figure 3.1-1 function F47 is the use product system function and the principal expansion of the customer need statement that includes operations and maintenance for the product system. This function F47 must be expanded for each program in the work associated with function F41 (Grand Systems Requirements) of that program.

The F prefix is used for all function block IDs because it clearly differentiates a function from other structures that may be used concurrently in computer databases. Functional flow diagram practice the author is familiar with does not generally call for a leading "F" in the function ID.

Where more than one functional flow alternative exists for a single system at the top level, the analyst may code these alternatives with different lead letters such as "F13," "G13," and so forth, until the alternatives are reduced to one as result of trade studies. At the lower functional flow levels, the analyst can use parenthetically included characters to achieve the same purpose. For example, let us say that we have identified a function F473 and it has not yet been determined which of three alternative functionalities are needed subordinate to that. These can be carried as F473(1), F473(2), and F473(3) with lower-tier functions identified as F473(1)1, F473(1)2, and so forth for alternative F473(1). These character sets will sort properly in databases programs used on desktop computers so long as you use only a single character in the parenthesis.

The function identification numbering system alternative that many customers prefer uses a base 10 place value structure. Levels of the flow diagram are separated by periods. This is necessary to mark the level (arithmetic base) boundaries. A typical Function ID would appear as "1.9.10.5." If this were written without the decimal delimiters, it could mean either 1.9.10.5, 19.10.5, or any of several other possibilities. The difficulty with this approach is that it does not sort well in a database that follows ASCII priorities. Strings are sorted one character at a time. So, the IDs 10.2 and 100.2.3 will be listed before the ID 2.5 since zero has a higher precedence than the period. By forcing each place value into a single character width, in the base 60 system, the periods can be eliminated and the sorting problem is overcome. The only disadvantage is that it is limited to 60 characters. In functional flow diagramming you want to avoid making a diagram with too many blocks, so there is really no disadvantage in the base 60 system applied to functional analysis. The author made one of these diagrams that had 70 blocks on one sheet during the pre-computer days of typewritten stickyback blocks connected with pencil lines. He used backward E and upside down A as well as other characters like #, $, %, and @ to identify all of the blocks uniquely. Twenty years later on the day the secretary who did the typing retired she used an expletive to refer to the author, but by that time in good humor.

The functions, depicted as blocks on the FFD, must be interconnected to define their intended sequence. This is accomplished by arrow-headed lines and three simple logic symbols. In all cases, the lines should be unidirectional in nature. Do not use lines with arrows at both ends.

It is uncommon when all of the functions of a system can be properly illustrated in a straight, serial fashion. Generally, systems possess a degree of parallelism such that more than one function must be simultaneously in process, or choices among alternatives are necessary. These characteristics establish a need to separate and rejoin the flow in accordance with two basic logical concepts, one of which has two alternatives. Figure 3.3-6 illustrates these

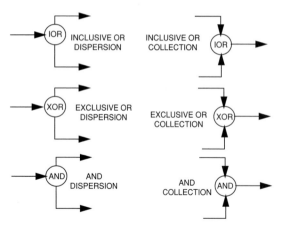

Figure 3.3-6 *Functional flow diagram style sheet, combinatorial symbols*

symbols. When flow must split, one of two possibilities exists: (1) it is necessary to follow all of two or more paths simultaneously or (2) it is required to follow one or more of two or more paths, including all of the illustrated paths. In the first case a logical AND symbol is used to branch the flow to appropriate points. In the latter case a logical inclusive OR symbol is used branch the flow that the author calls an IOR. If the logical OR relationship calls for taking only one of two or more paths, it corresponds to an exclusive OR relationship referred to by the author as an XOR.

These symbols are used to branch outward from one function to many successors or to collect flow from many precursors to one successor function. Where "OR" symbols are used without distinction or only inclusively, the flow can become confusing to those not present at the creation because path possibilities may not be clear. Some of this confusion can be overcome by separating the flow using a combination of "AND" and exclusive (rather than inclusive) "OR" symbols as illustrated in Figure 3.3-7. In Figure 3.3-7a functions F2 and F3 may be accomplished either in an exclusive OR relationship (one or the other but not both) or in an AND relationship because they are connected by an inclusive OR symbol. If it is important to differentiate

between these possibilities, the two functions can be shown connected by a combination of exclusive OR and AND symbols as shown in Figure 3.3-7b. Clearly, illustrating the latter kind of connections requires more work and does add complexity to the drawing while providing a more logically precise definition of permissible flow. By using a combination of AND, IOR, and XOR symbols one can achieve logical clarity with the simplest visual image.

It is convenient to construct flow diagrams in several levels as shown in Figure 3.3.8 derived from the first DSMC system engineering manual that in turn was derived from a Lockheed Space and Missiles manual. One of the reasons for this is to keep each diagram as simple as possible such that a person can visually take in the complete function illustrated on the diagram. It is generally true, however, that if too few blocks appear on individual sheets, then it will be necessary to prepare many individual flow diagrams to fully define system functions, and it will be difficult to interrelate all of these sheets. Therefore, the analyst must strike a compromise between the simplicity of individual diagrams (few blocks per level) and ease of following the lower-tier flow (few levels). The customer may seek to impose a standard that prescribes a minimum and maximum number of blocks per diagram but this is an arbitrary action that will generally work against the thoughtful development of system functionality.

Another important reason for using multiple levels is that it facilitates an iterative process that forces interaction in time with the performance requirements allocation and concept synthesis processes. The functional flow process must not be allowed to take on an independent life of its own. It must be forced to pace but not outstrip the other elements of the system requirements analysis process. By intelligently introducing levels into the FFD, the overall process can be better regulated by the system engineering community.

3.3.3.5 Detailed flow diagrams
The life-cycle master flow diagram is an expansion of the system need statement (top function). Detailed FFD have exactly the same characteristics as the master diagram and are expansions of specific blocks of the master diagram or other detailed FFD.

Perhaps the most critical system development task in the early phase of a developing program is the coordination and control of the preparation of the detailed FFD sheets. At the

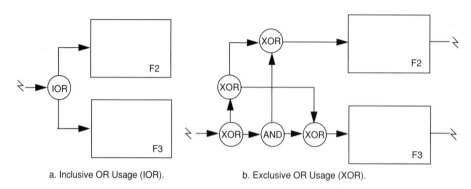

a. Inclusive OR Usage (IOR).   b. Exclusive OR Usage (XOR).

Figure 3.3-7 *Alternative OR symbol usage example*

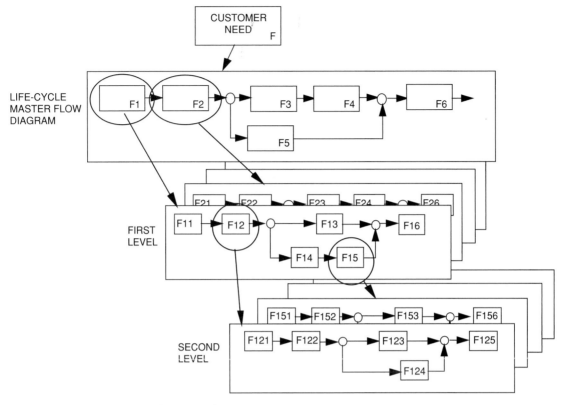

Figure 3.3-8 *Functional flow diagram levels*

time this process begins, there should already exist an approved mission analysis report (or equivalent) containing a clear definition of the key operational and logistics system characteristics and a mission model, a need statement and corresponding expression of it in terms of a top system function (these are either equivalent or one a paraphrasing of the other), a life-cycle model providing the level 1 FFD, and an appreciation by the project team of the character of top-level system assets (architecture elements) that will be used to satisfy the life-cycle model functions. The latter is derived by allocation of the level one functions to major system architecture elements.

In truth, there may exist two or more candidate mission models, logistics concepts, life-cycle model, and top-level architecture concepts, but each set should be documented as a package with their corresponding alternatives. The project team should conduct trade studies to select preferred alternatives before the detailed FFD expansion is allowed to proceed too far for each alternative, although it may be necessary to develop each alternative to some degree in order to fully understand important consequences of their selection.

The conduct of the functional analysis and preparation of the detailed FFD sheets may be too large a job for a single system engineer. If so, it may be necessary to form a special team to accomplish the work. The team name the author prefers is the Program Integration Team (PIT). Even if one engineer could do the whole job in terms of satisfying

schedule needs, a team should be formed to take advantage of a broad range of experience. More ideas are better than fewer ideas at this stage of the program.

The system engineers doing this work should work against a schedule approved by the Chief Engineer (PIT team leader or lead systems engineer, if delegated), giving priority to those functions where the greatest difficulty is anticipated. It may develop at some point that some blocks on the life-cycle model will not have been developed at all while others have been expanded to three or four levels of detail. In general, however, it is good form to develop the life-cycle model into detailed FFD in a balanced fashion. Ideally, the enterprise would develop a generic common process diagram depicting the life-cycle model functions listed below which are illustrated in the generic process diagram shown in Figure 3.2-1.

F41	Grand Systems Requirements
F42	Grand Systems Synthesis
F421	Design Grand System
F422	Material Operations
F423	Manufacture Grand System
F44	Grand Systems Verification
F46	Assure Product and Process Quality
F47	Use System
F48	Grand Systems Sustainment
F49	Manage Program

The actual task of expanding a function on the life-cycle model, or other detailed FFD, is a matter of systematically identifying all of the more detailed steps necessary to accomplish the function. This is facilitated if the analyst has access to the performance requirements that were derived from the higher-level function being expanded because these performance requirements and knowledge of the architecture to which they were allocated provide insights into lower-tier functions. Also, it cements together the functional analysis and concept synthesis processes. More on this in a moment when we discuss four pacing rules.

3.3.3.6 Functional N-Square diagramming

We will later apply a construct called an N-Square diagram to gain an insight into a preferred set of interfaces between architecture elements in a system. This tool is a square matrix of size N on a side. On the matrix diagonal you enter the set of objects you wish to establish relationships between. In the intersections of the matrix you indicate by some mark, symbol, or words whether or not the objects depicted by the diagonal boxes on the horizontal and vertical axes of that intersection need to interact in some way. The intersections on one side of the diagonal relate to one direction in the relationship and the intersections on the other side relate to the other directional sense.

This same construct can be used for the purpose of exploring the inter-relationships between functions shown on the functional flow diagram and, strangely enough, with the same motive we have for creating N-Square diagrams for interface purposes. We have said that we intend to allocate functions to architecture elements and in the process understand what the system is composed of in order to satisfy the needed functionality. The challenge is to allocate these functions so as to minimize what we will call cross-organizational interfaces. These are interfaces that cross from an element under the development of one engineering organization or team to one under the responsibility of a different development organization. These interface planes are where the system engineer can expect the greatest probability of inconsistency in the evolving system concept and design.

Theoretically, it is possible to use a functional N-Square diagram to prevent allocations from ever occurring in the first place that will create undesirable cross functional interfaces, but it is not obvious how to do this to most mortals. Figure 3.3-9 illustrates such a matrix. This matrix is based on the functional flow diagram illustrated in the same figure. The blocks on the diagonal, within which the function identifications are written, correspond to functional relationships internal to the blocks listed on the diagonal.

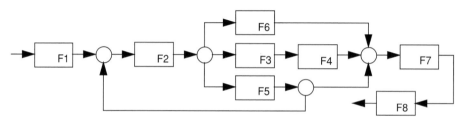

a. Flow Diagram

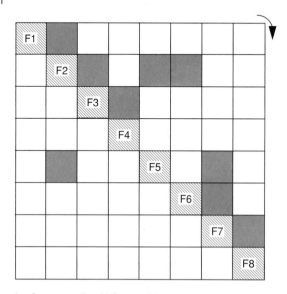

b. Corresponding N-Square Diagram

Figure 3.3-9 *Functional N-square diagram*

All of the interfaces in the system are predetermined by a pair of function allocations (in general, FX allocated to A1 and related FY allocated to A2 predetermines that A1 must interface with A2). In Chapter 3.6 a thoroughly structured approach to interface identification is covered.

The intersections below the diagonal are used to mark inter-relationships between two functions where a function on the diagonal produces an output applied to a function higher up on the diagonal. Likewise, the upper half of the matrix is used for the opposite case (a function produces an output applied to function lower on the diagonal). Thus a marked intersection in the lower left half means that the function in that row will create information or otherwise act as a source for some entity that is needed by the function in that column. This relationship is inverted for intersections marked in the upper half of the matrix. The analyst can reverse this pattern if desired, but in any case should use some way of indicating the directional sense (arrow at top right corner of the N-Square diagram) and get in the habit of using the same sense on all N-Square diagrams, at least all of those produced for a particular program.

As we attempt to allocate the expanding functional flow diagram functions to architecture, we can gain some insight into the resulting interfaces, even though the architecture elements are not yet conceptualized, by thinking through the consequences of the inter-functionality expressed in the functional N-Square diagram. Admittedly, functional N-Square diagrams are most effective in an organization that has applied this process several times to a common product line. In this situation, it often happens that a nearly one-to-one correspondence between functions and architecture items will evolve. The author has never been able to gain advantage from this technique but anyone who can has a distinct advantage because they may avoid bad mistakes in interface development that have to be solved on the architecture or solution plane.

3.3.3.7 Performance requirements analysis
Given that we have identified the needed functionality, it is necessary to derive performance requirements from the exposed functions. Chapter 3.4 picks up with this story. These performance requirements are then allocated to product entities identifying a need for certain physical objects in the system and linking the performance requirements with those items and the specifications that will contain the requirements.

3.3.3.8 Allocation pacing
There exist four different sequences we can apply in accomplishing functional analysis. They relate to the pacing we adopt in identifying functions, deriving requirements from them, allocation of those requirements to product entities, and developing design concepts for the entities. There are two extremes here that we should try to avoid and two intermediate approaches that offer promise. When using any of the four approaches, functions, or their corresponding requirements, may be allocated to an item that already exists on the architecture diagram or items added to which the functions may be allocated. If the analyst is not quite sure which he or she should do at the moment, simply allocate the function to some existing architecture item and as the picture becomes more clear ripple the allocation down that architecture branch to a more appropriate item as the analysis expands.

3.3.3.8.1 Independent mode
The functional flow diagram could be created independently of all other work with no allocation of functionality until the flow diagram is complete to perhaps 7 or 8 levels involving hundreds of blocks in total.

The supporters of the independent mode would remain in the functional analysis mode for some time while evaluating as many alternatives as possible in order to avoid selecting a concept that fails to pan out after considerable development effort has been expended. This approach has merit in the earliest phases of the development of an unprecedented system but can lead to analysis paralysis or runaway reduction without good leadership. Beware the system engineer with a lot of budget doing structured analysis. It may be more fun than they deserve. Eventually, you must break out of this mode, of course, and move from the functional (or logical) view to the physical view.

The team will find it increasingly difficult, once the allocation process does start, in allocating functions or performance requirements as the process continues downward into the flow diagram. It will probably be necessary to redo the lower-tier flow diagrams at some point. But the real problem is that it is difficult to know when you are finished with the functional analysis without the expanding physical facet also available.

3.3.3.8.2 Instant allocation mode
The other extreme is the see-function-allocate-function approach where functions are instantly allocated as they are identified. The problem with this approach is that once you have expanded the diagram a few layers you will find that the higher tier functionality could have been done differently and better. If the flow diagram is changed and it ripples into the architecture to which the changed functionality has been allocated, then there is a lot more error correction work to do that could have been avoided if the functionality had not been allocated so immediately.

3.3.3.8.3 Progressive allocation mode
There is a happy medium between these extremes where the functional analysis remains a few levels ahead of the allocation process. As the architecture is synthesized, higher-tier concepts are developed and fed back to effect lower-tier functional analysis. This steers the lower-tier functionality toward compatibility with higher-tier concepts. Some system engineers who support the independent approach feel this may result in single point design solutions and that might be a valid point if one is not careful. But the advantages of progressive allocation seem to the author to be worth the risk. There is an art form in doing this work that can only be mastered through doing it. There is not necessarily a right way and wrong way to implement a functional analysis. Three different contractors might produce three different functional analyses yet still come up with the same architecture.

3.3.3.8.4 Layered approach
Another intermediate approach as an alternative to progressive allocation is layered allocation pacing. In this approach, encouraged by Mr. Bernard Morias and Professor Brian Mar while applying their FRAT modeling alternative to traditional structured analysis, the analyst accomplishes the decomposition of one block on a higher tier diagram and then allocates the exposed functionality to architecture. This is repeated for other functions on the parent diagram. Then the analyst can begin with the functions on each of the expanded diagrams as the next layer or level.

3.4

Product and Process Performance Requirements Analysis and Allocation

Contents

3.4.1 Preliminaries

It is convenient to group these two kinds of requirements analysis work together (product and process) because they share a common methodology for gaining insight into appropriate requirements. This common methodology is functional flow diagramming. We shall begin by defining each requirements analysis category and then we shall apply the structured analysis approach described in Chapter 3.3 to product performance requirements analysis and to the several components of process performance requirements analysis. In each application we are interested in answering the question, "What does it have to do and how well does it have to do it?"

We have been careful to encourage a practice called concurrent engineering throughout this book, where engineers and analysts from the several different disciplines needed for a particular project, work together to first understand the problem capturing the results in a specification and then work together to solve the problem in design solutions. We might also require that this work be done in a simultaneous fashion to make an important point. We should develop the product and process requirements simultaneously cooperating across this gap intensely. Product requirements will influence the requirements appropriate for the manufacturing process. Similarly, the design of the product will influence the design of the manufacturing process. These facts are true between the product and all of the process components: manufacturing, quality assurance, logistics support, product operations and maintenance, verification, deployment, and disposal. Therefore, we must accomplish the product and process requirements analysis both simultaneously and concurrently and integrate and optimize at the grand system level.

Many system engineers and several system engineering standards and books offer an alternative to the author's method of breaking up the overall requirements task into pieces or tasks. Others describe a process they call requirements analysis, which is a precursor of functional analysis. The author believes this to be exactly opposite to a proper sequence. The author has tried to fit the pieces together, as described for example in MIL-STD-499B (never approved by DoD) and IEEE 1220, into a logical sequence and failed. The author's problem is that he believes that the ultimate function is the need statement and that this should be instantly allocated to the thing called system, the peak of the product physical model. This first act in the functional analysis and allocation process precedes everything that happens from an engineering or logical perspective on a development program. Therefore, there is no room on the front end of functional analysis for something called requirements analysis.

Now, the intent of the sequence often offered in standards is to respect the work that the user and acquisition agent may have to do to characterize the system they require as a prerequisite to entering into a contractual relationship with a developer to acquire the system. The understanding is that the contractor will then accomplish functional analysis to determine the composition of the system architecture necessary to satisfy the system requirements allocating lower-tier functionality to things that form the system and to accomplish the necessary trade studies to select the preferred architecture.

The author would prefer to recognize that functional analysis is a continuous stream of work from the first insight into a need through the time when we are sure of the preferred architecture and the necessary characteristics

those entities and relations must respect. Performance requirements analysis depends on functional analysis to expose needed performance related functions that will then be transformed into performance requirements suitable for inclusion in a specification for the item to which they are allocated. Ideally, users and acquisition agents would accomplish their requirements work leading to formation of an operational requirements document (ORD), concept of operations (CONOPS), or preliminary system specification as an adjunct to a well-executed functional analysis. The author offers Figure 3.4-1 as a vision of performance of this work early in the evolution of a new system where a large customer acquires systems through contracts with developers. The user and acquisition agent accomplish such requirements work as they feel is needed to control acquisition risks and populate a preliminary system specification. The contractors pick up with this work when working on their proposals to win the contract and each bidder couples a recommended system specification with their proposal submittal. When a contractor wins and is awarded a contract, they continue to apply structured analysis to perfect the system specification and develop lower-tier specifications. In the ideal situation, the bidding contractors would inherit the prior structured analysis work accomplished by the user but in the author's experience that information is most often a void because it simply has not been done. Ad hoc is alive and well in too many of these cases.

As suggested by Figure 3.4-1, there are many requirements analysis tasks involved in building a good specification, and functional and performance requirements analysis are but two of those tasks. It is hard to justify requirements analysis leading the process when you consider the whole process since functional analysis is the performance requirements identification mechanism. We will find that there exist requirements identification mechanisms available for the three kinds of constraints analysis (interface, specialty engineering, and environmental requirements analysis) also. In each case identification precedes definition. In identification we recognize a need for the control of a particular parameter and in definition we fully characterize that parameter.

3.4.1.1 Product performance requirements analysis
The results of product performance requirements analysis for a given system or item flow into the corresponding

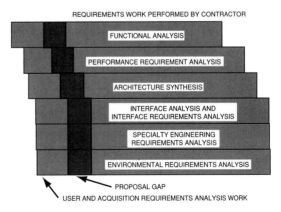

Figure 3.4-1 *Integration of user, acquisition agent, and contractor requirements work*

specification performance or entity capability requirements section. In this process we attempt to clearly define what the product system or item must do and how well it must do it, that is, how well it must perform. We wish to be complete while not over-specifying. This is a very difficult objective to satisfy because it is very hard to know when you have attained a condition of completeness. For this reason, we adopt a very organized methodology that encourages us, no forces us, to think of everything important while avoiding unnecessary characteristics.

You should already be aware that there is a fundamental contradiction in what we shall try to do. Creativity is encouraged by freedom from constraints. Yet, we shall attempt to be creative within the constraints of an organized approach. If we choose to apply too much structure, we run the risk of angry analysts forced to mindlessly work within a structure so rigid they are not able to think expansively. If we apply too little rigor, we run the risk of incompleteness. The author offers no clear direction here. Clearly, a condition of balance is required but the precise balance point for a given program is best selected through experience and knowledge on the part of the person who will perform the work.

A completely ad hoc approach represents one extreme while the overly organized approach covered in AFSCM 375 series probably represents the other extreme. In the latter case, analysts are required to fill out special forms called forms B that were actually requirements analysis sheets identified briefly in Chapter 3.3 and described in this chapter. The forms B work was based on a functional flow diagram. The forms B would be reviewed and approved and passed on to a key punch operator who would translate the form content into punch cards for mainframe computer database entry. The whole process was driven by the information needs of the main frame computer, using the then state of the art 80 column Holerith card I/O technology. Today, networked computers using database systems with well designed screens and keystroke patterns and more direct data entry methods have dramatically improved this whole process permitting us to focus on requirements writing rather than the rigid information controls of yesteryear.

3.4.1.2 Process performance requirements analysis
In process requirements analysis we seek to provide planning information useful in establishing program plans for: (1) material acquisition, (2) manufacturing, (3) quality assurance, (4) testing, (5) deployment, and (6) logistics support for the product system that are consistent with the evolving product system requirements and concept. One might add cost and schedule to this list. Cost is discussed as a specialty engineering requirements analysis component of design constraints analysis in Chapter 3.7 due to the allocable nature of cost. Schedule requirements analysis is simply the application of the product timeline methodology described in Chapter 3.8 for all of the process requirements analysis disciplines. Program schedule data can be presented in PERT or CPM artifacts, of course, as well as the simple Gantt chart format.

All of this work should be happening concurrently with product structured analysis and product performance requirements analysis such that the product requirements are influenced in intelligent ways by process requirements and vice versa. In some cases a product requirement that is essential will have to influence process decisions and in other cases process requirements will have to influence product characteristics. This process should result in

requirements optimized at the grand system level consisting of both product and process. If product and process requirements analysis work is not accomplished concurrently, this result is unlikely.

3.4.2 Requirements Development Strategies

We have identified four strategies for developing requirements, whether they be product or process performance requirements or design constraints. They are freestyle, cloning, Q&A, and structured analysis. The first three are not encouraged as the principal strategy though parent item cloning (copying) is an integral part of techniques employed by many specialty engineering disciplines applying mathematical models to their requirements analysis chore. The principal downside with the freestyle strategy is that it is neither comprehensive nor systematic. We will never know if we have identified all of the right attributes to control. The question and answer (Q&A) approach should be applied but it cannot be comprehensive unless applied in combination with some form of pictorial modeling.

The fact is that cloning is practically useless as a performance requirements approach. Performance requirements are necessarily wedded very closely with the kind of function and object involved. It is much more work to develop a boilerplate to cover all the possible performance requirements contingencies than it is to do each performance requirements analysis job in some organized way without benefit of a boilerplate.

The parent item cloning strategy gives little useful aid for performance requirements as well because the components of a parent item may be very disparate in their appeal to technology. If the parent item is a dump truck and we know the components to include a chassis, cab, dump body, engine, and running gear, we cannot always depend on the parent performance requirements directly providing us a comprehensive insight into the necessary performance requirements for these components.

The secret to the success of structured analysis, the recommended method for exposing performance requirements, is that it provides an exhaustive, comprehensive environment within which to pursue the identification of all appropriate performance requirements. It stimulates us to think about the central performance requirements question of, "What does it have to do and how well does it have to do it?"

3.4.3 The General Plan

This chapter stresses the use of functional analysis using functional flow diagramming as the means to gain an organized insight into the identification of product performance and programmatic process requirements. We will make a first pass with that as the mechanism. In Chapter 3.9 we will visit some other mechanisms and how they may be used to the same end.

Functional analysis, discussed in detail in Chapter 3.3, is the first step in a three-step performance requirements analysis process. It produces what we have called the attributes of primitive requirements statements, essentially the names of requirements. The second step is to add an appropriate value, units, and a relationship to this attribute creating what we called in Part 2 of this book a primitive requirements statement. The third step entails expanding the primitive statement into a fully specification-ready statement. An engineer very familiar with the product line, technologies applied, and customer base may be able to replace this three-step process with a single move from function name to performance requirement statement suitable for inclusion directly in a specification.

Since we have covered the general functional analysis process in Chapter 3.3 and expansion of primitive requirements statements and various methods of identifying appropriate values in Part 2, this chapter will focus on applying the general method to the several product and process performance requirements analysis disciplines listed above and the transformation of the functional statements into quantified primitive performance requirements appropriate to the specific discipline.

Figure 3.4-2 brings forward an illustration from Chapter 3.2 showing a generic life-cycle functional flow diagram that could apply to the development of most any system imaginable at the top level. Depending on the character of the system, in particular what its mission is, the product use oriented functions would decompose in a number of different ways using the approach described in Chapter 3.3.

The contractor's organization should be mapped to these major enterprise functions for the purpose of establishing the principal responsibility for requirements analysis for these steps. Table 3.4-1 offers one example of this map. Your organization's map will reflect how you are organized and may, therefore, be different. Functions F1 through F3 are focused on management of the enterprise and on acquiring new business. The responsible functional departments should cooperate in developing a common process diagram

expansion of Figure 3.4-2 for these functions. Function 4 is accomplished on each program of which there may be several in the house at any one time. Some of these functions will yield to common process development as well but some of them like F47 and F48 in the system employment overlay will have to be analyzed from an individual program perspective and will form the central focus for product functional analysis leading to identification of appropriate product performance requirements.

This table tells us that people from each of these organizations must be cooperating in the functional analysis effort described in this part. The organizations indicated must take the lead in decomposing the master flow diagram blocks noted in Table 3.4-1. For example, Verification Engineering (Department 250 in the generic organizational structure the author uses in his system engineering certificate program) should take the lead in decomposing function F44, Verify Product System. The lead organization must, in each case, lead the efforts of the supporting organizations in reaching their goals. The whole requirements analysis process for a program, including the performance and process portions, should be managed by a program system engineering function referred to in this book as a program integration team.

3.4.4 Transition to Process Analysis

As we decompose the operational functionality of the system, and allocate performance requirements derived from these functions to architecture elements, we should begin to

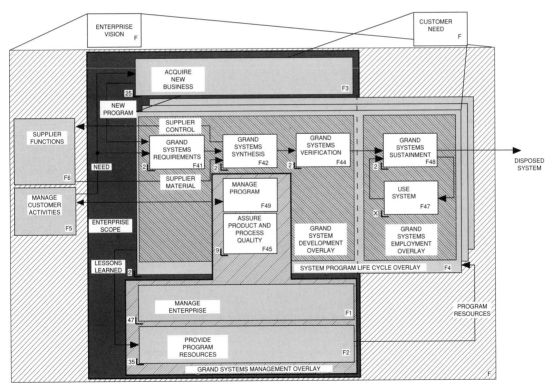

Figure 3.4-2   *Generic master function flow diagram*

Table 3.4-1 *Principal organization responsibility map*

| Process step | | Principal department | |
ID	Name	Dept	Name
F1	Enterprise Management	D000	Enterprise Integration Team
F2	Functional Management	D100	Business Functions Departments
		D200	Technical Functions Departments
F3	New Business	D900	New Business
F4	System/Program Life Cycle	D300	Enterprise Integration Team
F41	Grand System Requirements	D210	Engineering
F42	Grand System Synthesis	D200	Technical Functions Departments
F421	Design System	D210	Engineering
F422	Material Operations	D240	Material and Procurement
F423	Manufacture System	D220	Manufacturing Support
F44	Grand System Verification	D250	Verification Engineering
F46	Assure Product and Process Quality	D500	Quality
F47	Use System	D230	Logistics
F48	Grand Systems Sustainment	D230	Logistics
F482	Deliver/Deploy Product System	D230	Logistics
F484	Logistically Support System	D230	Logistics
F485	Modify System	D230	Logistics
F486	Dispose of System	D230	Logistics
F49	Manage Program	D400	Programs
F5	Manage Acquisition		Customer

see a correlation with the real-world operation of the system during its mission that can be focused using a more process-oriented view. We should not force this conversion too rapidly because it may act to shield us from good alternatives we might otherwise be stimulated to see from a functional perspective. But, as soon as the top-level items in the architecture are irrevocably selected so as to define the character of the system, we should make a conscious move away from functional analysis to process analysis.

This can be stimulated fairly painlessly if we progressively tune the functional analysis to the evolving product architecture solution. As functions or their derived performance requirements are allocated to architecture, we should identify design concepts for these items and feed those concepts back into lower-tier functional analysis. There is an art in knowing how to relate these two components of the structured analysis and it should be somewhere between the two extremes. One of the extremes is to allocate functionality as soon as it is identified. The other extreme is to complete the functional analysis before allocating any of it. As the functional analysis unfolds we will find that we could have done the higher-order functional analysis differently and better so there will be a degree of instability in a band of functional analysis above where the analysis is in progress. By allocating functionality instantly, we will not only have to maintain the functional analysis data in accordance with these beneficial changes but the allocations and

possibly the architecture as well. By delaying the allocation this instability will have time to dampen out resulting in a smaller scope to the rework. The other extreme suffers because when the allocation process is begun it will almost produce higher-tier architecture that is compatible with lower-tier functionality forcing rework of the lower-tier functional analysis.

Combine some delay of function allocation with architecture concept development and feedback to the lower-tier functional analysis to minimize the unnecessary work. In the process you will reap another benefit. The lower tiers of the functional analysis will start to take on the characteristics of a process flow diagram. The fundamental difference between a functional flow and a process flow diagram is in the meaning of the blocks and directed line segments as revealed in Figure 3.4-3.

The lower tiers as they relate to product maintenance and logistic support in the field provide the logistic support analysis (LSA) process with a vehicle for understanding appropriate requirements for support equipment and tools, technical data, personnel and training, and spares. The people responsible for operability should continue to expand the operations aspects in the process format to understand requirements for the human operational interface.

Manufacturing functional analysis should evolve into manufacturing process flow planning with the timeline being the manufacturing schedule. Functional analysis

SYMBOL	FUNCTIONAL FLOW	PROCESS FLOW
BLOCK	Identifies something that must happen	Identifies an analog of real-world activity
DIRECTED LINE	Indicates sequence only	Indicates sequence and flow of resources

Figure 3.4-3 *Diagramming comparison*

related to verification work should quickly transition into process planning for specific test activities and be correlated in time with the integrated verification schedule.

3.4.5 Primitive Statement and Transform

It is possible to allocate either the function name, the functional requirement stated in primitive format, or the complete performance requirement ready for inclusion in the specification. For example, if the function is to Navigate to Waypoint X, we could instantly allocate that function to a Guidance, Navigation, and Control (GN&C) Unit and then worry about how well it had to do this function. Alternatively, we could first determine that the performance requirement is to "Navigate to Waypoint X with less than a 10 meter circular error of probability from planned position." This performance requirement could then be allocated to the GN&C unit. Finally, the complete specification content could be phrased prior to allocation in the following form: "The flight vehicle shall arrive at any planned waypoint with a CEP of less than or equal to 10 meters over any distance between any two waypoints and shall not exceed a 20 meter cross track error during the leg between waypoints." The quantification is driven by a clear understanding of the degree of accuracy needed to satisfy a mission requirement defined by the user.

3.4.6 Value Identification

Functional analysis is very useful as a tool to gain insight into the identification of all of the needed performance requirements for an item. It helps answer the question, "What should I write performance requirements about?" This is not the complete answer, however. We also must define a numerical value for each requirement. We must not only answer the question, "What does it have to do?" we must answer the question, "How well does it have to do it?" Functional analysis is not helpful in establishing an appropriate value. We need other tools for this.

The functional requirements identified for each parent system element should be studied for allocation, or flow down, to child architecture elements. This allocation may follow one of several allocation methodologies: (1) apportionment, where a value is partitioned in accordance with a mathematical rule and the portions assigned to child elements, (2) identity, where the same value is assigned to all children, or (3) synthetic, where several parent requirements are combined in a complex way to drive the identification of a child requirement.

Some requirements can be interactively valued through parametric analysis very effectively. In parametric analysis, we relate one requirement to one or more other requirements by a mathematical equation or graphical presentation. We can then evaluate the utility of various alternative values for one requirement in the context of the effects on values for another requirement. You will commonly find through this technique that there is no combination that yields optimum values for all requirements included in the analysis, rather a combination that offers the best compromise.

It is especially helpful to define requirements parametrically in terms of cost while trying to define appropriate performance requirements values. If we can determine the relationship between a particular requirement, such as payload weight, and life-cycle cost (LCC), we have in our hands a powerful lever to focus on an appropriate value for payload weight. We may see that over a range of payload weights there is no significant effect on LCC, but beyond a particular point, cost begins to climb rapidly. We then have to ask ourselves how valuable additional payload is incrementally beyond this point in terms of LCC. These are questions, by the way, that the customer should be involved in answering early in the program.

Commonly, things are not this simple, however. Rather, it may be necessary to simultaneous evaluate several requirements inter-relationships over wide ranges of values. An analyst can handle a single parametric relationship, maybe three or four. But, it rapidly becomes inefficient for an analyst unaided by a computer model to manipulate these multiple relationships.

Parametric analysis carried to its ultimate blends into modeling and simulation where the relationships between several key requirements are linked together and can be manipulated to produce outputs indicative of system effectiveness in terms of mission success measures. Values for some of the most fundamental requirements for systems, such as guidance accuracy and engine thrust, must often be determined by complex models and simulations. A set of independent variables is established and they are related to a set of dependent variables. We plug in the independent variable values in different combinations using sound engineering judgment and experience and observe the effects on dependent variables. The data is studied for an optimum solution set.

3.4.7 Product Class Differences

The requirements analysis discipline discussed in this chapter can result in identification of requirements for objects in five uniquely different kinds of product elements: (1) hardware and equipment, (2) computer software, (3) personnel (procedures or tasks), (4) facilities, and (5) composite entities. Hopefully, any readers with a naval bent will forgive omission of ships as a unique category, but they will fit into the author's composite entity as will many other high-level items. All of the composite entities will, at lower levels of indenture, decompose eventually into one of the other four kinds of elements.

It may be convenient to initially allocate functions, exposed in the functional analysis that leads up to performance requirements analysis, to these five types and then to refine that allocation to particular items within those types. Some comments about four of these types follow.

3.4.7.1 Product computer software

In the development of product systems to satisfy very complex needs it is nearly unthinkable that the solution would not include computer software today. The aggregate of functions allocated to computer software become the "customer requirements" for the software engineer. Once all of these customer requirements are collected for a given software entity, the software engineer begins a software requirements analysis process using one of a number of very stylized methods. Several of these approaches are covered in Part 4.

3.4.7.2 Operational and logistics task analysis

Certainly every system will have some of its functions allocated to personnel. Operational functions allocated to personnel (aircraft pilot, maintenance technician, power plant control room operator, etc.) are operational tasks. Maintenance-oriented personnel tasks will normally be identified through some form of logistics support analysis (LSA). The same task analysis process used in LSA can also be used for operationally oriented personnel tasks. These maintenance and operational functions allocated to personnel are tasks that define the personnel actions required to accomplish the defined function and they should have a profound impact on the characteristics of the product through careful statement as requirements for the product. These tasks may require careful human factors engineering analysis and clear knowledge of the limits of normal human capability. Task analysis should consider the following:

a. Analyze multiple crew operational requirements.
b. Identify tasks associated with safety and security.
c. Detail and sequence tasks to eliminate possibilities of technically incorrect human performance and to provide continuity among all tasks required to accomplish a function.
d. Detail unique task considerations that would not otherwise be available through design data.
e. Analyze the series-parallel sequence relationship of tasks (and the associated elapsed times) within functions and among tasks of related functions at all appropriate indenture levels.
f. Complex task descriptions should include identification of the input stimulus, cues, signals, and indications that call for action or reaction, identification of the action or reaction (the individual's observable performance), and identification of the output stimulus, cues, signals, and indications that the action is (or is not) complete, correct, or accurate.
g. Where performance, in accordance with a particular standard is required, include that standard (with tolerances) in the task description if it is not stated in the functional requirements.
h. Where requirements relate to other functional requirements, personnel requirements, and recommended hardware, software, or facilities requirements, they must be correlated to insure the relationship is respected in future work.
i. Evaluate all normal alternate task sequences resulting from a user decision initiated by direct observation or input from audio or visual displays. The task analysis should be success oriented but do consider task sequences and related functional requirements resulting from degraded or no-go indications presented to a user by a system interface. This analysis should make maximum use of failure modes analysis data. Where possible, reference should be made to another functional or task analysis (rather than repeating the analysis) or to an existing publication.

Task analysis should continue through successively lower indenture levels until the following conditions have been satisfied:

a. All constituent perceptions, decisions, and motor actions required to accomplish a function have been determined.
b. All associated functional requirements have been defined.

c. The tasks have been completely characterized (e.g., all tasks and training characteristics above the lowest level have been determined).

Where the system or system element is to be operated by military personnel in the normal operational scenario, each functional requirement allocated to personnel entailing operations and functions should require the additional identification of the following requirements:

a. The appropriate military customer specialists codes or commercial equivalents.
b. Task elapsed times.
c. Quantity of personnel.
d. Teaming relationships.
e. Training characteristics and objectives.

3.4.7.3 Product facilities

Functions allocated to facilities must be transformed into facility design criteria in terms of space, floor loading, partitioning, and utilities. These factors are heavily dependent on the functions that will be performed within the facility. When the functional analysis process begins to allocate functions to facilities, the process should have already passed through an imaginary plane separating top-level functional analysis from lower-tier process analysis and the process flow diagram should begin to be overlaid with the facilitization scheme.

3.4.7.4 Composite product objects

Some functions will be allocated to objects in the architecture that include various combinations of hardware, software, facilities, and personnel actions. These top-level allocations will, on further decomposition, be refined into subfunctions that will eventually allocate uniquely to one of the other kinds of objects listed above. These high-level functional requirements will commonly be general in nature.

3.4.8 Guidelines

The following guidelines will be helpful in the identification of performance requirements regardless of the type of end product involved:

a. Establish parameter and quantitative boundary values within which element characteristics must fall.
b. Translate qualitative criteria into quantitative values for elements.
c. Provide reasonable assurance that design solutions will be acceptable and will be the most cost effective.
d. Prevent incorporation of undesirable characteristics.
e. Permit quantitative evaluation of element characteristics at technical reviews and audits.
f. Restrict requirements statements to a single requirement per statement (one requirement per specification paragraph).
g. State requirement in terms that are verifiable by test, analysis, demonstration, or examination as these words are defined in MIL-STD-961E.
h. Identify requirements to assure survivability of equipment and safety of personnel and equipment.
i. Explain qualitative criteria in terms of unique application to elements.
j. Avoid qualitative descriptions of design goals, objectives, and design solutions.

k. Use illustrations and tables to amplify or clarify requirements statements only if absolutely essential.

l. Insure that checkout requirements do not result in any destructive effects.

3.4.9 Verification Planning Analysis (VPA)

Refer to Chapter 6.6 for a broad discussion of verification requirements analysis, planning, implementation, and reporting as it applies to the three verification streams: item qualification, item acceptance, and system test and evaluation using four different methods of verification. In this section we will focus more tightly on item qualification accomplished through the test method while trying to link the product requirements with the verification process.

3.4.9.1 Overview

Verification planning analysis (VPA) has product verification requirements analysis, test article requirements, and program planning requirements aspects as illustrated in Figure 3.4-4. The product aspect focuses on identifying the verification process requirements for inclusion in specification Section 4, titled Quality Assurance Provisions or Verification depending on the specification standard. This is the portion of VPA that this book will primarily focus on. VPA must also develop any special requirements for test labs, facilities, ranges wherein testing will take place and test articles, test fixtures and adapters that are used to simulate the actual product systems elements during development testing.

All verification work is accomplished to verify achievement of some characteristic of the product design defined in a specification. As requirements are identified for system elements, we are obligated to devise a way to verify that each of them is satisfied by the design. These verification actions may be accomplished by analysis, testing, demonstration, or examination. Often, contractors try to do this by analysis because it is generally (but not always) less

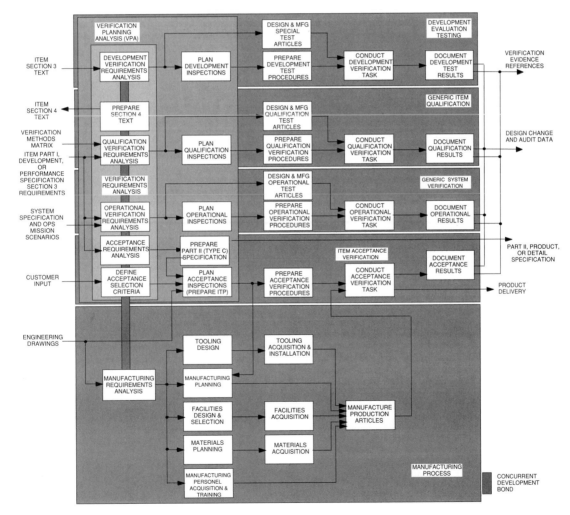

Figure 3.4-4 *VPA & MRA process flow*

expensive. The requirements that cannot or should not be verified by analysis, examination, or demonstration must be verified by test. In order to keep the cost of required testing as low as possible, the many verification requirements must be integrated into a complete program that minimizes the number of test articles and test tasks. Our objective in verification is the collection of the best evidence we can afford.

There are four kinds of program testing commonly accepted in the development of systems: development, item qualification, item acceptance, and development or operational test and evaluation. Development testing is accomplished to validate design concepts, mitigate design and development risks, evaluate alternative design concepts, and demonstrate availability of needed technology. The author refers to this testing as requirements validation testing because it is accomplished to mitigate against the risk of not being able to satisfy a requirement.

Item qualification testing is accomplished to verify that the design satisfies the development or performance requirements and is adequate for the application. Item acceptance testing is accomplished on each production article, or in accordance with some sampling technique, to verify that particular product items meet customer requirements and are suitable for use as planned. System or operational testing is essentially qualification testing performed to verify system operational requirements and to demonstrate system capabilities.

Our game plan is to apply functional or process analysis to these four kinds of product inspection activities. Flow diagrams are created and mapped to test articles, resources, materials, and facilities. These test activities are studied to determine if the planned special test and product articles to be tested require any special features for test compatibility. It may be necessary, for example, to locate a special bore sight target on an airframe to monitor alignment or special instrumentation to monitor strain during structural testing. The flow diagrams help us to recognize the needs for these test requirements.

We really should devote as much attention to systematic analysis, demonstration, and examination planning as we do to test planning, but in most organizations non-test planning is done in an autonomous environment within each engineering group with some oversight by engineering management. It is uncommon to include an analysis requirements planning activity in any formal way. Often, however, testing activities must be intertwined with analysis activities and they have to be planned as part of the test activities. A test may be preceded by an analysis and the test results analyzed prior to initiating yet another test. As a result, a test network may have gaps corresponding to these analyses if they are not integrated into the test network.

3.4.9.2 Development evaluation test requirements analysis

The basic process for development test requirements analysis is very similar to qualification test requirements analysis. DET is, of course, run much earlier than qualification testing because it is accomplished to validate the requirements giving confidence that it is going to be possible to satisfy them in the design process. The principal difference is that we also have to determine the requirements differences for special test articles. We may conclude that we need to perform a fuel system development test to evaluate fuel flow rates and pressures in different conditions of flight. It will

not be necessary, or even desirable, to perfectly reflect the planned physical design configuration of the system. We may, in fact, not be able to complete the detailed design until we have completed a development test. As a result, we mix the product article requirements with special test article requirements so as not to compromise test results applicability to the product article.

We may conclude that the ideal test article is a fluid flow bench laid out to permit us to walk among its parts for test monitoring and adjustment purposes. The final design may have to fit into an aircraft airframe in otherwise unused space. Our requirements would call for plumbing runs that mirror the planned effects of the design on fluid flow and pressure but the precise system geometry may be very different. The results of these tests will be useful in driving final design features and in validating the theoretically derived performance requirements reflected in the special test article design.

3.4.9.3 Item qualification verification requirements analysis

The requirements for Section 3 of our development, Part I, or performance specifications are fed into the block called Qualification Verification Requirements Analysis on Figure 3.4-4. These requirements place a demand on the integrated test plan that test engineers have to determine how best to satisfy within available resources. These requirements are coordinated with methods and collected in Section 4 of the same specification. The verification requirements are collected into bundles and allocated to verification tasks. The tasks are assigned to principal engineers who must prepare plans and procedures for assigned tasks.

Verification process flow diagrams are developed illustrating how the tasks will be sequenced. Schedules are developed and test resource acquisition planning completed. Where program technical risks have been identified that indicate difficulty satisfying a requirement, we evaluate planned activity and test requirements against the planned risk mitigation approach and make adjustments, if necessary. If a requirement is identified as a technical performance measurement (TPM) parameter, we ensure that the test activity supports the planned parameter control plans and schedules.

The test articles supplied to these tests should be as much like production articles as possible but it is seldom possible to really do this. You cannot gain a DoD customer acceptance of your production process until after you have proven that your design satisfies the requirements. You cannot provide the customer with convincing evidence of this until you have completed qualification testing. So, we have a catch-22 situation that is resolved by testing product articles that are as close to the planned production article as possible and have followed as closely as possible the planned manufacturing and quality practices as possible in their building. We then conduct analyses to assure ourselves that any necessary differences will not invalidate the results. Where differences develop that unavoidably do compromise the test results, we may have to run a supplementary test later using a real production article.

3.4.9.4 System test and evaluation requirements analysis

System test and evaluation is commonly broken into two major activities in the acquisition of large complex systems. First, the contractor conducts what is called development test and evaluation (DT&E) and reaches a conclusion about whether or not the system satisfies the contractual requirements contained in the system specification. Subsequently, the user conducts operational test and evaluation (OT&E) to determine if the system satisfies the user requirements. The requirements analysis process for DT&E is similar to qualification test requirements analysis except that we are dealing with the complete system or very large portions of it. System specification Section 3 content is transformed into appropriate methods for verifying product design compliance and verification process requirements. These statements are grouped into bundles and allocated to verification tasks of the four kinds respected. For each task identified, a principal engineer is assigned who has the responsibility to prepare a task plan and procedure that must be responsive to the verification requirements allocated to it. When the work is accomplished, the results are published in a verification task report. The reports are reviewed and conclusions reached about whether or not the evidence shows compliance with the driving product requirements.

Special requirements may have to be developed because of differences in the test article from the planned production article. This is especially true if phasing plans exist to evolve a final design through two or more product test phases. We may have to build a prototype vehicle and subject it to operational tests before we fully understand the needed characteristics of the final configuration. This is really a case of development testing, but since it involves the principal operational article in field trials, it fits better under system test and evaluation.

3.4.9.5 Item acceptance test requirements analysis

Product acceptance VPA defines requirements for product acceptance test stations and equipment that must be integrated into the production line, product design features ensuring acceptance testability, manufacturing planning requirements inputs, and requirements for acceptance test plans and test procedures.

The customer will commonly expect the contractor to select a series of critical system level parameters that cannot produce good results if anything in the article is not acceptable. VPA identifies these critical parameters for coverage in the Part II specification Section 3 and coordinates the verification process requirements in Section 4 with the appropriate verification requirements. This data is then used as the basis for fashioning the acceptance test plans and procedures.

Note that the content of the performance specifications is all design independent and must be crafted prior to the start of detailed design while the content of the detail specification must be determined after the detailed design has been determined and is, therefore, design dependent. This may seem like a case of design before requirements but the content of the detail specification does not apply to the design of the product, rather the design of the acceptance test process. So, this is still a case of requirements before design.

3.4.10 Logistics Support Analysis

Figure 3.4-5 illustrates a fragment of an unmanned aircraft turn-around process diagram. It will be used as an example of how logistics analysis can be used to drive product and support equipment requirements and design. This

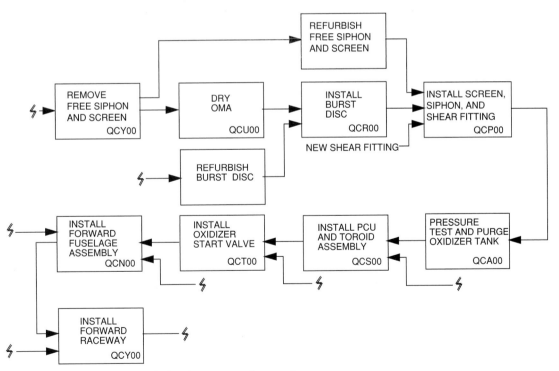

Figure 3.4-5   *Maintenance analysis using process diagramming*

particular vehicle was propelled by a hybrid rocket engine using rubber and inhibited red fuming nitric acid as propellants. Subsequent to flight it had to be decontaminated of sea water (if retrieved from ocean surface recovery) and nitric acid before it could be restored to flight status for its next flight. The process diagram reveals what steps must be accomplished and the sequence of steps appropriate. Each of these steps must be analyzed to determine required resources and any vehicle design considerations observed.

In accomplishing this work on the program the analyst found that step QCS00 could not be performed by one mechanic because it required three hands to hold all of the parts relative to each other while tightening bolts and tubing nuts to secure them. A requirement was established that two of the nose cone installation holes be precisely located rather than simply match drilled with the nose cone and these attach holes were used in the maintenance cycle to attach a special tool temporarily to hold and locate parts during task QCS00 so one person could do the work.

This same approach can be applied to the whole logistics support process for the product system and it can be done during the development process such that the results can be introduced into the design in an economical fashion. The results identify all of the needed resources. A few lines of text describing each task stimulate tech writers to craft the procedures needed for technical data. Logistics engineers can study the diagram for spares and parts demands as well as the outflow of faulted material requiring logistic action. Lists of tools, support equipment, personnel, and materials are maintained to reflect the aggregate resources needed to maintain the system.

3.4.11 Allocation of Functionality

We have two choices in allocation. We may allocate the function names directly to architecture entities and then consider what the corresponding performance requirements should be for that entity. Alternatively, we can first derive performance requirements from the function names and then allocate the performance requirements to architectural entities. Both cases are offered here but the author has after many years of thinking about it concluded that the latter is the better course because it is more often supports a one-to-one mapping situation.

The functional flow diagram (FFD) defines the approved sequence of functions that must be accomplished by the system in order to satisfy the system need statement (or system goal) within the context of the program mission analysis, system logistics concept, and system environmental analysis. That diagram is initiated, developed, and maintained as one portion of the functional analysis process. Two of the outputs of the functional diagramming steps are a FFD and function descriptions. The other output is a timeline that provides required timing information for the time-critical functions illustrated on the FFD. We will discuss timelines in Chapter 3.8 under environmental requirements analysis.

Figure 3.4-6 illustrates the functional analysis process where the output is a series of functional flow diagrams progressively expanding to lower levels of granularity. The function streams flow into a task where we define primitive performance requirements for each function. Each function should yield one or more performance requirements each of which may be entered into a requirements analysis sheet (RAS) ideally supported in a requirements database system.

Figure 3.4-7 illustrates a function allocation process that is team driven and relies upon brainstorming. On a small project, the work can be done by one person but there is value added by having more than one person involved in any case. The functional allocation activities provide a link between functional and performance requirements analysis processes and the architecture synthesis process. The importance of these links cannot be overemphasized as the means by which it is properly determined what the system shall consist of in terms of hardware and software entities.

The identification of performance requirements and allocation of them to architecture elements in this task should follow, in time, the functional flow analysis that results in the FFD, but not too far behind. The desired result is that these two activities proceed interactively with the functional flow analysis leading. When the timing between these activities is well coordinated by the responsible person, lower-tier functional flow analysis will take advantage of higher-level performance requirements allocations and development of design concepts for those higher-tier entities. This will result in a functional flow diagram that is progressively more clearly related to the preferred system concept as the analysis progresses more deeply into the flow. If the functional flow analysis progresses too far ahead of the functional allocation, then there is a danger that the lower-tier functional flows will be difficult to relate to the evolving architecture and concept.

The method employed to allocate (or map) performance requirements to system architecture elements can apply a structured team approach that assigns working level and mid-level program engineering and management personnel to interact with at least one skilled analyst. These teams bring their collective experience to bear on questions of preferred system composition using a structured problem-solving technique. That technique includes a brainstorming phase, a consolidation phase, and a selection phase. If the selection is sufficiently difficult, a formal trade study should be initiated as the selection phase.

The end result of the process is that performance requirements are formally allocated to particular system architecture elements that must perform specific functions. At the lower levels of indenture these elements can be assigned to a single design organization for concept development and preliminary design. The allocations are offered to the appropriate team leader (PIT or IPPT) for review and approval at an engineering review meeting.

The approved allocations flow to: (1) the architecture synthesis process for assembly into the system architecture block diagram (ABD), (2) back to the functional flow diagramming process as a basis for further breakdown of the functional flow diagram and development of timelines, and, if functions or functional rather than performance requirements are being allocated, (3) to the performance requirements analysis process for development of quantified performance requirements for the element to which the functions are allocated.

The architecture synthesis process assembles the approved output of the performance requirements allocation process into functionally related families of system elements respecting and adapting the customer Work Breakdown Structure (WBS), the enterprise engineering organizational structure, program contractual arrangements (sub-contractors, associates, and teaming partners) and company configuration management principals. Necessary interface relationships are developed and illustrated on schematic block diagrams (SBD) or N-square diagrams. These diagrams provide a systematic method for

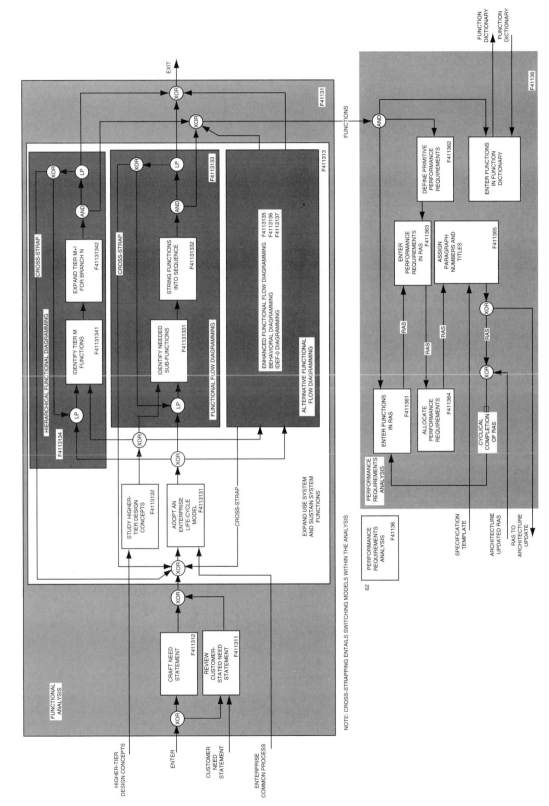

Figure 3.4-6 *Functional and performance requirements analysis*

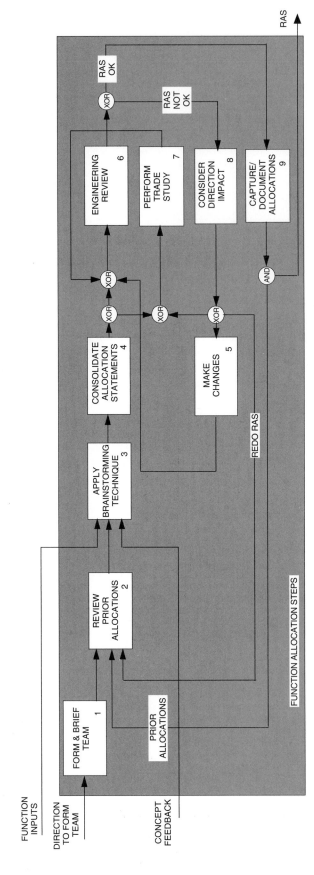

Figure 3.4-7 *Team oriented function allocation*

identifying all needed interfaces and assigning them to appropriate responsible engineers for requirements and concept development.

Once the architecture elements have been identified from performance requirements allocation, the requirements analysis work must continue to refine them into fully characterized performance requirements, resulting in the identification of how well the element must perform each assigned requirement. Concurrently, a constraints analysis is performed that results in a clear understanding of the (1) interface, (2) environmental, and (3) specialty engineering factors (including cost) that must be considered in design development.

The allocation of performance requirements to architecture elements is the first step in the concept synthesis process, and it is through this methodology that the fundamental twin system development tracks of requirements analysis and concept synthesis interact systematically. Later synthesis steps will result in a clearly defined concept for each system element. These steps include: (1) assembly of the architecture elements into the system architecture diagram, (2) development and selection of the preferred concept, and (3) development of the formal concept description for inclusion in the concept description document. In later program phases, these concepts will be more fully defined in engineering drawings prepared during preliminary and detailed design activities.

The design community will initiate action to develop alternative concepts for each system element that satisfy the requirements listed for that element. These alternative concepts are subjected to a trade study where no clear preferred alternative is easily identifiable or a concept predefined by the customer or current product line not available. The concepts approved in an engineering review meeting are included within the system definition document produced in that process keyed to the corresponding system element architecture code.

This cyclical process is repeated working downward through the expanding system functional flow diagram allocating the functions to system elements in the architecture (elements that already exist in the architecture or elements that must be added). The functional flow process provides a systematic way of understanding the system and subelement functions and their sequential relationships. The FFD provides the functional requirements analysts with inputs that help us identify viable architecture alternatives. The architecture conclusions that result are linked into family structures in the architecture synthesis that provides a growing framework for the growing definition of the system synthesis.

The process started with the need statement and is complete when the engineering team is satisfied that each of the system elements along the lower border of the ABD will yield to detail design to a specific set of written requirements by a single design agent or that it will be procured in accordance with a specific set of requirements.

3.4.11.1 Team briefing
A program may assign the PIT the responsibility to perform all function allocation work or assign the PIT top-level responsibility and the several IPT for all items below their own top-level architecture responsibility, all following the same pattern of behavior. If several teams are used, their work must be coordinated and their results integrated by the PIT. The first time a team meets, the team leader should brief the team on the process to be used with particular emphasis on allowing reasonably unrestricted input in a brainstorming process.

3.4.11.2 Review past allocations
Under the guidance of the team leader, the team reviews prior allocations made at higher levels of the functional flow and architecture. This will avoid the confusion resulting from allocating a lower-tier function to hardware when a higher-level function has been previously allocated completely to software. If a single team accomplishes the complete analysis, this need only be accomplished once since the team will retain the corporate memory derived from prior work.

3.4.11.3 Brainstorming and analysis
The next order of business is to develop a list of function allocations from the functional flow blocks under consideration. This is a creative, synthetic (rather than analytic) process. Since it is very rare that any one person has accumulated enough knowledge and experience that they can successfully accomplish the desired result in this activity, it is wise to use a group approach. While there are several models that could be used by the team leader to encourage the team to produce a creative work product, the one that most people can most easily relate to with little instruction is called brainstorming.

Brainstorming is a group process of encouraging participants to give voice to every idea that occurs to them germane to the subject. The objective of the brainstorming exercise in this application is to produce a list of functional requirements statements supportive of the functional flow block title and description. The result should be a list of functional requirements statements on a chalk board, plastic viewgraph sheet, computer projection, or paper tablet in full view of the group members. The leader should encourage energetic participation and discourage critical comment from team members about ideas expressed. Every idea should be written down, however foolish it may appear on first inspection. A creative environment requires the creation of an attitude among the group members that they can say anything that comes to mind with no concern for how their words will affect others. The team leader should inspire the members by example and with appreciation for ideas expressed.

Figure 3.4-8 offers one example of what the final product of this step could look like for function F1134, Separate Payload. The ID and FUNCTION/REQUIREMENT columns list the primitive functional requirements statements derived from the F1134 function numbered from 1 to N. These are a list of subordinate actions that must occur in order for the function F1334 to be satisfied. They may very well also be the next layer of functions subordinate to F1134 but need not be. The team, or analyst, must then allocate these functional requirements to system elements in the architecture. It may be helpful to identify some element character and class codes. These codes offer an intermediate level to which the allocations can be first made before deciding what specific item in the architecture it will be allocated to. Those codes are listed in Table 3.4-2. They are not included on Figure 3.4-8 but could be a sub-column under allocation information.

The character codes relate to the fundamental character of the allocation. Is that function going to be satisfied in hardware, software, facilities, or personnel (procedures)?

TEAM PIT	ANALYST B. Jones	DATE 12-05-98
FUNCTIONAL REQUIREMENTS INFO	PERFORMANCE REQUIREMENTS	ALLOCATION INFORMATION

ID	FUNCTION/REQUIREMENTS		ARCH ID	
F1134	SEPARATE PAYLOAD		A12678	
	1 Physically release locks		A1261A	
	2 Provide relative velosity		A1261E	
	3 Sense separation		A1452	
	4 Initiate release command		A1452	
	5 Verify successful separation		A183641	
	6 Monitor separation event		A12683	
	7 Communicate separation			
F1135				

Figure 3.4-8   *Typical requirements analysis sheet*

Table 3.4-2   *Intermediate Allocation Classes*

Character codes

Char code	Character
H	Hardware
S	Software
F	Facilities
P	Personnel

Class codes

Code	Analysis class	Responsible department
O	Operational Requirements Analysis	System Engineering
L	Logistics Support Analysis	Logistics Engineering
D	Deployment Planning Analysis	Logistics Engineering
V	Verification Planning Analysis	Verification Eng.
A	Assembly and Checkout Analysis	Logistics Engineering
M	Manufacturing Requirements Analysis	Production

Some functions may be allocated completely to hardware, software, or some to a combination of two or more character codes. Each allocation should be a separate line item on this form. If a function is allocated to three architecture items, three line items should be carried. Next, identify the appropriate class code from those listed above. These codes identify the primary cast of characters needed to complete the requirements analysis job.

Be careful to avoid excessively complex functional requirements statements as they eventually become very difficult to manage in the verification process. Include only one requirement per statement. This statement may be nothing more than the function name from the functional flow block diagram. If one function is allocated to two or more architecture items each should have a unique functional requirement statement focused on the item to which it is allocated.

The team defines what architecture element(s) the functional requirement statement is allocated to using the ARCHITECTURE ID CODES from the system architecture block diagram (ABD). If the team finds that there is no current architecture element to which they can allocate a requirement, they should identify a new element with a recommended architecture relationship.

The brainstorming phase may be completed in one meeting or in a series of meetings depending on the complexity of the system function and the degree of creativity of the team members. The team may find it useful to first meet to identify the allocation character (hardware, software, etc.) and the functional requirements statements followed by a period of individual study activity, and a second meeting to nail down the architecture elements.

3.4.11.4 Consolidation
The team leader will have to determine when the brainstorming exercise has reached a point of diminishing returns whereupon the team should switch to a consolidation mode. The team leader leads the team in a process of eliminating duplicates, eliminating clearly flawed alternatives, clarifying statements, and considering alternative allocations to architecture elements. The focus in the brainstorming phase is to capture as many ideas as possible. The focus in the consolidation phase is to organize the accumulated information. The two-step process disconnects the ideas from ownership encouraging everyone to contribute and then to evaluate all of the inputs fairly no matter who offered them.

If the alternative allocations do not yield to simple logic and inspection, the team should move the decision into a formal trade study process. In an early program phase the system engineer should discourage the program from choosing the subjects for trade studies independently from the functional flow and allocation analysis processes. It is precisely this point in the allocation process that offers the system engineer an insight into valid subjects for trade studies focused on how a particular function will be allocated. In the process, the trade study team will determine what element should be added to the system architecture to accomplish the function. The importance of this connection between the functional analysis and concept synthesis processes cannot be overemphasized.

At the conclusion of either the Consolidation process or the Perform Trade Study process the team should have a completed requirements allocation sheet (RAS) such as the typical fragment shown in Figure 3.4-8. This document should represent the aggregate views of the team. The team leader may not be able to bring about complete agreement by all team members. The option should be made available for a minority position paper or statement at the approving engineering review meeting by anyone not sharing the final allocation conclusions. The alternative views can be captured in the meeting minutes. We have discussed this as a paper process but today this work would generally be done in the context of a computer database tool using computer projection with group interaction with the projected image and data entry based on the group discussion.

3.4.11.5 New architecture identification
One of the important purposes of the functional allocation activity is the identification of new architecture elements needed to satisfy system functions. The analysis team may find that it is possible to allocate functional requirements to existing elements in the architecture, but they may also find that there is nothing currently identified that will fully satisfy some function. On large programs it may be necessary to provide a special form for use in declaring a new piece of architecture. Or, on a small program with physical collocation, it may be adequate to simply tell the engineer responsible for the ABD of the new item verbally.

3.4.11.6 Engineering review meeting
The RAS, at some level of indenture, and any new architecture identified since last reviewed should be presented by the team to the team leader (PIT or IPPT as appropriate) at an engineering review meeting. The team leader and his or her staff may conclude that the RAS is acceptable as is, completely unacceptable, or acceptable with changes. The team leader will make the final decision after considering the views of specialty engineering representatives and provide direction to the team. When the RAS is approved, the allocations will flow to the architecture synthesis activity for inclusion of any changes in the ABD. The performance requirements allocation statements in the RAS will be provided to the performance requirements analysis activity. The RAS will also be made available to the functional flow team for use in developing the next tier of the FFD. It is not suggested that each of these activities be accomplished by some separate team. All of this work may all be done by one team or, on a small project, by one person.

If the RAS is not completely accepted, the team will have to meet to determine how best to respond to direction and make the required changes either by recycling the process (if the changes are substantial) or simply make the changes if they are not complex. After the changes are made the team will reschedule the review or gain team leader approval otherwise.

3.4.11.7 Overall coordination
If many teams are employed, the PIT must provide a coordinating influence to control the flow of analytical work. This includes identification of team members and coordination of their availability with supervision, assignment of allocation tasks to the teams, maintaining status records and communicating them to management, maintaining the flow of paper and/or computer entries, coordination with the structured analysis teams, and coordination with the engineer responsible for architecture synthesis.

This process is essentially a creative one. Performing the task is difficult work requiring skills and experience not possessed in equal measure by all engineers, but the project team will have to complete the task with the personnel available. Therefore, it will be helpful for the team leader and members to have available to them some guidance from company procedures that continually benefit from critical feedback from the work experience of teams doing the work on programs. The engineering team must have a way to gain benefit from their experiences and fold the lessons of that experience into improvements for their next opportunity. All too often, companies attempt unsuccessfully to apply the structured approach once and conclude from that experience that autonomy is not so bad after all. It may take a leap of faith to escape from serial work performance habits and functional organization autonomy, but that faith will be rewarded if the organization has a built-in means for continuous process improvement and education.

Human foibles and group dynamics offer many ways for the team to fail. The team leader, and the members, should be aware of these potential traps in order to avoid falling victim to them. There are many good group dynamics books available that more generally cover this area. The professional system engineer, who must deal with groups as a regular part of the job, should master the fundamentals of group psychology.

3.4.11.8 Allocation criteria guidance
We should try to minimize system complexity. Attempt to satisfy system functions with the simplest possible system. This goal will often result in the least cost solution. Frequently, it will also result in a reliable system and one that is easy to maintain.

Minimize the interfaces between elements under the development responsibility of different organizations. This is another way of saying "minimize system complexity." What complexity that must exist in the system should reside in the system elements with the simplest possible interfaces. It is among the system interfaces where the greatest opportunity for system development difficulty exists because of the difficulty of communications between different design agents. It may be necessary to employ N-Square diagrams and other analytical tools to achieve this objective.

Identify architecture elements that can be assigned to a single design agent for concept development and design. This will result in simplified interfaces. It may not be possible at the higher levels of architecture, but should be possible at lower levels.

Minimize the appeal to new but proven technology. Some choices may result in a new technology requirement in hardware, some in software. It is important to select an alternative that balances the technology demand in the aggregate. It may also be true that the development team may be equipped to handle either software or hardware technology development better.

3.4.11.9 Additional performance requirements analysis examples

As an alternative to the matrix shaped RAS, some engineers prefer a sheet RAS. Several examples of this approach are included below. This depiction of the process leads you to a clear understanding of why engineers leaped to a mainframe computer application for requirements analysis long before computer hardware and software technology was ready for the application. Note in Figure 3.4-8 that if a function were allocated to multiple architecture entities it would lead to a logistics problem in a paper based analysis process. We would like to pile up the sheets of paper in order of functions to accomplish the analysis. But, we would then like to pile up the sheets of paper in architecture order to support specification development because we write specifications about architecture entities.

The beauty of a computer, even a mainframe computer in 1965, is that it can index or sort data in more than one way. If you are interested in functionality, you can sort it in function order. If you are interested in architecture, you can sort it in architecture order.

3.4.11.9.1 Performance Requirements Analysis Example 1

In this first example, exposed in Figure 3.4-9 in the form of a sheet rather than tabular type RAS, we have a space transport system upper stage delivering a satellite payload to orbit. This function (F475) has been allocated to the upper stage, A12. We need to define requirements for the design of the separation system located on the upper stage.

REQUIREMENTS ALLOCATION SHEET

FUNCTION IDENTIFICATION

FUNCTION ID:	F475
FUNCTION STATEMENT:	SEPARATE PAYLOAD
FUNCTION DESCRIPTION:	When the Upper Stage completes the transport of the Payload to its planned orbit, it must separate from the Payload allowing the Payload to enter its final transport or operational phase. Stability of mated entity assured by other means.
ENTRY CONDITION:	On-Board computer issues the payload separation command.
EXIT CONDITION:	The Payload physically moves away from the Upper Stage mating surface and develops a Payload Separated signal.

ALLOCATION

ARCHITECTURE ITEM:	A12, Upper Stage

PERFORMANCE REQUIREMENTS IDENTIFICATION

RQMT 1 ATTRIBUTE:	Relative Separation Velocity.
RQMT 1 VALUE:	Less than or equal to 5 feet per second.
RQMT 1 RATIONALE:	Engineering judgment.
RQMT 2 ATTRIBUTE:	Separation Distance.
RQMT 2 VALUE:	Equal to 4 inches plus or minus 1 inch.
RQMT 2 RATIONALE:	Selected to clear potential interference with Upper Stage unit extending 1 inch above the interface plane.
RQMT 3 ATTRIBUTE:	Separation Time.
RQMT 3 VALUE:	Less than or equal 30 seconds.
RQMT 3 RATIONALE:	Engineering judgment.
RQMT 4 ATTRIBUTE:	Payload Separation Acceleration.
RQMT 4 VALUE:	Less than or equal to 5 Gs.
RQMT 4 RATIONALE:	Limit the separation acceleration applied to payload components to 5 Gs to control component cost.
RQMT 5 ATTRIBUTE:	Relative Angular Velocity.
RQMT 5 VALUE:	Less than or equal to 1 degree per second.
RQMT 5 RATIONALE:	To ensure no re-contact occurs between the Payload and Upper Stage components during separation.

Figure 3.4-9   *RAS for example 1*

The question is, "What attributes does the system have to possess in order for it to perform this function (payload separation)?" In this example, we have identified five specific attributes and quantified each requirement based on a rationale that is public and open to criticism and, where good sense dictates, changed for the better. These requirements can be fashioned into fully compliant specification statements very easily by tagging them with a paragraph number and paragraph title and adding English phrases and punctuation as covered in Part 2. Requirement 4 converted to a specification statement looks like this:

"3.2.5.7 *Payload separation acceleration.*   Payload separation acceleration shall be less than or equal to 5 Gs."

This completes the cycle of performance requirements analysis for this requirement: (1) functional analysis, (2) allocation of functions to architecture, (3) identification of attributes driven by these allocated functions, (4) establishing appropriate values for these attributes, (5) establishing traceability to the process (rationale statements), and (6) spinning these primitive requirements into full specification language. We will keep working away on this pattern in the remainder of this chapter with more examples.

3.4.11.9.2 Performance Requirements Analysis Example 2
Given that a Naval oilier must replenish U.S. Navy ships at sea, it must be able to string lines and pump fuel while safely maintaining course relative to and distance from the receiving ship. Let us assume that it has already been determined that the oilier must only be responsible for maintaining a steady heading and that the receiving ship is responsible for maintaining relative distance. Figure 3.4-10 reveals the results of the functional and performance requirements analysis. One performance requirement that falls out of this analysis is the following:

"3.2.5.1 *Course error.*   Own ship's course error shall be less than or equal to 0.5 degrees throughout underway replenishment operations."

3.4.11.9.3 Performance Requirements Analysis Example 3
In this example, we seek to protect an aircraft on the ground from attack by other aircraft. The best implementation of this function may be to never get caught in this situation, but that may have to be ruled out for operational reasons because it is not 100% effective. Figure 3.4-11 shows the corresponding RAS.

3.4.11.9.4 Performance Requirements Analysis Example 4
The pilot of a remotely piloted vehicle must be able to fly the craft without the sensation of flight using basic flight instruments and an XY plotter indicating geographical position. See Figure 3.4-12 for the RAS.

REQUIREMENTS ALLOCATION SHEET

FUNCTION IDENTIFICATION

FUNCTION ID:	F4783
FUNCTION STATEMENT:	REPLENISH SHIPS WITH FUEL
FUNCTION DESCRIPTION:	The oilier must come along side, string lines to secure the fueling hoses, and pump fuel through lines.
ENTRY CONDITION:	Oilier approaches target ship from aft at a relative speed of 5 knots at a distance of 1000 yards along track and 50 yards abeam.
EXIT CONDITION:	Disconnect last line and steer a diverging track.

ALLOCATION

ARCHITECTURE ITEM:	A1, Oilier

PERFORMANCE REQUIREMENTS IDENTIFICATION

RQMT 1 ATTRIBUTE:	Course Error.
RQMT 1 VALUE:	Less than or equal to 0.5 degrees.
RQMT 1 RATIONALE:	Engineering judgment.
RQMT 2 ATTRIBUTE:	Course Instability.
RQMT 2 VALUE:	Less than or equal to a mean of 0.5 degrees with a variance of TBD degrees and a rate less than TBD degrees per second.
RQMT 2 RATIONALE:	
RQMT 3 ATTRIBUTE:	
RQMT 3 VALUE:	
RQMT 3 RATIONALE:	

Figure 3.4-10   *RAS for example 2*

```
┌─────────────────────────────────────────────────────────────────────┐
│ REQUIREMENTS ALLOCATION SHEET                                         │
├─────────────────────────────────────────────────────────────────────┤
│ FUNCTION IDENTIFICATION                                              │
│                                                                      │
│ FUNCTION ID:              F47442                                     │
│ FUNCTION STATEMENT:       PROTECT AIRCRAFT FROM ATTACK               │
│ FUNCTION DESCRIPTION:     An aircraft on the ground which cannot get │
│                           airborne to fight or escape in the time    │
│                           available from enemy aircraft detection    │
│                           must be protected from the effects of      │
│                           known enemy weapons the attacking aircraft │
│                           may employ.                                │
│ ENTRY CONDITION:          Aircraft unpowered on ground with          │
│                           approaching enemy aircraft.                │
│ EXIT CONDITION:           Enemy aircraft leave the area.             │
├─────────────────────────────────────────────────────────────────────┤
│ ALLOCATION                                                           │
│                                                                      │
│ ARCHITECTURE ITEM:        A431, Bunker, Aircraft Protection          │
├─────────────────────────────────────────────────────────────────────┤
│ PERFORMANCE REQUIREMENTS IDENTIFICATION                             │
│                                                                      │
│ RQMT 1 ATTRIBUTE:         Overpressure.                             │
│ RQMT 1 VALUE:             TBD psi.                                   │
│ RQMT 1 RATIONALE:         The largest bomb expected to be employed   │
│                           by the enemy is their MK32 2000 pound bomb.│
│                                                                      │
│ RQMT 2 ATTRIBUTE:         Shrapnel and Gunfire Protection.           │
│ RQMT 2 VALUE:             Penetration denied for flying fragments    │
│                           from bomb blasts and gun-fired shells up   │
│                           to 20 mm.                                  │
│ RQMT 2 RATIONALE:         Largest guns on enemy attack aircraft are  │
│                           20 mm.                                     │
└─────────────────────────────────────────────────────────────────────┘
```

Figure 3.4-11 *RAS for example 3*

3.4.12 Performance Requirements Analysis Preceding Function Allocation

As noted previously, we have the option of completing the performance requirements analysis before or after allocation of the functionality. There are actually three entities that could be allocated: (1) function titles from the blocks, (2) functional requirements which in the author's view are technically complete requirements but not in complete sentence form, and (3) a performance requirement which is in the complete form for inclusion in the specification. If we had a function navigate to waypoint, we could simply allocate that function to a Guidance and Navigation set and subsequently decide how to phrase the complete statement and how accurately the navigation must be. Alternatively, we could determine how well the navigation must be accomplished before allocating the complete performance requirement.

Completing the performance requirements prior to allocation is generally a good idea because there tends to be a one-to-one correspondence between performance requirements and architecture but there may be a one-to-many relationship between the driving function and architecture entities.

3.4.13 RAS-Centered Requirements Analysis

In Chapter 3.10 we will take up a highly coordinated way to accomplish all of the requirements work such that every requirement, not just the performance requirements, is captured in the requirements analysis sheet. Performance requirements are derived from functions and allocated to product entities (architecture). But so are interface, environmental, and specialty engineering requirements treated with this same discipline relative to driving models and alloca-

tion to architecture. Requirements in each class are identified using a model and each requirement is identified with a model ID (MID). For each MID, one or more requirements are defined and allocated to a product entity. The requirement is linked to a selected template paragraph number for the product entity to which it is allocated. In this context, the RAS ties together the whole requirements analysis process.

3.4.14 Process Summary

The traditional structured analysis process has been explained in the context of functional flow diagramming with hierarchical functional analysis, process flow, and IDEF 0 alternatives for functional analysis as described in Chapter 3.4. As a result of applying this process, we should have a clear understanding about system functionality, what items the system shall consist of, the interfaces that must exist between the things in the system architecture, and the seeds from which performance requirements will spring in the form of allocated functions, functional requirements, or performance requirements.

Interfaces will exist in the new system as a function of relationships between function-architecture allocation pairs as all interfaces are predetermined by the way you choose to allocate functionality to the things in the system. Commonly, interfaces are identified in an unstructured fashion based on the experience of the person or persons doing the functional analysis and architecture definition. These interfaces that are identified ideally would be coupled to function allocations to the interface terminal items. That is, a case can be made for encouraging interface traceability to pairs of lines in the requirements analysis sheet

REQUIREMENTS ALLOCATION SHEET	
FUNCTION IDENTIFICATION	
FUNCTION ID:	F4754
FUNCTION STATEMENT:	PILOT CRAFT TO MISSION AREA
FUNCTION DESCRIPTION:	The remote pilot monitors remote flight instruments and geographical position on an XY plotter. Preprogrammed events are observed to occur on time and maneuvers at the planned points. Corrective action is taken if errors threaten to degrade mission results.
ENTRY CONDITION:	Unmanned aircraft arrives within radar range of controller as indicated by carrier light and beacon decode.
EXIT CONDITION:	Aircraft is handed off to another station or commanded into recovery
ALLOCATION	
ARCHITECTURE ITEM:	A742, Remote Pilot
PERFORMANCE REQUIREMENTS IDENTIFICATION	
RQMT 1 ATTRIBUTE:	Monitor flight vehicle altitude, airspeed, roll, pitch, and engine throttle setting.
RQMT 1 VALUE:	Once per minute for 10 seconds total.
RQMT 1 RATIONALE:	These parameters provide a basic indication of health. Other activities will force the controller to focus on them for a total of 50 seconds per minute.
RQMT 2 ATTRIBUTE:	Monitor flight vehicle position for planned track and mission timing.
RQMT 2 VALUE:	Detect error equal to or greater than 1 nautical mile cross track or along track.
RQMT 2 RATIONALE:	The radar plotting system will not permit more accurate detection. Errors up to 3 miles are acceptable before corrective action is necessary.

Figure 3.4-12   *RAS for example 4*

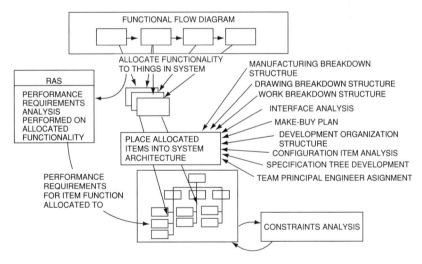

Figure 3.4-13   *Functional analysis summary*

(RAS). Chapter 3.6 offers a structured pair-wise analysis procedure for linking function allocation pairs to interfaces.

Figure 3.4-13 provides an overall view of the process discussed in this chapter. We first defined the needed functionality using a functional flow diagram. We allocated this exposed functionality to things via a RAS that did two things. It identified the element responsible for implementing the function and defined one or more performance requirements for that item based on the allocation. We next synthesized all of the allocations into an organized hierarchical structure of the product called an architecture. This architecture was built by a team recognizing the needs of several different specialized views.

The remaining problem from a requirements perspective is that we need to identify the design constraints in the form of interface, environmental, and specialty engineering requirements. It is important to note that we can define the performance requirements for an item in the functional plane before we know what item is going to perform it but the design constraints must be identified on the physical plane. We have to know what the things are before we can define their constraints.

3.5

Architecture Synthesis

Contents

3.5.1 Introduction to Architecture

The word *architecture* takes on many meanings. The author uses the word to cover the physical entities of which the system consists piled up in a hierarchical breakdown diagram. Others picture architecture in a more dynamic sense including the physical things in the system as well as a scenario view of how those things will be used. Some system engineers refer to the functional or logical and the physical architecture. Department of Defense Architecture Framework (DOD AF) entails 26 different diagrams. So, when discussing this matter with others, particularly people from the customer's ranks, try to reach an understanding of the other person's concept of architecture early in the conversation.

As the structured analysis process proceeds, it generates allocations of the needed functionality of the system, or performance requirements derived from those functions, to specific physical entities through trade studies, an appeal to historical precedent, good engineering judgment, and respect for customer direction to use specific resources in the system. Government furnished property is an example of the latter when dealing with the government, but any customer may have residual property from other systems they wish to make continued use of in a new or updated system.

The engineering community, in concert with the production and logistics communities must decide how to organize these allocations into families of things. We must map the evolving architecture to: (1) organizations responsible for the development of each item, (2) the evolving planned manufacturing process flow, (3) the customer's interests in managing the development process through configuration or end items, (4) make or buy considerations, (5) specification needs, and (6) the customer's work breakdown structure (WBS) that will be used to organize the cost and schedule performance of the contractor.

It is possible that in the process of allocating system functionality to system architecture elements that resultant interface relationships are not optimum only to be discovered later in the analytical process. The system engineering community must continually evaluate the evolving architecture for optimum interface relationships because most system development problems occur at the interfaces where functionality differences coincide with organizational responsibility differences. Optimum interface is defined here as that interface condition characterized by minimized interface complexity between elements under development responsibility by different design agents.

The results of this interface analysis must be fed back into the architecture definition process to close the loop on this system optimization activity. It is possible in the process that a need for many other changes will be triggered, but this is a part of the iteration, or churning activity, that must occur as we become smarter about this system we are creating. We proceed from the simple to the complex, from the general to the specific. On the route we may find that ideas we had when less knowledgeable were not good ones. We must have the courage to change those ideas as early as we possibly can since it is much less costly to do so early in a program than later.

On a small system one system engineer can do the whole job of functional analysis as well as architecture and interface development. On larger systems, a system team should perform the whole process or at least the system-level work down to include to top-level architectural entities that will

be the responsibility of IPPT formed to develop the design. Fewer allocation mistakes of an interface nature will occur if the team doing the work is sensitized to the truth about development problems occurring at the development organization's interfaces where they coincide with product system interfaces. Simplify those interfaces and you will build a more successful system at lower cost. The most powerful way to simplify those relationships is to form the team about the product system architecture as cross-functional teams physically collocated on the program. This encourages that the product interfaces align perfectly with the human communication paths that must be exercised between the teams.

Figure 3.5-1 illustrates the process we will be discussing in this chapter. The discussion opens with architecture block diagramming followed up with mapping this architecture to several important hierarchical aspects of the system: customer configuration items, work breakdown structure, and specifications.

Configuration items are identified by the customer as the major elements through which they would prefer to manage a development program. In early program phases the customer may ask the contractor to supply a candidate list of configuration items. A sound approach for configuration item list building is to draw a line across the architecture at some level of indenture, not necessarily at the same level in every branch, and select items on that line as configuration items. Make sure that everything below the line is in one of the items that will be a configuration item so that one can say honestly that everything in the system is being managed through the configuration items.

The Work Breakdown Structure (WBS) is a cost-oriented view of system structure defined for the Department of Defense at one time in MIL-STD-881A. The specification tree defines the elements in the ABD that require formal documentation of their requirements in a customer specification format, procurement specification format, or in-house format.

In Chapter 3.6, we will return to the most important relationship of all, interface analysis. We will use a tool called N-Square diagramming to gain insight into optimum interface relationships among the evolving architecture arrangements and schematic block diagramming to define the interface solutions. Chapter 3.10 shows how to tie all of the requirements analysis work into a single integrated analytical model and in particular how interface analysis really should work. Finally, in Chapter 3.11, we will discuss how to collect all of the work products of the tasks depicted on Figure 3.5-1 and others into a single system-defining document during the early program phases.

3.5.2 Architecture Block Diagramming

We seek first to build a process for creating a system architecture block diagram (ABD) that depicts all of the elements of a system arranged in a physically oriented hierarchical tree, to map the elements on that tree to the principal design agent, and to provide a framework for linking all development work focused on the system elements to a central, approved understanding about system composition. We are concerned in this endeavor in translating the results of the functional analysis and performance requirements work that resulted in the derivation of performance requirements from those functions into a recognition of the most appropriate physical entities to accomplish the indicated

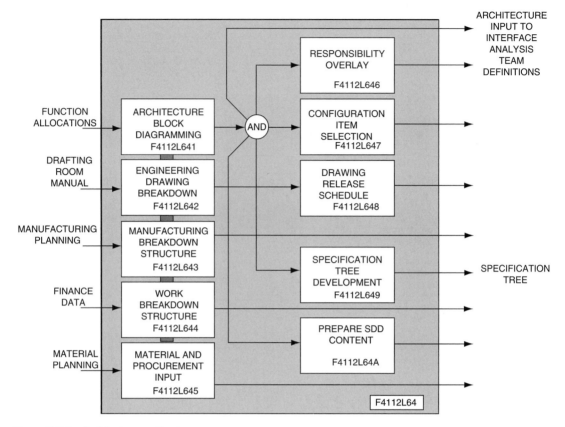

Figure 3.5-1   *Architecture synthesis process*

functionality through allocation of those performance requirements to product architecture entities. The requirements analysis sheet (RAS) started in functional analysis and extended in performance requirements analysis, will be used to capture these relationships. We will then synthesize the evolving list of architecture entities in the RAS into an ABD to illustrate the resultant architecture. Finally, we may build a tabular listing of all of the architecture elements illustrated on the ABD called an architecture dictionary.

We should keep in mind that the RAS is also going to be used by the principal engineer responsible for the item we place on the architecture diagram as an insight into performance requirements for that item. As a result of the architecture synthesis activity we will define who that principal engineer is. This architecture diagramming step is at the very core of the system engineering process because it brings together at one point the functionality of the system, the physical organization of the system to be, and the organizational responsibilities for developing the elements.

We will create two principal products in this activity: an architecture block diagram and an architecture dictionary. The ABD defines system composition. Everything that is part of the system is illustrated in its family relationships. It uses simple blocks and interconnecting lines to define the existence of elements and their relationships. The Architecture Dictionary is a tabular listing of system architecture elements illustrated on the ABD (one tabular line

item for each block on the ABD) containing the element name, a description of the element, and other reference information too extensive to place on the face of the ABD. This information includes the name of the principal engineer and/or the responsible design agent and/or their department or team.

3.5.3 Diagramming Fundamentals

Our Architecture Block Diagram (ABD) is formed from simple boxes, connecting lines, some text, and a minimum of other symbols. This diagram is intended to visually convey the ordered arrangement of all of the parts or sub-elements of a system and is thus purposely not cluttered with extraneous data however useful it may be for specific purposes. The diagram provides the eye and mind with system structure information pertaining to how the system is put together to the end that the details about specific system sub-elements will fit into that mental framework or pattern and result in greater understanding. Figure 3.5-2 illustrates a typical ABD that happens to be for the Titan-Centaur System, a large expendable space transport vehicle, built by Martin Marietta at the time using a General Dynamics Centaur upper stage.

While employed by Ryan Aeronautical, the author encountered difficulty trying to release system drawings. He found out that the drawing number structure included a

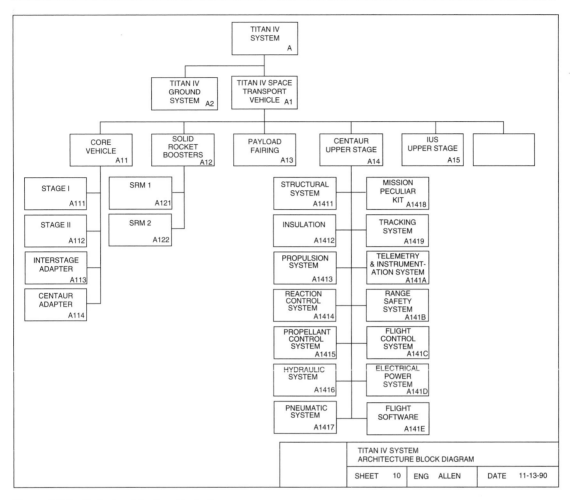

Figure 3.5-2 *Typical architecture block diagram*

letter that was linked to functional departments. System Engineering did not exist when the letters were handed out but the author discovered that there was one letter not assigned. The next time he tried to release one of these drawings he included this letter and the release clerk asked what department letter "D" was for. The author replied it was for system engineering. The clerk responded, "Oh yeah," and released the drawing. One can get a lot of good system engineering done if not hindered by rules. Figure 3.5-2 appears as an engineering drawing from this heritage but the author has since concluded that these drawings would be better documented in a system definition document (SDD) explained in detail in Chapter 3.11.

On programs which begin with little but a customer need statement, this diagram is constructed from the top down in step with the evolving system definition derived from performance requirements identification and allocation. The structured allocation of system functions or performance requirements to particular sub-elements in the ABD is then followed by identification of design constraints and a synthesis of all these requirements collected together for a par-

ticular item to form one or more design concepts. If there is more than one potentially useful concept developed, a trade study is performed to select the preferred concept from the alternatives. The approved preferred concepts in combination with the approved requirements provide the information needed to begin preliminary design work on an item in the architecture.

3.5.4 Architecture Element Coding

A system is constructed of objects arranged in a hierarchical structure based on useful ways of partitioning the gross system function (stated in the customer's needs statement or system specification) into sub-functions that can be accomplished by product entities upon which development work will be performed by a limited and manageable set of disciplines on a program team. Each sub-element of the product system has associated with it a further sub-structure and so on down to the elemental system parts. This pattern follows what is called a breakdown structure where each object is composed of two or more identifiable entities.

This pattern can easily be represented by a conventional block diagram where subordinate elements are depicted connected to the parent by lines, as in drawing breakdown structures, and this is the pattern followed here. This is quite adequate for pictorial display but does not satisfy the needs of computer listing of the system objects. A computer listing of the architecture is needed to support the management of the evolving architecture, its requirements and concept development. The tabular document to satisfy this need will be referred to here as an architecture dictionary. Therefore, some method of coding the system objects is required to facilitate computer sorting and indexing.

This breakdown pattern can be represented by a place value system of coding where each place is associated with a particular level in the block diagram and each unique symbol in a particular place is associated with a branch from its parent block. The product architecture coding symbols used here are formed by the ten Arabic digits 0 through 9 (the character zero is not normally used), 25 upper case English alphabet letters (the letter "O" omitted to avoid confusion with the number 0), and 25 lower case English alphabet letters (the letter "l" omitted to avoid confusion with the number 1). This set forms a base 60 system that is more than adequate so long as any one architecture branch contains no more than 59 elements at the same level.

As an alternative, some customers prefer a decimal delimited structure composed of only numbers in the place values delimited by periods. When more than nine elements exist on any one branch, two digit numbers are used. Periods separate each architecture level in the code. The difficulty with this arrangement is realized when trying to use a simple microcomputer database to generate the architecture dictionary. The normal ASCII sorting/indexing priority results in faulted ordering of the architecture elements. There are other computerization benefits derived from each architecture level being defined by a single character in the code structure as well.

The codes formed from the characters identified above are referred to here as the Product ID, or PID. This same basic coding structure is also used in this book for functions, interface elements, WBS, and other hierarchical system entities. The top-level system architectural entity, the system, is identified with the Product ID "A." It is sometimes useful to identify cooperative systems with other first letters such as "B," "C," and so forth. The A Product IDs are assigned to subordinate elements in any order selected by the system engineer responsible for the architecture synthesis process, but generally the most important or largest should be identified with a "1."

During the early phases of system development, particular sub-elements may survive for considerable periods of time in an indeterminate status where several alternative configurations or concepts are being pursued in a trade study or parallel development (for risk reduction). These alternative concepts or configurations can be identified in the Product ID notation through the use of parenthetically included characters. For example, the ID A12(1), A12(2), and A12(3) identify three possible alternative syntheses of element A12 requirements.

3.5.5 Sheet Cross-Referencing

Movement between ABD sheets is facilitated by cross-referencing notation at the upper left or lower left corner of a block. A number at the top left corner identifies the ABD sheet number where the block appears integrated with its parent block. A number at the lower left corner of a block identifies the ABD sheet where the block is exploded. The author also uses a bold angle at the lower left corner of a block to show that the block so designated is expanded on another sheet given by a sheet number next to the angle.

Where one single structure cannot be continued on a single sheet, triangular symbols may be used with contained cross-reference codes to note continuation on another sheet. It is not necessary that the codes be the same on both sheets and, in fact, sometimes cannot be. The continuation reference note included with each triangle symbol tells what sheet and reference symbol on another sheet provides the extension. For example, the cross-reference connection of a triangle 5 on sheet 6 and a triangle 2 on sheet 3 is accomplished by a note next to the triangle 5 on sheet 6 as follows: "SH 3-2." At the triangle 2 on sheet 3, a note reading "SH 6-5" would appear.

3.5.6 Alternative Organizational Structures

The ABD is structured primarily from a functional perspective in terms of systems, subsystems and components required to satisfy functions depicted on the system functional flow diagram. As the system definition matures, it is necessary to overlay physical and cost organizational perspectives onto this functional view. The overlaying structure is implemented on the diagrams by enclosing related objects within dashed lines. Overlay structures should not be carried any lower than necessary to expose and resolve structural conflicts.

This can be illustrated in the case of the physical perspective, by dashed lines enclosing system objects (depicted by blocks) that collectively belong to particular contractor-defined end item description (EID) structures (or whatever a particular company uses as the basis of an engineering drawing structure), with notation within the bounded area identifying the EID. Ideally, Drawing Breakdown Structures (DBS) would be compatible with Manufacturing Breakdown Structures (MBS) that partition the system into sub-elements that will be manufactured in coordination with a particular facilitization of the production process. One other kind of physical partitioning scheme is the customer-directed Configuration Item (CI) structure that identifies particular system sub-elements that will be managed with high visibility. It is important that every effort be made to cause alignment between architecture codes and the WBS, using the WBS as the architecture model wherever possible.

The WBS is frequently developed early on a program, in accordance with a customer's cost structure, when only the functional perspective is sufficiently mature to provide an organizational basis. The EID and DBS are generally structured along the lines of the developing contractor's organization. As a result, the physical structure may cross the WBS, where the division's organization is in conflict with the customer's preferred cost organization of the system. The crossing of these structures should be minimized as the system architecture matures. The ideal arrangement would be to use the ABD hierarchy to refine or replace the generic customer WBS. Unfortunately, the generic WBS has, in the past, been respected more vigorously than good sense dictates resulting in a mismatch between the ABD oriented toward the contractor organization structure and the generic WBS. The reason for this is generally two-fold:

(1) inflexible computerized cost models cannot be easily adjusted to a different structure, and (2) the WBS does line up with the way the customer is organized and wishes to manage the program. Where possible, the ABD and WBS should use exactly the same structure and ID codes. Where this cannot be achieved for reasons indicated above, the system engineer must map the ABD to the WBS in the Architecture Dictionary.

3.5.7 Implementation Notes and Responsibility

Given that the developing company has a system engineering organization, it makes a lot of sense for engineers from that organization to be made responsible for the overall architecture synthesis process and the architecture block diagramming in particular. System engineering should at least be held responsible for the architecture down to and including the level where there will be multiple design organizations involved. In a program organization structure employing functional department teams, once the decomposition process has arrived at a level on a branch that a single design group is responsible, engineering management could require those design groups to expand the architecture beneath that level within their own domains down to the component level. If the lower-tier responsibilities are distributed, system engineering should collect the results into a system ABD, audit the interface implications derived from the design group conclusions, and ensure a common format, style, and symbol set. A much more satisfactory way to establish architecture responsibilities is by organizing cross-functional teams coordinated with the product architecture.

Every system may be illustrated on an ABD as a single block with the system name contained within the block with an Architecture ID of "A." The system name will generally be given by the customer, but it will not always be the only name used to identify the system. It is important that the contractor team settle on a single name for the system. The final authority in this matter within the program is the Program Manager, but system engineers should work through the Chief Engineer or program integration team (PIT) leader to get the name firmly resolved.

Where alternative systems at the top level are under consideration, they may or may not all have the same name depending on their character. Each should be designated by a different Architecture ID parenthetical identifier as covered earlier. At the system level, the System Architecture may alternatively use different letters for the different alternatives, that is "A" for the principal alternative and "B" and "C" for others.

Based on the anticipated system complexity, identify the hierarchy levels that will be needed in the ABD. Figure 3.5-3 identifies and names lower-tier levels that may be used in very large systems. In smaller systems all of these levels may not be necessary. In very large systems, it may be necessary to identify multiple levels within the levels listed.

Figure 3.5-3 addresses the hardware entities that populate a system. A system can be composed completely of software as can its segments and elements. More often a large system is composed of a mixture of hardware, software, and people accomplishing procedures on the hardware and software. At the bottom end software is composed of lines of code corresponding to hardware parts like resistors. These lines of code may be grouped into software entities in any number of ways all the way up to the system level.

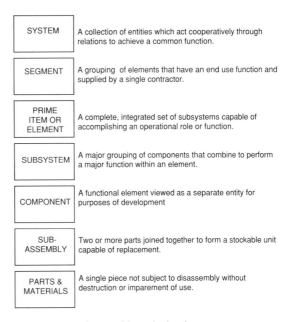

SYSTEM	A collection of entities which act cooperatively through relations to achieve a common function.
SEGMENT	A grouping of elements that have an end use function and supplied by a single contractor.
PRIME ITEM OR ELEMENT	A complete, integrated set of subsystems capable of accomplishing an operational role or function.
SUBSYSTEM	A major grouping of components that combine to perform a major function within an element.
COMPONENT	A functional element viewed as a separate entity for purposes of development
SUB-ASSEMBLY	Two or more parts joined together to form a stockable unit capable of replacement.
PARTS & MATERIALS	A single piece not subject to disassembly without destruction or imparement of use.

Figure 3.5-3   *System hierarchy level names*

The system engineer with ABD responsibility should coordinate with the Program Office and Chief Engineer/PIT Leader for insights into planned teaming and associate possibilities in the early phases such that architectural entities may be set up to simplify interfaces at corresponding points. Interface simplification is an important underlying goal in the architecture definition, actually.

Coordinate the developing architecture with the design groups for compatibility with their Drawing Breakdown Structure and End Item Description needs and with manufacturing for their Manufacturing Breakdown Structure needs. Coordinate the developing architecture with Economic Analysis for compatibility with the Work Breakdown Structure and their work on life-cycle cost. Coordinate with the Make-Buy Committee or the Program Office for procurement decisions and Logistics Engineering for government or customer furnished property. A particular program may not include these organizational functions and some or all of these functions may be satisfied by the program office.

The ABD, and the system development decomposition stroke, is complete when all of the elements at the lower fringe of every branch: (1) may be developed by a single design department or project team coordinated by a Principal Engineer, (2) will be purchased from a supplier, or (3) will be supplied by the customer or an associate.

3.5.8 Architecture Crossing Conditions

The previous discussion assumes that we will choose to organize the architecture around the physical entities into which the system parts will be packaged. That is the author's first preference but there are other possibilities that appear more favorable in particular situations. For example, if the system is a new petrochemical plant or space launch site, it might be advantageous to organize the whole or some part of the architecture around the functional subsystems included. In a petrochemical plant the gasoline system may be plumbed through several physical plant structures. Rather than organize a space launch complex by pad, propellants, and blockhouse, we might choose to organize it into the several functional subsystems like liquid oxygen, liquid hydrogen, electrical control, and structures.

A large space launch vehicle like a Lockheed Martin Atlas vehicle consists of several major end items including the primary stage with various names such as booster or first stage. This stage may include several physical elements manufactured in different facilities but they contain parts of functional subsystems supplied by vendors spread across two or more physical entities.

The reality is that all systems include within their physical architecture some mix of functional subsystems interacting within some number of physically packaged entities. It is seldom possible to focus all of the physical architectural organization on functional subsystems or physical packaging. Thus, we will commonly have to accept that there will be crossing conditions between the overlays of the architecture. This is especially true when we consider all of the overlays addressed in paragraph 3.5.6. But, the question we must answer as a priority is which perspective will be used to primarily depict the architecture. The answer is often a compromise and properly so.

The author would recommend that the physical structure (packaging) be selected as a first attempt with some consid-

eration for laying out the physical architecture such that functional subsystem content crossover can be overlaid laterally across the physical branches.

Ideally, all of the different overlay perspectives would respect the same architectural structure selected for the definition of the system. In this case there are no crossing conditions between different overlays, which tend to cause development concerns. The Lockheed Martin Atlas Centaur upper stage offers an example of crossing conditions as well as several other interesting architectural relationships. Figure 3.5-4 illustrates the architecture of this vehicle that is an upper stage for the Atlas and Titan space launch vehicles used in the 1990s to transport payloads from Earth surface to low Earth orbit and/or beyond into geosynchronous orbit or into deep space.

The reader will note that in Figure 3.5-4 the lower-tier architectural entities are functional subsystems with WBS numbers, that is, the system architecture was laid out originally from a functional perspective. In fact, this pattern went all the way back to the development of the USAF/General Dynamics Astronautics Atlas ICBM in the 1950s. The original contract for that system included a functionally defined architecture aligned with the program WBS. The whole division was spun off from Convair and functionally organized around this customer driven WBS arrangement. When the Centaur upper stage was developed years later for a space transport application of the Atlas

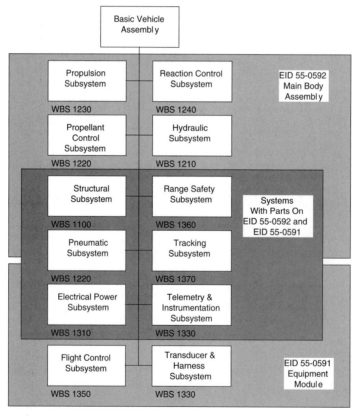

Figure 3.5-4 *Centaur upper-stage architecture*

rocket, it was natural to employ essentially the same WBS and functional subsystem pattern.

The reader will note that there are several subsystems that have all of their parts on one of the two major physical entities of the vehicle: (1) a main body end item containing the liquid hydrogen and liquid oxygen tanks and (2) an equipment module end item upon which was mounted the avionics equipment. Six subsystems included parts that were physically located on both physical entities. Thus it was not perfectly clear how the engineering drawing tree related to the WBS structure. At least one engineering group, all of which were organized around the functionally organized WBS elements, developed a top drawing for their subsystem, which included three blocks. The top block was for the subsystem and each of the two lower blocks identified a drawing number covering the parts located on one of the two end items.

The design team for a new warship concluded early in its development that it would depart from traditional ship construction relative to installing electronic equipment aboard during construction by pre-packaging the equipment in large steel modules that could be stuffed with equipment in a factory rather than at the ship yard piece by piece into the various compartments. The architecture therefore had to be viewed from the perspective of functionally organized segments, elements, and components as well as how this equipment was to be packaged in these physical modules. There were many considerations in making the packaging decisions, of course. It was not necessarily desirable to locate all of the radio equipment in one module, for example, because one hostile hit could disable all or a large part of the radio capability of the ship. There were also interests in balancing the mass above the waterline, power consumption, and other aspects of ship design. Figure 3.5-5 shows the architecture

packaging rules adopted. All of the elements in a segment did not have to be located in the same module. All of the components of an element did not have to be located in the same module. But, all of the parts of a component had to be in the same module.

A third correlation between architecture decisions and physical packaging considerations is observed in what some car manufacturers did to reach a goal of not having to replace any engine parts in the first 100,000 miles. They discovered that all of the parts could satisfy the goal except the high tension wires for spark plugs. The ignition system was re-designed to replace the single coil (that created the high tension jolt passed on to the distributor) with a small encapsulated coil at the top of each spark plug. The high tension circuit was shortened from several inches for each plug to zero and all of the wires leading to the spark plugs could be crafted in low tension wiring that would easily last for 100,000 miles.

There is a powerful connection between system architecting, interface definition, and procurement. Witness the automaker's movement to purchasing complete brake systems from suppliers rather than the parts to assemble them from. The final assembling company must be careful in making these decisions to avoid out-sourcing key technical disciplines that they must retain, however. Very often none of the available options are optimal in these cases.

3.5.9 Reversing Traditional Structured Analysis

As discussed in Chapter 3.1, traditional structured analysis works well following Sullivan's encouragement (form follows function) when the problem being attacked is relatively unprecedented but it is not necessarily the right approach for solving a heavily precedented problem such as major

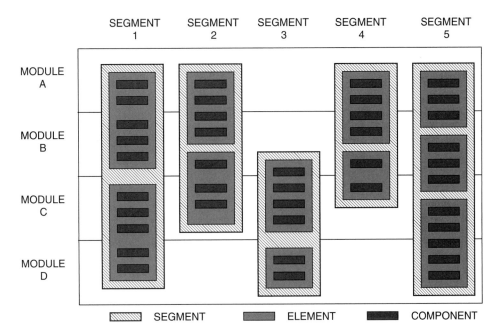

Figure 3.5-5   *Warship packaging architecture*

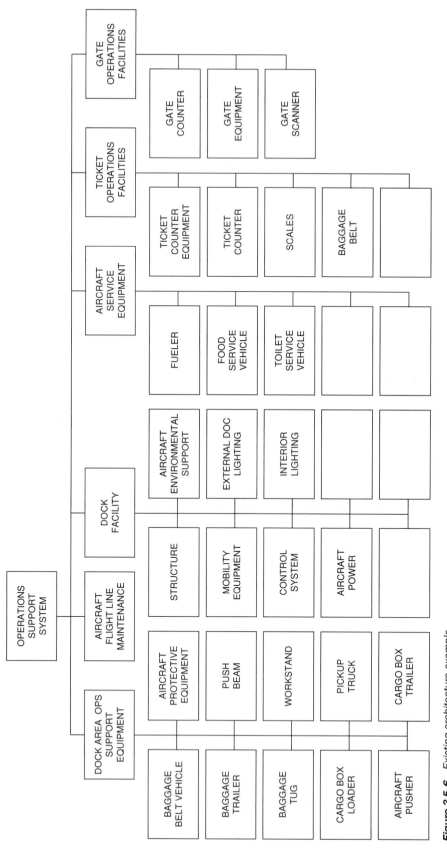

Figure 3.5-6 *Existing architecture example*

modification of an existing system. In this case there is likely an existing system that can be observed operating in its intended environment. The system has an established architecture and there is a complete description of the engineering and manufacturing baseline in existence.

The traditional structured analysis process can be revered in this case entering the problem space on the architecture plane with the existing architecture. If there is no existing architecture diagram, it can be drawn from the engineering drawing structure or from observing the system functioning in the field. Then one can evaluate the existing architecture against the functionality differential between that which the system was built to satisfy and that which is now required. Figure 3.5-6 shows the architecture for an airline terminal interface at a gate created by the author from too many boring hours sitting at terminal gates

waiting for a flight. The reader can check it out on his or her next flight.

Given that we must make a significant change in the existing system, the question is, How shall the functional differential be buttered onto the existing architecture? The existing elements of the architecture will end up being identified as okay as is, deleted, or modified and one or more new items may be added. The new and modified elements can be subjected to structured analysis.

Generally, one cannot expect to find the work products from the structured analysis work performed, if it was, from a previous development life. So, this work will commonly have to proceed without that resource, another reminder that one should have a way of capturing the structured analysis work products. One way to do this is explored in Chapter 3.11.

3.6

Interface
Identification
and Definition

Contents

3.6.1 Introduction to Interface Analysis

Systems consist of entities, referred to as elements of the system architecture in the previous chapter, and relations between those entities. The relations between the entities are referred to as interfaces. Interface analysis is one of the methods by which we arrive at an optimum architecture for a system before we commit to design it. We define precisely where the interface planes are in the system, who is responsible for them, and ensure that the teams and principal engineers responsible for the interface terminals and the related media all have precisely the same understanding of the requirements and design concept for each one. In the process we may affect changes to the system architecture derived from allocation of system functionality. The desired goal is to refine the match between how the design teams are organized and how the product system is organized in order to simplify the system development process and assure we have accounted for all needed system functionality. Systems with the simplest interfaces relative to system design responsibilities result in programs with the fewest risks.

This book encourages the use of n-Square diagramming techniques augmented by specialized analytical techniques corresponding to the several engineering disciplines (avionics, fluid dynamics, and so forth) to determine the optimum interface relations between system elements. Generally, the analyst seeks to maximize the capacity for interaction between system elements while minimizing the system need to interact. Generally, this results in minimizing the number and complexity of interfaces.

Given that we have identified an optimum interface condition within the system through interface analysis, we need to communicate the results of that analysis very effectively to all development team personnel to ensure they are all working to this model across their own portions of the system. Schematic block diagramming and interface dictionaries can be very effective in satisfying this need. Schematic block diagramming is a graphical methodology for reporting upon the preferred interfaces defined through a separate but closely related process referred to here as interface analysis. It should be pointed out that n-Square diagramming, schematic block diagramming, and interface dictionaries all communicate the same message in different ways and one could economize by only using one of them.

The author prefers to collect all system and hardware item requirements into two subsets: (1) performance requirements (dealt with in Chapter 3.4) that define what an item must do and how well it must do it, and design constraints. The latter come in three varieties: interface requirements, environmental requirements, and specialty engineering requirements. This chapter opens the door on the study of interface constraints. Chapter 3.7 continues with specialty engineering requirements and Chapter 3.8 completes the trio with environmental requirements analysis. The reason the author uses these categories is that the work and special models effective in identifying requirements in these categories align with them. Interface constraints analysis quantifies the interface relations identified through interface analysis and results in clear interface requirements statements for inclusion in specifications and interface control documents.

3.6.1.1 Interface defined

An interface is a plane or place at which independent systems or components thereof meet and act or communicate with each other. An interface is characterized by two terminals, each touching one element in the system architecture or environment, and a media of communication. An interface is completed between these terminals via an interface media such as physical contact, electrical signals in wiring, fluid flow in plumbing, or a radio signal in space. The interface media is provided by either an element of the system architecture or the system environment. The interface is not the media itself, rather the functionality facilitated by the media during the development period when there is no physical reality yet established. The need for particular interfaces is pre-determined by the way we allocate functions or their derived performance requirements to architectural entities that are joined by the interfaces. Interfaces extend the functionality of one item across those interfaces to another. It is through the interfaces that the entities of which a system is composed enter into synergistic behavior.

It is well known that the most difficult development problems often occur at a particular kind of interface the author refers to as cross-organizational (two different organizations responsible for the two terminals). It would be grand if we could simply build systems with no interfaces and avoid these development difficulties but, then, we would have to remember that a system is composed of two or more things that interact to achieve a common purpose and the interaction is via the system interfaces. We must learn how to develop the interfaces well because they are an unavoidable part of every system.

As the physical design matures, we must define the interfaces in terms of physical measurements. Thus we have exactly the same situation in interface development as item development. First we identify the development requirements in a design independent fashion and then define the product requirements in a design dependent fashion. The first set is the basis for design and verification and the second set is the basis for acceptance and technical data.

It is common practice to associate an interface with one of several types: functional (or logical), physical, fluid, electrical, electromagnetic (optics, radio, radar, UV, and infrared), or environmental. A physical interface involves the form and fit of mating parts. Examples of physical interfaces include: mounting bolt patterns, drive shaft flange connections, mating wire harness connector physical attachment, and the tires of a fighter plane resting on the apron. As a space launch vehicle rises off the launch pad, a large volume of liquid oxygen from a rocket propellant system flows via a duct to the engine. Another fluid interface is observed in a hydraulic braking system.

Some examples of functional interfaces are: a 28VDC signal passing from a solenoid driver in an on-board computer I/O to a valve solenoid in a pressurization control unit and a digital data stream flowing from an instrumentation control unit to a flight data recorder input port. As a function of being connected by a wire harness, an aircraft on-board computer may send a command signal to the aileron actuator of an aircraft that causes the aileron to move interacting with the air mass of the environment, in relative motion with respect to the aircraft, to roll and turn the aircraft such that the command signal is nulled out in the on-board computer as a function of the guidance set detecting approach to the direction commanded.

Electrical power must commonly be distributed throughout the product via wires that connect the sources to the loads. Electromagnetic relationships provide a wide variety of functional extensions involving light, radio waves, heat, and x-ray.

Environmental interfaces can exist between two items when the natural environment communicates environmental stresses between two items. A reconnaissance camera company once argued that one of their cameras could take excellent pictures through the engine exhaust of an unmanned reconnaissance aircraft. If this were true it would have made it possible to outfit a very inexpensive unmanned photo reconnaissance bird. Unfortunately, no one questioned the environmental interface between the jet engine exhaust and the camera system before considerable money had been spent to prove the existence of an irreconcilable environmental interface incompatibility. During the first ground engine run the problem was obvious to most while looking through the exhaust stream at the blurry buildings in the background. During the engine run, however, the camera marketeer was still trying to convince the Air Force program manager that it really was feasible.

3.6.1.2 The interface dilemma

Included in the definition of a system is the existence of two or more elements of a system that cooperate via interfaces. It is through these interfaces that a system attains its superiority over an unorganized collection of things. We must conclude that a system must have some interfaces. The richer the interface complexity, the greater the potential for synergism between the components.

We know from Chapter 3.3 that we must decompose a system functionality into system elements that can be designed by teams of specialists. We commonly organize our engineering departments according to how we apply these specialized engineers to designing elements in our company's product line. It is a fact that the majority of system problems will occur at the planes in a system where different specialized organizations are responsible for the opposite terminals of an interface. The way to minimize these problems is to minimize the number of such interfaces.

We must have interfaces between elements designed by different agents and we desire to minimize these interfaces. This is perhaps the most important task of a system engineering community, to control the development of the interfaces in a system to satisfy these conflicting demands— system richness versus development simplicity. Obviously, the solution will be a compromise between two extremes, a rich interface in a system that will drive you to the poor house for crazy people on one extreme and an uncoordinated collection of independent things that do not form a system on the other.

3.6.1.3 The solution

The principal technique for arriving at a reasonable solution to the interface dilemma is careful development of the architecture to respect the known development team organizational boundaries. We allocate the functionality and aggregate the architecture elements so as to minimize the need for cross-organizational interfaces. The program team is in control of two levers to steer the process. First, the team can control how they are organized relative to the architecture and second, they can control how the functionality is allocated to the architecture. It is how we allocate functionality to architecture that predetermines what interfaces must exist and this is a key element in identifying the needs for interfaces between the system elements.

As the architecture is created, we continue to analyze its expanding interface implications within the system itself and between the elements of the design team in order to assure that we remain in a condition of optimum interface definition. An effective tool for this purpose is the n-Square diagram. An n-Square diagram is a square matrix with size "n."

The diagonal corresponds to internal interfaces for each item of architecture so we can simply identify the items down the diagonal. The other squares correspond to interface possibilities between the elements. Figure 3.6-1 illustrates such a diagram. The squares are marked with interface names where there is an interface between the elements identified on the diagonal. You will note that there are two squares, one above and one below the diagonal, for each pair of elements. We choose the cells above the diagonal for interfaces that have their source in a block on the diagonal above the destination block and vice versa for the other side of the diagonal. There is nothing wrong with making the opposite selection. Just be sure that you use the same convention in any one analysis. Mark the matrix with a directional arrow like that at the top right corner of Figure 3.6-1 showing in this case clockwise rotation.

As we can see in Figure 3.6-1, there is an interface requirement between the actuators that are a part of the hydraulic system and the DCU (on-board Digital Computer Unit) that develops the command signal. There are also interfaces between the actuator and the hydraulic control unit that provide hydraulic power to the actuator in response to the control exercised by the DCU. We can use this diagram to evaluate the functional allocation choices we have made against the organizational responsibilities.

Let us assume that each of the subsystems we have identified on the diagram is the responsibility of a separate design department or engineering team. Therefore, the shaded squares on the diagonal represent interface situations between system elements within a design group responsibility. Any interfaces identified outside of these shaded blocks represents a cross-organizational interface of the kind we wish to minimize but the kind that makes our system a system.

As one means of reducing cross-organizational interface, we can reassign an architecture element from one subsystem to another. We may also change an existing interface condition by re-visiting the functional allocations. For example, we could conclude in the case of the actuator that we should have allocated the steering function to a separate steering propulsion system that does not require actuators, only exhaust jets that are turned on and off under the control of the on-board computer. In the process we would eliminate the hydraulic system if it was not required for any other function. Finally, we could change the design concept from a hydraulic actuator solution to an electrically driven actuator solution. Now the complete control system is in the avionics design domain, the computer and the actuator. We may have removed the need for a hydraulic system as well. These three approaches are effective in refining the system architecture to simplify system architecture while assuring that we are satisfying the needed system functionality.

3.6.2 Interface Identification

3.6.2.1 Intuitive interface identification

The first task in interface analysis is to identify the interfaces that are necessary in a system. The author conducted an unscientific survey of how some experienced system engineers felt they identified interfaces in developing systems while at the 2003 INCOSE Symposium in Crystal City

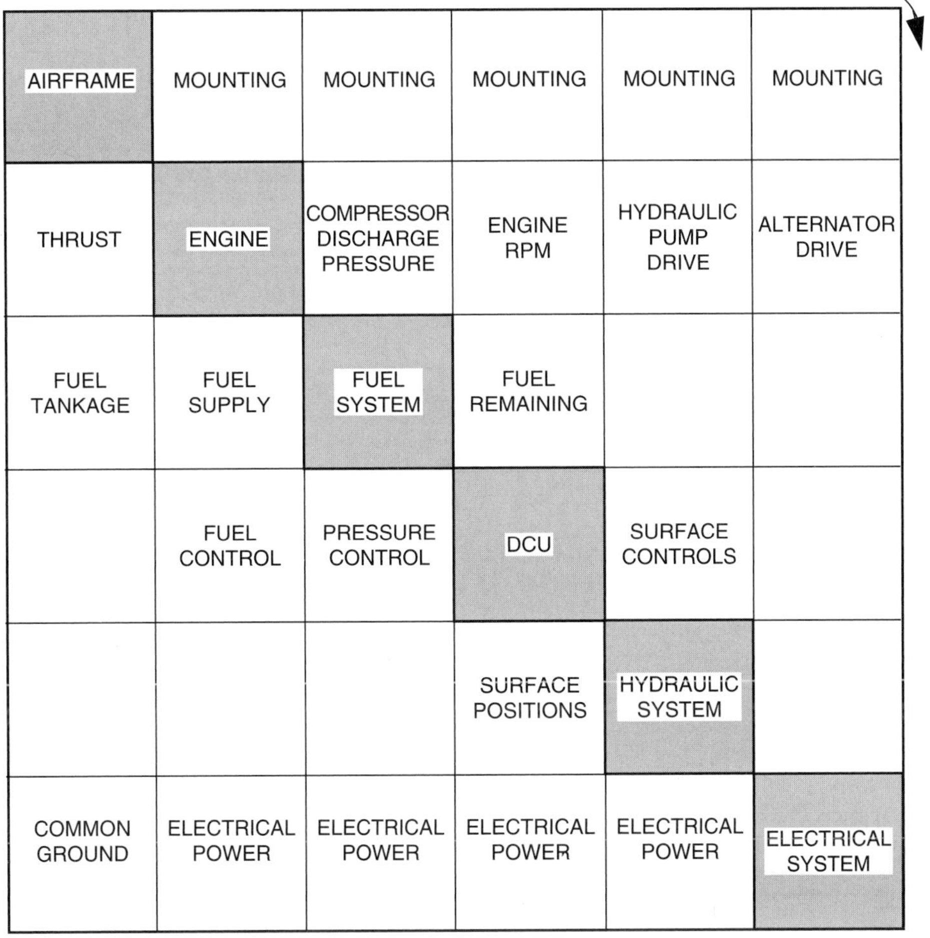

AIRFRAME	MOUNTING	MOUNTING	MOUNTING	MOUNTING	MOUNTING
THRUST	ENGINE	COMPRESSOR DISCHARGE PRESSURE	ENGINE RPM	HYDRAULIC PUMP DRIVE	ALTERNATOR DRIVE
FUEL TANKAGE	FUEL SUPPLY	FUEL SYSTEM	FUEL REMAINING		
	FUEL CONTROL	PRESSURE CONTROL	DCU	SURFACE CONTROLS	
			SURFACE POSITIONS	HYDRAULIC SYSTEM	
COMMON GROUND	ELECTRICAL POWER	ELECTRICAL POWER	ELECTRICAL POWER	ELECTRICAL POWER	ELECTRICAL SYSTEM

Figure 3.6-1   *Typical n-Square diagram*

across the Potomac River from Washington, DC. The author found that everyone asked did it the same way the author did—experience with certain product lines makes it very easy to intuitively identify needed interfaces based on the elements the system contains. This is essentially the approach suggested in paragraph 3.6.1.3 above. The question is, "Is there a more structured approach to accomplishing this work?"

3.6.2.2 A thoroughly disciplined method

We know that the way functionality is allocated to architectural entities pre-determines the interfaces needed but it is hard to conceive the interfaces while thinking in terms of the functional plane alone. It is relatively easy to do so on the architectural plane after the functionality has been allocated. The tools we use (n-Square and schematic block diagrams) to identify interfaces are effective on the architectural plane. Yes, one can build a functional n-Square diagram but it tells you no more than the functional flow diagram. It does not give you insight into needed interfaces in the product.

In Chapter 3.10 we will construct a merger between the n-Square diagram and a requirements analysis sheet (RAS) discussed in Chapter 3.4 that ties this together. Figure 3.6-2 offers an alignment between the n-Square diagram and a graphical form of the RAS where the diagonal of the n-Square diagram coincides with the architecture axis. The figure illustrates this construct suggesting that particular pairs of trios of functions, performance requirements, and architecture jointly establish the need for an interface between the two items to which particular performance requirements derived from particular functions have been allocated. In Chapter 3.10 you will see an extension of Figure 3.6-2 to expose a complete RAS for all of the requirements that have to be exposed in the system and hardware requirements analysis work.

If requirement R_a is derived from function F_g and the requirement is allocated to architecture entity A_m and requirement R_b is derived from function F_h and the requirement allocated to architecture entity A_n, then we should inquire if the functions F_g and F_h extend across the gap between the two architectural entities to require a synergistic relationship I_p.

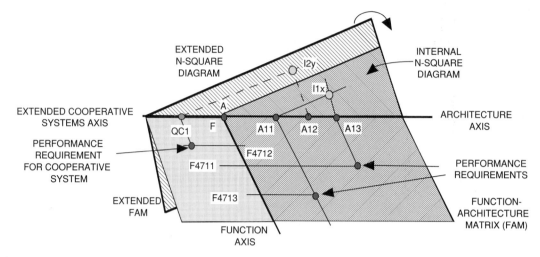

Figure 3.6-2   *Extended RAS*

So, given all of the functions corresponding to the performance requirements allocated to a particular pair of architecture entities, we should pair-wise evaluate if there is a synergistic relationship between those architectural entities driven by the allocation pair. If, for example there are three performance requirements allocated to item A_m and three allocated to architecture entity A_n, then there are 9 pair-wise evaluations to consider. In the general case, if there are x allocations for item A_m and y allocations for item A_n then there will be xy pair-wise evaluations that have to be made for the pair of items.

If there are 5 items in a particular team level entity (the highest level item for which a particular team is responsible) there are 20 pair-wise relationships that may yield an interface need in that the number of non-diagonal intersections is n^2-n). The number of pair-wise interface evaluations for the internal interfaces for this team item will be $\Sigma x_i y_i$ where the range of i is 1 through 20. If x and y were both 3, then in this example, the number of pair-wise evaluations needed for one item would be $9 \times 20 = 180$. The system engineer who wanted to be perfectly disciplined would follow this path to determine what interfaces are needed. It is possible that the system engineer could reach erroneous conclusions on the need for an interface based on one or more evaluations of function-architecture pairs but, barring that, the result would be a perfectly disciplined and correct evaluation of needed interfaces. If the number of team level items in the system were 15 and they all shared the same number of internal items (5), then it would be necessary for the 15 teams to each make 180 internal interface evaluations for a program total of 180×15 or 2700. This number excludes the number of possible higher-tier cross-organizational interfaces. If the 15 items are all at the same level and each pair has same number of allocation intersection cases to deal with as the lower level teams (9), then the number of program interface evaluations mounts to 2700 lower-tier plus the higher-tier evaluations which would number $9 \times 15 \times 14 = 1890$ for a total number of system interface evaluations of 4590. Granted, some (perhaps all) of these higher-tier interfaces would actually be expressions of lower-tier

interfaces but we should identify them separately in the top-down definition of system interfaces. This total does not include external interface possibilities that might very well put us over the 5000 mark.

The discipline needed to accomplish this evaluation is tremendous and the time required could be long and expensive. The question is, To what degree might we miss a needed interface if we accomplished the work the way it is more commonly done where the system engineer intuitively identifies needed interfaces based on the character of the items in the system from prior experience? The answer probably depends on the experience of the system engineer doing the work. If the system engineer was one of those interviewed at the 2003 INCOSE Symposium about this matter, the author would be content with his or her evaluation but everyone probably does not have the experience to make a success of this intuitive pattern. Programs commonly cannot afford the cost and time needed for a comprehensive pair-wise evaluation of interface possibilities. So, the lesson is that programs require experienced and well-trained system engineers to do this work well and affordably. This is not really news but, hopefully, the logic by which we arrived at the conclusion is of value.

3.6.3 Identification Work Products

We will describe two diagrammatic tools for the development of a clear definition of the interface relationships needed between the system elements defined on the architecture block diagram. These tools may be used together with n-Square diagramming used as an analytical tool and schematic block diagramming used to publish the results. Either tool can also be used independently to accomplish the complete interface definition task, the analytical and the exposition portions. An interface dictionary implemented with a computer database can also be very useful in controlling the development of interfaces throughout the system.

Our methods will focus on producing some combination of three work products:

a. An n-Square diagram that identifies interface relation-ships between a set of system elements in a way that forces us to consciously consider every possibility.

b. A schematic block diagram (SBD) that defines system interfaces. It uses simple blocks and interconnecting lines to define the existence of interface elements and their terminal relationships.

c. An interface dictionary that is a tabular listing of sys-tem interface elements illustrated on the SBD (one tabu-lar line item for each line on the SBD) containing the element name, a description of the element, its two ter-minals and media identification form the architecture dictionary, and other reference information too exten-sive to place on the face of the SBD.

3.6.3.1 N-Square diagramming methods

One tool that has been found effective in exposing interface requirements between the many elements of a system is a matrix called an n-Square diagram illustrated in Figure 3.6-1. The "n" refers to the dimension of the square. To make an n-Square diagram the analyst makes a square marked off on each side with a space for each system element under consideration. These spaces are then annotated with the architecture ID or element names. The analyst then marks the intersections within the square to note the required inter-faces between the elements. The diagonal represents the internal interfaces required for each of the elements and need not be explored in this analysis.

At first glance it is easy to see that there may be a degree of ambiguity in this matrix since it includes two intersec-tions for each pair of architecture elements. This allows us to use the matrix to define interface directivity as well as existence. It is common practice to select a clockwise rota-tion but one could pick the other direction. Identify the direction you have picked on your diagram for the ease of use of the diagram by others.

The method of noting the existence of an interface in the intersections of the matrix can be as simple as marking the intersection with an "X," using a code with an explana-tion in an accompanying dictionary, or by simply writing cryptic notes within the intersections that describe the inter-faces needed.

When illustrating complex interfaces within a system ele-ment a triangular matrix can be used where the same list of elements appear along the vertical and horizontal axes or down the diagonal. At each intersection where an interface is required, an "X" is assigned. A dash in an intersection indicates no interface requirement between these ele-ments. A blank means that a decision has not been made. Only one half of a square is required because the other half is redundant (directivity not needed).

It is possible to arrange a series of n-square diagrams along a diagonal so as to represent the interfaces not only within particular system elements but between those ele-ments and others as well. Figure 3.6-3 illustrates such an application. You could create this diagram on a large grid

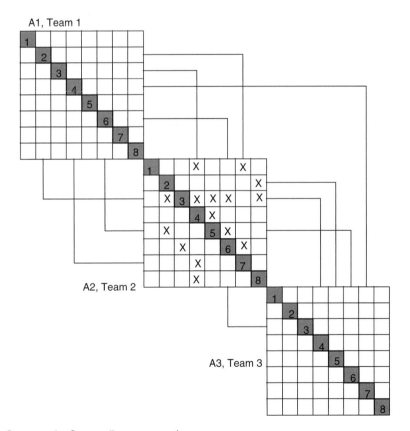

Figure 3.6-3 *Compound n-Square diagram example*

simply marking the interfaces in boxes for external and internal interfaces in the same way. But, assuming that each of the three subsystems is under the responsibility of three different teams, note how the cross-organizational interfaces jump out of the page so obviously. For example A14 must send something to A37 meaning that people from Team 1 and Team 3 are going to have to negotiate this interface.

3.6.3.2 Schematic methods

Schematic block diagram illustrations are structured from very simple symbols, blocks, and lines. The blocks are the objects, and only the objects, illustrated on the ABD and the lines that join them represent the interfaces between those objects. Figure 3.6-4 defines all of the symbols used on the SBD.

Interface lines are either of the bundled or elementary type. At the higher levels, bundled lines are used denoting the existence of an interface requirement between two system elements. These lines do not include arrows to denote direction because they very likely include elementary interface elements of many kinds with both directions. An elementary line, which represents a single specific interface, must show direction with an arrow at one end (unidirectional) or both ends (bi-directional).

At the lower tiers where an interface element represents a single elementary interface, interface element names may be included on the line. Also at the lower tiers, different line types may be used to illustrate different classes of interface media such as solid lines for electrical, long dashed lines for pneumatic, and alternating dashed lines for fluids, for example. This diagram should be created by the system engineering community in order to ensure continuity. For example, left to their own devices, the specialized design disciplines will each use the solid line to represent the line they have to use most often because it is easier to draw than any other kind. On electrical schematic block diagrams you will see the solid line for wiring and on hydraulic and pneumatic schematics it will represent plumbing.

It is vital that only those blocks used on the ABD be considered for use on the SBD. If in the functional analysis, one finds that it is not possible to satisfy system functions cur-

rently allocated to system elements through the process of constructing the SBD and the solution is to add a system element, then the analyst must cause the object to be added to the ABD before it may be used in the SBD.

Each interface can be coded for unique identification in computer databases. These codes use the same structure applied to the product ID system. The complete set is identified as "I." It need not be used in cases where the context is clear, but the author retains it for use in computer databases. At the top level the interfaces are arbitrarily numbered I1, I2, I3, though the author prefers to use the ID I1 for all internal interface at the system level and I2 for all external interface where one terminal touches the system. The author reserves I3 for all interfaces that are interesting for the program but neither end touches the system under development. For example the system communicates with a satellite that is not part of the system, the satellite relays the signal to another satellite and then the signal is downlinked to a ground station that is part of the system being developed. The signal between the two satellites would be identified with an interface ID beginning with I3. At the next level each of these interfaces are broken down to the next level as I11, I12, ... I1n. This process is continued until the component architecture level is reached in the SBD.

A base 60 ID system is used in this book employing the 10 Arabic numerals, 25 capital English alphabet letters (O excluded), and 25 lower case English alphabet letters (l excluded). The number of characters in an interface ID indicates the interface level. Within each level and branch, up to 60 different interfaces can be coded using this system. If more than 60 are needed they can be broken up into subclasses each no more than 60 in number.

Some people prefer a decimal delimited ID system allowing an unlimited number of identifications within any one level. The code 1.10.4.17 is an example of this method. The problem with this ID system is that it does not sort and index properly in simple computer database systems available for microcomputers. The ID 1.10.4.17 will appear before the ID 1.2.5.7 because a simple database implementation sorts character-by-character recognizing ASCII character precedence and the 1 in 10 of the first ID has a higher priority than the 2 in the second ID.

The ultimate, or level zero, SBD is exactly the same for every system and is illustrated in Figure 3.6-5. It is identical to the fundamental system abstraction. Every system interacts with its environment to achieve its top-level function (expressed in the system need statement). That environment includes the natural environment and may include hostile systems that the system is intended to destroy or avoid, cooperative systems that provide the system with useful services, and non-cooperative systems that may adversely interact with the system unintentionally (by creating electromagnetic interference, for example). We have now experienced the ultimate system block diagrams under traditional structured analysis. The ultimate function is one block inscribed by the system need with ID "F." This function is instantly allocated to the ultimate architectural entity, one block inscribed with the title "System" with ID "A." Figure 3.6-5 illustrates the ultimate system interface view. This trio of views happens to be identical for every system that has ever been and every one that will be.

The system environment is labeled by the author with an Architecture ID of "Q" and the system with an Architecture ID of "A." The interface between the environment and the system is labeled here with a level one Interface ID I2 and

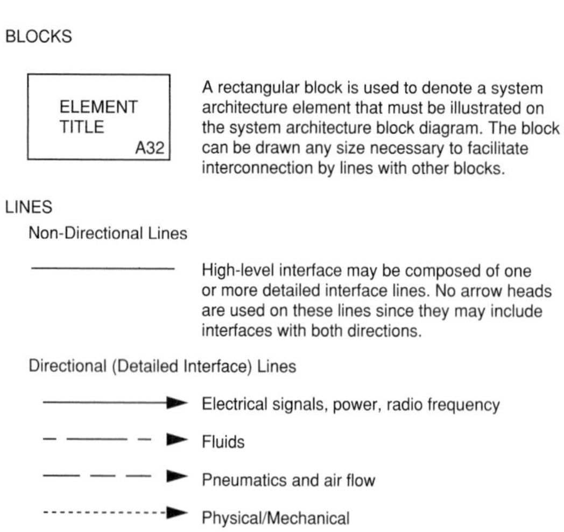

BLOCKS

ELEMENT
TITLE
A32

A rectangular block is used to denote a system architecture element that must be illustrated on the system architecture block diagram. The block can be drawn any size necessary to facilitate interconnection by lines with other blocks.

LINES

Non-Directional Lines

High-level interface may be composed of one or more detailed interface lines. No arrow heads are used on these lines since they may include interfaces with both directions.

Directional (Detailed Interface) Lines

Electrical signals, power, radio frequency

Fluids

Pneumatics and air flow

Physical/Mechanical

Figure 3.6-4   *Schematic block diagram symbols*

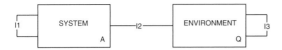

Figure 3.6-5 *Universal ultimate schematic block diagram*

the system innerface with a level one Interface ID I1. In order to reduce the number of levels in the Interface ID, the analyst may choose to assign level one Interface ID to the major interfaces within the top-level system elements as well as to the environment interface.

Figure 3.6-6 illustrates a typical top-level (level one) SBD for a space transport system. As you can easily see, this diagram could be fitted into the form of Figure 3.6-5. The top-level SBD is created by laying out each of the top-level system ABD elements (as a minimum, those architecture elements immediately subordinate to the system block on the ABD) on a sheet of paper and connecting pairs of them appropriately with interface lines.

Lower-tier diagrams must expand the interfaces defined on the top-level SBD by creating an expanded SBD for each interface line on the diagram. This process is carried from the top downward progressively in step with the advancing ABD definition. The product of the interface analysis work should be fed back to help define the expanding architecture responsibility assignments.

Note the use of the lines that are attached at both ends to the same block. These interface lines signify the internal interface for that item. These lines correspond to the diagonal on an n-Square diagram. It is useful to illustrate them on a schematic block diagram as a part of the discipline in assigning interface ID codes. The internal interface is coded at the same level as other interfaces at that level. If a schematic block diagram is developed for that item, all of the internal interfaces have the prefix of the internal interface ID. This is the gimmick that assures that interface ID codes can be expanded from the top down in step with the expanding architecture.

Every interface in a system can be said to have a source, a destination, and a media (wires, pipes, attaching bolts, etc.). The intensity of interface problems in a developing system are directly proportional to the percentage of system interfaces that possess development organization responsibility differences among these three interface aspects of source, destination, and media. Therefore, one of the principal objectives in laying out the system architecture is to structure the functional elements of the system, driven out by the functional analysis, into subsystems that are cleanly related to the engineering design organization or team structure intended on the program.

Interfaces that have different design agencies responsible for source, destination, or media are defined as cross-organizational interfaces. These interfaces must be tracked by system engineers to insure they are properly developed by the two or three responsible agents. Those interfaces that involve different associates or teaming partners should have interface control documents (ICD) prepared to manage the interface development. Interfaces that have all three aspects (source, destination, and media) under a single design agent's responsibility are non-critical and will be the complete design responsibility of a single design agent.

This objective is satisfied by the concurrent development of the ABD and the SBD under the responsibility of system

engineering members of the program integration team (PIT). As elements are added to the architecture from functional analysis action, they are fitted into the SBD with a watchful eye on alignment between interfaces and design organizational boundaries. This may be done by coloring the SBD blocks different colors for the different design organizations, by overlaying the design responsibilities onto the SBD with dashed boxes, or by simply encircling elements under each design group's responsibility with a line of a particular color (color computer graphics and automated identification of critical interfaces could be a great help in this area).

It is of crucial importance that the analyst respects the discipline that the SBD use no blocks that are not already identified on the system ABD. The reason for this is that architecture elements can have assigned to them organizational development responsibilities, which, in turn, can be used to define interface development responsibility pairs. If these are not clearly connected, then management of the development process will suffer.

Later we will refine our definition of system interfaces to identify three subsets: innerface, crossface, and outerface as a function of the principal engineer's responsible for architecture development. For the time being let us just say that innerface is those interface elements that have both their terminals at the same architecture element. A crossface interface element has two different responsibilities for the two architecture terminals. It is useful to have a name for interface elements that are of no interest to a particular development agent, thus the outerface class.

The top-level SBD is expanded in step with the expanding ABD. There are two principal expansion approaches: (1) innerface (internal interface) expansion, and (2) crossface (external interface) expansion. These two subsets of the complete interface set for a system encompass all system interfaces so it is unnecessary to focus special attention on the third class of interface called outerface with the one exception that there may be a few interfaces which neither end touches the system but in which the program has interest.

Innerface is identified on any SBD as a line with both ends on the same block. This line will be coded like any other with an Interface ID. To expand one of these innerface elements, the analyst should prepare a SBD sheet in either the block diagram or triangular matrix form that illustrates each of the system elements subordinate to the block that has both ends of the line attached. The subordinate blocks are derived from the ABD. The analyst will find that initially it may be useful to simply make a circle on a piece paper for each block, an example of which is shown in Figure 3.6-7. Then draw a line joining each pair that must have an interface. Add an innerface line to each block. Add the Interface ID to each line in an arbitrary fashion by appending 1, 2, 3, through z (using our base 60 system) to the Interface ID for the line expanded in this SBD.

From this preliminary sketch, it will be possible to see a pattern of lines and blocks that can be arranged in an artistically pleasing way as well as technically accurate. The final illustration, shown in Figure 3.6-8, should have as few line crosses as possible. Prepare the final diagram on a computer using a graphics package or using templates and a pencil.

If the rough sketch becomes too cluttered consider using a triangular matrix, or semi n-Square, format to illustrate the SBD. The triangular matrix format corresponding to

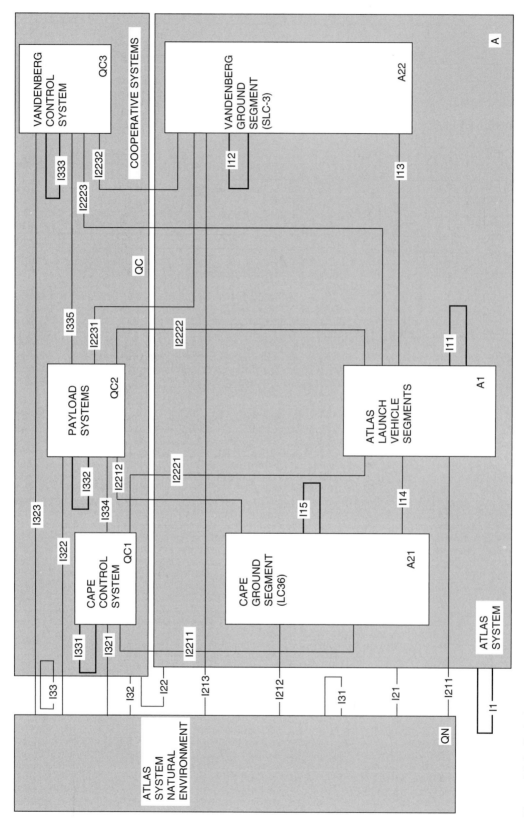

Figure 3.6-6 *Typical system schematic block diagram*

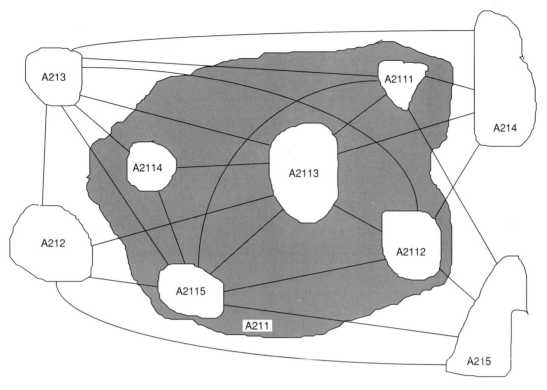

Figure 3.6-7   *Primitive schematic block diagram*

the one shown in Figures 3.6-7 and 3.6-8 is illustrated in Figure 3.6-9. It is formed by simply listing the architecture IDs for the set of elements on both axes. For each intersection corresponding to a needed innerface, mark the square with an Interface ID expanded from the original Interface ID.

A slightly different technique is necessary for expansion of crossface lines on a SBD. A crossface line will, of course, always have its two ends each on a different block. To expand one of these lines, draw a rough sketch with all of the blocks subordinate to each of the two parent blocks illustrated on one side of the sheet as suggested in Figure 3.6-10. One way to get this sketch is to use the innerface expansions for each of these elements, if they have already been prepared. Next connect pairs of these blocks, one end on a sub-element block on one side of the page and the other end on a sub-element block in the other side of the page.

When all of the line pairs have been entered, number them all expanding on the Interface ID that is being worked as the root for each of lower-tier Interface ID. Use the sketch to see the non-crossing patterns of interest in creating the final drawing. Figure 3.6-10 illustrates the initial top level, primitive, and final form for a typical diagram expansion. Note that this diagram would expand as you allocated functionality to the system. First one would recognize that it was necessary to fly the aircraft in the Earth's atmosphere and that we would been sensory data about the immediate environment provided by a sensor suite. Continued analysis might reveal that it would be necessary to fly relative to the Earth surface as well as the atmosphere in order to preserve a fairly uniform V/H ratio for photo reconnaissance pur-

poses resulting in selection of a radar altimeter to control height (H) above ground level with the understanding that aircraft velocity (V) would be controlled by engine thrust. Finally, a more detailed understanding of needed functionality would have suggested the details shown in Figure 3.6-10c.

2.6.3.3 Interface dictionary

It is not possible to use the SBD to retain all of the information of interest about the interfaces illustrated there. The SBD actually only illustrates the existence of interfaces and identifies unique names for each interface. The analyst should also prepare an interface dictionary which provides an inventory of all system interfaces in the form of an alpha-numeric listing by Interface ID corresponding to all of the interface lines on the schematic block diagrams.

Ideally, the dictionary is retained in and printed from a computer database that is a part of a computerized requirements system. It includes information about the source, destination, and media of each interface element plus other data of interest. The database allows system engineering to map all interfaces to design responsibility groups for the purpose of identifying the critical interface subset as well as providing interface definition and statusing data. Figure 3.6-11 illustrates the interface dictionary columns that define each interface in terms of its source, destination, and media. The interface connectivity fields list the architecture IDs corresponding to the entities that provide the indicated functions. Other columns can be included for other purposes. The interface dictionary could be used to also capture

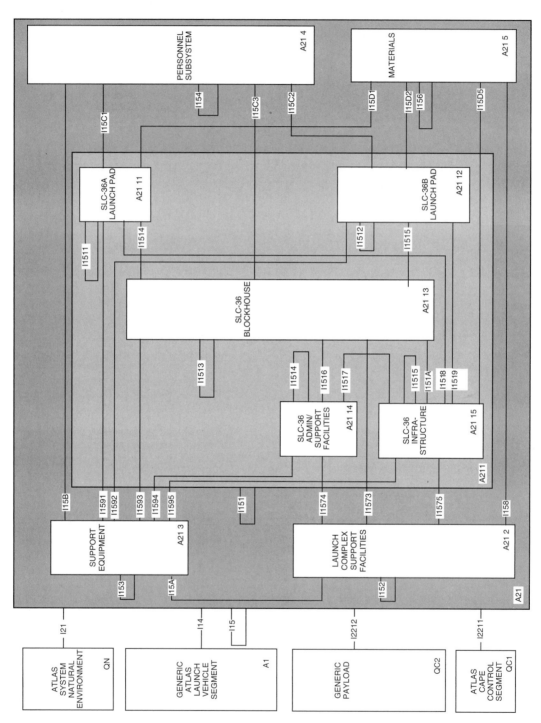

Figure 3.6-8 Finished schematic block diagram

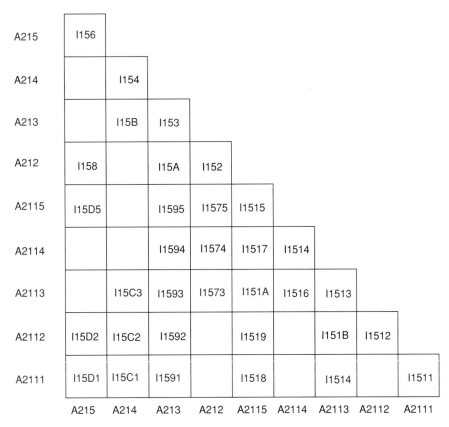

Figure 3.6-9 *Triangular matrix schematic block diagram example*

the function-architecture pairs that encouraged the identification of the interface in the first place but this data will be difficult to maintain.

3.6.4 Interface Media and Requirements Definition

The interface identification methods covered in Paragraph 3.6.3 were specifically developed to enable a top-down interface development approach that could keep pace with the top-down architecture development process. As the development process penetrates deeper into the system architecture, we are able to progressively refine the interface definition between system elements until we are dealing with single, primitive interface elements about which it is possible to write detailed requirements for the terminal items.

Interface definition involves writing appropriate interface requirements coordinated with the lines on the schematic block diagrams or marked intersections of the n-Square diagram. It is necessary to first know the interface media technology to be applied by the terminals. This is a reality in the development of requirements in the three design constraints categories (interface, environmental, and specialty engineering). Performance requirements can be determined prior to their allocation to an architectural entity but one must know what the items are before writing design constraints and have some idea about the design concept for the

item. This is not a case of design before requirements, however. Generally, the design concepts are at a higher tier than the requirements work is focused on.

For each interface defined for an item, the item principal engineer must cooperate with the opposing terminal principal engineer within a prescribed responsibility rule, as discussed shortly, to define one or more requirements by first deciding what aspects of the interface they should attempt to control. Sometimes this will be very obvious. In other cases it may prove very difficult to identify the attributes that must be controlled at a given interface to ensure that the system will satisfy its overall function.

But this should not come as a total surprise to either principal engineer or require a new beginning. The team should have captured already on a schematic block diagram, interface dictionary, or n-Square diagram exactly what interfaces are required. They should have participated in the give-and-take process of deciding how the elements must interface to satisfy the functionality allocation decisions already made. It is, after all, the way we allocated functionality to the things in the system that drives the interfaces that will have to exist between the things. The only thing that should remain is to define precisely how those interfaces are characterized before the design of the terminal elements is contemplated.

This matter can best be discussed in terms of some examples. In each of these examples read first only the attribute that must be controlled (you might place a sheet of

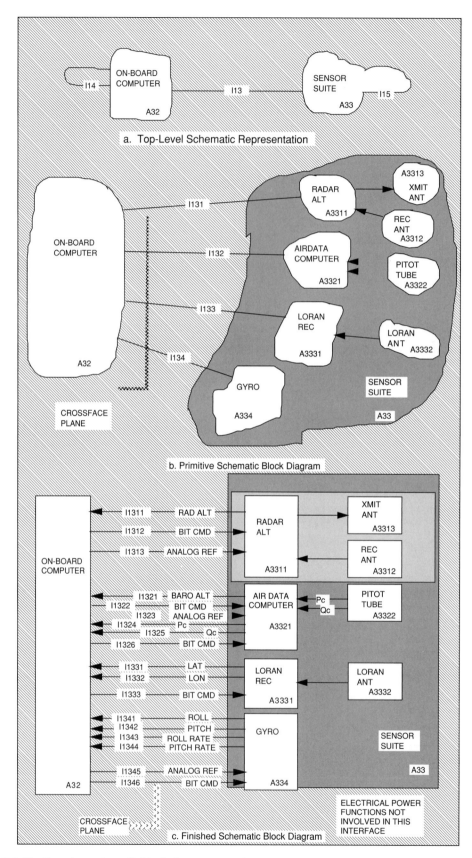

Figure 3.6-10 *Crossface schematic block diagram expansion*

INTERFACE IDENTIFICATION		INTERFACE CONNECTIVITY		
ID	NAME	SOURCE	DEST	MEDIA
I1511	Fuel On Command	A21	A252	A32
I1512	Fuel Flow Light	A251	A21	A32
I1513	Fuel Level Tank 1	A252	A251	A32
I1514	Fuel Level Tank 2	A252	A251	A32
I1515	Fuel Alert	A252	A251	A32

Figure 3.6-11   *Typical interface dictionary listing*

paper across the page and move it down only far enough to expose the title, for example) and think through how you would phrase the requirement before exposing the example given. Each title is intended as the name of an interface element appearing in an interface dictionary between generic terminal elements X and Y. The Interface ID codes included in an interface dictionary are not printed with the interface requirements in these examples, but could be parenthetically included.

3.6.4.1 Electrical power example
3.1.7.4 *Electrical Power*.

3.1.7.4.1 *Power Level and Duty Cycle*.   Unit X shall supply to Unit Y 400 watts of electrical power intermittently over a 24-hour period of time with a duty cycle of at least one period of 2 continuous hours off and a total of 6 hours on spread across the 24 hours.

3.1.6.4.2 *Voltage*.   The voltage level shall be 28 VDC plus or minus 0.5 VDC.

3.6.4.2 Electrical signal example
3.1.7.8 *Radar Altitude Signal*.

3.1.7.8.1 *Signal Magnitude and Sense*.   The radar altitude signal supplied by Unit X to Unit Y shall be 0 to +5 volt DC in magnitude with 0 volts corresponding to 0 feet altitude and +5 volts corresponding to 5000 feet above local terrain.

3.1.7.8.2 *Signal Power*.   The radar altitude signal supplied by Unit X shall be capable of driving a 2000-ohm resistive load in Unit Y.

3.6.4.3 Physical attachment example
3.1.7.3 *Physical Interface Plane*.

3.1.7.3.1 *Attaching Bolt Pattern*.   Attaching bolt centers shall appear on a 36-inch radius circle centered as defined on Figure 3.6-13. A total of 36 bolts shall be used at even 10 degree spacing around the circumference.

3.1.7.3.2 *Attaching Bolt Size*.   Attaching bolts shall be 1/4-inch diameter, 3 inches long.

3.1.7.3.3 *Bolt Hole Size*.   Attaching bolt holes shall be 0.248 inches in diameter plus or minus 0.002 inches.

3.1.7.3.4 *Mating Surface Smoothness*.   The mating planes shall each be surfaced to a smoothness of **TBD-13**.

3.1.7.3.5 *Bolt Head Location*.   The bolt heads shall be on the unit X side of the interface.

3.1.7.3.6 *Use of Safety Wire*.   Nuts used to secure the pressure plate in place shall be safety wired in accordance with company Mechanical Practices Manual 33-120.

3.1.7.3.7 *Mating Orientation Control*.   Mating surfaces shall be keyed to ensure that only one orientation is possible to preserve unit X pressure test port orientation for alignment with the maintenance access door.

3.6.4.4 Fluid transmission example
3.1.7.8 *Fluid Coupling Interface*.

3.1.7.8.1 *Connections*.   Mating fluid connectors shall satisfy the requirements of QPL-25427-18(T). It must be possible to make and break these connections without a subsequent requirement for bleeding the system.

3.1.7.8.2 *Fluid Flow Rate*.   Unit X shall supply Unit Y with fluid at a rate over the range of 10 to 82 liters per minute.

3.1.7.8.3 *Fluid*.   The fluid shall be MIL-H-19457D hydraulic fluid.

3.6.5 Interface Documentation

3.6.5.1 Capture in the requirements analysis sheet and database system
As the requirements analysis work is on going, the interface requirements must be captured in some fashion that is compatible with the eventual publication of these requirements in a specification. Throughout this book the author has encouraged the use of the requirements analysis sheet (RAS) as the engineering means to capture requirements as they are identified and defined during the requirements analysis work. Ideally, this RAS would reside in a computer requirements application like DOORS, SLATE, or CORE, for example, but it can be done with paper, spreadsheet, or word processor. Figure 3.6-12 illustrates a fragment of the RAS with one representative requirements pair included.

MID	MODEL ENTITY	PID	PRODUCT ENTITY	RID	REQUIREMENT
I127	Radar Altitude	A21	Radar Altimeter	G74H	$0 \leq$ Altitude $\leq$ 5000 feet
I127	Radar Altitude	A22	On-Board Processor	QJU7	$0 \leq$ Altitude $\leq$ 5000 feet

Figure 3.6-12 *Requirements analysis sheet capture of interface requirements*

Note that the interface is listed twice as every interface between two items should be because an interface has two terminals. The data captured in the RAS includes three pieces of information. The first column pair deals with the model from which the requirement was derived in terms of a model ID (MID) and entity name. The second column pair names the product entity that will implement the requirement (product ID or PID, and item name). The third column pair captures the requirement in terms of the unique requirement ID (RID), commonly assigned by a computer database in the background, and the actual requirement statement that will be captured in the specification or interface document. If we use a specification pair, the interface requirements for the On-Board Computer (item A22) will pull these interface requirement into the list to be published with that document. If we capture the requirements in an interface control document (ICD), the single interface will be captured in that document. Commonly, the development or performance interface requirements will be the same for the two terminals. This is often not the case for detail (part II) documentation because a plug and socket kind of sense is important. So, both terminals will have to be specified and controlled.

3.6.5.2 Interface definition publication
Interface requirements have the characteristic of two terminals making it difficult to clearly establish responsibility, which is the next topic. But, we must capture the requirements in one of two ways. The specification templates preferred in this book are MIL-STD-961E for systems and hardware and EIA J STD-016 for software and both of these documents cover internal and external interface requirements. In this documentation mode we would, of course, have to capture and define a particular interface in a specification pair and ensure that the two requirements sets were compatibly captured.

An alternative capture approach is to document the interface requirements in a document focused on the interface rather than on the terminal item specifications. This document is variously called an interface control document (ICD), interface requirements specification (IRS), or interface requirements document (IRD). No matter the name, when we select that method of documentation we should avoid double booking by including the requirements in both the interface document and the specification pair. It is nearly impossible to ensure that both references will remain coordinated amidst changes that will often be necessary.

Where an interface document is used, the specification pair should include the following statement in place of a requirement statement and interface control document 23-43567 then clearly defines each interface that exists between the items covered in the specification pair:

> 3.2 Interface requirements.   Interface requirements shall be defined in interface control document 23-43567.

Whenever capturing interface requirements using an ICD, never also capture those requirements in the specification pair across that interface. You will never be able to maintain the coverage between the specification pair and the ICD in synchronism.

3.6.6 Interface Responsibility

3.6.6.1 Program organization
In a matrix-organized enterprise, there is no valid counter argument for organizing a program into cross-functional teams oriented about the product and physically collocated by team rather than functional departments. The advantages are that the assigned team coordinates with a contiguous collection of planning data for the WBS item assigned to the team including the WBS, SOW expansion data, the IMP/IMS expansion data, the corresponding budget and schedule responsibilities, and a specification previously prepared by the higher-level or system team. It is, therefore, clear who is responsible for the development of each architecture item. It is much more difficult to determine who is responsible for the development of the interfaces because there is a one-to-two relationship between an interface and its terminals but this method of organization causes perfect alignment between the product interfaces and the human communication patterns that must be effective to integrate those interfaces resulting in good interface accountability.

3.6.6.2 Three views of interface
Interfaces exist between objects of systems for the purpose of introducing synergism between the objects. A system is said to be a collection of two or more entities that interact in purposeful ways to achieve an objective that no subset of the objects could otherwise achieve and it is through the synergism provided by interfaces between the system objects that this takes place. Interfaces communicate information from one object to another, which the other would have no way of obtaining through its own unaided capacity. They provide energy with which to function as in the case of electrical or hydraulic power. Interfaces act to physically tie a system together where objects are bolted or otherwise joined together.

An interface element has two fundamental characteristics: (1) a terminal at each end and (2) a media of communication. One terminal will always be some element of the system depicted on the system architectural block diagram. The other terminal will always be an element of the system or an element of the system environment. This includes the elements of the system of interest, possibly some cooperative and hostile systems within the system environment, natural environmental elements, and no other possibilities. Likewise, the media of the interface element will be provided by some element of the system of interest, or the system environment (natural environment or cooperative systems). Wire harnesses, fluid plumbing, and the space through which radio signals pass are examples of interface media. The first two media are provided by the system architecture and the last example by the system environment.

We take three different views of the system interface here as a function of the perspective of the observer. These three views take into account the variable intensity of interest and responsibility for interface by the design community. The basis of these differing views is the relationship between the terminals and media of an interface element and the responsible design organization. It is a great systems management truth that the interfaces that will result in the greatest system problems are those where different organizations are responsible for two or three of the following: source terminal, destination terminal, and media. Relatively little difficulty can be expected where the same organization is responsible for all three.

The traditional view of interface is that an interface exists where two different contractor organizations are responsible for the design of the two terminals of the interface. We will take a more general view where an interface is any means of relating one system element to another at any level of system indenture. Interface thus imagined may be partitioned into subsets of interest as a function of organizational responsibility by establishing three classes as follows:

a. OUTERFACE - Interface elements with neither terminal or media under the design responsibility of organizational element X will be considered in the organization X outerface class. Organization X has no immediate interest in this class of interface and is not responsible in any way for its proper design.
b. INNERFACE - Interface elements with both terminals and the media under the design responsibility of organization X will be in the organization X Innerface class. Only organization X is immediately concerned with this class and is completely responsible for its proper design.
c. CROSSFACE - Interface elements with some subset (other than the null set) of two terminals and media under the design responsibility of organization X and

the remaining subset (not the null set) under the design responsibility of a different organization are in the organization X Crossface class. Organization X and some other organization are jointly responsible for the design of this class. This is the interface class where system engineering energy and budget must be focused because it is where system problems will develop. These are the cracks between the specialized knowledge and experience of the engineering organizational elements. They are inevitable in any system and are determined by how the organization is structured with respect to the structuring of the system architecture. It is important that the system engineering community be capable of sorting all of this class of interface from the whole and assigning responsibility for its development.

3.6.6.3 Interface responsibility model
How shall we assign responsibility for interface development? It is clear there is a fundamental difference between architecture development responsibility and interface development responsibility. The former can be completely assigned to a single development agent with little difficulty. Interface on the other hand has two, and with the interface media possibly three, agents that should be interested in any one interface element. We need a way to make interface responsibility unambiguous.

The previous discussion of the three classes of interface as a function of your outlook offers a solution. First let us create a hypothetical schematic block diagram in Figure 3.6-13. It includes seven components in three subsystems. Our challenge now is to determine who, among the seven principal engineers at the component level and the three principal engineers at the subsystem level, is responsible for the each of the interfaces illustrated here.

The system illustrated in Figure 3.6-13 is composed of a set of architecture consisting of A={A11, A12, A21, A22, A31, A32, A33} arranged into three subsystems (A1, A2, A3)

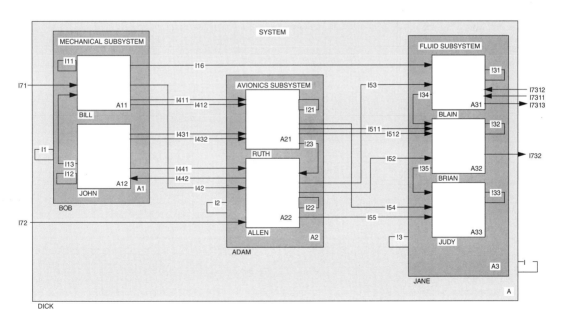

Figure 3.6-13 Interface responsibility model

and a set of interfaces defined by I={I11, I12, I13, I21, I22, I23, I31, I32, I33, I411, I412, I42, I431, I432, I441, I442, I511, I512, I52, I53, I54, I55, I6, I71, I72, I7311, I7312, I7313, I732}. Let us now assign some human beings as principal engineers for these elements who you see named at the lower left corner of each block.

It is easy to assign architecture responsibility. Let us assume that A1 is a mechanical sub-system with Bob the principal engineer supported by Bill for A11 and John for A12. Similarly, we will assign Adam the principal engineer for the Avionics Subsystem, A2, with Ruth responsible for A21 and Allen for A22. Jane is the Fluids Subsystem principal engineer, A3, with Blain responsible for A31, Brian for A32, and Judy for A33. We must have a system engineer responsible for the system as a whole, assign Dick to that task. Now, how do these people share development responsibility for the interfaces between the elements for which they are responsible?

To answer this question let us partition the set of interfaces into the three views of innerface, outerface, and crossface from the perspective of each one of the subsystem principal engineers. Figure 3.6-14 provides a Venn diagram of all of the interface possibilities for the system based on an input partition of the set I, an output partition of the set I and the superimposition of these two sets. This diagram includes subsets for each of the three subsystems and one for the system environment.

Figure 3.6-15 maps all of the 29 interface elements illustrated on Figure 3.6-13 into the 16 subsets defined on Figure 3.6.14 as a function of subsystem level input/output relationships. Some subsets are voids with none of the 29 interface elements involved. Now let us, in Figure 3.6-15, see how each of the three subsystem principal engineers views these interfaces from their innerface, outerface, and crossface perspective. This same pattern can be continued through all layers of system interface in a top-down development of interface in step with the top-down development of the system architecture.

The problem remains to decide which of the subsystem principals should be responsible for interfaces identified as crossfaces on Figure 3.6-13. We have three choices: (1) make the architecture principal engineer on the source end responsible, (2) make the principal on the destination end responsible, or (3) let them share responsibility. Clearly they have to work together, but one of them should be made responsible for the proper development of each interface. It is possible to select either rule 1 or rule 2 above to avoid interface respon-

sibility ambiguity. In the interest of "total quality management," it may be popular to select rule number 2 since the destination principal engineer can be thought of as the customer for that interface. This is a case of a demand-based interface development and is encouraged by the author.

Whichever rule you select, you will find many good reasons for exceptions based on: (1) architecture elements that are grandfathered into the system as customer furnished equipment derived from previous systems, (2) existing equipment designs that are only modified for this application, and (3) biased engineering judgment. Since it is so difficult to pick a foolproof interface responsibility rule, you may conclude that it is foolish to try, but try to resist this feeling.

Where a system is being developed from a clean sheet of paper and will use no existing designs (a very uncommon situation), we could pick rule 1 or 2 without fear of compromise. In a real world situation, where exceptions are possible, the rule could be 1 or 2 as the default with exceptions decided on a case-by-case basis. Given that we have an interface dictionary, we can include interface responsibility information in that dictionary and remove all doubt about who is responsible. Either the default rule applies or the default responsibility rule is reversed as noted in the dictionary.

Several years ago, a student at UC San Diego approached the author at a break during a requirements course he was teaching and whispered that there was an error in the figure shown above. The author made a note about the error and thanked the student. As the student walked back to his seat, the author felt a little sad that he had caused the student to ruin a perfectly good evening tracing through these relationships. It finally dawned on the author that this was an interesting curiosity but not an effective practical tool.

This set oriented view of interface is not intended to suggest a practical way of interface management. It is only intended to communicate to the reader that everyone looks at interface differently and a system engineer must be aware of these differences in order to approach and encourage compatibility across all program interface planes. If you apply the team approach oriented toward the product architecture, there is an easier way to focus on the cross organizational interface than the set theory approach noted above. Figure 3.6-16 provides a schematic block diagram of a Lockheed Martin Centaur upper stage of an Atlas Centaur space launch vehicle (when it was a General Dynamics Space Systems Division Centaur) circa 1993.

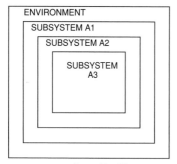

a. Input Partition

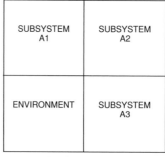

b. Outout Partition

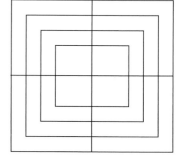

c. Superimposition

Figure 3.6-14 *Interface partitions*

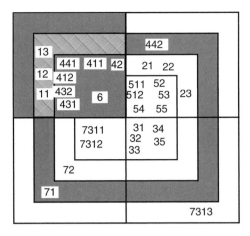

a. Subsystem A1 Perspective

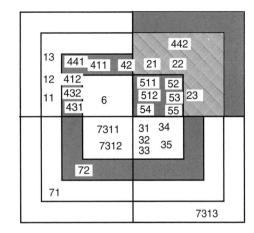

b. Subsystem A2 Perspective

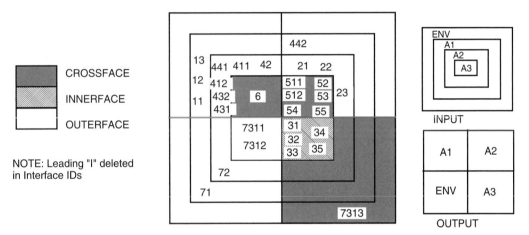

c. Subsystem A3 Perspective

CROSSFACE

INNERFACE

OUTERFACE

NOTE: Leading "I" deleted in Interface IDs

Figure 3.6-15 *Subsystem principal engineer views*

If we assigned teams to the things on the schematic block diagram as noted, it is clear at this level of interface indenture where the cross-organizational team (IPPT) interface planes reside. Any interface line that crosses one of these planes is a potential problem for the system agent. So, the PIT can craft a clear view of cross organizational interface by building a schematic block diagram with each item under the responsibility of each team included as one or more blocks. Where one team has more than one item on the diagram draw a border around all of the things that team is responsible for. This will take some artistic skill to locate the blocks so that all of the things under one team's responsibility can be located in one contiguous space.

3.6.6.4 The special need for external interface development

Interfaces between elements that are both under the same company's design responsibility should never be a problem in a development activity, but they commonly are even where different design departments of the same company are involved. Interfaces between elements a contractor is developing in-house and those that are procured from an outside vendor should never be a problem since the contractor controls the vendor through a subcontract or purchase order. Interface definition is a common area of dispute between contractors and their suppliers, however. If interfaces that are fully under the control of a contractor are difficult to develop flawlessly, those between two contractors who have no contractual relationship could be nearly impossible. Because it is so difficult to develop these interfaces, we need some special controls at these very important interface planes.

Interface Development is an engineering management and technical approach to the development of interfaces where the contractors responsible for the design of the elements at the two terminals are different and not linked through a contractual arrangement. Commonly, the contractors are called associates, and each has an independent contract with a common contracting agency. This management approach seeks to introduce discipline and precision

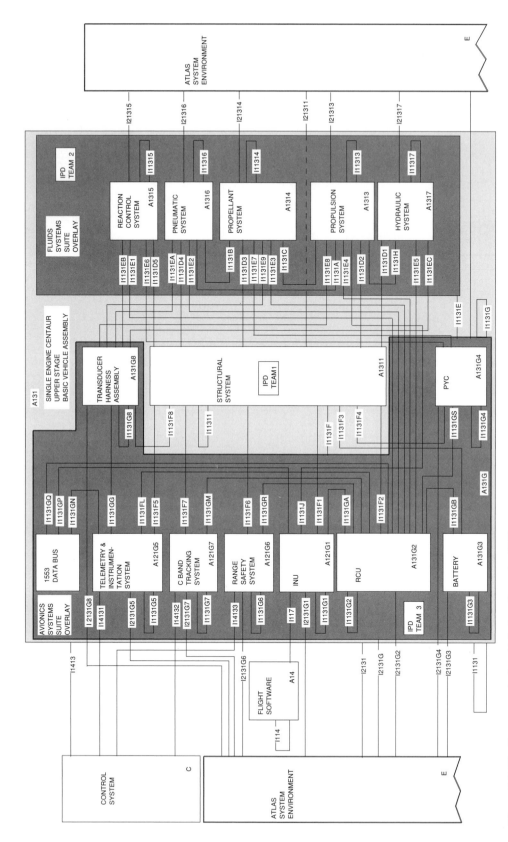

Figure 3.6-16 Cross-organizational interface through a SBD

into the technical communications between the contractors and to formally resolve problems across the interface between their products.

A customer will commonly require two interfacing associate contractors to reach an agreement on how they will jointly work a mutual interface and to jointly sign a memorandum of agreement stating the means by which they will cooperate. They may also be required to develop an interface management plan to expand on the memorandum of agreement and both sign that plan. One of the associates will be required to prepare an Interface Control Document (ICD) and they and the common customer will manage the development of the interface through that ICD.

An Interface Control Working Group (ICWG) will be formed with the customer chairing its meetings. The ICWG will meet periodically and take up interface issues over time the interface definition between the two elements managed in this way. Interface issues will be resolved by joint action and approved as evidenced by approving signatures on the ICD and revisions by ICWG members.

Two possibilities exist for this ICD. One possibility is that it will only be used as a means to an end, that being to provide each associate with the same interface definition for use in their design work. In this case, each associate uses the ICD as the source of the interface requirements for the specification on the element on their side of the interface plan. When these two specifications are authenticated (signed) by the common customer into the program baseline, the interface can thereafter be managed though engineering change proposals (ECP) against the two specifications. The ICD goes away at this point having served its purpose.

The other alternative is that the ICD remains a living document throughout the program. In this case, the interface requirements contained in it may never be copied over into the specifications at the terminals. Each terminal specification will reference the ICD for the interface requirements. The ICD is taken up in Chapter 6.5 in association with the special management challenges of associate contractor management.

3.7

Specialty Engineering Requirements Analysis

Contents

3.7.1 Serial Versus Parallel Work Pattern

The system engineering process as initially conceived involved close cooperation between design engineers charged with synthesis of requirements into a practical design solution responsive to those requirements and an expanding cast of specialty engineers contributing to the conversation. Unfortunately, some companies allowed their specialty engineering and design groups to become isolated from each other. They evolved a condition where design drawings were reviewed in a serial way by each specialty engineering discipline after the drawings were completed. The feedback from the specialty groups had to be reacted to by design, if there was any budget left at that point in the program, followed by another review cycle. This serial process is a blueprint for cost and schedule overrun.

On one program involving nuclear survivability requirements, a designer had his design rejected three times by the survivability analyst before asking how he could get it right and complete the job. The analyst replied that he wasn't sure, but he would know the correct design when he saw it. Many designers have had similar experiences with specialty engineering groups and engineers and it drives them to distraction. Interest in concurrent engineering during the late 1980s focused on restoring a condition of real-time teamwork between design and specialty engineers such that the design drawings included the specialty concerns before they would go off-board. The designers and specialty engineers were encouraged to work concurrently to develop a design solution that satisfies the performance requirements and specialty constraints.

Eventually, the computer will restore much of the designer's freedom from specialty engineering dependence by embedding specialty engineering constraints in the computer-aided design tools used to create design drawings. When the designer creates a design feature using CAD that is in conflict with an automated specialty check, he or she will be informed of the conflict and will be able to use help features to find out how to avoid the conflict. Prior to the arrival of this utopia, we humans will have to continue to try to work closely together during the design process. In this book we are interested in a cooperative requirements analysis process, but will initially cover the overall generic specialty engineering process within the environment most companies operate in today.

3.7.2 The Generic Specialty Engineering Process

All specialty engineering disciplines follow a similar pattern of behavior with respect to design engineers focused on identifying requirements, encouraging understanding of the requirements, concurrently supporting the designer while the designer synthesizes requirements into a concept followed by expression of the concept on design drawings, and assessment of the design solution for compliance. There is a spectrum of implementation possibilities here from a simple focus on compliance assessment only to effective concurrent engineering. We will assume here that the organization is capable of concurrent engineering meaning that the design and specialty engineering team members cooperate concurrently to develop requirements followed by cooperative work developing a design solution responsive to those requirements.

The first specialty engineering challenge is to identify specialty requirements for system elements, each of which has a principal engineer or integrated design team leader assigned overall responsibility. Once the requirements are identified, the specialty engineer must help the designer/team understand the requirements and ways that these requirements have been complied with in the past. During the concept and design development process, the specialty engineer must interact with the designer to concurrently assess compliance while the design unfolds.

Specialty engineers responsible for logistics support, operational employment, and product manufacturing should be simultaneously designing the logistic support system, operational employment process, and manufacturing processes (tooling, facilities, manufacturing flow, etc.) while the product design is in progress. As a result, the design engineer functions as a system engineer with respect to their activities trying to understand these sometimes conflicting requirements and fashioning a synthesis of the totality of those requirements. Taken together, we see that all of these people must forget about who is a designer and who is a specialty engineer and form effective cross-functional teams producing a coordinated product design, manufacturing process, and employment process. The goal for the team should be to become the equivalent of one all-knowing engineer. Mankind's success in building new knowledge precludes the possibility that any one engineer will ever be able to understand a truly complex problem so we will forever have to create the equivalent of one great mind, that can master a problem space, by teaming. The ultimate function of the system engineer from the system development process perspective thus becomes integration and optimization across the several team member perspectives.

3.7.2.1 Requirements identification responsibility aid

We must have a foolproof way to communicate to the several specialty engineering analysts that they have a specific requirements analysis task to perform. We must have a way to place a clear demand upon them to provide a specific service. One method for doing this is to use a specialty engineering scoping matrix (SESM). The model used here is based on the design constraints scoping matrix (DCSM) used in Air Force Systems Command Manual AFSCM 375 series. The matrix, an example of which is illustrated in Figure 3.7-1, correlates the different engineering specialties on the left margin with the architecture elements across the top of the matrix. An appropriate symbol in a matrix intersection means that the responsible engineering specialty discipline (left margin) must identify one or more requirements for that discipline against that architecture element (top margin).

The top, architecture identification border may be structured to permit wide flexibility in form use. The example identifies components using product or architecture ID codes to designate system elements, but WBS could be used instead of, or in addition to, these codes. Obviously, one matrix for a complete system may become completely unwieldy because of the number of elements in the system. This can easily be handled through multiple pages suggested in Figure 3.7-1 each with the same vertical axis defining the constraint categories. We can use this approach in a progressive way as the architecture expands through the application of structured decomposition. As a new layer in the architecture unfolds, we can make a new page of the matrix.

We place a symbol in each intersection to denote whether or not the constraint category applies to that architecture

PRODUCT ARCHITECTURE

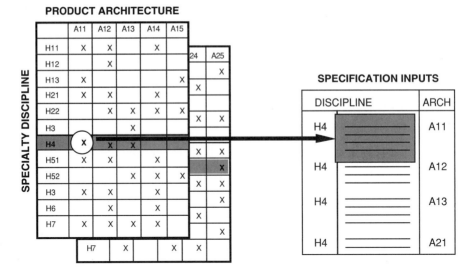

Figure 3.7-1   *Specialty engineering scoping matrix*

element. We can use a simple "X" to indicate yes, an "O" to indicate no, and a blank to indicate we don't know yet. In this case, the specialty engineer responsible for a given specialty constraint category simply responds by writing one or more constraints against each item marked "X" in his or her row.

But, we may also use the matrix to capture a more complex reporting mechanism in addition to tasking the specialty engineering analysts. The system engineering function could first mark up the matrix with "Xs" as indicated above to identify what disciplines will be required on the program. The matrix (some subset of the total matrix) is then passed around to the specialty groups for review. Each specialty group annotates the matrix to indicate the anticipated impact or difficulty using one of the codes listed below and returns the marked-up matrix to system engineering for integration and database update. This action could be performed more efficiently in a fully automated, on-line fashion, of course.

X   The constraint applies to this element. No difficulty
         is foreseen in implementing the constraint.
Y   The constraint applies to this element. Additional
         study or analysis is required to determine the
         impact of implementation.
Z   The constraint applies to this element. A serious
         impact is foreseen in implementing the constraint
         in terms of cost, schedule, and/or performance risk.
O   The constraint has been reviewed against this element and found not to apply.
         A blank indicates that the analysis related to this
         intersection has not been completed.

The SESM (referred to at the time as a DCSM) was used as part of a paper-oriented analysis approach many years ago and later during mainframe computer database implementation before networked systems became possible and popular. The author, in his book *System Requirements Analysis* published by McGraw-Hill in 1993, offered praise for this tool but confesses to later becoming less enthusias-

tic about its use. More recently, however, since trying to evolve the RAS Complete reported upon in Chapter 3.10, the author has accepted the use of this device as a means for system engineering to report upon the results of its analysis at the system level on the need for these many specialty engineering discipline views in system specifications.

3.7.2.2 Requirements capture
In earlier days, the engineering specialty analyst responded to the final DCSM by applying his or her specialized tools and procedures to identify design constraints for which he or she was responsible, often by reference to applicable documents appropriate to the development. These documents must have been screened earlier (preferably in the proposal phase) for any needed tailoring and the tailoring agreed upon with the customer. Alternatively, the specialty analyst might appeal to a system mathematical model to allocate a requirement value from parent to child through each level and branch of the architecture. This response could occur within the environment of any of the four basic requirements analysis strategies (structured analysis, cloning, question and answer, or freestyle).

3.7.2.2.1 Freestyle Approach
In the freestyle strategy, the analyst responsible for a specialty discipline simply ensures that he or she writes one or more constraints against each architecture item defined by the matrix and provides it/them to the appropriate principal engineers. This could be in the form of a tabular printout of architecture versus requirement values, an applicable document referencing requirement statement that applies to all architecture elements, or a series of statements with blanks that are filled from a tabular listing as a function of architecture.

3.7.2.2.2 Cloning Approach
In the cloning strategy, the specialty analyst might provide the person responsible for preparing boilerplates with the generic information that applies across all of the elements

that might use a particular boilerplate and a tabular list that contains the specific values for each architecture element. Each principal engineer using the boilerplate would simply fill in the value from the table for that item into the blank provided in the boilerplate text.

3.7.2.2.3 Question and Answer Approach
The Q&A approach does not have a really useful connection with specialty engineering because it implies that someone other than the specialty engineer might know more than the specialty engineer does about his or her specialty and after all they owe their existence to their specialized knowledge.

3.7.2.2.4 The Structured Strategy in Years Gone By
In the structured approach encouraged in AFSCM 375-5, each specialty engineering analyst would be required to define one or more specific design constraints for each intersection marked with a letter character on each specialty engineering scoping matrix, as in the other cases. The difference is that he or she would be required to respond in a definite time frame for given architecture items on a special form provided for that purpose. Figure 3.7-2 illustrates a typical design constraints identification form (DCIF) from mainframe database days. This form could be used directly in a manual implementation of design constraints identification or as a computer data-entry form.

It was intended that any one particular form be completed by one specialty discipline, such as Maintainability. The specialty analyst enters the constraint number from the design constraints scoping matrix, followed by the constraint category name. Some typical source documents that could be referenced in the form are: (a) higher-tier specifications, (b) military specifications, standards, and handbooks, (c) trade study and analyses reports, (d) associate contractor specifications and ICD, (e) design review and technical interchange meeting minutes, and (f) internal correspondence and documents

The criticality code (CRIT) column could be used to solicit specialty engineering criticality information. For example, if Safety Engineering concluded that failure to satisfy a particular requirement could result in personnel death or catastrophic damage to system elements, they might identify the requirement as safety critical.

Enter the constraint in quantitative engineering terms with tolerances where appropriate. Use specification phraseology (as defined in your specification standard or style guide) such that the constraint may be used directly in the appropriate specification. Specify those essential characteristics that must be controlled during the detailed design and development process. The same constraint may apply to more than one system element. Number each unique constraint identified for each constraint number and category starting with the number 1.

Enter the system Architecture code to which the constraint corresponds and the accepted element name or acronym. If required, and configuration item numbers have been assigned, enter the configuration item number. Alternatively, a program could require use of WBS numbers as the means of identifying architecture items. Remember, there should be one or more line items on this form for each intersection marked on the specialty engineering constraints scoping matrix. Figure 3.7-2 calls for identification of the paragraph (PARA) number that the requirement will flow into in the specification associated with the architecture item identified.

3.7.2.2.5 Structured Analysis in the Twenty-First Century
Today the analyst might be called upon to enter the specialty requirement directly into a program requirements database or into his or her own specialty database which would ideally be linked to the requirements database such that the specialty database content could be either brought into the requirements database at specific benchmark times or the model values linked into the specification paragraph structures when spinning a specification from the requirements database.

In many programs the specialty engineers applying a mathematical model will maintain their model in a special database tool and have to transfer the requirements to the separate requirements database with additional key-strokes suffering the potential for failure to maintain the two sets of information in synchronism. Any time this prescription is followed, a double booking arrangement is being applied and there is a good chance that the two databases will become uncoordinated. A better arrangement is to link the two databases such that the specialty database is accessible

ANALYST John Jones		DATE 05-13-1986	LSA REFERENCE		ACTION			
SPECIALTY ENGINEERING DISCIPLINE CODE H11		SPECIALTY ENGINEERING DISCIPLINE NAME Reliability			REV A	DATE 06-03-1986		
LINE ITEM	REQUIREMENTS INFORMATION					ALLOCATION INFORMATION		
	DESIGN CONSTRAINT STATEMENT		SOURCE REFERENCE		CRIT	ARCH ID		PARA
1	Item mean time between failures shall be greater than 2000 hours.		Program RAM AAA Report.			A11 Computer		3.1.6
2	Item mean time between failures shall be greater than 2500 hours.		Program RAM AAA Report.			A12 Guidance Set		3.1.6
3	Item mean time between failures shall be greater than 2400 hours.		Program RAM AAA Report.			A13 Altimeter		3.1.6

Figure 3.7-2   *Design constraints identification form*

from the requirements database and can be used to update the requirements values.

A paper implemented requirements analysis sheet (RAS) can be used to capture specialty engineering requirements but the sheet can better be implemented in a requirements database system. Table 3.7-1 illustrates an example of a few specialty engineering entries in a program RAS. The requirements flow into the specification corresponding to the product entity identified in the right hand pair of columns. The RAS maps a model entity (model ID and name) to a product entity (product ID and item name) and a requirement entity (requirement ID and requirement text). A requirement ID (RID) is a computer assigned unique code used for establishing traceability and other relationships inside the database.

3.7.2.3 Constraints integration
The architecture element principal engineer should be held accountable for integrating the effects of the specialty constraints. The principal engineer will normally be a system engineer at the highest system levels, possibly an Integrated Product and Process Team (IPPT) leader at intermediate system levels between segment and subsystem (on programs that use teams), or a subsystem or component design engineer at lower system levels. All of these principal engineers will have different degrees of skill and experience with specialty engineering integration and some may need support from a system engineer.

There are two alternative applications of a system engineering function in the integration process. First, they could be used to audit the performance of the principal engineers in this process and provide feedback to principal engineers about their conclusions. Second, system engineers could be assigned to teams to do the integration work for the principal engineers. The particular program team must evaluate the skills of its personnel and select an appropriate approach for their situation.

In addition, the system engineering community should be tasked with coordinating the specialty engineering requirements analysis activity to (1) ensure all required disciplines have access to adequate budget to perform their task, (2) monitor that each required discipline has qualified personnel assigned to specialty requirements tasks, (3) verify that each discipline is responding in a timely way to the SESM content (or other mechanism of direction), and (4) ensure that the specialty requirements integration process is working. This requires energetic pursuit on the part of the specialty engineers and eager interaction on the part of the design engineers. All too often both parties tend to withdraw from this relationship. What is required is that both plunge across the abyss and engage in real teamwork.

The integration agent must also look across the different specialty domains to seek out any cases of conflict between two or more requirements. All of these specialty engineers will be working within the narrow confines of their specialty engineering model and may choose to allocate values in a pattern that creates conflicts with other value allocations from other engineers. For example, the reliability engineer allocates a reliability figure of 0.9999 to an item while a maintainability engineer allocates a remove and replace time of 0.1 hour to the same item. These numbers come together in the maintainability and availability models and will represent a waste of resources. If the item is going to be so reliable, it will not hurt aggregate system maintainability or availability to allocate a longer remove and replace time to it. It is never going to fail so why worry about a speedy replacement? The maintainability engineer could be tasked with auditing this condition because his/her model includes R and M numbers.

3.7.2.4 Specialty constraints communication
Many specialty engineering requirements are contained in often-extensive applicable documents. These are imported into the program through a specialty engineering requirements statement such as, "Pressure vessel design shall conform to MIL-STD-1522A." This simple statement places a constraint on the responsible designer to comply with the many requirements in sections 4, 5, and 6 of that document to the extent that they apply to pressure vessels if they are not tailored out.

One of the most difficult tasks in system engineering is the efficient communication of the meaning of the contents of the pile of applicable documents referenced in specialty engineering requirements statements to the designers. Designers who have a great deal of experience will have acquired a good understanding of many of these requirements, but not all. Recent college graduates will have little experience with the content of any of these documents nor the concept of importing requirements through the referencing mechanism.

Each engineering specialty discipline must work assertively and interactively with the design community to insure that the designers understand the specialty requirements and their consequences. This interaction could take the form of any combination of four principal initiatives on the part of each specialty discipline: (1) specialty checklist, (2) person-to-person discussion, (3) organized interaction meetings, and (4) participation in trade studies and engineering review boards. This very difficult process could eventually be replaced by integrating specialty requirements into computer aided design packages.

3.7.2.4.1 Checklist Approach
In the checklist approach, each specialty domain engineer who invokes an applicable document containing a voluminous list of design constraints must translate that listing into a checklist focused on the specific product line and element, if possible. The checklist must be formatted such

Table 3.7-1 *RAS specialty engineering entries example*

| Model entity | | Requirement entity | | Product entity | |
MID	Model name	RID	Requirement	PID	Item Name
H11	Reliability Math Model	DF6T	MTBF > 2000 Hours	A11	Computer
H11	Reliability Math Model	HY7R	MTBF > 2500 Hours	A12	Guidance Set
H11	Reliability Math Model	J9I8	MTBF > 2400 Hours	A13	Altimeter

that the designer can clearly and rapidly understand the important attributes that must be satisfied. This checklist must be made available to each design principal engineer affected by the requirements. The checklists could be communicated with printed material or via a networked computer system.

Each engineering specialty creates a specialty checklist under this approach that is responsive to the applicable document(s) called out for that discipline. The checklist simply lists each specialty requirement contained in an applicable document and provides a space for each item to be checked off. A specialty checklist may have several columns for different kinds of elements (AGE, flight vehicle, etc.). The checklists should be created during the early period of the program prior to the beginning of design work. Specialty requirements often apply to all system elements across several layers of architecture but can best be dealt with in association with assembly or component level design work. The checklists must be available by the time it is necessary to identify requirements for component level elements that will yield to detailed design.

If the checklist approach is applied to all specialty disciplines, there are many disciplines on a program, there are many requirements per discipline, and all of the specialty disciplines use the checklist approach effectively, the design engineers can easily become completely overloaded just dealing with these checklists.

It is, however, important to make every reasonable effort to help the design engineers understand the complete set of requirements that must be complied with. One element of that assistance could be a joint checklist peer review. This is an organized review of all of the specialty checklists by the specialty disciplines. A representative from each specialty discipline reads all of the checklists of the other disciplines with an eye for spotting any conflicts with their own checklist.

One way to implement this review is to meet for one hour per day for "n" days (where n equals the number of specialty disciplines on the program). Each day, a different one of the specialty disciplines should act as the host, providing copies of his or her checklist, explaining the content, and reacting to critical comment. A system engineer should set up these meetings and oversee them as necessary to ensure issues are clearly resolved. The resultant joint checklist can then be briefed to some or all of the designers with follow-up interaction on a personal basis by the specialty engineers.

Another checklist simplification approach is for each specialty to partition their aggregate checklist into subsets that provide all of the constraints related to particular kinds of elements common to the company product line or particular program elements. This may eliminate many checklist items from the list that a designer, interested only, for example, in valves, must use. This becomes very important when you consider that a designer may have to deal with checklists from ten or more specialties.

The checklist approach commonly breaks down in implementation for several reasons. There is often an open hostility between design engineers and specialty engineers. The design engineers tend to be very positive about their emerging design concept while the specialty engineer appears to the design engineer as very negative in outlook searching for ways that the beautiful design may fail or kill someone. There are other reasons that help to nurture this hostility and it should be recognized by management that these atti-

tudes have to be stamped out on the program if there is to be success in specialty engineering work. Too often, the engineers simply wish to get the specialty engineer out of their sight in association with the checklist review. So, the checklist is completed in record time with little or no improvement of design engineer understanding of the specialty requirements. In a badly run engineering organization, the specialty engineer only cares about getting a signed list back which can be trotted out when the design fails and waved as evidence that the engineer said he or she understood the requirements. This is clearly not specialty engineering integration or concurrent engineering.

3.7.2.4.2 Individual Person-to-Person

In the person-to-person approach, each specialty discipline interacts on a one-on-one basis with each designer, with or without benefit of a checklist, to help the designer understand their requirements. If checklists are used, these conversations can focus on the checklist items and whether or not the designer understands the listed items. Where checklists are not used, the specialty engineer must develop a feeling for whether the designer understands the raw content of the applicable document or allocated specialty requirement.

In these discussions, the specialty engineers may be able to offer advice about which potential design solutions under consideration would have the better result in satisfying their requirements. The specialty engineer may also give the designer some examples from similar programs about how these requirements were satisfied.

3.7.2.4.3 Organized Interaction Meetings

The designers and specialty engineering representatives could be brought together periodically for informal discussions about current problems and shared insights. This may be most useful when a development team must respond to a specialty engineering discipline very uncommon to their experience. For example, if a private aircraft company won a U.S. Army contract to provide an air vehicle that could survive in nuclear battlefield situation, they would have to master some very uncommon design solutions. A periodic meeting may be very effective in this situation until the designers have come to understand appropriate design approaches. These meetings may be more a matter of instruction at first eventually turning into effective technical exchanges.

This approach can also be worked into periodic team meetings such that one or more specialty engineers are asked to give a brief presentation on their discipline encouraging designer understanding of the fundamental principles. In the case of the nuclear survivability and vulnerability story included earlier, the designer of the advanced cruise missile fuselage section angry about having his design rejected repeatedly by the specialty engineer in a fashion that did not help the designer understand how to comply, went to the Chief Engineer for relief. What was discovered was that the designer had been responding to the Chief Engineer's direction to get the weight out of the design because of a missile overweight condition that exceeded the max gross weight of one of the carrier aircraft locations. The program was managing the specialty discipline it understood and ignoring the one they did not.

In his zeal to cooperate with weight reduction, the designer had thinned out the design to the extreme that there was a danger of the high current flows that can occur in metallic objects in close proximity to a nuclear blast might damage the thin skin acting like a fuse rather than a fuselage. The designer understood weight reduction but did not understand nuclear survivability and the specialty engineer did not or could not explain it to him. The answer here was a meeting where each discipline (structures design, mass properties, and nuclear survivability and vulnerability) had to explain their concerns and solution space. A compromise was worked out very close to the limits set by the laws of physics and the design proceeded but the weight problem was sufficiently severe that missile max gross weight growth did occur requiring carrier aircraft modification. If this problem had been identified earlier through the kinds of interaction meetings encouraged above, a considerable amount of program risk might have been avoided or at least reduced. A complicating factor in this case was a failure to provide for a weight margin early in the development effort.

3.7.2.4.4 Decision Support

Another way the specialty engineers may interact with the design process is to participate in trade studies, design reviews, and engineering review boards. In these forums, the specialty engineer is responsible to ensure that the preferred solution offered or the decision arrived at in the selection among alternatives has properly taken into consideration the relative valuation offered by the specialty engineer and any specialty concerns.

3.7.2.5 Specialty design assessment

From the 1950s through the 1980s many companies allowed growing specialized engineering design, analysis, and specialty disciplines to form departments and build walls such that development work progressed in a serial fashion. Specialty groups were allowed to review finished drawings without concurrent interaction during the creative concept and design development process. This phenomenon is called stovepipe engineering by some people. It is encouraged in many companies by virtue of the functional departments tending to be more stable than the program organizations and over time longevity wins the relative power game. Functional managers would generally prefer that all of their people be collocated with them. Barring that, the functional managers tend to set up reflections of themselves on each program if allowed by the programs.

As discussed earlier, the specialty and analysis groups must be employed in a concurrent way in order to create around the creative design engineer the effect of a single, complete all-seeing engineer effective in initially designing-in all product specialty engineering requirements simultaneous with the design of the product manufacturing and employment process. This requires the formation of effective, physically collocated teams of design, specialty, and analysis engineers and their manufacturing, logistics, and operational specialists oriented about the product architecture.

Even in an effective concurrent design environment, however, we should not fail to independently and formally assess the design for requirements compliance. If a project organizes by integrated product and process teams (IPPT), this can be done through cross-IPPT assessment of each other's designs to preserve a degree of independence or by the specialty engineering function referred to as a program integration team (PIT) in this series. The results from this assessment should be captured in some enduring way so that it may be used as specialty requirements verification data at a customer-mandated functional configuration audit (FCA). If checklists are used, a set of signed-off checklists may be adequate for this purpose. This checklist could have one column for pre-design interaction and another for post-design assessment.

3.7.2.5.1 Non-Compliance Identification

During the design process, the engineering specialists must interact with the designers, study the preliminary sketches, and finally study the detailed engineering drawings offered for formal check and release to assess the degree of compliance with their requirements. If a checklist was used as a means to inform the design engineer then the specialty engineer should use the same checklist as a systematic aid in the assessment.

The specialty engineer must have some understandable rationale for a conclusion of non-compliance. It is not enough for the specialty engineer to simply conclude that a design fails to meet the requirements. The designer must be offered an understandable statement of the reason in terms of the requirement. A checklist will help in this regard where the specialty engineer may refer to a specific checklist item and describe the design characteristic that is at fault.

It is less costly for the specialty engineers to identify non-compliance issues before design release than after and the earlier the better. But, situations do arise where it is necessary to view several drawings and associated analyses before the specialty engineer can form an educated opinion about compliance. In these cases, some drawings may have been released in the process of waiting for the others needed for the analysis. This is an unavoidable problem the severity of which can be reduced if the specialty engineer gives these kinds of compliance assessments top priority when the complete set of data does becomes available. A policy of drawing package release is one way to soften this problem. The drawings are withheld from release until a complete package for the item or some major element of it is complete and has been thoroughly reviewed by team members.

3.7.2.5.2 Non-Compliance Correction

When the specialty engineer finds a design feature that fails to comply with a valid specialty requirement, that engineer must first try to resolve the issue with the designer directly. It will be helpful to refer to the specific checklist item or documented requirement and the specific way in which the design fails to satisfy the requirement in this conversation. The experienced specialty engineer can also offer fairly specific ways the problem could be overcome in the design. The specialty engineer should recognize that the designer may very well feel protective and defensive about the results of his or her creativity. A little tact mixed with assertiveness will generally be repaid in this conversation.

If the direct approach to the designer fails to achieve a satisfactory solution, the specialty engineer must bring the non-compliance issue to the attention of the designer's supervisor, manager, IPPT leader, or the Chief Engineer. Where the specialty discipline is on the drawing signoff (approval) list, recognition of a problem is guaranteed, but correction is not. Where the engineer is not on the drawing signoff list, he or she must ensure that the non-compliant issue does not pass unnoticed. A program issue system can be useful in such cases. A program issue is a special kind of action item that must be resolved by the designer of the element against which the issue is written. Like all engineering problems, the team meeting is the court of last resort for these issues.

It is a healthy thing for engineers to recognize that designers and specialty engineers look at the design process very differently. Designers take a very positive view of their design looking for ways to make it work and satisfy the requirements. Specialty engineers, by the nature of their job description, tend to be negative in outlook. They look for ways the design can fail. This attitude difference is one of the reasons that hard feelings can develop between specialty and design engineers. In some companies, as a result of a constant drumbeat of hostility from designers, specialty engineers withdraw from the fray contenting themselves with drawing signoff as a means of dealing remotely with a disagreeable adversary. System engineers and team leaders have to be watchful that these conditions do not develop within their team and program.

3.7.3 Engineering Specialty Activities Overview

This section offers the system engineer a summary understanding of each of the many specialty disciplines that may be required on a program. Not every one of these disciplines will be required on all programs.

Also, the Integrated Master Plan (IMP) for a given program could include an identification of the specialty disciplines required. It is possible that a program may conclude that one or more specialty engineering disciplines must be applied even though the contract does not specifically call for it in order to ensure the design is adequate for the requirements defined in the customer's system specification.

An alternative to IMP listing is to include this in the system definition document (SDD) as indicated in Chapter 3.11. Appendix E of that document as encouraged in this book provides the system engineer a place to list all of the specialty disciplines coded with the identifier H. This same list can then be used as the vertical axis of the specialty-architecture matrix, which in Chapter 3.11 takes the place of the specialty engineering scoping matrix referred to in this chapter.

3.7.3.1 Reliability engineering

The reliability program will commonly be conducted in accordance with some customer-mandated standard such as MIL-STD-785B, as tailored by the program statement of work (SOW). Reliability planning data specific to the program may either be contained in a deliverable Reliability Program Plan or only in a section of the System Engineering Management Plan or IMP narratives. There are many good textbooks on reliability since it is a very mature and widely recognized specialty engineering discipline. You may also refer to MIL-STD-756B for a thorough treatment of how to perform the mathematics associated with reliability engineering.

Reliability is frequently expressed as the probability that a system (or item) will perform its intended function for a specified interval (normally one mission) under defined conditions. It is stated as a number between 0 and plus 1. A typical reliability requirement would read, "Item reliability shall be greater than or equal to 0.985." Alternatively, the reliability requirement could be stated in terms of a failure rate commonly referred to as lambda (λ). The failure rate for an item is defined as the ratio of the number of failures

expected divided by the total operating time (commonly in hours). If an item were to have no more than 3 failures per 10,000 hours of operation, a failure rate requirement statement for this item would read, "Item failure rate shall be less than or equal to 0.0003." A third way to state a reliability requirement is in terms of mean time between failure (MTBF), which is the reciprocal of the failure rate (MTBF=1/failure rate). For the item above the reliability requirement statement would read, "Item MTBF shall be greater than or equal to 3333.3 hours."

These three reliability figures of merit are all related to each other as you can see in these three equations:

$$\text{Failure Rate } (\lambda) = \frac{\text{Number of failure}}{\text{Total Operating Time}}, \text{MTBF} = \frac{1}{\text{Failure Rate}}$$

Reliability = $e^{-\lambda t}$, where the exponential distribution is appropriate, λ is the failure rate, and t is the amount of time measured in the same units as λ.

Reliability engineering serves as an integral part of the system engineering process by (1) performing apportionments of quantitative system reliability requirements to lower levels, (2) estimating the basic reliability and mission reliability of the system, subsystems, and equipment, (3) determining whether these reliability requirements can be achieved with the proposed design, and (4) assessing whether specific designs do comply with reliability requirements. Since reliability failure rates drive many other analyses and trade studies, it is important to have these figures established early and given wide distribution.

The following summaries of 11 reliability program tasks describe the basic management techniques, analysis methods, and analysis tools used to satisfy common program requirements and ensure product reliability.

3.7.3.1.1 Task 1, Reliability Program Plan

This plan identifies and describes task elements that fulfill program reliability requirements and also reflects policies and practices. It may be an integral part of a System Effectiveness Program Plan which incorporates additional plans for related disciplines such as maintainability, quality assurance, human engineering, parts and materials, processes, and so on. Reliability activities are planned to meet contract objectives and revised as required when engineering change proposals are generated.

3.7.3.1.2 Task 2, Subcontractor and Supplier Control

Reliability requirements, as they pertain to management of subcontractors and suppliers, are flowed down through item procurement specifications and statements of work. Reliability maintains control by (1) evaluating the subsequent data submittals, (2) assessing reliability through test data, (3) participating in subcontractor reviews, and (4) by discretionary visits to the subcontractors facilities to monitor their reliability program.

3.7.3.1.3 Task 3, Failure Reporting, Analysis, and Corrective Action System (FRACAS)

A FRACAS is a closed-loop system that provides the primary means to report failures, evaluate causes, and implement decisive corrective actions. A FRACAS applies to all equipment that the contractor is responsible for, commencing with the lowest indentured assembly subjected to test. Closed loop problem reporting is applied to suppliers with reports provided to the prime contractor. Records are maintained for use in future problem solving support.

3.7.3.1.4 Task 4, Failure Review Board (FRB)

FRACAS provides the mechanism to acquire and update failure and reliability data. A Failure Review Board is established to review failure analysis reports to ensure that corrective actions are implemented in a timely way. It should include representatives from the program office, reliability, design engineering, risk management, system safety, producibility, and quality assurance.

3.7.3.1.5 Task 5, Reliability Modeling

A reliability block diagram is established and maintained to show the functional reliability interrelationships of the system to the subsystem and/or black box level. The model provides input for reliability allocation calculations, and when filled with prediction failure rate data, produces basic system reliability requirements and compliance data. This task should make use of computer tools to enable rapid "what-if" manipulation and data collection. Tasks 5 and 6 are the principal requirements analysis activities performed by the reliability engineer.

Figure 3.7-3 illustrates a typical simple reliability model with only a single case of redundancy. The blocks represent the architecture items and should come only from the system architecture diagram. They are arranged in a network that reflects whether they are related in a serial or parallel fashion. Parallel networks indicate redundancy while serial relationships indicate potential single point failures. This model corresponds to a system that is required to have a reliability of 0.98. This means that the probability of a failure during a prescribed period of performance of the system (commonly during a mission) is 1–0.98 = 0.02.

The reliability mathematical model corresponding to Figure 3.7-3 is:

$$R_A = R_{A1} \times R_{A2} \times [R_{A3} + R_{A4} - (R_{A3} \times R_{A4})] \times R_{A5}$$

You would expect to see system reliability figures on the order of 0.92 to 0.98 except in very high reliability systems. There is, as you can imagine, an unavoidable conflict between very reliable systems and affordable systems. In order for the system reliability figure to be met, the lower-tier elements of the system must have an increasingly reliable character as you can see from the way reliability figures (probabilities) have to be combined. When this flows down to the lowest order elements, it can require a higher degree of reliability than can be provided within cost constraints. You can see a need for a very careful balancing between these two requirements values throughout the hierarchy of a system.

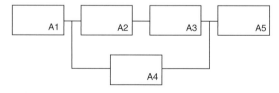

Figure 3.7-3 *Typical reliability model*

3.7.3.1.6 Task 6, Reliability Allocations

A system reliability requirement is defined and allocated down through the hardware of the breakdown structure to give an initial developmental reliability goal for each of the elements. Where there is some chance of reliability compliance risk, margins are assigned to permit intelligent management. The allocation calculations are entered into the reliability model and hooked in to the requirements capture process. Figure 3.7-3 provides a model for allocation of system reliability R_A to elements A1, A2, A3, A4, and A5.

Given the system in Figure 3.7-3, with a system reliability requirement of 0.92 and a mission duration of 3 hours, we might allocate the reliability to these items in accordance with Table 3.7-2 based in part on the design concept selected for these items. The analyst might choose any of the three reliability figures shown based on customer requirements, analyst preferences, and available computer tools. A computerized tool should be capable of rapid manipulation of alternative values.

If you plug the lower-tier reliability figures into the math model given above, you will see that they result in a system reliability of 0.92. Failure rate and MTBF numbers can be calculated from the equations previously listed.

3.7.3.1.7 Task 7, Reliability Predictions

Allocation is accomplished prior to design. As the design evolves, it is reviewed and reliability predictions made based on a parts count, part reliability figures, and redundancy patterns. Prediction data should be entered into a computer database permitting periodic status reports, rapid analysis of compiled data, and comparison of predictions with allocations. Prediction data is, like allocation, analytical. The only way to acquire actual reliability data is from tests or actual use within the context of a method of collecting failure data.

The reliability analyst should have some way to record reliability figures, both allocated, predicted, and measured or demonstrated for each system item. Table 3.7-2 provides an example of such a list. This table can be combined with a specification boilerplate approach to provide principal engineers with the basic reliability statement (boilerplate) and specific values corresponding to each architecture item. The principal engineer picks out the value corresponding to his or her assignment and fills it into the boilerplate reliability statement blank.

3.7.3.1.8 Task 8, Failure Modes, Effects, and Criticality Analysis (FMECA)

FMECA can be performed at the part, component, subsystem, or system level. The product of this analysis is a series of statements about what can fail, the effects of failure, and the criticality of the failure for each failure analyzed. The results are used to concurrently support the designer in the selection of parts, design of networks, and redundancy decisions.

The results are also used by system safety to trigger hazard analyses to determine safety consequences of failure. These failures become safety hazards, which safety engineers discuss with designers seeking ways to mitigate the probability of occurrence or serious consequences of the occurrence. In some cases, the problem is mitigated by procedural coverage that prohibits operation of the system in certain ways but in other cases hardware or software design changes will be necessary to reduce the problem severity.

Table 3.7-2 *System reliability data table*

Arch ID	Allocated reliability			Predicted MTBF	Demonstrated MTBF
	Reliability	*Failure rate*	*MTBF*		
A	0.92				
A1	0.98				
A2	0.97				
A3	0.95				
A4	0.94				
A5	0.97				

This data is useful in determining support equipment needs and requirements by showing what can fail and the modes in which it can fail. It is the basis for determining the degree of success in the current built in test (BIT) design by providing the total number of faults and forming a ratio of this figure with the number of faults detected by BIT. If, for example, there are 21,241 faults that can occur and 18,235 faults can be detected by BIT, then the BIT effectiveness is 18,235/21,241 = 0.8585. If the requirement is for 90 percent, the program has a little ways to go to satisfy this requirement. It may close the gap by reducing the failure modes or increasing the number of failure modes that can be detected by BIT.

3.7.3.1.9 Task 9, Reliability Critical Items and Critical Item Control Plan

Items identified by FMECA as being critical, within the context of a particular definition of criticality included in the Reliability Program Plan, are controlled by the application of provisions of that plan. Critical components might be defined as those whose failure will cause loss of the item during its mission, failure to satisfy mission requirements, heavy damage to material or facilities, or serious injury or death of personnel. The control of critical components might include recording serial numbers installed versus product end items and collection of component history data.

This task may also include participation in the Government Industry Data Exchange Program (GIDEP) described in MIL-STD-1556. Alerts processed through GIDEP provide immediate visibility to users concerning parts and materials problems in industry usage.

3.7.3.1.10 Task 10, Reliability Development, Growth, and Test (RDG&T) Plan

This is the primary reliability testing program implemented during the development phase, where the majority of reliability problems are identified and corrective actions implemented and verified. The principal tool used is Test, Analyze and Fix (TAAF) testing commonly conducted in accordance with MIL-STD-781D. This is a very costly process not often implemented.

3.7.3.1.11 Task 11, Sneak Circuit Analysis

Sneak Circuit Analysis (SCA) is a computerized technique for analyzing a network (commonly electrical but it could be a computer software, fluid, or mechanical network) for ways the circuits, as designed, may complete unintended paths that result in unplanned events. Almost always these unplanned events will lead to bad results so sneak analysis can be thought of as an extension of FMECA to detect subtle, unintended, hidden circuit design defects.

3.7.3.1.12 Reliability References

Table 3.7-3 lists several reliability references primarily from Government sources. It is said that you can tell the maturity of a discipline by the length of the references list. Some of these have been discontinued by DoD but they contained useful information of continuing value as guides.

3.7.3.2 Parts, materials, and process engineering (PMP)

Standardization of parts, materials, and processes (PMP), together with rigorous control over their selection, qualification, application, procurement, and verification, is a mandatory activity on many programs. The standardization and PMP program management ensures that the minimum number of electrical, electronic, and electromechanical (EEE) part types, all reliable, will be used in the system design and promotes maximum commonalty of PMP across all systems and subsystems.

A parts engineer provides design resolution involving parts selection, application, procurement, manufacturing, and field service problems. Design and specialty engineering people are provided full access to supplier data, military parts documentation, GIDEP, parts standards, and parts history files that are maintained current for optimal effectivity and reliability. DOD-STD-1686 provides requirements that can be established, with possible tailoring, for every subcontractor and supplier of electronic equipment. A PMP Control Plan is commonly required by DoD customers. Other references for parts, materials, and processes include MIL-STD-1546A, MIL-STD-1547A, and MIL-STD-965A.

A parts control board (PCB), materials and process control board (MPCB), or joint board may be established early in a program to manage and control PMP activities, and establish the program parts, materials, and process selection list (PMPSL), a common data deliverable. This list is generally developed from an existing approved parts list and approved materials and processes list for a company's product line updated as necessary and possibly reformatted for customer delivery requirements.

The PMPSL provides the basis for design studies and alternative designs in early engineering and manufacturing development (EMD) and will commonly be expanded to include other approved PMPs for subcontractors. PMP engineers review and approve company and subcontractor design drawings, specifications, and change documentation for compliance with PMPSL and PMP Control Plan requirements. Electronic parts/circuit tolerance analysis and parts application review (PAR) is applied to equipment designs to verify part application compliance with part electrical, thermal, and derating requirements.

Table 3.7-3 *Reliability references*

Document number	Title
3235.1-H	Test & Evaluation of System Reliability, Availability, and Maintainability
4245.7-M	Transition From Development to Production
5000.40	Reliability and Maintainability
AFP 800-7	USAF R&M 2000 Process, January 1989
AR 702-3	Army Material System Reliability, Availability, and Maintainability
GIDEP	Government Industry Data Exchange Program, Summaries of Failure Rates
LC-78-1	Storage Reliability of Missile Material Program, Missile Material Reliability Handbook Parts Count Prediction
MIL-HDBK-189	Reliability Growth Management
MIL-HDBK-217E	Reliability Prediction of Electronic Equipment
MIL-HDBK-251	Reliability/Design Thermal Applications
MIL-HDBK-781	Reliability Test Methods, Plans, and Environment for Engineering Development, Qualification, and Production
MIL-STD-280	Definitions of Item Levels, Item Interchangeability, Models, and Related Terms
MIL-STD-721	Definition of Effectiveness Terms for Reliability, Maintainability Human Factors, and Safety
MIL-STD-756B	Reliability Modeling and Prediction
MIL-STD-781D	Reliability Testing for Engineering Development, Qualification and Production
MIL-STD-785	Reliability Program for Systems and Equipment Development and Production
MIL-STD-1388	Logistics Support Analysis
MIL-STD-1543B	Reliability Program for Space and Missile Systems
MIL-STD-1556	Government/Industry Data Exchange Program (GIDEP)
MIL-STD-1591	On Aircraft, Fault Diagnosis, Subsystems, Analysis/Synthesis
MIL-STD-1629A	Procedures for Performing Failure Mode, Effects and Criticality Analysis
NAVORD OD 44622	Reliability Data Analysis and Interpretation, Volume 4
NAVSO P-6071	Best Practices: How to Avoid Surprises in the World's Most Complicated Technical Process
NPRD-3	Non-electronic Parts Reliability Data
RADC-TR-73-248	Dormancy and Power On-Off Cycling Effects on Electronic Equipment and Part Reliability
RADC-TR-74-269	Effects of Dormancy on Non-electronic Components and Materials
SECNAVINST 4490.2	Transition from Development to Production, 13 March 1987
NO NUMBER	RADC Reliability Engineer's Toolkit, July 1988

The objective of the PMP plan and related activities is to ensure integrated and coordinated management of the selection, application, procurement, control, and standardization of PMP for the program. PMP generally identifies a program requirement to use only those items listed in an approved parts, materials, and processes list. Requirements for PMP are commonly identified in specifications by reference to an approved parts list and materials and process specifications.

PMP is an area that will easily adapt to automation to avoid serial behavior in your engineering organization. PMP data is fairly simple to integrate into a CAD capability such that the designer will get a negative message when trying to use a part, material, or process not on the approved parts, materials, and processes list.

3.7.3.3 Maintainability engineering

Maintainability management is commonly conducted in accordance with program-tailored MIL-STD-470A, MIL-STD-2080A, MIL-HDBK-472, and AFR 800-18 or commercial equivalents. The management objective of the maintainability program is to integrate maintainability requirements with system requirements and equipment design thus ensuring that the system and equipment are readily maintainable at the designated maintenance levels at the lowest possible system life-cycle cost. Management of the maintainability program is designed to provide required maintainability activities throughout the system life cycle to assure attainment and retention of the desired maintainability characteristics.

An effective maintainability program emphasizes integration of maintainability attributes into the system and its components. The integration management procedure must ensure a condition of concurrent development exists between maintainability, design, system safety, reliability, life-cycle cost, manufacturing engineering, and logistics engineering personnel during the program effort to result in maximum exchange of information common to all disciplines.

Maintenance planning provides the necessary integrating input to the logistics engineering function to identify support requirements and resources for individual end items, subsystems, assemblies, and components. These requirements are then synthesized into maintenance plans for the overall system and they outline maintenance functions, flows, responsibilities, and actions. Initial maintenance plans identify system level servicing requirements, on-equipment repairs, and planned maintenance levels. Resulting plans describe functional system design in sufficient detail to identify configuration, construction, interfaces, and features of repairables. Maintenance plans can be developed for a variety of operations including: (1) test support, (2) interim contractor support, (3) contractor logistics support, (4) depot, (5) organization and intermediate, and (6) automatic test equipment.

Initial iteration of maintenance analysis activity provides the management framework for detailed task analysis within the logistics support analysis activity. Successive iterations will identify maintenance actions on removed repairable assemblies and will define the appropriate levels of repair pertaining to specific program requirements.

The Maintainability Program Plan, if required, will identify management activities necessary for integration of maintainability characteristics into the system design, otherwise the SEMP, System Effectiveness Program Plan, or Logistics Support Plan may contain this information. Examples of nine common maintainability tasks are summarized below.

3.7.3.3.1 Task 1, Maintainability Analysis
Maintainability analysis includes all the maintainability tasks addressed in the program plan. It translates overall system operational and support requirements into detailed quantitative and qualitative maintainability requirements and evaluates how many established maintainability requirements have been achieved. It is an iterative process that begins with preliminary concepts, proceeds through the development of a preferred maintainability model for the Preliminary Design Review (PDR), and culminates in a detailed design with well defined quantitative and qualitative maintenance support requirements to the lowest repairable item level.

3.7.3.3.2 Task 2, Document Maintainability Requirements and Criteria
Maintainability engineers provide inputs to the design engineers to identify and integrate specified maintainability requirements. The design criteria constitute specific maintainability goals such as modularization, standardization, accessibility, interchangeability, repair versus discard guidance, quantity and placement of test points, and degree of self-test features. These criteria are stated qualitatively or quantitatively and are used as guidelines by the design engineer. A maintainability design checklist may be provided to the design engineers and used by maintainability engineers to record their evaluations of maintainability features within the design.

The allocations and predictions are performed using one or more of the maintainability parameters listed in Table 3.7-4. These parameters are measured commonly in hours but the time figure could be selected based on the time scale of the system. Each of these parameters has a very precise definition that may have slightly different meanings to different people. It is important to get a clear understanding with the customer about the parameter they wish to use as a basis for maintainability requirements.

These and other parameters can be included in a maintainability model using mathematics common to queuing theory that permits manipulation of maintainability figures in "what-if" analyses helpful in resolving maintainability concerns by adjusting values and consuming maintainability margins in alternative ways. The model can also be arranged to compute an aggregate maintainability figure from the many lower-tier figures to validate that allocations made are consistent with top-level figures. Note: MTBM is similar to MTBF (a reliability parameter), but is equivalent only when there are no preventive maintenance or maintenance-induced failures.

The maintainability engineer should develop and maintain a maintainability requirements model that also can produce a report with content driven by the customer's data needs (expressed in a data item description). Commonly, you will be required to report detailed corrective maintenance data and a summary mean time to repair figure plus preventive or turn-around maintenance data.

There are many ways to measure maintainability and many ways to count time in maintenance events, so the corrective maintenance times must be very carefully defined and agreed upon by the customer. These times could include only remove and replace time of line replaceable units or they could include access, check-out, and logistic delay times. The maintainability engineer should keep a tabular list (in a computer database model if possible) of allocated corrective maintenance times for each element accepted into the maintainability program.

Table 3.7-5 provides a fragment of an example of a corrective maintenance requirements list and reporting format using system MTTR as the maintainability parameter. Allocated failure rate numbers from the reliability analysis are multiplied by allocated repair times for each item included in the maintenance analysis. These products are added and divided by the sum of allocated failure rates to yield a measure of MTTR. In this case the MTTR figure would be 0.3500 meaning that it takes on the average

Table 3.7-4 *Maintainability parameters*

Parameter acronym	Parameter full title	Meaning
MTTR	Mean time to repair	How long it takes to repair the item on average
MTRR	Mean time to remove and replace	Item removal and installation time
MTBM	Mean time between maintenance	Time between maintenance actions
Mct	Mean corrective maintenance time	Time to repair a failure on average (S/A MTTR)
Mpt	Mean preventive maintenance time	Time required for preventive action

Table 3.7-5 *Corrective maintenance requirements list*

Arch ID	Allocated failure rate	Allocated repair time (HR)	MTTR	Other data
A121	0.02	1.3400	0.0268	
A122	2.24	0.1000	0.2240	
	⋮	⋮	⋮	⋮
	⋮	⋮	⋮	⋮
A31	10.43	7.0805	3.6505	
	20.23		7.0805	

$0.35 \times 60 = 21$ minutes to repair the item when it fails. MTTR is computed here from the component figures based on a weighted averaging technique.

The design engineer may extract the maintainability figure from this table and plug it into the requirements for his/her item. If the design respects these allocations, the system MTTR will be as advertised. The maintainability engineer may include a margin in the allocated numbers such that a target figure is actually given to the designer and a portion is withheld at each architecture level. This margin is dispensed grudgingly to resolve designer difficulty in meeting the allocated values.

The equation applied in Table 3.7-5 is:

$$\text{MTTR} = \frac{\sum\limits_{i=1}^{n} \lambda_i \text{MTTR}_i}{\sum\limits_{i=1}^{n} \lambda_i} = \frac{7.0805}{20.23} = 0.35$$

Preventive or turn-around maintenance requirements are developed for a system or end item using a process flow diagram and companion time line. The times for each task are determined through analysis, estimate, or appeal to history. The times are combined in accordance with the flow diagram pattern to develop the total time for comparison with a system requirement. If the analytical time exceeds the required time, the maintainability engineer may have to assign specific time requirements for time-critical item tasks that apply to specific items involved in parts of the overall use process.

3.7.3.3.3 Task 3, Maintainability Quantitative Analysis to Assure Requirements Are Met

Maintainability allocations are assigned in a top-down fashion from a top level quantitative maintainability figure to provide a quantitative goal for the designer to satisfy and to permit us to evaluate cost-effective alternatives in the integration of maintainability and other requirements into the system, subsystem, and component concepts and designs. It may be necessary to develop allocations for both corrective and scheduled maintenance actions as noted above.

As the design matures such that it is possible to imagine the maintenance actions that must be performed, the maintainability engineer adds predicted failure rate and maintenance time columns to his or her data table and computes predicted MTTR. If the system level MTTR is within the required value, all is well. If it is not, then it is necessary to identify the items that are contributing the excess maintenance time and evaluate with the designer how the times can be reduced. Later, when actual failure rate and maintenance time data become available from reliability and maintainability testing, manufacturing testing, or the user, the engineer can develop a figure for achieved MTTR.

System or end item scheduled maintenance time is computed from detailed scheduled maintenance tasks by combining them in accordance with the process flow diagram and timeline diagram as noted above. In a deterministic environment, the estimated times are simply added where they are serially accomplished and critical path times added where parallel tasks are involved. In a probabilistic environment, mean and variation numbers are required and they are introduced into a queuing model to determine aggregate values. In both the deterministic and probabilistic approaches, margins should be used to provide for risk management opportunities.

As the design concept matures, maintainability engineers study the evolving design and make scheduled maintenance predictions within the context of the planned maintenance concept. Where predictions suggest a failure to satisfy allocated values, the maintainability engineer must first determine which area of the design offers the best avenue for reduction of this figure.

There may be several alternatives that may have to be studied in a trade format. The maintainability engineer then works with the selected item principal engineer for scheduled maintenance time reduction. As a last resort, maintainability figures may be re-allocated and maintainability margins attacked or margin values exchanged in other parts of the architecture.

3.7.3.3.4 Task 4, Design Surveillance/Assessment

Maintainability design inputs are formally transmitted to design engineers by conversation, tabular maintainability data, and design specifications where flow down is applicable. The design effort is monitored by attending design meetings, reviewing preliminary design data such as sketches and drawings, and through concurrent engineering discussions with the design engineer. Issues that cannot be informally solved are resolved by engineering management in internal reviews.

3.7.3.3.5 Task 5, Participate in Design Tradeoff Studies

Maintainability engineering provides inputs to alternative design concepts, support concepts, subcontractor/supplier proposals, and analysis of the effect of alternative manufacturing processes on maintainability. Normal criteria for assessing the compatibility of concepts or alternatives include relative MTTR and/or M_{pt}, access requirements, skill levels and number of personnel, special tools and test equipment, impact on facilities, and the relative life-cycle maintenance cost.

3.7.3.3.6 Task 6, Participate in Design Reviews

Maintainability engineering actively participates in all formal and informal in-house reviews and is also included in the distribution of all proposed design changes. Each change is evaluated for impact on quantitative and qualitative maintainability requirements. Upon approval of a design change, prediction parameter values, design criteria, or maintenance procedure documentation is updated as appropriate.

3.7.3.3.7 Task 7, Subcontractor and Supplier Control

Subcontractors and suppliers providing newly designed equipment are subject to the maintainability constraints allocated to them by the prime contractor. Based upon the results of the maintainability analysis, quantitative parameters are allocated to and incorporated in subcontractor or supplier specifications. Supplier progress is monitored and maintainability concerns highlighted for resolution.

3.7.3.3.8 Task 8, Failure Reporting, Analysis, and Corrective Action

Management may review failure trends, significant failure, delinquent actions, and corrective actions at a Program Reliability Review Board. A maintainability engineer participates on the review board in the maintainability analysis process. In-plant corrective action data may provide useful maintainability data including maintainability action timeline information.

3.7.3.3.9 Task 9, Conduct Maintainability Demonstration

The achievement of system maintainability requirements is evaluated by analysis and formal and/or informal maintainability demonstrations. The demonstration tests are conducted to support verification of contract requirements. Often the contractor will be required to demonstrate the system MTTR, which is comprised of the time for fault detection, fault isolation, removal and replacement, and repair verification.

3.7.3.3.10 Maintainability References

Table 3.7-6 lists several maintainability engineering documents primarily drawn from government sources. Some of these military standards have been dropped but they still contain useful background information.

3.7.3.4 Availability

There are several possible expressions of availability, but in general it is a probabilistic measure of the probability that an item will be in an operable condition and committable state at the start of a mission initiated at a random time. The range of values is, of course, 0 to 1. Inherent availability is computed from reliability and maintainability measures as follows:

$$A_i = \frac{MTBF}{MTBF + MTTR}$$

Mean time between failures (MTBF) and mean time to repair (MTTR) are normally measured in hours. Therefore, this factor is based on a ratio of operating time to the sum of operating and maintenance time (total time).

The development of availability requirements is a test of the ability of an engineering organization to perform concurrently within its specialty engineering functions in that it requires cooperation between a reliability and maintainability engineer or group. Some organizations combine these three disciplines into one RAM department. The similar probabilistic methods applied are supportive of this approach. Such a department may have to hire more finely specialized engineers but should be able to cross train them into more general engineers without their suffering from loss of detailed understanding of the details.

Availability is normally defined only for end items and systems since it relates to use of the system in the context of its intended missions and environment. All of the reliability and maintainability data for the components rolls up into these figures. If the engineering team employs a computer database approach to reliability, maintainability, and availability data, the availability numbers can be maintained automatically as a function of the independently and separately determined reliability and maintainability figures. Other availability factors, such as achieved availability and operating availability, may also be contract requirements.

3.7.3.5 Producibility engineering

The producibility program develops a synthesis of hardware design requirements and manufacturing methodologies with objectives for reducing cost and improving the producibility of a product. It provides the link between product design development and manufacturing process development. These two activities should proceed concurrently such that product and process are consistent and the combination represents an optimized solution at the grand system level resulting in least cost, maximum quality, and maximum customer value.

Producibility requirements will commonly be qualitative in nature as they apply to the product, but may be driven by quantitative manufacturing rate requirements based on planned customer delivery rates. These requirements may also include needed tooling points, features supportive of factory handling in different stages of production, and materials and processes compatible with local environmental constraints where the factory in or will be located.

A manufacturing characteristics matrix with a relative weighted system can be used to evaluate design concept producibility. Each concept for a given piece of product

Table 3.7-6   *Maintainability references*

Doc number	Document name
MIL-HDBK-472	Maintainability Handbook
MIL-I-8500C	Interchangeability and Replaceability of Component Parts for Aerospace Vehicles
MIL-M-23681	Manual, Technical, Periodic Maintenance Requirements, Preparation of
MIL-M-24365A	Maintenance Engineering Analysis: Establishment of, and Procedures For Formats For Associated Documentation; General Specification For
MIL-STD-280	Definitions of Item Levels, Item Interchangeability, Models, and Related Items
MIL-STD-470A	Maintainability Program For Systems and Equipment
MIL-STD-471	Maintainability Verification/Demonstration/Evaluation
MIL-STD-721	Definition of Effectiveness Terms For Reliability, Maintainability, Human Factors, and Safety
MIL-STD-780	Work Unit Codes For Aeronautical Equipment; Uniform Numbering System
MIL-STD-1388	Logistics Support Analysis
MIL-STD-2080(AS)	Maintainability Plan Analysis For Aircraft and Ground Support Equipment
Booher, Harold R.	Manprint, An Approach to Systems Integration, Van Nostrand Reinhold 1990
Blanchard, Benjamin	Maintainability, A Key Effective Serviceability and Maintenance Management, Wiley and Sons, 1995. Co-authors are Dinesh Verma and Elmer Peterson.
Blanchard, Benjamin	Logistics Engineering and Management, Wiley, 5th Ed, 1998

architecture is comparatively evaluated relative to its peculiar characteristics. The objective of the matrix is to facilitate rapid comparisons of alternative concepts to identify potential risks. For example, a matrix comparison between candidate materials (such as aluminum, graphite-epoxy composite, and titanium) would weigh the relative advantages and detriments based on material cost, manufacturing processes, tooling, manufacturing resources, facilities, available technology, production rate capabilities, and risk.

Manufacturing/producibility specialists employ several methods for producing inputs to this analysis. The methods involve cost and schedule tradeoffs and help to identify new methods or to verify or demonstrate existing production methods. These include:

a. Break-even analysis for evaluating projected volume or demand when comparing alternative processes or design.
b. Sensitivity analysis to determine the effects of system requirements on producibility, or the effects of producibility changes on other system requirements.
c. Value engineering analysis to identify the function, establish a value on that function, and provide that function at the lowest possible cost without degrading performance or quality.
d. Pareto analysis is used for ranking components in terms of cost and focusing the engineering effort on relatively small percentage of components with a large impact on problem causes and cost.
e. Tolerance analysis to develop compatibility between the capability of the manufacturing facility and design demands. Where incompatibility exists, a decision is made to upgrade capability or modify design demands or incur cost penalties for special handling of components.

3.7.3.6 Design to cost/life-cycle cost (DTC/LCC)

Cost unfortunately is not often considered a product requirement. MIL-STD-490A, Specification Practices, cautioned that it should not appear in a specification because it was a programmatic requirement properly covered in the contract documents. That was a simpler time when performance was everything. Today, cost is often the driving requirement perhaps referred to as affordability. One may find a unit cost figure in a specification based on a certain lot size. This figure may be a system life-cycle cost in a system specification.

The value of the dollar, and all other currencies, fluctuates over time so it is advantageous to identify cost figures against some reference value. This is more important during a period of high inflation as in the 1970s and 1980s in the United States. During a period of relative stability, this is not so important but one never knows when conditions will change driving the relative value of currencies.

A design to cost (DTC) figure establishes a unit cost target. The cost engineer allocates total cost assigned to the system down through the architecture like weight or any other quantified requirement following the sum rule. The top level cost target may be determined somewhat differently as a function of the kind of system and customer. In a military system development program, the customer will likely be involved in the definition of the DTC logic involving a projection of the cost per item in a certain lot size. Where an enterprise is selling a product commercially and applies a DTC program, the target cost may be related to the projected market price minus desired profit.

Life-cycle cost (LCC) is the sum of all of the cost figures across the life of the system. This includes non-recurring development cost, total recurring cost over the production run, spares cost, operation and maintenance (O&M) cost over the life of the system, and retirement cost. This cost too may be allocated to items in the system and used as a decision-making development measure but it may not make a lot of sense in relation below the level of end items.

3.7.3.7 Human factors engineering

Human factors engineering integrates human functions into the system. The objectives are to develop the crew/equipment and crew/software interfaces to achieve the required effectiveness of human performance during system operation, maintenance, and control. Normal human capacities, measurements, and performance capabilities are well known and recorded in manuals for use by human engineers for use in assessing designs for consistency with these figures.

Human engineering requirements are often qualitative but may include reference to documents that provide objective requirements such as MIL-STD-1472C or some other customer-mandated standard. Designers may have difficulty applying these requirements to their design work so the human factors engineer should work closely (concurrently) with the designers to evaluate design concepts and evolving designs. For example, human access to components for remove and replace actions will require access. The actions necessary to accomplish this work may not be obvious to the designer in terms of the necessary screwdriver grip or wrench swing space and consequent access door features and size.

The human factors engineer may apply process flow diagramming as a means to understand the actions people must accomplish. This is done by decomposing processes down to the level that human imagination can be effective in detailing human actions in terms of tool use; use of visual, hearing, and touch senses as information sources; and manipulation of operational and support item features.

This work may be relatively simple for a military system where the population you have to deal with is assumed to be in good physical condition and the variation of capabilities is limited by rules controlling recruiting military personnel. In commercial systems, you may have to account for physical and mental disabilities as well as a wider range of dimensions, capacities, and capabilities.

In addition to maintenance actions that result in man-machine interfaces, operator actions are often much more intense and time-dependent. The operator must be presented with information easily understood and controls that permit easy implementation of actions related to the performance and status information presented to the operator.

Operational sequence diagrams can be used to identify needed human actions coordinated with machine and software actions. Figure 3.7-4 shows an example of this kind of diagram often referred to as a "swim lanes" diagram for obvious reasons. Time runs down the page. The columns correspond to the several entities the analyst wishes to establish coordinated action between. Each entity action is followed by a subsequent action by the same entity or some other entity with connecting lines showing the sequence. Concurrent operation can be illustrated in these diagrams. This is similar to the sequence diagram of UML.

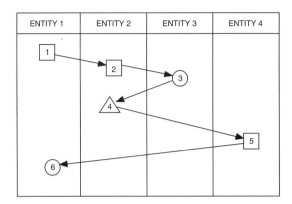

ENTITY 1	ENTITY 2	ENTITY 3	ENTITY 4

Figure 3.7-4 *Operator sequence diagram*

3.7.3.8 Corrosion prevention and control (CPC)
Corrosion prevention and control is commonly accomplished by a materials and processes engineering function, which reviews designs for dissimilar metals in contact, material properties when exposed to corrosive environmental factors, and other sources of corrosive degradation of the product.

3.7.3.9 System safety engineering
Safety requirements are commonly identified in a statement referencing a customer-selected document that lists required safety characteristics. These are often difficult to transform into requirements that a component designer can easily understand and apply. The safety engineer must interact with the designers to make sure these requirements are clearly understood and monitor the on-going design work to ensure that the design is compliant. Divergence between requirements and design must be identified as early as possible because the cost of correction is directly related to the amount of program time that passes before correction. Early correction is always relatively inexpensive.

The currency of the safety engineer is the safety hazard. These are characteristics of product and process of use that will or may result in damage to the system, its environment, or personnel. The safety engineer studies system features and a process model looking for ways that the system can cause or encounter these hazards. At the time that this study is accomplished, system use is in the future and there is likely not any physical product with which to work as well. So, we are dealing with future actions that must be imagined within the context of a model of system use. Hazards are therefore risks that can be characterized by two parameters: probability of occurrence during system operation and severity of the consequences of occurrence. The safety engineer prioritizes the hazards by rating them in the context of these parameters as suggested in Figure 3.7-5a. The hazards that are the most worrisome are those with a high probability of occurrence and result in seriously adverse consequences. These are the high-risk safety hazards that must be corrected through design changes, procedural controls, or other methods.

Where the design cannot be changed to address the perceived hazard, the safety engineer should pursue the following precedence of actions to remove or mitigate the safety risk:

a. Introduce a protective system such as a fire protection system, radiation shielding, blast shied, or protective personnel equipment.
b. Provide a warning of a hazardous condition with clear instruction on how to respond to this warning or an automatic response that relieves the condition.
c. Provide procedural coverage that when followed precludes the development of a hazardous condition.

Figure 3.7-5b, which charts hazard index versus program time, indicates that one should be able to aggregate the safety hazard index for the whole program and use it as a program metric. The methods offered in some references do not encourage this result because the higher risks have lower index numbers.

3.7.3.10 Electromagnetic compatibility (EMC) engineering
EMC engineering is concerned with the suppression of electromagnetic interference (EMI), electrostatic discharge, destructive lightning effects, and manmade equivalents. EMC engineers also identify related requirements and evaluate system designs for compliance. Where EMI concerns exist, the design must either be changed to eliminate the internal source, suppress the external source, or harden the design.

This is a two way street. Systems should produce no more than a required strength over a specified spectrum and should be capable of operation within an environment characterized by EMI up to some level. EMI is an example of a non-cooperative environmental stress applied to systems. Our system must endure these stresses applied to it unintentionally by other systems and natural sources. Where these stresses are purposely applied with malice, they can be characterized as elements of electromagnetic warfare (EW) and part of the hostile environment.

3.7.3.11 System security engineering
The objective of security engineering is to eliminate any system characteristic that could result in the deployment of the system with operational security deficiencies. System security assessment should begin during the initial program work defining security requirements. These requirements should be driven by an assessment of possible security threats. These may have to be defined by the customer and the customer may be slow to do so placing the development process at some risk of engineering changes that are more costly than would have been the case if the requirements had been defined in a timely way.

The principal security issue is physical security, which may be addressed by protective features or suppression of the hostile stimulus. In systems containing computers there is a great danger of the computers being attacked in subtle ways that have potentially disastrous results. This may include destruction or corruption of the data or the introduction of viruses or altered programs that result in erroneous operation.

3.7.3.12 Mass properties engineering
Mass properties commonly include weight, space, and center of gravity control. These requirements should be allocated to all items in the system and designs evaluated for compliance. Weight is commonly defined in a computerized mass properties model that may include the ability to manipulate the data to examine alternative allocations and the assignment of margins through which

PROBABILITY OF OCCURRENCE

	1	2	3	4	5
1	1	2	3	4	5
2	2	4	6	8	10
3	3	6	9	12	15
4	4	8	12	16	20
5	5	10	15	20	25

SERIOUSNESS OF OCCURRENCE

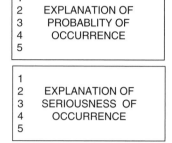

1
2   EXPLANATION OF
3   PROBABLITY OF
4   OCCURRENCE
5

1
2   EXPLANATION OF
3   SERIOUSNESS OF
4   OCCURRENCE
5

a. Safety Index

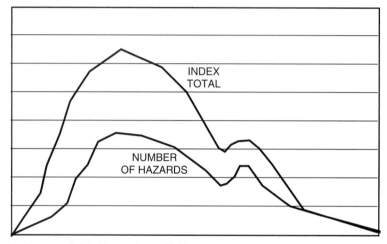

INDEX
TOTAL

NUMBER
OF HAZARDS

b. Program Safety Hazard Index Metric

Figure 3.7-5   *Safety hazard diagram*

product weight may be more easily managed as the design evolves. During the design process, weight is estimated or predicted based on the design and materials. As the product becomes a reality, the items are weighed and actual weights compared with predicted. Where space is critical, it should be managed every bit as energetically as weight. This may include controlling item form factor as well as volume.

3.7.3.13 Environmental impact engineering

The environmental impact engineering process establishes constraining boundaries on the ways the system may be allowed to impact its natural and man-made environment throughout its life cycle, including disposal. The objective is to identify all of the environments within which the system will function as part of the system environmental requirements analysis and ways that the system may cause damage to elements in those environments. The principal threats to the environment will come from product energy sources, chemicals, propellants, and explosives. The specific threats must be identified and the path through which environmental damage is possible. Where the impact is sufficiently adverse, this pathway must be interrupted, the design

changed to eliminate the impact source, or the impact otherwise mitigated.

3.7.4 Science Projects and Natural Systems

3.7.4.1 The ultimate system diagram

One of the most fundamental system concepts is the relationship between a system and its environment. Every system that has every been and ever will be but one is structured as shown in Figure 3.7-6. The system interacts with its environment. The environment consists of everything in the universe except the elements that are in the system. Most of the environment thus defined is not remotely involved in influencing the system of our interest.

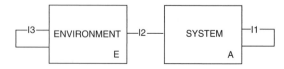

Figure 3.7-6   *The system and its environment*

The challenge is to identify precisely which elements will influence the system.

The universe consists of things that interact. It may be stretching it a bit to say that these parts interact to achieve a common purpose unless you are a very religious person but that can be covered by saying the purpose is simply in defining reality or existence. Now, this system has a very unique feature in that it consists of everything that is so it violates the concept shown in Figure 3.7-7. There is only one block. One could say that this block is all system or all environment but it is one or the other, perhaps both at the same time but fundamentally different that the view offered in Figure 3.7-6. This book is primarily concerned with the man-made systems noted within the Earth boundary.

Obviously, there are other kinds of systems, namely natural ones some of which have been around a very long time. All of these systems do consist of two or more things that interact. At the galactic level, the interfaces are gravitational as far as we know but there are particles that navigate the either sometimes striking one body or another. When man steps onto this grander stage, as he has in a very local way relative to our solar system, he and his machines must play by the laws of physics. The forces and time necessary to move through the enormous space between adjacent objects are too great to contemplate on our scale of existence. It is amazing that man has come to understand this grand view of his environment (or is it his system) with only this little outpost from which to view the whole reality.

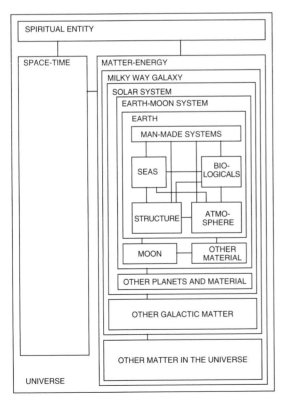

Figure 3.7-7 *The ultimate system*

We don't know if our current knowledge of the Universe is the truth or only a stopping point on the way to full understanding. What we know of this natural environment/system has been captured in study reports and standards and generally available for every conceivable place we may seek to operate a man-made system. The principal use of this information is to provide clear insight into the stresses that will be applied to man-made systems operating within particular subsets of the natural environment. But, in the process of characterizing these natural forces, we have come to understand how some collections of these entities interact to form natural systems.

We are rapidly closing in on a detailed understanding of the climatic system at work on, under, and above the Earth's surface influencing global weather patterns. We have broken the code for the movement of the tectonic plates giving great insight into geological systems. We are knocking at the gates of the genetic secrets palace a short distance from understanding life and its many normal and abnormal forms. The most complex natural system of all still eludes us in the form of our own mind but we are beginning to have an appreciation for some of the electrochemical operations at the micro level.

3.7.4.2 Give us the sense to know the difference

In all of our natural system exploration work you will note that we have not been a designer but a spectator examining these systems that we did not create, trying to characterize them, describe them. Sometimes the scale of our man-made systems and the cumulative effects of our lack of environmental concern do interpose on natural systems. In the 1940s when the author was growing up along the Penobscot River in Old Town, Maine, the river was an open sewer as were many others and getting worse through the 50s. Most of these problems in the United States have been reversed because we found ways to relate man-made systems to natural systems so as not to overload the capacity of the latter to absorb punishment and restore themselves.

The Yucca Mountain nuclear storage project in southern Nevada is a man-made system that is very tightly interfaced with natural systems. One TRW engineer working on that project confided to the author that he perceived that the mountain was GFE. This acronym has for decades been used to designate government furnished equipment on contracts to identify things that the government would provide. In this case the engineers used the acronym to refer to the mountain as God furnished equipment. Given a particular material being stored that could contaminate the surrounding water table, the chief interface requirement for this man-made system was to deny an interface between that material and the water table for thousands of years until such time that the material was safe for human contact. The team approached this problem by characterizing the natural protection formed by the geological features of the area chosen (GFE) and then determining what degree of isolation was thought to be necessary. The difference was the added protection that the man-made component was going to have to be capable of satisfying. This aggregate protection, of course, would have to be provided within the context of earthquakes and every other reasonable natural calamity that might occur over a very long time.

It is very easy to get caught up in specifying natural environmental characteristics as if we were in control of them. The reality is that we are not. We can be selective about ranges of variables within which our system will have to endure but we cannot define reality from any other perspective than experience. There is little room for creativity here. A safe way to proceed is to treat the natural environment as an interface element interacting with the man-made system while recognizing that we have to appeal to reality for parameters and values that will apply stresses to our product system.

3.7.4.3 Characterizing reality

Man has characterized his environment extensively. Most of the possible Earth bound environments have been characterized and documented in environmental standards. We have extended far from Earth as well with space environmental standards and acquired a great deal of knowledge about the Moon and other planets in our solar system as well as deep space.

We have developed theories that in the main work for many physical phenomena to the extent that we can predict outcomes given certain information. We understand how many natural systems act and can predict with fairly good accuracy many very complex outcomes such as the weather in Madison, Wisconsin, several days in advance.

We have cataloged many of these living natural systems and arranged them in an architecture in a sense. We have mapped the Earth and explained how the continents were formed and evolved. Some ultimate questions continue to elude us but we have theories that permit us to carry on while trying to develop better theories of everything.

3.7.4.4 Specific scientific development programs

Some large science projects of the immediate past include Superconductor Super Collider which would have created a huge atom smasher if it had not proven to be so costly, international thermonuclear experimental reactor (ITER) program which will build the first economically viable fusion reactor, Hubble space telescope, and the many NASA space shuttle science projects to name a few. One could argue that the international space station is a science project but it has many goals, which include some that are scientific in nature.

The Superconductor Super Collider was initially designed by a consortium of universities and the Department of Energy then let contracts to manufacture it. General Dynamics Space Systems Division won a contract to manufacture the CD magnets that would have formed the collider circle only to find that the design was not producible. Luckily the division was saved from a financial problem with an impossible fixed price contract by the government's inability to fund the program. ITER, another program in financial trouble from time to time, must also solve an extremely difficult problem involving a control system attempting to maintain a plasma inside an intense magnetic field. ITER, like many science projects, in the process of developing the solution do invent or expose new science that often leads to new regulatory controls that have the effect of increasing the cost of the solution.

Hubble will forever have to carry the development decision that resulted in the optics not being capable of perfectly focusing when initially deployed, despite the brilliance of the program's recovery. This program was more aligned with aerospace methods than university science effects so it would be unfair to paint the Hubble with that brush.

The danger in science projects appears to the author to be the methods applied in science that establish a hypothesis and seek to determine if it is valid or not. This translates into picking a particular preferred design solution not necessarily accomplished within the context of the kinds of controls that engineers insist upon to optimize the selection process.

3.8   Environmental Requirements Analysis

3.8.1 Overview

Environmental requirements actually define the requirements for interfaces between one of the environmental components appropriate to the system and an element of a system. They are, therefore, a special case of interface requirements in that sense. The environmental components are: natural, self-induced, cooperative, non-cooperative, and hostile. We seek to define the environmental stresses that our system and its parts shall experience in each of these categories so that when the system is placed in service it can withstand the combined environmental stresses placed upon it while satisfying its performance requirements.

A naval aircraft's wing must be designed to withstand not only the static stresses experienced in flight due to atmospheric drag in clear air and the effects of gravity on the supported vehicle, but for the dynamic forces experienced in conditions of wind shear, carriage and release of wing stores (including possible pyrotechnic shock during explosive separation of stores), some level of battle damage, possible overloading stresses a pilot may have to apply in tactical maneuvers, and the bone-jarring impact with an aircraft carrier deck arresting and catapult systems. Here we begin to see some of the fundamental differences emerging between military and commercial requirements for aircraft.

We are also vitally interested in not only what all of these environmental stresses are but how these many stresses combine in time to affect the wing structure over its life cycle. Must the wing endure all of these stresses simultaneously throughout its life or are some of the stresses applied in different patterns without simultaneity. If we design for the aggregate worst-case scenario, we may introduce unwanted characteristics and more capability than necessary. Most often this results in added weight and cost. One very effective tool to help piece this puzzle together is an environmental use profile where we attempt to characterize the normal combined stresses applied to the product in its normal and aberrational life cycle elements.

3.8.2 Environmental Categories

Figure 3.8-1 illustrates the several environmental partitions that are useful in developing environmental requirements. They may all be partitioned into two major sets to recognize that the cooperative interfaces are developed as system external interfaces because there is someone with whom one may talk to make agreements and evolve an agreement satisfactory to the two parties responsible for the terminals. All of the other environmental elements shown in Figure 3.8-1 are developed as environmental stresses that the system must be able to endure in operational use.

3.8.2.1 Natural environment (QN)

The natural environment is defined as the composite of space, time, and physical phenomena naturally characteristic of the space within which the system operates. Where the space is the Earth's surface and atmosphere to 30,000 feet, this includes Earth gravity, atmospheric pressure, humidity, temperature range, cyclical solar heating and light, and seasonal climatic conditions. If the space involves Earth orbital space at an altitude of 150 nautical miles, it includes alternating solar energy impingement, vacuum, and particle impact. Other environmental parameters can be imagined for an Earth undersea, Earth surface seacoast, Earth surface desert, or Martian surface environment.

The natural environment provides wonderful opportunities to write silly requirements. Our natural environmental requirements must reflect what the system must withstand rather than attempt to redefine reality. Over the period 1966 through 1975 the U.S. Air Force Strategic Air Command (SAC) operated an unmanned photo reconnaissance aircraft program flying over North Vietnam with Teledyne Ryan Aeronautical special purpose aircraft. This aircraft had some fiberglass leading edges that at one point were being damaged during flight at high altitude through the heavy seasonal weather common to the area. SAC Headquarters responded in alarm to a message from the field reporting this damage with a message to the local command saying, "In the future, there shall be no hail storms in the path of the vehicle." Requirements are written against the product system and no matter how wonderful the apparent effect, the words in our requirements will not change or influence the natural forces applied to the product system. We must have the good sense to know what we can control.

The analyst must identify the natural environmental parameters to be considered in the design based on the planned system operating spaces. MIL-STD-210, Climatic

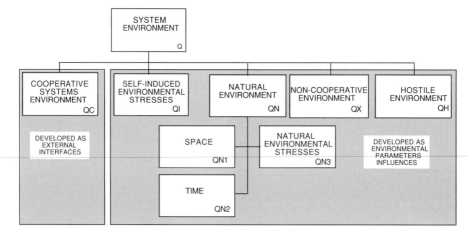

Figure 3.8-1 *System environmental categories*

Extremes for Military Equipment, is a useful source for worldwide land surface, sea surface, and atmosphere parameters to 262,000 feet and values for these parameters that should be considered for equipment. Table 3.8-1 lists several other documents that define environments corresponding to these and other spaces. Depending on the planned system employment environment, the analyst should select one or more of these references as a source of natural environmental parameter definition.

3.8.2.2 Self-induced environment (QI)

When elements of a system interact with the natural environment, it can cause other environmental effects that are not present in the absence of the stimulus provided by the system itself. These environmental stresses are commonly called induced environmental effects. For example, the rocket engines on a space launch transport booster create a tremendous amount of acoustic energy that can be heard by human observers for miles.

This same energy also pounds the components of the transport rocket itself. Some engine vibration is transmitted through the basic structure itself (an internal interface between engine and airframe communicated through the engine mounts), but the engine thrust also creates a very powerful acoustic noise that communicates through the air of the natural environment back to the other components of the launch vehicle and payload resulting in additional stress. Were it not for the presence of the system, there would be no acoustic noise in this environment. This is a characteristic of an induced environmental parameter.

Some other examples of induced environmental effects are:

a. An external store (bomb, missile, fuel pod, etc.) is attached to an external wing pylon and it is located such that the exhaust from one of the aircraft's engines impinges upon the store. The engine exhaust creates locally heated air in the natural environment that a component of the store is exposed to causing possible damaging thermal effects.

b. An enterprising aerial reconnaissance camera company once convinced an Air Force procurement officer that their camera could take good pictures through the exhaust stream of a jet engine greatly simplifying the camera installation in the airframe of choice. When this was actually tried in a test vehicle (yes, science was an insufficient basis for a decision in this case), the engine exhaust created such turbulence in the atmosphere (natural environment) under the camera window that there was very little photographic value. The light energy from the photographic target was so confused in transit that it produced a useless image on the camera film.

c. An aircraft flying through the air generates static electricity that builds up a charge on the airframe. This static electrical charge can interfere with sensitive navigation instruments like Loran C. In order to combat this phenomenon, it is common practice to install static dischargers on wing trailing edges to bleed off the charge. In this case, a special piece of equipment (a sharply pointed discharger) is installed on wing trailing edges to interface between the natural environment (atmosphere) and the airframe to equalize a charge introduced by the rapid movement of the airframe through the natural environment.

d. The U.S. Air Force A10 aircraft, which mounts a Gatling Gun in its nose, was plagued during flight testing with a high frequency of failure of electronic components. Many of these components were used in other aircraft and were not failing at this same high rate in other applications. It did not take long to determine that the cause was excessive noise, shock, and vibration generated by the Gatling Gun.

A common source for induced environmental stress is obviously found in energy sources and potentials within the system: engines, weapons, and nominally very hot or cold elements are examples. A good teammate in identifying these sources is a safety engineer because they also often result in safety hazards.

3.8.2.3 Non-cooperative environment (QX)

Other systems operating within the same space as our system may generate and apply stresses to our system without intent to interfere. A good example of this kind of influence is electro-magnetic interference. A friendly radar system that does not have any direct relationship to our system may generate RF energy that reduces our system's effectiveness.

A mischievous Marine Sergeant in a Marine Air Control Squadron radar shop once greatly enjoyed himself by directing a young woman Marine photographer to a point where she wanted to take some pictures of radar equipment used by the unit and watching the reaction when all of her flash bulbs went off in her bag in route. A Sergeant in the communications organization loved to order new troops to replace neon lights in the communications shack and watch them go into a state of shock as he keyed a BC-610 transmitter while they passed by the long wire antenna connections of the transmitter. These same energy sources can have a similarly deleterious effect on systems more complex than these, probably including the human beings radiated in these cases.

The USS Ranger had difficulty leaving port in Subic Bay one morning in route to Yankee Station in Tonkin Gulf during the Vietnam War because a sailor with critical

Table 3.8-1   *Natural environmental parameter references*

Document number	Document name
MIL-STD-810	Environmental Test Methods
MIL-STD-1540	Test Requirements for Space Vehicles
NASA-TMX-53139	A Reference Atmosphere for Patrick AFB, Florida, Annual
NOAA-S/T 76-1562	U.S. Standard Atmosphere, 1976 for NOAA, NASA, and USAF
MTP-AERO-61-78	Surface Wind Statistics for Patrick AFB (CCAFS) Florida
Special Report 264	Smithsonian Astrophysical Observation Special Report 264

knowledge of the pier fueling system was late returning to ship and the ship departed about an hour late. The ship carried with it an unmanned reconnaissance aircraft system scheduled to fly one last training mission by a small target range on the West coast of Luzon before going to war for an operational evaluation. The aircraft was launched about an hour late and flew into the range time of a target mission prelaunch checkout. The two systems shared a common command control system with different command assignments. When the target operator issued a remote command to exercise the target on its ground launcher, the reconnaissance aircraft would respond as well but in a different way. By the time the reconnaissance aircraft remote pilot got control, the bird was nearly to Clark Air Force Base in central Luzon. It didn't help that the remote control officer in an E2A was not very experienced as a controller.

There was nothing hostile in this situation (you could argue there was in the previous two examples of shocked humans). It was a case of two systems operating normally while sharing a common space and interfering as a function of non-cooperation.

These kinds of effects could be treated as interfaces, but since it is not always clear where these effects will come from, there is no one to interact with and negotiate a mutual agreement.

Therefore, non-cooperative system effects are commonly considered under and lumped into environmental effects. During the development of a new system, we need to try to foresee as many of these kinds of effects as possible. Some of these problems will not be possible to solve through specific system configuration features and may require procedural methods involving cooperative action between potentially interfering systems.

3.8.2.4 Hostile environment (QH)

Hostile environmental effects are also generally accepted under the title of environment rather than interfaces since we cannot interact with anyone to gain a mutual agreement. This may include effects of weapons such as: the blast and pressure generated by explosives, mechanical forces of impact, electrical currents induced in conductive components, and radar jamming energy. The complete list of hostile environmental effects must be produced and realistic values defined. A threat analysis is commonly performed to systematically develop this list. Classified intelligence reports are the principal source of information needed to characterize these requirements.

3.8.2.5 Cooperative environment (QC)

When we define a system, we say that everything is either part of the system or it is in the system's environment. Therefore, any systems that cooperate with our system but are not part of our system are in the system's environment. We can treat any interactions between these cooperative systems and ours as environmental effects if we choose to, but it is more common to treat them as external interface requirements between associate contractors responsible for the development of the systems.

3.8.3 Environmental Requirements Models

The structured analysis model for environment is the most complex of the traditional structured analysis processes and in fact three different ones are required, one for the system level, another for end items, and a third for component level.

3.8.3.1 System environmental requirements analysis

At the system level, we must define the aggregate environment that influences the major system elements that operate within the system space. It may be possible to define a single system environment or it may be useful to define a set of environments to which system elements are exposed in some pattern. This starts with identification of all of the spaces within which the system will have to function. Then we can select one or more standards corresponding to each of these spaces. Finally, we need to select the parameters from these standards that apply to the product and determine an appropriate range of values for each as covered in Figure 3.8-2.

3.8.3.2 End item environmental requirements

An Environmental Use Profile is a clearly defined relationship between the major items of a system often called end items, system processes in time, and environmental stresses effective during those processes over the life of the system. It is captured in a map between the system elements, system processes defined on a process flow diagram and companion timelines, and environmental subsets affecting those processes. This calls for a three-space vector: process, architecture, and environment. Figure 3.8-3 illustrates a process for assembling these data in a form useful for defining aggregate effects of environmental stresses.

Process flow diagrams are covered in Chapter 3.10. They look a lot like the functional flow diagrams covered in Chapter 3.2 but the blocks contained on them represent analogs of real world activity and we can imagine the system architecture flowing through them being transformed in some fashion. The architecture definition process was discussed in Chapter 3.5. The other axis of interest for end item environmental requirements analysis is the system environment defined as described in paragraph 3.8.3.1. Figure 3.8-3 illustrates the service use profile process.

As an example of performing this analysis consider the development of a hypersonic unmanned target air vehicle. Based on an understanding of the mission profile derived from mission analysis, we can construct a list of environmental subsets: (a) truck shipment, (b) rail shipment, (c) naval ship shipment, (d) air carrier shipment, (e) logistics storage, (f) operating site storage, (g) indoor maintenance, (h) outdoor maintenance, (i) hoisting and handling, (j) indoor transport on a tug-towed trailer, (k) outdoor transportation on a tug-towed trailer, (l) captive carriage on a launch aircraft, (m) launch, (n) free flight, (o) recovery, (p) mid-air helicopter retrieval, and (q) boat retrieval. Then we can define appropriate values for these subsets for a range of mission situations. Table 3.8-2 only includes two situations (air transport shipment and free flight) in the interest of space, but the analysis for this end item would have to include all of those listed above. Table 3.8-2 refers to some figures (could refer to tables as well) that capture environmental relationships that are too complex to be easily put into text form. These referenced tables and figures would have to be included in the document providing the environmental subset definitions.

Now we can map these subsets to each process on the system process diagram in a matrix like that shown on Table 3.8-3. The Process ID column lists every process at some level of indenture, not necessarily the same level

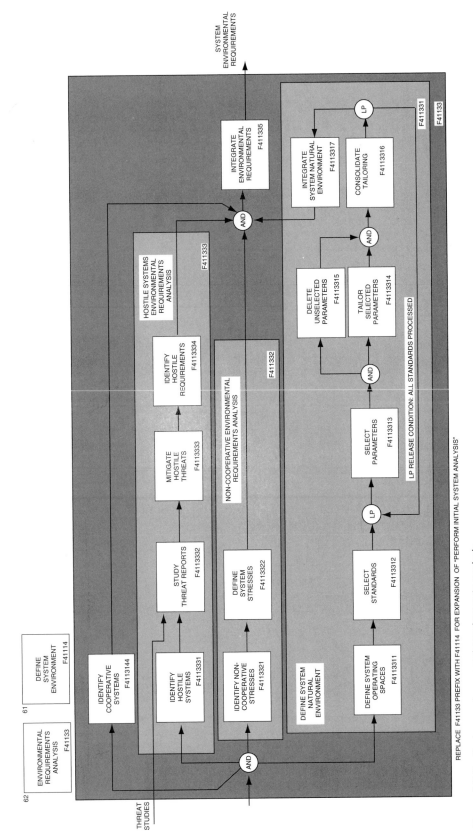

Figure 3.8-2 *System environmental requirements analysis*

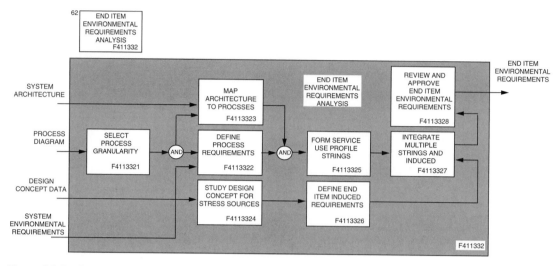

Figure 3.8-3   *Service use profile analysis*

Table 3.8-2   *Sample environmental subset definition table*

Environmental parameter	Air carrier shipment (D)	Free flight (N)
Altitude	0-8000 ft	Defined in Figure 2 with pressurization loss 8000 to 40,000 ft change in 0.5 sec
High Temp	140 degrees F	Air mass temperatures Figure 1
Low Temp	0 degrees F	See Figure 1
Rel Humidity	95% at 85 degrees F 3% at 130 degrees F 100% at 0 degrees F	5% at −100 degrees F 65% at −70 degrees F
Solar Radiation	Not Applicable	Table 1 of MIL-STD-210
Rain	Not Applicable	Not Applicable
Sand & Dust	Not Applicable	Not Applicable
Fungus	Not Applicable	Not Applicable
Vibration	1.5 g, 5.5 to 200 hz	As defined in Figure 3
Acceleration	−1.5 g forward, 3.0 g aft, −2.0 g down	Plus or minus 5.0 g vertical plus or minus 1.5 g lateral, acting independently

Table 3.8-3   *Process environment matrix (PEM)*

Process ID	Environmental subsets														
	A	B	C	D	E	F	G	H	I	J	K	L	M	N	O
F4711	X	X	X	X											
F4712			X		X										
F4713					X		X		X		X	X		X	
F472											X	X			
F473													X		
F474														X	
F475														X	X
F476					X	X	X	X	X	X					

throughout the list. The analyst then places a mark in each environmental subset that applies for each process. In the interest of simplicity, a time column that would show how long (from the process timeline) system elements would be exposed in each process, has been left off the matrix.

Next, we have to map the processes to the architecture that is the basis for writing specifications. Table 3.8-4 carries the example this one step further. The item environmental requirements are now defined but can be made more obvious by implementing one more matrix.

Table 3.8-4 *Process architecture matrix (PAM)*

| | Architecture ID | | | | | | | | | | |
Process ID	A1	A2	A3	A4		A5	A6	A7	A8	A9	A
F4711	X	X	X		X	X			X		
F4712		X		X	X			X		X	
F4713			X				X			X	
F472	X							X			
F473		X	X	X	X						
F474	X								X		
F475			X				X		X	X	
F476					X	X	X			X	

We can complete this analytical process by now merging the results of the two prior analyses producing a map between architecture items and environmental sets partially defined in Table 3.8-2. We can now see the complete natural environment that system element A1 will be exposed to. It includes the integrated effects of the environments defined for processes F4711, F472, and F474. From Table 3.8-5 we can see that A1 will have to endure the integrated environmental effects defined by some combination of environmental subsets that comprise environmental set $\alpha=\{A, B, C, D, L, M, O\}$. In this particular analysis it has happened that each end item will require a unique environmental set in Σ. Often the end items will require some number of subsets less that the number of end items.

We may apply the service use profile concept to a complete system as discussed here. As an alternative, we could create a separate service use profile matrix for each of two or more end items so that we do not have to use two matrices (three dimensional data) simultaneously to understand the total effect on a particular item. For example, we might create one matrix for a flight vehicle and another for support equipment used on the flight line to check out and prepare the flight vehicle for flight.

The final step in environmental use profile analysis is to integrate the effects of environmental subsets for a particular item where it may be used in two or more processes each of which may call for two or more environmental subsets. One approach to this integration process is to simply take the worst case for each parameter. Where this conclusion does not drive the design team into new, high-risk technology or high-cost solutions in search of a design solution, it may be an acceptable approach. This is also one of the easiest and most insidious ways that you can add to system cost without corresponding customer benefit. In this integration process, you must take into account what is physically possible or likely in our World.

There are other sources of potential conflict that the analyst must be watchful for. A famous one is the case where a structure must withstand the heating effects of direct sunlight in the daily cycle as well as a seasonal snow load of four feet. If the designer takes the worst case union here, he or she will have designed for a snow load in July (Northern hemisphere assumed). We have to be careful the way we combine environmental effects so as to reflect real world possibilities and minimize cost.

The process described above may appear to be somewhat tedious, but it is very systematic and it ensures that nothing will be left to chance. It forces the analyst to think through every possibility. A systematic process can be corrupted by bad analysis and bad engineering decision-making, of course. Much of the tedium can be erased by a computer database approach that encourages independent development of the process flow, architecture mapping to processes, environmental subsets, and mapping the environmental subsets to processes. Once each of these steps has been accomplished, the database should be capable of displaying the aggregate environmental stresses applied to each architecture item. The requirements analyst must then apply his or her human mind to the environmental requirements integration job to develop a realistic requirements set for each item.

Table 3.8-5 *Architecture environment matrix (AEM)*

| | Environmental subsets | | | | | | | | | | | | | | | | | |
ARCH ID	A	B	C	D	E	F	G	H	I	J	K	L	M	N	O	P	Q	S
A1	X	X	X	X								X	X		X			a
A2	X	X	X	X	X									X				b
A3	X	X	X	X		X	X	X	X	X	X			X		X	X	c
A4				X	X									X				d
A5	X	X	X	X	X	X	X	X	X	X	X			X				e
A6	X	X	X	X		X	X	X	X	X	X							f
A7						X	X	X	X	X	X					X	X	g
A8				X	X							X	X					h
A9	X	X	X	X										X		X	X	i
AA				X	X	X	X	X	X	X	X					X	X	j

You will note that we have used the term process flow diagram here rather than functional flow diagram in relation to the service use profile. The reason is that a functional flow diagram does not necessarily reflect the real world scenario that a process diagram must. When creating a functional flow diagram we do not necessarily understand the physical situation into which we are interjecting the new system. We do not necessarily even know of what the system consists. That is why we create the functional flow diagram. A process flow diagram is a model of the real world situation. The blocks are analogs of planned system activities that we picture occurring in time at known places or in particular spaces.

We have no difficulty attaching a real world environment to the process blocks, but it is not always easy to do so with the blocks of a functional flow diagram. We may begin to understand an environmental use profile in terms of the functional flow diagram, but one should avoid completing the profile in that framework. Spend the energy in transforming the functional flow diagram into a top-level process flow diagram as early as possible so that the environmental use profile can be related to the planned physical situation. The process diagram is needed by the logistics people anyway to complete their analyses. In fact, logistics people could be assigned this responsibility with some help to cover the operational processes.

3.8.3.3 Component environmental requirements
Now that we completely understand the natural environment a system and its end items will be exposed to, we must apply that knowledge to the problem of defining the environments that components will be exposed to within one of the end items. Let us use a fighter aircraft as an example. A radar altimeter internal to the fighter aircraft fuselage will be affected by environmental stresses that are influenced by the natural environment, induced environment (engine acoustics, for example), and the design of the airframe and on-board environmental controls.

The first step is to select the element or elements that will be subjected to a detailed internal environmental analysis. Most system elements will commonly expose all of its components to essentially the same environment that can be defined based on the integrated natural environment defined as covered previously. Some elements, like a fighter aircraft,

will include components installed in significantly different environments as a function of the different spaces available within the airframe and their proximity to engines, surfaces exposed to aerodynamic heating, or proximity to a 20 mm cannon, for example. These are the kinds of elements that require a detailed internal environmental analysis.

Our goal is to avoid over-designing components of a system for environmental stresses because it adds unnecessary cost that a customer could gain better return from through other applications. How then can we be sure that we clearly understand the system environmental requirements as they relate to each component in an end item of a system? Is it necessary to design every element of a system to the same set of environmental requirements or can we safely reduce the environmental stresses for some components over others?

One approach that has been found effective is called environmental zoning. We analyze the end item for spaces of similar environmental characteristics based on the way the structure is organized and the proximity of environmental stress generators (engines, guns, pyrotechnic devices, and other high-energy devices). We partition the element into zones of common environmental characteristics. Figure 3.8-4 illustrates such a zoning diagram for our fighter aircraft.

The next step is to map the item's components to the zones for our fighter aircraft. Initially our map will focus on a physical installation scheme driven by interface patterns. We should accept that this map is a variable that will have to be adjusted as the analysis proceeds. It may develop that we can avoid environmental controls in one or more zones by redistributing the components. This is essentially a packaging exercise.

Next we must define the environment for each zone shown on Figure 3.8-4. This is done by integrating the effects from our understanding of the natural environment derived through the environmental use profile study, induced environmental effects, and the physical structure of the item. We should first explore the environment that will result based on no special environmental control efforts. In those cases where the zone un-controlled environment exceeds that which can accommodate the kinds of components we will want to locate within them, we should consider first moving components that will have difficulty. Barring that possibility, we must consider environmental controls.

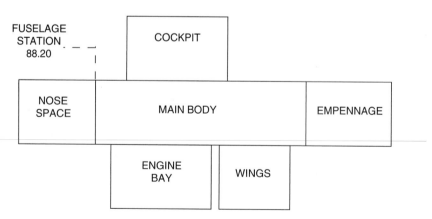

Figure 3.8-4   *Sample zoning diagram*

Once a decision has been made to control one or more spaces, the location of components should be re-evaluated to see if any may be located within this controlled environment thereby decreasing problems in other zones. At any one time we must have a clear definition of the component locations by zone and an effective way of communicating that baseline. Changes to this baseline should be carefully evaluated within the development team in a cross-functional way to check impacts on manufacturing, maintainability, reliability, and other specialty factors.

Strictly speaking, the environmental requirements analysis portion of this task is isolated on defining the environmental parameters and their ranges. You can see, however, that this task is inextricably entwined in the design process associated with the components, the containing spaces (airframe in our example), and the environmental control system we find necessary. It is essentially a packaging design problem.

3.8.4 Time Analysis

System timing is part of the environmental requirements analysis work but it is perhaps more convenient to discuss it in association with functional analysis because we are interested in associating time with the functions identified. Time lines may also be useful in connection with the design and the same techniques discussed here related to functions can be applied in connection with design features. Time appears tacked onto the tail end of the environmental requirements discussion because the author previously included this material in the chapter dealing with functional analysis because of the strength of the relationship between functional flow diagramming and time lines.

It is vitally important that we understand system timing. Timeline diagrams (TLD) give us a way to define the time requirements corresponding to functions identified on the system functional flow diagram (FFD). The principal product produced in this modeling environment is one of the following:

a. A timeline diagram (TLD) that graphically portrays the timing relationships between the actions depicted by selected blocks on the functional flow diagram (FFD).
b. Mean and variance time entries in fields of the Function Dictionary database (which is a tabular listing of system functions illustrated on the FFD) forming a tabular timeline.

3.8.4.1 Diagramming fundamentals

It is very important to understand how much time is available to accomplish functions illustrated on the FFD because the kinds of resources allocated to the functions, their cost and development difficulty may be driven by these times. It may be a far simpler problem to transport a cargo or a weapon 1000 miles in a three-week period than in a 30-minute period. We, therefore, resort to timeline analysis to systematically evaluate the appropriate times that should be associated with system functions. Once the top-level times for system operation are determined from the mission analysis, the timeline analysis process is essentially one of organized allocation, not unlike weight allocation. Throughout this process, the evolving set of timelines provides the engineering team with a systematic picture of how these times inter-relate and helps to force appropriate decisions on lower tier time allocations.

The TLD is created on a sheet of paper (or computer equivalent) marked off with a time axis in some unit of measure appropriate to the actions to be illustrated. Different time scales may be used for different portions (different sheets) of the timeline, but only one time scale should be used for any one timeline. Since the TLD will use the FFD breakdown, which will result in progressive refinement of the times, most of the potential difficulties involving relatively long and short times should be avoided. Where there is only one very long function, a break where everything continues undisturbed for a period of time may be used to foreshorten diagram length.

Timebars, corresponding to the blocks of a particular FFD sheet, are placed on the selected time scale of the blank sheet in a staggered or echelon formation where each block has a length proportional to the amount of time it will require to accomplish the corresponding function.

3.8.4.2 Timeline diagram symbols

The simple symbols used on timeline diagrams are defined on Figure 3.8-5 and these are illustrated on a typical Timeline Diagram in Figure 3.8-6 using the flow diagram in Figure 3.3-9a. The principal symbol is the timebar. The timebar is another expression of the FFD block. The analyst must respect the discipline that no timebar shall be used on a TLD that does not appear somewhere in the FFD. The length of the timebar is defined by how much time is allotted to the function. The height of the timebar is purely an artistic matter, but timebar height should be uniform throughout all of the timelines.

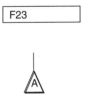

The time bar is a rectangular box of a length proportional to the length of time allowed or required by the corresponding function. The function ID should be noted in the box or in the left margin along with the function name.

A triangle identifies an event marker used to mark an major event associated with the times related to its location on the diagram. Identify the marker with a number or letter explained somewhere on the face of the diagram or in related text material.

An elastic point signified by a circled "E" can be placed at a point on the time axis to identify a plane where the time axis may be stretched without disturbing the staus quo of functionality on either side of the plane. A variable storage time is an application example.

Figure 3.8-5   *Timeline diagram symbols and conventions*

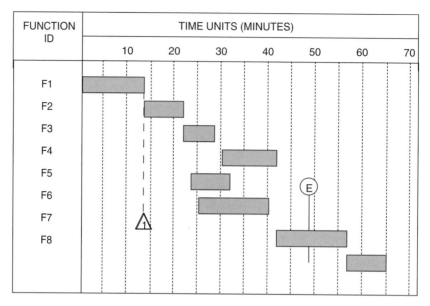

Figure 3.8-6 *Typical timeline diagram*

The event marker is self-explanatory. The elastic point marker is a useful device for overcoming timelining difficulties dealing with variability. There are functions that naturally vary in time such as storage (store until needed) and temporary indefinite holds in a missile launch process that result from planned holds that have a length based on past events. These are very difficult to illustrate on a graphical timeline because one does not know how long to make them. The elastic point allows the analyst to simply show the time as zero at the elastic point with a note about how long the longest "stretch" is at that point.

Philosophically we should imagine that we may stretch the timeline at an elastic point to reflect time passage up to the maximum indicated value. Beyond that point the timeline will breakdown for certain reasons at other points. For example, it may be that when an element is in storage longer than 24 months it may not be used without 12 specific servicing actions that are not necessary when stored for less than 24 months.

3.8.4.3 Variability

In every system there exists a degree of flexibility, variability, and uncertainty that makes it difficult to identify a single specific time for each function, but it must be done. Some of these sources of indecision resulting from variability and uncertainty may be satisfied by an appeal to probabilistic rather than deterministic time statements and others by using elastic points. Both of these will be discussed below.

Flexibility is a characteristic of a system that permits it to be used for a range of applications or offers the user options or alternatives under certain conditions of use. A flexible system will generally have a FFD with a lot of logical OR symbols permitting alternative flows. Problems caused by system flexibility will generally not yield to the use of probabilistic or elastic techniques and will require a more sophisticated methodology. That methodology could

entail a conscious selection of a design reference mission, or set of missions, as a result of the mission analysis. System operational possibilities might be infinite in nature but the understanding reached with the customer is that if you can show that the design reference mission(s) is(are) satisfied then it has been proven that all other combinations have also been satisfied as well.

The design reference mission set should provide the functional analysts with guidance in answering questions about what set of alternatives and assumptions apply to the creation of the timelines. The set should present a worst case scenario from a timing perspective such that all other permissible missions could be accomplished with the system designed to satisfy timing requirements based on the design reference mission set.

3.8.4.4 Selectivity

All functions illustrated on the FFD may not be time-critical and therefore need not be accepted into the timeline analysis. The timeline analysis should therefore be preceded by a filtering process to weed out those functions that need not be considered. This does two things: (1) saves time and money, and (2) allows the analysts to focus the use of available budget on the most pressing need. A sample set of criterion for use as a guide in selecting time-critical functions is listed below. They are presented in the form of questions to ask about the function in question. If the analyst finds the answers to many of the questions are "Yes," then the function is probably time-critical. If the answer to all questions is "No," then the function is probably not time-critical. These questions may stimulate you to think of other criteria more precisely tuned to your specific situation.

a. Does the function influence system reaction time to hostile or competitive actions?
b. Is the function involved in a critical event involving a countdown?

c. Are there concerns for human ability to perform needed actions within a limited time span?
d. Does this function have to be coordinated in time with other functions?
e. Is the item to which this function has been allocated involved in multiple functions such that a decision is needed whether to provide one shared unit or multiple items specialized to particular situations?
f. Is this function on a critical or only path in the flow diagram?
g. Does the function involve operational activities where relative timing of friendly and hostile action is critical?

3.8.4.5 Tabular timelines

As an alternative to diagrammatic treatment, the timeline can be included in a tabular format. This is especially indicated where probabilistic times are needed. For any one function time, two figures are given, a mean and variance. The mean, of course, tells the anticipated or required normal time increment while the variance gives a measure of variability about the mean.

The ultimate FFD generality can be expressed by a flow diagram consisting of one block titled with an expression derived form the system need statement. Corresponding to this function, there is a top-level timebar that identifies the gross time available or required for accomplishment of that function, commonly one cycle of the system mission. It will usually be possible to determine the gross time available to accomplish the complete system function before its decomposition into the major functions illustrated on the life-cycle model if the system functions are constrained by the functions of other systems. If the top system function (expressed in the system need statement) breaks new ground, it may be necessary to assemble the top timebar from an appreciation of its life cycle model components.

During the preparation of the system concept diagram in support of the mission analysis is an appropriate time to quantify the length of the top-level timebar. Commonly, a system will be observed to function in a cyclical fashion, repeating a series of functions in time. The top timebar generally corresponds to one period of this system cycle, which may include some preparation and termination function times common to each cycle.

Once the top-level timebar is quantified, determining the times for subordinate functions is for the most part a matter of time allocation not unlike weight or cost allocation. There is only so much available at each level and it must be allocated to subordinate elements with some margin retained for future problem solving during the development period.

Time allocation works well where the time available must be split among two or more serially connected functions. If a function has seven subfunctions all serially connected on the FFD, and there is 30 minutes available to accomplish the parent function, then the analyst may allocate say 27 minutes of this time to the seven functions retaining a 10 percent margin. Because of differences in function complexity and physical factors, the analyst may determine a distribution of the 27 minutes as shown in Table 3.8-6.

This is a simple land transportation problem where all of the subfunctions are clearly connected in a serial fashion. The times are offered in a deterministic frame but could be probabilistically stated with a mean of 27 minutes. The table provides a good example of a time analysis summary where the rationale for specific times is recorded and preserved for review and possible later re-evaluation.

Determining appropriate function times becomes more complicated when the functions are illustrated in parallel on the FFD. One effective approach in this case is to select what appears to be the pacing or critical path (the path that will require the longest time) among the parallel paths and perform the serial analysis, as discussed above, using that path only. Next, the analyst must analyze the time requirements of the paralleling paths to determine if they can be completed in the time identified for the path used in the analysis. It is possible that the one selected was not the longest path and the analysis will have to be adjusted accordingly.

In any case, the analyst will have systematically arrived at a sound conclusion with documented results, having completed a time analysis sheet as suggested by Table 3.8-6. The times will be in the public domain where they can be reviewed, criticized, and debated to the education and improved and common understanding of everyone on the project team.

Time margins should be used throughout the timeline analysis to provide a buffer for the solution of timing problems during the decomposition process and subsequent design phases. In critical path and scheduling circles these margins are frequently called "float." These margins could be captured in another column in Table 3.8-6. To form a time margin the analyst sets aside a percentage of the time available at each level of the analysis and puts it in the bank, so to speak. As problems arise in the development process, these margin times provide for crisis-free solution by re-allocating the margins to function times where the

---

Table 3.8-6   *Typical serial time allocation example*

Function number	Function name	Time (min)	Time rationale
F23	TRANSPORT LOAD TO PLANT	19.0	27.0
F231	Inspect Load	2.0	Simple visual/touch
F232	Exit Security Gate	2.0	Drive from Bldg 5 200 feet and show papers
F233	Move to Freeway	4.0	Two stop signs and two traffic lights
F234	Freeway Run	12.0	Six miles at 30 mph
F235	Move to Gate	3.0	Distance of 3/4 mile, two stop signs and a traffic light
F236	Clear Security Gate	2.0	Queue and paper check
F237	Drive to Dock and Park	2.0	Distance of 2/10 mile Park
	TOTAL ALLOCATED TIME	27.0	
	MARGIN	3.0	
	TOTAL TIME	30.0	

design groups are experiencing difficulty satisfying timing requirements.

The customer may require that times be developed in probabilistic terms for all functions or only for logistics activities such as maintenance. Probabilistic times provide a more realistic model of system operation in that the operation of any system entails a degree of uncertainty that cannot be expressed in deterministic terms. Availability of probabilistic times also makes it possible to perform queuing analyses, which are probabilistic analyses of arrival and service times.

The times listed in Table 3.8-6 example are of a deterministic nature and in reality they are an estimate of the mean value of the time required for the function. If one measured the time required for the system to perform the function later in the system's life cycle, they would find that there was a scatter of times, hopefully centering on the predicted figure shown in Table 3.8-6. The scatter would be introduced by all of the factors in the system environment that cannot be controlled by the system (timing of traffic lights on the route and traffic density in this example) in combination with the sources of variation in system operation.

If probabilistic times are required in the timeline analysis, it will be necessary to collect two time figures. The first is the mean or average time expected for the function. This is essentially the same as the one figure captured in the deterministic case. There may be subtle differences, however. In the deterministic case, we will sometimes be more interested in the worst case time figure than the average figure and this worst case time may find its way into our analysis consciously or otherwise. In the deterministic case, always clearly establish what the time figures will represent: nominal, worst case, or what. In the probabilistic case there is a conscious discipline at work to force us to select the average time in each case.

The second figure needed in the probabilistic case is an estimate of the possible time variation. This may be stated as a range of variation (for example, a mean of 132 feet per second and range of variation of plus or minus 30 feet per second) or a mathematically more precise deviation or variance. Refer to a good statistics textbook for a discussion of these terms.

The first number (the mean) will provide information about the central tendency and the second (the deviation) will provide information about the variation about that central value. These two numbers together can be used in mathematical queuing models of system performance. Figure 3.8-7 offers a suggested time analysis sheet providing for capture of both the mean and deviation figures. If the latter is not required on a given program, simply leave that column blank or prepare a form without that column.

3.8.4.6 Timeline reporting

The tabular timeline can be integrated into the function dictionary such that every time-critical function has a time requirement recorded in its function dictionary record in paper or computerized application.

Functional analysis, including the timeline analysis, is best implemented within the context of a computerized requirements analysis tool. In the good ones, the graphical image is automatically linked to the database content. In other tool sets it may be necessary to accomplish the graphical analysis independently of the database content. If it must be implemented in a totally manual way, the analysis should focus on the development of a TLD and a set of time analysis sheets using computer word processors, graphics and databases if possible. Figure 3.3-7 offers a simple format for recording this data.

These data could be included in appendices of the system definition document for formal reporting purposes. This report should include the complete FFD, a Function Dictionary, and a set of time analysis sheets if the times are not included in the function dictionary. The author prefers to include the timeline in Appendix B, dealing with system environment, as described in Chapter 3.11. In the computerized implementation, the time figures should be listed within the function dictionary and the time rationale saved within the dictionary database.

As in the case of all of the data obtained through analysis, the times that result from the timeline diagramming process must be reflected in system concepts and designs. There are two ways to insure this outcome:

a. *System Definition Document Reference.* Formally reference this report in the system specification and require that times contained within it shall have the force of requirements.

b. *Performance Requirement Time Inclusion.* Include the time identified through the timeline analysis within performance requirements statements where appropriate.

The reason that the timing of detailed FFD development is so critical is that the concept synthesis should not lag too

ANALYST		DATE		REV
FUNCTION NUMBER	FUNCTION NAME	TIME (UNITS)		TIME RATIONALE
		MEAN	DEV	

Figure 3.8-7 *Time analysis sheet example*

far behind the functional analysis process. The functions depicted on FFD must be translated into performance requirements and allocated to system architecture elements. This is the methodology for the orderly identification of system elements. If the FFD becomes too rapidly developed to a great depth far ahead of the concept synthesis process, there is a danger that it will be very difficult to allocate the lower level functions to the lower tier architecture elements identified by allocating higher-level functions to them.

Bringing about good timing between these activities (functional analysis and concept synthesis) is as much an art form as an engineering activity. One systematic way of ensuring good timing is to insist that the functions on each FFD be allocated to architecture elements before any of them are expanded to lower levels. This will result in the definition of specific functional requirements for each function illustrated and the allocation of those requirements to architecture elements. A natural dynamic will take over in an experienced system engineering organization to ensure that these identified items are assigned to concept development teams and that the resulting concepts are used by the functional analysis team to guide subsequent functional analysis. Close management guidance will have to be applied to this process if the teams are not composed of people who have succeeded in the application of a structured approach on earlier programs.

3.8.5 Environmental Requirements Capture

The stresses applied to the product system by the environmental effects discussed above must be captured in appropriate specifications. Ideally, the organization accomplishing the requirements analysis work would be employing a computer requirements database system but should capture the requirements in a requirements analysis sheet (RAS) however implemented. Figure 3.8-8 illustrates a fragment of a RAS containing some environmental requirements. A RAS includes three column pairs related to the model from which the requirement was derived (model ID or MID and model name), and the requirement (the RID is assigned by a computer database and the actual requirement statement), and product which will have to implement the requirement and into which specification the requirements will fall (PID and product entity). The requirement may be captured in primitive form as shown here, a referenced figure, or in a compete specification sentence or paragraph.

3.8.6 Environmental Impact

In the distant past, man had no means by which to apply significant stresses to the natural environment as would cause the quality of life on our planet to be damaged. Today, it is possible for man to apply at least locally damaging stresses to the natural environment. The author can remember as a grade school kid the Penobscot River being fouled by sewerage and industrial waste on its way through Old Town, Maine, and upstream points and by the time it arrived in the salmon pool in Bangor, Maine, being so bad as to fail to support the life of salmon. Today, that river runs fairly clean and many others as well. To avoid similar problems, a specification may include controls on the amount and kinds of stresses the system being developed will be allowed to apply to the natural environment.

If we are dealing with a military system, the system commonly will have a very harsh relationship with the natural environment in its normal successful use. Often military supplies contain very toxic materials and have a very destructive effect by design. One cannot avoid those effects but such a system can still be designed to be disposed of at the end of its life in a reasonably low stress fashion. An example of systems that were not designed in this fashion is nuclear weapons, which entail use and production of very dangerous by-products and chemicals. For many years these were simply stored in tanks that developed leaks. Only fairly recently has the money and talent been applied to this problem to avoid further buildup of these materials as well as the safe processing, packaging, and storage of these materials for the several thousand years that may be required.

There are materials used in commercial systems that can be damaging to people and our quality of life. Some simple examples are lead-acid batteries, anti-freeze, brake fluid, and brake pad residue. The author remembers being a high school kid working at a gas station in Centerville, Virginia. Another gas station along the highway was built on a mound of dirt that was black from all of the auto fluids that had drained into the dirt over the years. Many years later the author happened to be driving through Centerville hungry and noticed that there was a restaurant on that plot of ground. He kept on driving for a while.

Each environmental stress that is identified should be listed as a requirement and the design response to these requirements will be a mitigation policy that will reduce the probability of the stress being a problem or reduce the severity of the impact. So, the environmental impacts identified may be treated as safety hazards and mitigated in a similar fashion.

MODEL ENTITY		REQUIREMENT ENTITY		PRODUCT ENTITY	
MID	MODEL NAME	RID	REQUIREMENT	PID	PRODUCT NAME
QH7	Explosive Force	U83G	TBD-9	A1	Air Vehicle
QI3	Vibration	5TYU	1.5 g, 5.5 to 200 hz	A1	Air Vehicle
QN1	Temperature	FR5Y	$-40° \leq$ Temperature $\leq 120°$	A1	Air Vehicle
QX5	Electromagnetic Interference	K957	TBD-3	A1	Air Vehicle

Figure 3.8-8   *RAS containing environmental requirements*

3.9

Functional Analysis Alternatives

Contents

3.9.1 Variations Covered

This chapter covers several variations on simple functional flow diagramming applied in Chapter 3.2. These variations fall into six categories: (1) functional analysis variations, (2) state diagramming in several forms, (3) mathematical methods, (4) scenarios, (5) process analysis, and (6) quality functional deployment (QFD). Part 4 covers several computer software models including a new Department of Defense standard on complex information system architectures called DOD AF.

There are several models that include a functional sequence but also apply a kind of lateral axis as well representing data or commodity flow. These two-axis models are substantially richer in modeling a problem space than simple functional flow diagramming but the reader is cautioned that they are also not as effective in communicating modeling results into the human mind through the sense of vision because they are more cluttered on the page—they are visually more complex.

3.9.2 Functional Analysis Variations

3.9.2.1 Hierarchical functional analysis

This book encourages the use of functional flow diagramming as the preferred method for gaining insight into needed system functionality. It is, however, possible to decompose the system need using a hierarchical function structure as illustrated in Figure 3.9-1. Two levels of the diagram are illustrated to show how it can be decomposed. Each block on this diagram represents the same thing described for functional flow diagram blocks earlier but the diagram does not disclose intended sequence, simultaneity, or relative timing.

Where a system, or system element, is characterized by a fairly static situation, this model may be adequate. Where the system has an active space-time dynamic, this approach will probably fail the user by not providing insight into sequence-driven functions. It is difficult to understand how to translate this diagram into a timeline to evaluate system timing requirements, an activity that simply falls out of the functional flow approach in a very natural way.

One might think that hierarchical analysis would be an improvement after having applied functional flow once or twice. After all we wish to end up with a hierarchical structure called architecture. Why shouldn't we do both the functional and physical analysis in the same medium? The fact is that it is easier to do hierarchical analysis and that is the problem. The difficulty of allocating from a sequence-oriented structure to a hierarchical structure protects the analyst from the danger of leaping to point design solutions. You have a tendency to be more thoughtful about the possibilities.

System engineers who apply hierarchical analysis tend to develop systems with a one-to-one correspondence between the two structures and often build a functional hierarchy based perhaps subconsciously on the last physical

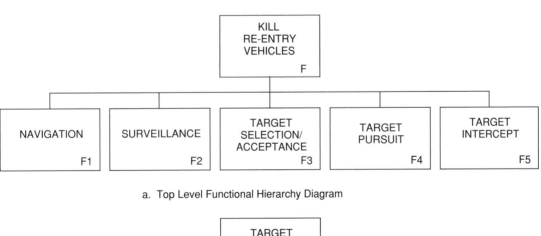

a. Top Level Functional Hierarchy Diagram

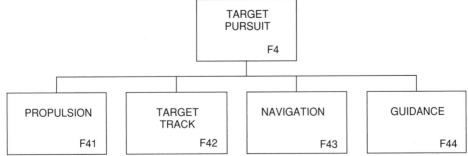

b. Second Level Functional Hierarchy Diagram

Figure 3.9-1 *Functional hierarchy diagram*

architecture. The result is that hierarchical analysis may retard technological advancement in an enterprise.

Despite this concern, it is possible to apply hierarchical functional analysis in a very effective way particularly on systems developed for commercial purpose where new models are periodically created with relatively little physical architectural differences.

3.9.2.2 Enhanced functional flow block diagramming

The AND and OR symbols used in the Chapter 3.2 do not offer the analyst a complete set of logical treatments to model some complex relationships functionally. Enhanced functional flow block diagramming as applied in the system engineering tool CORE by Vitech uses an extended set of logic symbols. These extensions also merge some old flow chart ideas with simple functional flow diagramming.

3.9.2.2.1 Trigger Construct

In functional diagramming described in the last chapter, the understanding was that a function is enabled when the prior function is completed. The Trigger construct offers a more complicated relationship. In Figure 3.9-2 function A or B will be enabled only if the prior function has been completed and the trigger is applied. In the case of the inclusive OR arrangement depicted, both functions may be completed if both are triggered, one or the other if only one is triggered or neither until one or both are triggered.

3.9.2.2.2 Multiple Exit Function

Commonly only a single output line is shown for a function on a flow diagram but enhanced flow diagramming recognizes a multiple output construct as illustrated in Figure

3.9-3. Whichever output is enabled in function A, the following function B or C is accomplished.

3.9.2.2.3 Iteration

Iteration allows a repeat of a specific function or "thread" of functions over a given "domain set" where a domain set represents: (1) a certain number of times, (2) a rate, or (3) a set definition. A "thread" of functions can be anything from one function within the iterate to a complicated set of functions controlled with multiple control constructs. Execution of Reference 1 in Figure 3.9-4 enables function A, function A is executed, and control is returned to the first iteration node which then enables the function A again and this is performed as directed by the domain set. If the domain set is 3, then function A is executed three times. When the iteration has been completed, Reference 2 is enabled.

3.9.2.2.4 Loop

The loop construct permits a repeat of a function or 'thread' of functions until a specific loop exit condition is achieved. In the figure, Reference 1 is executed and enables function A, function A is enabled and executed. If the loop exit condition is achieved, function B is enabled. If the loop exit condition is not satisfied, control is transferred back to the first LP node and function A is enabled again. The loop will continue to transfer control and enable function A indefinitely until the loop exit condition is achieved.

3.9.2.2.5 Kill Branch

The kill branch is used in conjunction with the AND construct and is illustrated in Figure 3.9-6. An AND construct with one or more branches designated as a "kill" branch

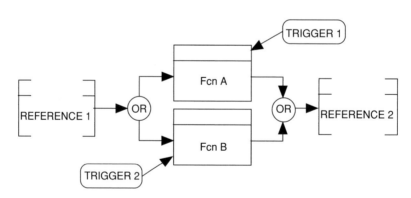

Figure 3.9-2 *Trigger construct*

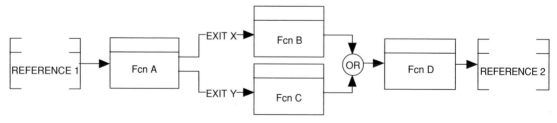

Figure 3.9-3 *Multiple exit construct*

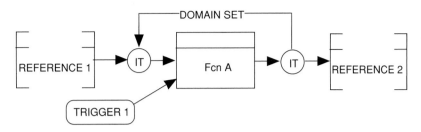

Figure 3.9-4 *Iteration construct*

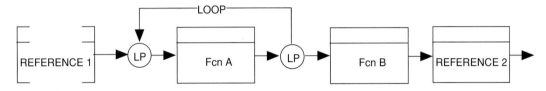

Figure 3.9-5 *Loop construct*

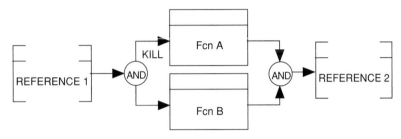

Figure 3.9-6 *Kill branch construct*

operates in the same manner as a regular AND construct until the "kill" branch is completed. Upon completion of all functions on a "kill" branch, control is transferred to the function or construct immediately following the closing AND node. If there are functions within the AND construct that had not been executed prior to completion of the "kill" branch functions, they will not be performed. In the "kill" branch figure, the branch that function A is on has been designated a "kill" branch. If function A is executed prior to function B, control will be transferred to Reference 2 and function B will not be performed.

3.9.2.2.6 Lateral Data or Commodity Flow
In enhanced functional flow block diagramming we can overlay another flow onto the left to right functional flow that may represent data needs by the functions or commodity flow where a physical process is being analyzed. Figure 3.9-7 illustrates this arrangement.

3.9.2.3 Behavioral diagramming
In Part 4 a model called IPO is discussed which has a vertical axis formed by the old software flowchart augmented by a lateral flow for data needed by the functional blocks. This

process was borrowed by Ascent Logic in the development of RDD-100 and re-named behavioral diagramming. The behavioral diagram is illustrated in Figure 3.9-8. The reader will note that behavioral diagramming and enhanced functional flow diagramming are essentially the same, each employing a two-axis diagram, but that they are rotated 90 degrees relatively.

Mack Alford, the developer of several computer aided software and system engineering products, including DCDS for the U.S. Army and RDD-100 for Ascent Logic, asserts that the traditional functional flow diagramming process is flawed. He proposed an alternative method called requirements driven development (RDD). RDD is rooted in a merger of the conventional functional flow diagram, arranged in a vertical orientation, with a treatment of data interfaces overlaid in the horizontal axis very similar to the input-process-output (IPO) model.

The Ascent Logic tool RDD-100 could be made to produce functional flow, IDEF-0, state transition, or N-Square diagrams, as required by a customer or internal company preference, from the information entered to complete the behavior diagram definition for a given system. The RDD method results in an executable definition of system functionality.

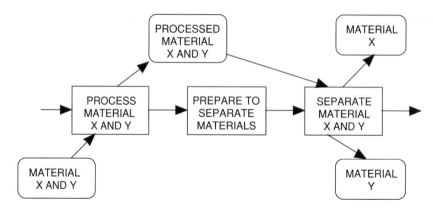

Figure 3.9-7 *Commodity flow in enhanced functional flow*

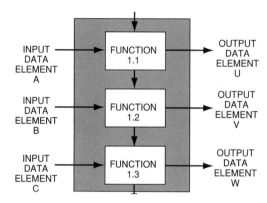

Figure 3.9-8 *Behavioral diagramming*

Simulations can be run using the same data upon which the specifications will be based as the functionality is allocated to system components. Mr. Alford's ideas along these lines are expressed in the Proceedings from a joint conference of ASEM and NCOSE (now INCOSE) at Chattanooga, Tennessee, October 1991 and in IEEE Computer, April 1985. At the time this is being written, Ascent Logic had gone bankrupt and been purchased by another company which may at some point return it to market.

3.9.2.4 IDEF-0

The U.S. Air Force, in a study completed by Air Force Systems Command, applied the Structured Analysis Definition Tool (SADT) model, developed at Softech, to an interesting expansion of the functional or process diagram called an IDEF. The expression IDEF is one of those wonderful compound acronyms that are so much adored by people dealing with complex systems. Acronyms offer a useful shorthand that makes it possible to communicate complex ideas efficiently between knowing members of a project team. Acronyms can also extend the time it takes an outsider or new team member to understand the project. The use of acronyms in aerospace and other industries is akin to the use of mathematics in all fields. Mathematics offers an economy of thought that is absolutely necessary in order to think and communicate complex mathematical ideas. A series of

acronyms mixed with some common English can economically communicate a powerful meaning in much the same way to those who understand them. Compound acronyms, however, are on the fringe of this otherwise useful technique.

The acronym IDEF originally meant ICAM DEFinition, where ICAM means Integrated Computer Aided Manufacturing. The methodology was developed in an attempt to provide a standard way of characterizing a development or manufacturing process. The Air Force went on to identify other models in the IDEF study so this original one, which is a functional analysis model, was re-identified as IDEF 0. As the Department of Defense rushed out of the standards business, the Air Force gave the Department of Commerce the funding to create a standard for IDEF 0, FIPS PUB 183, *I*ntegration *DEF*inition For Function Modeling (IDEF 0). Figure 3.9-9 illustrates an example of this kind of diagram. The blocks represent process steps. They are joined by directed line segments as in the process flow at the left and right ends of blocks. The difference is that directed lines, with special meaning, also are used at the top and bottom of a block. An entity corresponding to a line entering at the top of a block provides a controlling influence on the process. A line entering at the bottom of a block provides a supporting mechanism needed by the process. Lines entering at the left end and exiting the right end of a block correspond to principal inputs and outputs as well as show the sequence of flow.

Thus, with the IDEF 0 methodology, we can model a much more complex situation than with simple functional flow diagrams. We can, for example, illustrate a situation where one process produces a controlling influence on another in the form of a directed line from the right side of one block to the upper edge of another. These diagrams are of course more difficult to prepare than simple functional flow diagrams. They also tend to become very cluttered with interconnecting lines. This problem brings to the surface a great truth about the diagrams that are used for structured analysis. One has two choices. You can apply simple models that communicate powerfully visually with the human mind but do not offer a rich modeling environment. Alternatively, you can select a very powerful and rich modeling tool set that communicates poorly with the human mind. Functional flow diagramming is at the simplicity extreme on this continuum and IDEF 0 is close to the opposite extreme.

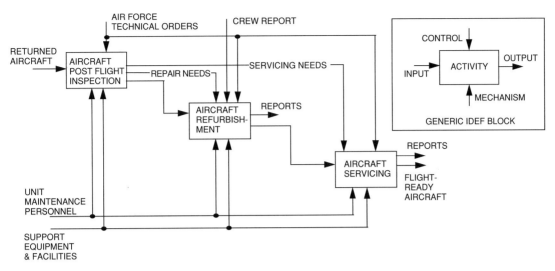

Figure 3.9-9   *Typical IDEF diagram*

3.9.2.5 FRAT

The acronym letters refer to the principle artifacts on which this model is focused: functions, requirements, answers (architecture in the context of this book), and test (verification in this book). FRAT was developed by Mr. Bernard Morais, President of Synergistic Applications, and Professor Brian Mar, Professor Emeritus, University of Washington in Seattle, WA. As suggested in Figure 3.9-10, the method splits the definition of what the system must do into functions that tell what the system must do and requirements that describe how well. The requirements are associated with answers that form the architecture of the product. Finally one must determine how to test the product (answers) to ensure that its design satisfies the driving requirements.

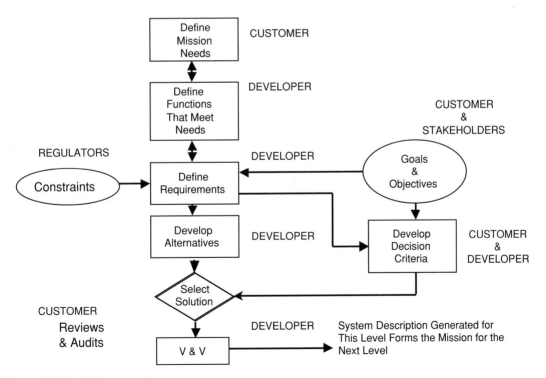

Figure 3.9-10   *FRAT sequencing*

The developers of FRAT recognize three interacting systems, the product system, the program system (that system which creates the product), and the rest of the world (or universe). Each of these systems must have a clear definition of its boundaries. One of the program system outputs is the product system, and inputs from the rest of the world provide resources and constraints for both the product and program system. Each of these three systems has different functions, requirements, answers, and test. For example, answers for the program system tend to be tasks or activities needed to create and operate the product system. WBS descriptions are answers associated with a program system; program risks are outputs of the program system and describe problems associated with meeting cost, schedule, or product quality requirements for the program. Each of these systems require the definition of all four FRAT views as well as graphical diagrams that define the parent-child hierarchy (vertical trace), the RAS relationships (horizontal trace), and all the interface, sequence, and processing dynamics (flow diagram). Thus the FRAT views and the FRAT pointers support the generation of all forms of systems engineering descriptions for both the product definition (specifications) and program management (SEMP and WBS).

The author of this book considers FRAT a nearly equivalent alternative to what he calls traditional structured analysis explained in this Part. The developers believe their intellectual construct to be a separate reality and of broader scope than simply supporting an effective requirements analysis process. The author is indebted to the developers of FRAT for opening his eyes years ago to a more satisfactory view of the allocation of needed functionality to things in the system through their derived performance requirements rather than the development of performance requirements subsequent to allocation of functionality to the things.

3.9.3 State and Event Analysis

3.9.3.1 State transition diagram analysis

Product lines characterized by a time-space dynamic will generally yield to functional flow diagramming analysis. Where the system or element is not characterized by a physical dynamic, it is sometimes difficult to apply functional flow diagramming to useful result. In these cases, state transition analysis may produce the desired insight into system functionality more easily.

State transition diagrams consist of two simple symbols connected to show what states (modes, forms, or conditions of existence) are possible and in what way the system may change from one to another. These symbols are a directed line segment (line with arrow at one end) and a circle or bubble. The circles indicate states in which the system can exist. The directed lines indicate allowed changes from one state to another.

A state is a condition of a thing with respect to circumstances or attributes. A system or an element of a system may exist in many different states commonly. At any one time, a single thing may exist in only one state. It therefore must be possible for a system to transition from one state to another in some predefined pattern based on specific stimulus. A system may be characterized or modeled by defining the states that are possible and the transitions that are possible between these states. The resultant information when combined with quantities defining values with some precision acts as a set of performance requirements.

Figure 3.9-11 offers a simple and typical state diagram. It consists of bubbles identifying states in which the thing may exist and transitions depicted by directed line segments that identify the permissible transitions from state to state. Since we do not want to clutter the diagrams, we should augment the diagram with a state dictionary and a transition dictionary, each of which lists the items and provides a clear definition for each state and transition. Table 3.9-1 offers a state dictionary and Table 3.9-2 a transition dictionary. The X and Y factors in the state transition definitions in Table 3.9-2 might be a function of relative traffic flow and time of day.

This diagram offers an application of state transition diagramming found in the traffic light system you encounter every day while driving your car. Any one combination of red, green, and yellow lights, at one instant, constitutes a system state. A simple two-way intersection with no left turn or walk signals and no emergency modes has only four possible states: (1) East-West green and North-South red, (2) East-West yellow and North-South red, (3) East-West red and North-South green, and (4) East-West red and North-South yellow. The system simply repeats upon this cycle of states indefinitely. There is no physical movement in time and space; no physical dynamic.

Note that the transition lines are lettered so they may be defined in an accompanying transition dictionary. Similarly, the states are numbered for correlation with an accompanying tabular definition of the states. The state and transition definition dictionaries for this example are left to the reader to work out. Simply make a list of state and transition symbols and after each include a definition of that entity. In the case of the transitions you may wish to use so many ticks of a 60-Hertz clock or monitor traffic intensity in some fashion to adjust operation to favor the principal pathways.

In the case of the design of the traffic light system, we would have to decide how we were going to implement the indicated states and transitions with people, hardware, and software. Colored lights or mechanical signals might suggest themselves right away, even if we had never seen a traffic light but other alternatives for communicating with the human drivers are possible. Old traffic light systems relied on time as the transition stimulant. Newer ones rely on magnetic sensing of the metallic vehicle mass in combination with time to stimulate a state change.

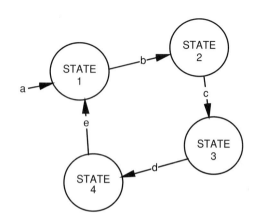

Figure 3.9-11   *State diagram*

Table 3.9-1 *State dictionary*

State	State name	State definition
1	North-South Traffic	Traffic is allowed to flow in the North-South direction with left turns permissible when North-South drivers see adequate space between opposing traffic needed for the turn. East-West traffic is not permitted to flow.
2	North-South Caution	North-South traffic is signalled that they must stop within a short period of time while East-West traffic remains halted.
3	East-West Traffic	Traffic is allowed to flow in the East-West direction with left turns permissible when East-West drivers see adequate space between opposing traffic needed for the turn. North-South traffic is not allowed to flow.
4	East-West Caution	East-West traffic is signalled that they must stop within a short period of time while North-South traffic remains hallted.

Table 3.9-2 *State transition dictionary*

	Transition	
NBR	Name	Transition definition
a	INITIAL	Power initially turned on with all lights showing caution switching to North-South traffic in 30 seconds.
b	NORTH-SOUTH GO	(STATE 1 FOR X SECONDS) AND (EAST-WEST TRAFFIC AVAILABLE)
c	NORTH-SOUTH CAUTION	(STATE 2 FOR 20 SECONDS)
d	EAST-WEST GO	(STATE 3 FOR Y SECONDS) AND (NORTH-SOUTH TRAFFIC AVAILABLE)
e	EAST-WEST CAUTION	(STATE 4 FOR 20 SECONDS)

The DoD data item description (DID) CMAN 80008A for system and segment specifications expanded on state transition analysis to identify a subordinate modes analysis approach. States are particular conditions of existence while modes are ways or methods of accomplishing these conditions of existence. In the context of CMAN 80008A, the analyst is supposed to identify possible system states and for each state, identify different modes to accomplish the purpose of that state. Then the analyst would have to identify one or more system characteristics (requirements) driven by that mode. So, one would have a hierarchy of states, modes, and characteristics in the performance requirements section of the specification.

Many system engineers find the CMAN 80008A approach too obscure to usefully apply in total to the general system development process as a decomposition tool. The state transition analysis portion of this DID has a very valid application in the instances noted above. But, the way modes are explained in CMAN 80008A, they appear to have little distinction from sub-states. With this interpretation, one could satisfy the DID with multi-level state transition diagramming. It may, however, be necessary to tailor CMAN 80008A in order to implement this variation depending on the specific customer program office interpretation. Another tailoring alternative would be to simply delete the modes analysis references and apply single or multi-level state analysis.

The more recent military standards for specifications (MIL-STD-961E and MIL-STD-498, the later replaced by EIA J-STD-016) refer to states and modes in a more general way, do not distinguish between them and simply encourage that there be an organized method for uncovering the content expressed in the specification.

The author has difficulty thinking of system or item modes disconnected from specific design concepts. With this inhibition, modes analysis becomes more useful later in the development cycle as an aid in defining the design solution concepts. In this chapter we are dealing with mechanisms for gaining insight into what the system is composed of and appropriate requirements for the items.

MIL-STD-498, which replaced MIL-STD-2167A and several others for software development on an interim basis until adequate commercial standards appeared, offered a much more sensible requirement for state analysis. It called for some form of modeling where the term *state* could be satisfied with states, modes, functions, or objects. The words *states* and *modes* could also be used interchangeably. Whatever the objects used, performance requirements (renamed entity capability requirements) would be identified that were associated with these objects.

Figure 3.9-12 provides a more complicated situation than the simple traffic control application illustrated earlier. It defines the permissible states for the superconductor super collider, a very large atomic research instrument for which development was discontinued by Department of Energy in the early 1990s due to cost. The diagram defines how the machine would be transitioned from ambient condition to a beam storage state where experiments can be conducted and back down to ambient conditions. This huge instrument, which was to be buried in Texas, involved the application of a tremendous amount of energy to drive the superconducting magnets to create the field intensity necessary for atomic particle experiments. The author had a great deal of difficulty expressing the needed functionality for this system with a functional flow diagram but an attempt to examine states and transitions spilled out onto the page readily.

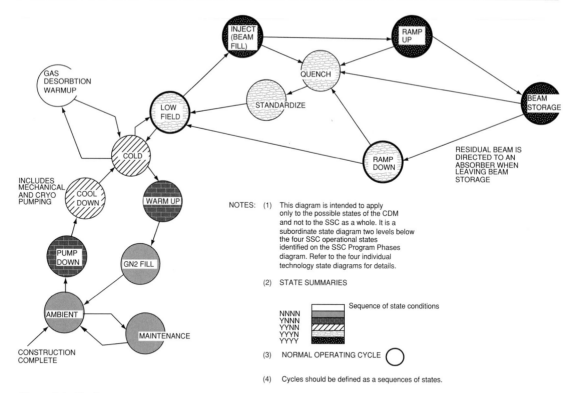

GAS
DESORBTION
WARMUP

INCLUDES
MECHANICAL
AND CRYO
PUMPING

CONSTRUCTION
COMPLETE

RESIDUAL BEAM IS
DIRECTED TO AN
ABSORBER WHEN
LEAVING BEAM
STORAGE

NOTES: (1) This diagram is intended to apply
only to the possible states of the CDM
and not to the SSC as a whole. It is a
subordinate state diagram two levels below
the four SSC operational states
identified on the SSC Program Phases
diagram. Refer to the four individual
technology state diagrams for details.

(2) STATE SUMMARIES

NNNN
YNNN
YYNN
YYYN
YYYY

Sequence of state conditions

(3) NORMAL OPERATING CYCLE

(4) Cycles should be defined as a sequences of states.

Figure 3.9-12 *Superconductor super collider state transition diagram*

3.9.3.2 Finite state machines

Problems that can be solved by computing machines and
many other kinds of devices can be modeled by several
kinds of state machines where the understanding is that the
machine will be in one state at a time and transition occurs
between these states in some pattern that can be discerned.
A finite state machine consists of a finite number of states
and a set of transitions that permit changes from one state
to another. These machines are discussed here because they
do come up in conversations with computer software and
hardware engineers. Also, at least one author in the systems
engineering field has used them as a means of defining the
system development process.

The simple traffic control system illustrated in Figure
3.9-11 could be described as a finite automaton FA = {S, A,
t, S_o, T} where S is a finite set of states $S = \{S_o, S_1, S_2, S_3, S_4\}$
in this case; A is a finite alphabet A = {a_o, a_1, a_2, a_3, a_4, a_5, a_6,
a_7, a_8}; S_o is an element of S called the initial state; and T is
the set of all terminal states. Function f describes how the
next state is determined, given a state that the automaton us
currently in. In the case of Figure 3.9-10, T is a null set in
that once initiated the automaton runs forever theoretically.
If we were using this model to design a real traffic control
system, we would want to identify specific terminal states
based on power failure and a need to maintain the system.

The reader interested in these machines is encouraged
to consult the literature recognizing that the reading will
likely be mathematically challenging. One introductory
book *is Introduction to Automata Theory, Languages, and*

Computation by John E. Hopcroft and Jeffrey D. Ullman.
Those familiar with Dr. Wayne Wymore's work and writing
will recognize a connection between state machines and
Dr. Wymore's *Model-Based Systems Engineering.* We will
take a brief look at three finite state machines that are often
referred to in the literature.

Alan Turing was a mathematician who worked during
WW II for British intelligence as a code breaker. Code breaking played an important part in the evolution of the digital
computer because of the tremendous complexity of the
work and urgency attached to it. In 1936 Turing developed
what came to be known as the Turing Machine. The Turing
Machine provides a mathematical model, M, of a digital
computer defined formally as M = (Q, Σ, Γ, δ, q_o, B, F),
where:

Q   is a finite set of states,
Γ   is a finite set of allowable tape symbols,
B   is the blank, a symbol of Γ,
Σ   is the set of input symbols, a subset of Γ not
     included in B,
δ   is the next move function, a mapping from $Q \times \Gamma$ to
     $Q \times \Gamma$
q_o in Q is the start state,
F   is a subset of Q and is the set of final states.

The Moore and Mealy machines are similar and in fact are
equivalent. That is, for every particular Moore machine there
is a Mealy equivalent and vice versa. The Moore machine is

named after E.F. Moore and the Mealy machine after G.H. Mealy. These machines and their related rich subject matter are useful in developing compilers and text editors as well as in support of many other useful activities. The professor for the automata theory course that the author attended in the 70s was using automata theory at the time to understand the language of dolphins for the U.S. Navy.

Both Moore and Mealy machines are defined by the six tuple $(Q, \Sigma, \Delta, \delta, \lambda, q_o)$ where:

Q	is a set of states and q_o is the initial state,
Σ	is the input alphabet,
Δ	is the output alphabet,
δ	is a transition function
λ	is a mapping

In the Moore machine, λ maps Q to Δ whereas in the Mealy machine λ maps $Q \times \Sigma$ to Δ.

3.9.3.3 Petri nets

State diagrams can be used to explain many fairly complex problems but cannot be used to model problems characterized by concurrent action because the understanding is that only one transition is permitted at any one time. Petri nets offered by Carl Adam Petri in his PhD. dissertation in 1962

at the Technical University of Darmstadt, Germany, solved this limitation. A Petri net is composed of nodes or places identified by bubbles such as 1 through 5 in Figure 3.9-13a interconnected by directed lines corresponding to transitions as an automaton lettered a through f in Figure 3.9-13a. Tokens are added in what are called particular markings of the net as in Figure 3.9-13b. A transition fires a token into another node or place under certain conditions defined for the net. One set of rules that could be specified are that a transition fires if all of the source nodes contain a token moving a token from each source node. With the net and marking shown in Figure 3.9-12b, the next net expression would be as shown in Figure 3.9-13c.

Note that in this case we have modeled a situation where five transitions occurred simultaneously. Transitions b and e fired one token each into node 4 from nodes 1 and 5 respectively. Transitions a and b each fired one token into node 5 from nodes 1 and 2 respectively. Finally, transition f fired one token from node 4 to node 3. There are many other possible marking pairs that this Petri net is capable of modeling and an infinite number of other nets that could be built.

Petri nets pose a problem in that it is not convenient to compose or decompose them, which makes it difficult to break large problems into sets of smaller problems. In many

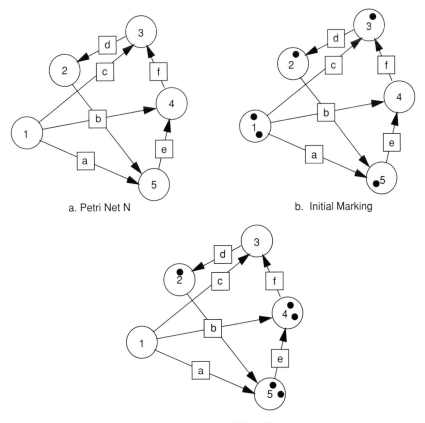

a. Petri Net N

b. Initial Marking

c. After Indicated Transitions

Figure 3.9-13 Petri nets

other models it would be possible to expand bubble 5, for example, into another detailed diagram showing more details. This techniques does not work very well for Petri nets. Event structures and state charts are possible extensions of nets that would overcome this objection. The evolution of models from automata theory through Petri nets to state charts is an example of the creative genius of man to progressively come up with methods that remove restrictions in current models. This same magic can be seen in everything man does. It is a long flight from the Wright brothers' flier to the F22 but each step along the way was a small advance to overcome some observed problem.

3.9.3.4 Event traces, lists, and trees

Events are what cause a transition from one state to another in state analysis but the events can be considered independently of states. As an alternative to listing the activities that occur in sequence, we could list a series of events where the understanding is that an event requires zero time. An event is, therefore, not an activity but an instant in time correlated with a condition of the system resources. An event list may be used to help identify the conditions under which states transition from one to another condition as well as define the entry and exit conditions of functions. Events and action lists are two themes that run through many structured models. They correspond to action on line and action on node respectively. In action on line modes, the action takes place in the transition from one condition to another and events signal the transitions. In action on node the transitions are assumed to occur instantly based on some event and activities occur in the bubbles.

An event tree is a graphic expression of an indentured event list. For any one event, we can define sub-events to as many levels as make sense for the problem. Since most systems have a cyclical pattern driven by some reusable aspect

to the system, these event lists or trees should close upon themselves. Unidirectional systems will be characterized by open-ended termination of the list or tree.

3.9.4 Mathematical Models

3.9.4.1 Mathematical equations

Some problems lend themselves to characterization in the form of a list of equations that must be solved for a list of independent variables. Military fire control problems fall into this category. Positions, distances, velocities, and rates are determined based on sensor inputs and projections made of future positions where the weapon can be placed coincidentally and detonated. In Figure 3.9-14, given an own ship (o) position x,y and target (t) position $(x + R\mathrm{SinB}, y + R\mathrm{CosB})$ at time T, we can integrate the velocities in x and y to define a moving predicted point where the target will be at $T + \Delta T$. Relative velocity in x is $V_o\mathrm{SinH}_o + V_t\mathrm{SinH}_t$. Relative velocity in y is $V_o\mathrm{CosH}_o + V_t\mathrm{CosH}_t$. In 1960 the requirements implied by these equations might lead analog engineers to implement a design using electromechanical servo loops with syncros, resolvers, and potentiometers. Ten to twenty years earlier, analog engineers would have implemented these same equations in mechanical resolvers, linkage multipliers, and ball and disk integrators. Today, these same equations would stimulate software engineers to write lines of code for a general purpose digital computer algorithm that accomplished the indicated mathematics.

At one time the author worked for Librascope on underwater fire control systems like the Mk 111 and Mk 114 ASROC and Mk 113 SUBROC systems. It was not uncommon in the development of the key operational performance requirements for a mathematician/scientist to write the equations for the problem in terms of own ship, target ship, and weapon orientations, positions, and motions as a basis

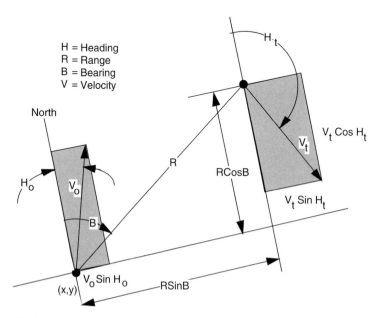

Figure 3.9-14   *Example of a mathematically specified problem*

for the design. The design would include a mechanical, electro-mechanical, electronic, or software implementation of a mathematical computing solution, depending on where you were in technology and time, which implemented the strings of equations with the required accuracy. Designers then designed or selected and interconnected the components that implemented the mathematics and packaged the equipment. In looking at the design solution you could clearly see the equations it implemented.

3.9.4.2 Formal methods

One of the problems in developing computer software, or systems in general, is the ambiguity of our spoken and written languages, such as English. Even computer programming languages can be used incorrectly. The problems that can occur with these languages have resulted in unacceptable consequences in some systems. One of the most infamous was the X-ray system that concentrated a lethal dose in the patient due to software flaws. Researchers trying to find a less error prone way in which to develop critical software evolved something called formal methods.

Formal methods apply logic and simple mathematics to computer program development. We all know that mathematical formulae unambiguously define situations whereas an English sentence may have several alternative interpretations. If a problem could be described completely in a series of mathematical relationships, as in the case of the fire control problem in the earlier paragraph, it should be possible to write error free code to implement these relationships. The mathematics need not be limited to real analysis, of course, but may also include mathematical logic, so many problems can be completely described in mathematical terms.

There exist several formal methods and one of those is Z (pronounced zed). Refer to "the way of Z" by Jonathan Jacky published by Cambridge University Press in 1997 for a detailed discussion of Z and reference to books on other formal methods.

3.9.5 Scenarios, Strings, and Threads Analysis

3.9.5.1 Scenario depictions

A scenario is an imagined sequence of events. There are many ways to place this imaginary vision into more concrete visual terms such that a human may read or view and think about them. These methods include descriptive text as well as various pictorial techniques.

3.9.5.2 Icon flow

One simple way to depict a scenario is a series of pictures or icons that clearly communicate ideas to the viewer. Figure 3.9-15 illustrates a scheme that places payloads in Earth orbit with post launch refurbishment of recovered components.

3.9.5.3 Descriptive text

The scenario can be communicated through pure text describing what must happen and in what sequence. This could be a tabular list or a narrative. If the problem attacked is very complex, it will require a very large text tract to communicate the same image conveyed by a picture. As they say, a picture is worth ten to the third words, and it is true. There is also an ambiguity in words and expressions that is difficult to avoid.

3.9.5.4 Strings or threads

A scenario can be characterized by listing a string of actions that must occur in the context of the system resources. A large system may be characterized by a 1000 different strings or threads where each one is an end-to-end series of actions possibly related to human actions, machine actions, and forms completed and communicated. We are tracing the flow of activity through the system under development as a way of understanding what the system must accomplish, an order in which it must accomplish these actions, and a fashion in which the actions will occur or be stimulated.

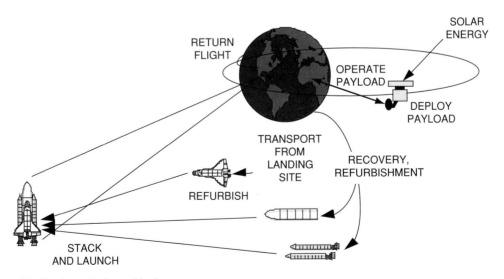

Figure 3.9-15   *Scenario formed by icons*

If we are analyzing the actions of a bank when a depositor arrives, we could string together the following actions by depositor and teller:

1. Depositor arrives at bank and proceeds to a table filling out a deposit slip.
2. Depositor enters the single teller queue.
3. Upon arrival at the head of the queue and clearing of a teller window, the depositor steps up to the window.
4. Depositor presents teller the deposit slip and financial instrument.
5. Teller reviews deposit slip and financial instrument making sure they are coordinated and associated with the correct bank and branch.
6. Teller enters account number in computer and selects transaction type (deposit). Teller completes entries required on computer screen from deposit slip. Teller closes transition.
7. Teller places the financial instrument and paper deposit record in appropriate locations.
8. Printer creates transaction record for depositor. Teller gives depositor transaction record.
9. Depositor exits window area and bank.

Now we may use this string or thread of activity to determine requirements for our bank system that responds to depositor needs. The result is very similar to flow diagramming, scenario writing, an event trace, or state modeling but it focuses on a series of actions oriented toward a form or single activity. The picture becomes more complex, of course, because the bank is interested in more activities than just depositors. The analyst would have to go on from depositors to check writing and deposits as well as many other actions in the normal life of a bank and its employees. In the process, the analyst may find some sub-strings that are common to many of these activities and places that strings cross and uncross that will help to simplify bank responses and associated cost.

3.9.5.5 Synthesis of functional threads
Some analysts find it difficult to make the initial transition between the need and the lower tier functions.. The author prefers to apply the life-cycle model as the initial breakdown but this same difficulty could flow down to the Use System function. Some analysts apply an interesting analytical approach for this immediate expression in the form of what a software analyst might call a context duagram. Figure 3.9-16 illustrates the beginning of this process. It has been determined that there are three functional inputs and one output. The analyst next develops a functional thread expansion for each input. Often the output thread will be

included in one or more of the input threads and can suggest a way to integrate the threads into a single flow diagram which is the desired outcome of the method in the first place as an alternative to simple expansion of the need or "use system" function into that resultant flow diagram.

3.9.6 Process Analysis

In the beginning of a development activity, when it is not yet clear what the system architecture should be composed of, the functional flow diagram commonly will not clearly relate to the physical, real world situation that eventually evolves. If the functional analysis process has been properly timed with respect to the evolving concept development and preliminary design activities and the concepts for higher-level elements is fed back to the lower-tier functional analysis process, the functional flow diagram will become adjusted progressively toward a model of the physical reality envisioned for system operation.

The functional flow diagram will gradually become a system process diagram where each block represents an analog of some physical process. In Chapter 3.1 we discussed a system relation where the process set is a set of relations that map the cross product of the power sets of architecture, interface, and environment to the function set. This phenomenon of the functional flow diagram evolving into a process diagram results in the functional flow diagram and the process relation set being in a one-to-one relationship, at least at the lower levels of the functional flow diagram. This is a cause for celebration because it gives us a lot of assurance that the system function set will be covered as a result of system operation in accordance with the planned process flow.

The potential danger in following this conversion course too early in the development process is that the team will become too closely wedded to a particular solution for the system need and fail to consider possible alternatives to satisfying high level system functionality. Since process analysis is an analog of some physical reality, it is very easy to allow yourself to become trapped in a rerun of past solutions with which you are familiar. That may not be a good thing if new technology and methods have evolved since the time frame of your prior experience that, if applied to this program, could result in better customer value. Since functional analysis is an abstract activity, it helps to force the participants to evaluate a broad range of options and that is exactly what is needed in the early development work.

Those familiar with logistics support analysis as well as manufacturing process planning can relate to process analysis very readily. But, it is very easy for anyone else to do so as well. It can be thought of as an analog of reality. While the process flow diagram can be drawn using the same symbols and methods common to make functional flow diagrams, the blocks correspond to physical activities that take place at prescribed physical locations or in particular facilities and under defined environmental conditions. We can imagine the objects of the system operating in particular ways (either under normal or faulted conditions) and what the consequences are.

3.9.6.1 Process fundamentals
3.9.6.1.1 Diagramming
Figure 3.9-17 offers two versions of a process flow diagram. The one shown in 3.9-17a was created using a computer graphics package and the one in 3.9-17b was created using

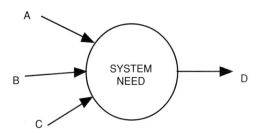

Figure 3.9-16 *System function depiction*

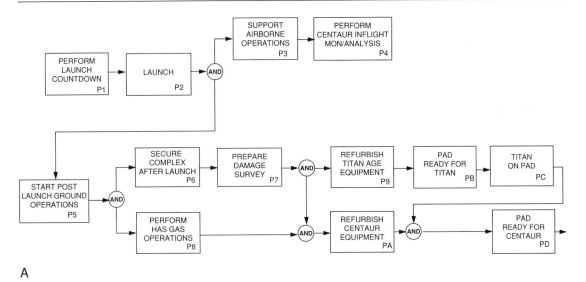

A

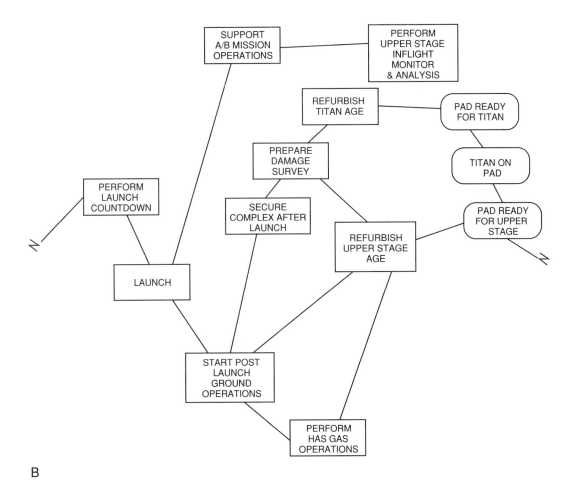

B

Figure 3.9-17 *Typical process flow diagram (a) Computer graphics process diagram; (b) Project planning software process diagram*

a computer planning software application (either a PERT or CPM approach will work). The advantage of computer planning software is that timelines and integrated resource lists can be created from the same data entered to create the flow diagram. Most project planning software applications also identify the critical path.

The diagram symbol set used in your process diagram is a function of the tool used to create the diagram as you can see from these two examples. As you can see from Figure 3.9-17a, it is possible to use the same symbol set used for functional flow diagrams, but the author believes the technique illustrated in 3.10-17b better relates to the physical model represented by the process diagram. You will note that where the flow joins and separates it does so at the blocks with multiple inputs and outputs. This is preferred to the use of AND and OR symbols found on the functional flow diagram because these lines can be carried directly to the appropriate lower-tier block on the flow diagram expanded from the blocks illustrated on this diagram.

The blocks on the process flow diagram represent physical reality, so it should be possible to overlay the diagram with facilitization plans to indicate in what facilities or spaces each process will take place. This can be done by shading the blocks or drawing outlines around blocks that take place within specific facilities. This could also be done by including this information in a process dictionary so as not to clutter the diagram.

3.9.6.1.2 Process-Resource Linkage

The process diagram provides an ideal mechanism for performance of logistics support analysis (LSA). The non-operational process blocks are exactly what the LSA process seeks to ventilate. As the process flow matures, the system hardware, software, procedures and people skills, and facilities can be mapped to the process blocks using a series of matrices. Where you see a system item not called for in any process, you want to question its need. Where two or more very specialized items have a tenuous application, it may suggest alternative design solutions involving a union of the functionality of the several items into one item. Where there is a great and diverse demand for particular items, it may stimulate an appreciation for appropriate quantities to include in the system definition.

A process-architecture matrix (PAM) is a simple structure with architecture items on one axis and processes on the other. An "X" in the intersections indicates that that item is needed in that process. Other symbols can be used to indicate a more specific application than simply "needed in this process." Figure 3.9-18 illustrates a PAM that corresponds to the process diagram shown in Figure 3.9-17 and the architecture illustrated in Figure 3.5-2 using a simple "X" at the intersections as discussed above.

Knowledge of the application of architecture to processes through the PAM can also be used to define the incidence of interface connections between these architecture items. Where the PAM indicates that two items are needed for a particular process and we know from the system schematic block diagram or N-Square diagram (both discussed in this chapter) that these two items share an interface, we should ask ourselves, "Is this interface completed or disconnected during this process, not connected throughout the process, or connected throughout the process?"

3.9.6.1.3 Process-Environment Linkage

In Chapter 3.8 we discussed an environmental requirements analysis methodology, called a service use profile, involving a process-environment matrix (PEM) that correlates system processes with system environmental forces. Given that you have defined a list of environmental forces, you can form a matrix of these forces with the system architecture. Then it is possible to form a three-dimensional matrix of process, architecture, and environment that provides insight into the

PROCESS		ARCHITECTURE						
ID	TITLE	A11	A12	A13	A14		A2	
P1	COUNTDOWN	X	X	X	X		X	
P2	LAUNCH	X	X	X	X		X	
P3	A/B OPS	X			X			
P4	CENTAUR OPS	X			X			
P5	P/L GRND OPS						X	
P6	SECURE COMPLEX						X	
P7	DAMAGE SURVEY						X	
P8	HAS GAS OPS						X	
P9	REFURB TITAN						X	
PA	REFURB CENTAUR						X	
PB	PAD READY						X	
PC	TITAN ON PAD	X	X				X	
PD	CENTAUR READY	X	X	X	X		X	

Figure 3.9-18 *Process-architecture matrix*

total environmental envelope that given items will have to endure throughout their life cycle.

3.9.6.2 Process analysis applications

A lot of requirements analysis work can be performed well with process analysis as a basis for structured analysis. This is true of logistics support analysis (LSA), verification planning analysis (VPA), manufacturing requirements analysis (MRA), deployment requirements analysis (DRA), and disposal planning analysis (DPA). But it is also true for the ways in which the enterprise itself chooses to function.

3.9.6.2.1 Generic Enterprise and Program Planning

There are four processes of interest in the development of systems as suggested in Figure 3.9-19. The enterprise which would develop systems should have a clearly identified generic process (1) which it applies as a template while creating each new or modification program process (2). During the program creation process (a proposal effort for a government contractor) the program unique product development process must be developed based on the generic process definition. During the program unique system development process (3) implementation there are two process developed or refined. First, we must refine the process for product manufacturing and test. Second, a fourth process should be exposed in the form of the product use process (4). Both of these result from the application of IPPT concepts to jointly develop a product and its process of use including both operations, logistics support, and disposal preparation.

3.9.6.2.2 Generic Process Analysis

As discussed in earlier coverage of the structured analysis process, functional flow analysis and process analysis differ in the meaning of the blocks and lines that compose these diagrams. In a functional flow diagram the blocks represent things that have to happen and through allocation of the objects to things in the physical model of the system we gain insight into the real things that will constitute the system architecture. The directed lines indicate the sequence in which the functions must be dealt with in the system implementation. The blocks of a process flow diagram represent analogs of real world activity. We can picture the tasks associated with these blocks being accomplished in the context of a list of resources by people with particular skills and knowledge within specific time and cost constraints.

Many organizations employ IDEF 0 to accomplish this work rather than the simpler process flow diagramming. The advantages of IDEF 0 are that it forces us to identify not only the sequence of activities but the resources needed and controlling influences and where these elements come from. The disadvantage of IDEF 0 is that it is graphically very complex and the profusion of lines can make it difficult for the analyst and users of the diagrams to effectively use the power of mind working through one's vision.

These same models can be applied to the enterprise to develop a generic process definition. The process identifiers used by the author in this book were given in functional context starting with the letter F rather than P in order to be consistent across the whole life cycle including the product

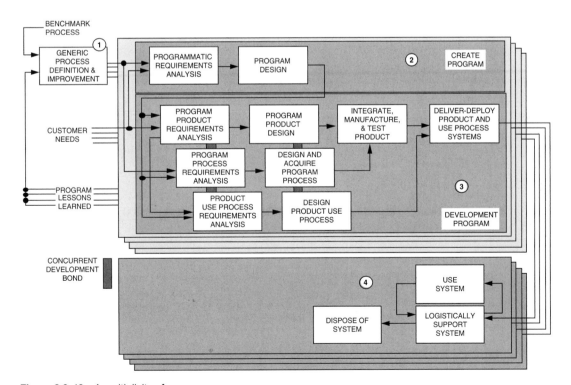

Figure 3.9-19 A multiplicity of processes

use for which the author commonly applies functional analysis. The ultimate process requirement can be defined as the enterprise vision statement. This can then be decomposed into lower-order processes to identify relatively smaller tasks that could be accomplished by one or more people working together to achieve a work goal. These lower order tasks are then allocated to functional department responsibilities which in the aggregate for each department form its charter.

Each process block in this process model is then associated with not only the source department for personnel to do that work on programs but also the practices and tools corresponding to that task. Cost and schedule estimates can be established for each of these tasks at any level of indenture desired.

When planning a particular program, we may connect functional planning strings composed of process step and functional department identifications (F411-D230 for example identifies a piece of generic work defined under process F411 accomplished by people from department D230) with program planning strings defined by a program and stage or phase identifiers, a program event, and an architecture ID (P05-S02-E02-A14 for example). The fusion of these two strings clearly identifies a specific package of work that must be accomplished for item A14 ending at event E02. That composite planning string would be P05-S02-E02-A14-F411-D230.

3.9.6.2.3 Program Specific Process
The program specific process is thus formed by combinations of generic work strings and architecture derived program strings. These work elements should be collected into a network application like Mac Project, Primavera, or Microsoft Project yielding a program schedule, cost profile, and critical path for use by program and team managers during program implementation.

There are three program processes that cannot be planned in detail from generic information sources. They are the product specific material and procurement, manufacturing and quality, and verification processes. The requirements for these processes must be developed concurrently with the development of the product requirements. Subsequently, the designs for these processes should be developed concurrently with the design of the product elements. They are then put into operation to implement the product development.

3.9.6.2.4 Continuing Cost and Schedule Requirements Analysis
Cost is a commodity like weight and reliability that can best be defined in the context of an allocation process where total cost is decomposed into component cost elements with margins distributed for management flexibility and cost risk avoidance. The total program cost of development (non-recurring cost) should be partitioned into subsets based on all of the program work elements defined in the architecture and linked to the program work breakdown structure for finance purposes.

Detailed work planning within each statement of work paragraph should be captured in a detailed task plan and linked to cost and schedule requirements. Ideally, this planning work will respect a cross-functional team structure overlaying the product architecture such that there is a clear map between available program funding, system items requiring development, and the organizations responsible

for development work. Refer to the specialty engineering requirements analysis section of Chapter 3.7 for a brief discussion of life-cycle cost and design to cost analyses.

Schedules are the timelines of the development process. They provide the same function for the development process as timelines linked to system functions do for performance requirements analysis. Schedule requirements are time figures for the amount of time available to accomplish a development task. Program schedule requirements are commonly identified through the application of Gantt charting or application of CPM or PERT networking techniques. One of the latter are recommended since these tools provide powerful what-if and resource management features.

3.9.6.3 Program product-oriented processes
3.9.6.3.1 Specialty Engineering Integration and Concurrent Engineering
The architecture block diagram, schematic block diagram, process flow diagram, process-architecture matrix, and process-environment matrix together provide a common set of core system engineering data that should be used by all specialty engineering people in their analyses of the system from their specialized perspectives. The link to LSA is obvious since this data provides an intellectual model of system operation long before the physical system exists (while the design may still be economically influenced). But these data also provide a common system definition for failure modes effects and criticality analysis (FMECA) performed by reliability engineers and the hazard analysis performed by system safety engineers. If you can cause all of the specialty engineers to share a common understanding of the system model with the design community, you are a long way up the road to sound concurrent engineering goals.

3.9.6.3.2 Program Material and Procurement Process Analysis
As engineering identifies the things which will comprise the product in a family tree breakdown, each of these things must be placed in one of four subsets. A decision must be made whether the item shall be purchased from outside sources, developed and manufactured by the enterprise, supplied by an associate not controlled by the enterprise, or supplied by the customer. This is often called a make-buy decision. If the item is going to be supplied from outside sources, it may be necessary to prepare a specification covering the item, a purchase order for an off-the-shelf purchase, or a set of drawings telling a supplier how to produce the item covered by a purchase order or contract.

So, each item, at some level of indenture, has a defined source connected to a schedule of availability. This work must be done in a concurrent fashion with participation by material and/or procurement. Procurement is a business discipline and department that has the responsibility on programs of acquiring material defined by programs and providing it to receiving in accordance with purchase orders and contracts as appropriate. Material is a manufacturing department that receives the material, stores it if necessary, kits it for manufacturing use, and routes it to the right place on schedule.

As manufacturing begins to lay out their program specific manufacturing process, material must participate to interface material feeds to that process. This includes identifying what material must flow and in what configuration. In some cases, intermediate storage may be necessary for subassemblies previously partially assembled by manufacturing. This is generally not a desirable condition but it may be useful to provide manufacturing float in the schedule.

Given these intersect points and conditions with manufacturing, material may then lay out their upstream requirements for material processing in order to feed the manufacturing process as defined in the joint analysis. This will determine need dates for procurement action.

3.9.6.3.3 Program Manufacturing and Quality Process Analysis

As the design concepts are formed on teams, manufacturing personnel should participate in order to ensure that the evolving design is consistent with a manufacturing process that either already exists or one which can be achieved by the date manufacturing must begin. Manufacturing can offer timely interaction with the design community if they build a process model as the high-level design concepts begin to flow. The lower-tier elements of these processes then represent relatively small manufacturing actions which can be envisioned in the imagination possibly aided by quick sketches. Manufacturing engineers can map manufacturing resources to these process blocks such as facilities, machine tools, special tools and test equipment, manufacturing aids, production line equipment, and personnel with specific skills and certifications (if needed).

As additional design details flow in, manufacturing engineers can use this manufacturing process model as a background in analyzing whether or not the product and the production process are mutually compatible. A good manufacturing engineer will spot design features that:

a. require a tube bend radius that cannot be dependably accomplished using planned methods,
b. human attachment action within a space inconsistent with any normal human's hand and forearm,
c. close human work to a very sharp surface that will cause injury,
d. ambiguous operating features that hide the actual product state from the manufacturing technician,
e. tolerances that stack up to defeat any reasonable manufacturing accuracy strategy, and
f. the use of several different kinds of fasteners in close proximity requiring several tools where one would serve for all.

The reader can imagine many other ways the manufacturing and engineering people can innocently create inconsistencies between their plans. In some cases, the manufacturing plan can be changed with little effort but in other cases it will be necessary to alter the design concept to correct them. The use of process models when there is not yet any physical reality has a powerful effect on risk reduction. It will encourage the identification of problems before they have any chance of causing discontinuities in production. This is another example of the low cost of finding and eliminating problems early and the relatively lower cost of correcting them earlier rather than later.

Quality engineering requirements must be identified in concert with the manufacturing requirements analysis so that appropriate inspection stations are located in the manufacturing flow, product design features permit adequate inspection of the product during production, and the design solution as well as the manufacturing and testing processes minimize the need for quality inspection through wide product tolerances relative to process tolerances.

3.9.6.3.4 Program Verification Process Analysis

Chapter 6.6 of this book offers a complete generic process for accomplishing product item qualification and acceptance planning as well as system test and evaluation and that process is encouraged. But, we must apply it as a template against which product specific planning is accomplished. That template encourages identification of the development requirements which drive the design of the product and the corresponding qualification verification requirements which drive the design of the qualification verification process. The verification process and corresponding plans and procedures are the verification design and they should be responsive to the verification requirements.

This same process is repeated for acceptance verification based on product or detail specification content driving the identification once again of acceptance verification requirements. These, in turn, should be the basis of acceptance test process flow, plans, and procedures. The third tier in this process involves system testing based on requirements in the system specification.

In all three verification cases, there is a specification (item or system) that contains the item (performance or detail) or system requirements that should be used as the basis for the definition of corresponding verification process requirements which, in turn, should drive the development of a process design captured in plans and procedures. Implementation of those plans and procedures results in evidence of compliance (or otherwise) that should be reported in verification reports which should be audited to ensure that the evidence included is best evidence and true.

Part 5 encourages preparation and maintenance of a Verification Management Data Report which contains a verification process flow diagram, verification process schedule, and a set of three management matrices (compliance, task, and item).

3.9.6.3.4.1 Test Planning Analysis (TPA)
TPA has product requirements analysis, test article requirements, and program planning requirements aspects as illustrated in Figure 3.9-20. The product aspect focuses on identifying the need for product features that are driven by the necessities of testing and defining verification requirements for inclusion in Section 4, Quality Assurance, of program specifications. This is the portion of TPA that this book will primarily focus on. TPA must also develop any special requirements for test articles and test fixtures that are used to simulate the actual product systems elements during development testing. Finally, the program requirements aspect focuses on providing inputs to the integrated test planning activity.

All testing is accomplished to verify achievement of some characteristic of the product. As requirements are identified for system elements, we are obligated to devise a way to verify that each of them is satisfied by the design. These verification actions may be accomplished by analysis or testing. Wherever possible, contractors try to do this by analysis because it is generally (but not always) less expensive. The requirements that cannot or should not be verified by analysis must be verified by test. In order to keep the cost of required testing as low as possible, the many verification requirements must be integrated into a complete program that minimizes the number of test articles and test events.

There are four kinds of program testing commonly accepted in the development of systems: development, qualification, acceptance, and operational. Development testing

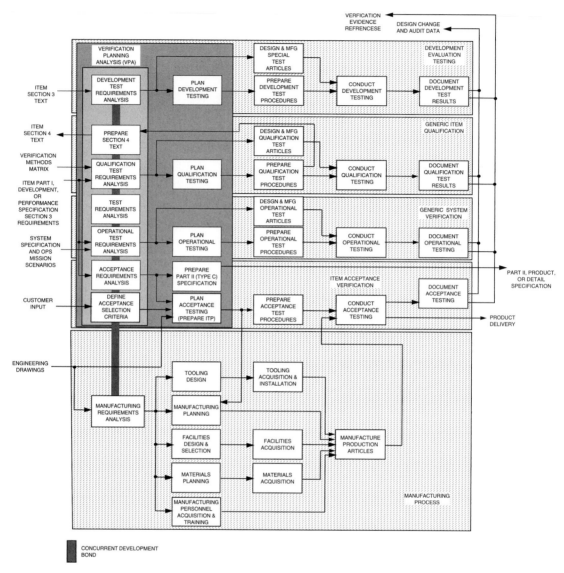

Figure 3.9-20   *TPA and MRA process flow*

is accomplished to validate design concepts, mitigate design and development risks, evaluate alternative design concepts, and demonstrate availability of needed technology. Qualification testing is accomplished to verify that the design satisfies the requirements and is adequate for the application. Acceptance testing is accomplished on each production article, or in accordance with some sampling technique, to verify that particular product items meet customer requirements. Operational testing is essentially qualification testing performed to verify system operational requirements and to demonstrate capability.

Our game plan is to apply functional or process analysis to these five kinds of test activities. Flow diagrams are created and mapped to test articles, resources, materials, and facilities. These test activities are studied to determine if the

planned special test and product articles to be tested require any special features for test compatibility. It may be necessary, for example, to locate a special bore sight target on an airframe to monitor alignment or special instrumentation to monitor strain during structural testing. The flow diagrams help us to recognize the needs for these test requirements.

3.9.6.3.4.2 Development Test Requirements Analysis   The basic process for development test requirements analysis is very similar to qualification test requirements analysis described above. The principal difference is that we also have to determine the requirements differences for special test articles. We may conclude that we need to perform a fuel system development test to evaluate fuel flow rates and pressures in different conditions of flight. It will not be necessary, or even desirable, to perfectly reflect the planned physical design configuration of the system. We may, in fact, not be able to complete the detailed design until we have completed a development test. As a result we mix the product article requirements with special test article requirements so as not to compromise test results applicability to the product article.

We may conclude that the ideal test article is a fluid flow bench laid out to permit us to walk among its parts for test monitoring and adjustment purposes. The final design may have to fit into an aircraft airframe in otherwise unused space. Our requirements would call for plumbing runs that mirror the planned effects of the design on fluid flow and pressure but the precise system geometry may be very different.

The results of these tests will be useful in driving final design features and in validating the theoretically derived performance requirements reflected in the special test article design.

3.9.6.3.4.3 Qualification Test Requirements Analysis The requirements for Section 3 of our development (Part I) specifications that have been assigned to a test verification methodology are fed into the block called Qualification Test Planning on Figure 3.9-20. These requirements place a demand on the integrated test plan that test engineers have to determine how best to satisfy within available resources.

Test flow diagrams are developed identifying specific tests with clear test objectives and resources. These tests are knit together into an integrated whole. The tests are mapped to specific item Section 3 requirements. Schedules are developed and test resource acquisition planning completed.

Test planning data corresponding to each individual test is transformed into a test procedure. The test plans are digested in Qualification Test Requirements Analysis and summary statements made available for inclusion in Section 4 of the item specification supplementing the verification matrix already available. Where program technical risks have been identified that indicate difficulty satisfying a requirement, we evaluate planned activity and test requirements against the planned risk mitigation approach and make adjustments, if necessary. If a requirement is identified as a technical performance measurement (TPM) parameter, we ensure that the test activity supports the planned parameter control plans and schedules. In Part 5, we will pick up on this story and tie it together with the requirements verification process.

The test articles supplied to these tests should be as much like production articles as possible but it is seldom possible to really do this. You cannot gain a DoD customer acceptance of your production process until after you have proven that your design satisfies the requirements. You cannot provide the customer with convincing evidence of this until you have completed qualification testing. So, we have a Catch 22 situation that is resolved by testing product articles that are as close to the planned production article as possible. We then conduct analyses to assure ourselves that any necessary differences will not invalidate the results. Where differences develop that unavoidably do compromise the test results, we may have to run a supplementary test later using a real production article.

3.9.6.3.4.4 Operational Test Requirements Analysis Operational test requirements analysis is similar to qualification test requirements analysis except that we are dealing with the complete system or very large portions of it. Special requirements may have to be developed because of differences in the test article from the planned production article. This is especially true if phasing plans exist to evolve a final design through two or more vehicle test phases. We may have to build a prototype vehicle and subject it to operational tests before we fully understand the needed characteristics of the final configuration. This is really a case of development testing, but since it involves the principal operational article in field trials, it fits better under operational testing.

3.9.6.3.4.5 Acceptance Test Requirements Analysis Product acceptance TPA defines requirements for product acceptance test stations and equipment that must to be integrated into the production line, product design features ensuring acceptance testbility, manufacturing planning requirements inputs, and requirements for acceptance test plans and test procedures. Also, as noted on Figure 3.9-17, TPA filters the resultant test procedures for each test plan to determine which of the requirements should be included in the item Part II specification.

The customer will normally not wish to base acceptance on the complete acceptance test procedure. Rather, they will commonly expect the contractor to select a series of critical system level parameters that cannot produce good results if anything in the article is not acceptable. TPA identifies these critical parameters for coverage in the Part II specification Section 3 and extracts the test requirements from the appropriate test procedures for inclusion in Section 4.

3.9.6.4 Deployment planning analysis (DPA)

Deployment planning analysis applies the same process covered previously to identifying requirements for system deployment. We create deployment functional flow (or process) diagrams and allocate these functions to specific product system elements and we apply the results to program planning activities to ensure that as product flows out of the production process it can be moved directly to its intended area of operation prepared in a state of readiness for immediate use.

This may include developing public acceptance of operation of the system in its planned space through an environmental impact analysis and abatement process. In the case of intercontinental ballistic missile systems developed in the 1980s, this was a greater obstacle than the technical, mission-oriented problem of hitting assigned targets. Some of these systems required an expansive network of bunkers above ground or underground rail lines. Few of the geographical areas with sufficient uninhabited space could also be cleared for use by the civilian population through their state and local governments.

The reality is that we may develop a very fine product through a very difficult development experience only to find that the result cannot be deployed. Systems have been developed and never deployed because conditions had changed during the development period mooting the original need. Teledyne Ryan Aeronautical developed a fantastic Model 154 unmanned reconnaissance aircraft in the early 1970s designed to over fly over Russia and China. The program stimulus was the downing of the Francis Gary Power's U2 flight. By the time it was ready for operational use, however, President Nixon had ushered in a period of detente and use of the system was politically unacceptable. Also satellite reconnaissance was becoming more effective and the SR 71 was coming on the scene.

During the Vietnam War, Teledyne Ryan Aeronautical produced an unmanned reconnaissance aircraft using a color television camera rather than a film camera. After flight test, this vehicle was air shipped to Bien Hoa Air Base for operational use by the 100th Strategic Reconnaissance Wing detachment based there but when the C-141 transport landed at Clark Air Force Base in the Philippines in route it was prohibited from flying on to Vietnam because it had not been cleared into the theater by the general in charge of the air war. After this incident a couple of years passed before the television capability could be fielded and it was with a black and white camera with a lot less capability. The advantage of a television camera, of course, was that its signal could be radioed to an aircraft operating outside of the enemy surface to air missile capability and pictorial data evaluated while the mission was in progress leading to possible changes in the mission plans under way based on targets of opportunity and near real-time data could be used to encourage strikes nearly coincident with observations of enemy activity.

The deployment process can be modeled using process diagrams that stretch from the availability of a deliverable capability to deployed operational capability. We start with one box labeled "Deploy the System." This one block can then be decomposed into some number of activities that must be accomplished to achieve the desired results represented by the single block. This process can be continued until the lower-tier elements will yield to detailed planning work involving lists of required resources summed across all of the process steps and specific tasks that must be accomplished. The process steps can be translated into a Gantt chart with dates noted when specific actions must be started and completed. Alternatively, the deployment process can be modeled using PERT or CPM diagramming.

Part of this whole process may entail an environmental impact analysis requiring special skills not available within the developing organization necessitating employment of one or more consultants. This analysis must identify any adverse environmental influences resulting from the deployment of the system and describe planned actions to be taken to mitigate their impact.

The final result of deployment planning analysis consists of programmatic requirements that drive program planning work that ensure that the developed system can be deployed into its planned operating environment. If this work is delayed too long, it may not be possible to deploy the system as planned leading to additional program cost to store manufactured material until it can be deployed or program changes to reshape the production process. Note that the effect of late planning of deployment can have the same effect as poor material planning.

Deployment of military systems or distribution and sale of commercial products should not be a foregone conclusion based on how tremendous the developer believes the product to be. Early in the development activity one must explore potential deployment problems and work out strategies to overcome observed problems.

3.9.6.5 System sustainment process analysis

Four major process steps exist related to the use of the product by the customer and all of these processes must be developed concurrently with the development of the product itself. These are: logistics support, operational use (some would fold this into LSA but the author does not), modification development, and disposal.

3.9.6.5.1 Logistics Support Analysis Overview

LSA is a systematic series of activities that help us focus our attention during the development period on supporta-bility features a system needs to provide maximum customer satisfaction during their later operational use of the system. It relates to architecture items and to processes within which these items are used in a customer operating environment. It seeks to identify product requirements driven by field use of the system long before the system sees the light of day. As a result, it is an analytical activity requiring some imagination.

We take away some of the theoretical nature of the task by creating system employment process diagrams that clearly relate to our physical operational facilitization plans, operational employment scenarios, maintenance concept, and logistics support concept. To the logistics engineer these flow diagrams combined with the architecture diagram and the map between them plus an understanding of design the concept make the system come alive long before the first rivet is bucked on the production line.

With this background it is possible for the logistics engineer to imagine the principal product element (e.g. aircraft, tank, rocket ship) passing through these processes acted upon by maintenance and operations personnel and checkout equipment. It is possible to determine if adequate resources are presently identified to satisfy the goals established for each process and the characteristics of these resources for successful system performance.

LSA focuses on creating information in the several categories noted on the blocks of Figure 3.10-21 derived from the LSA process defined in MIL-STD-1338-1A. This information is created through the indicated analytical activities conducted on the current system requirements and system design concept. As a result of these analyses, we expand the requirements and concept baseline exposing additional facets for iteration of the analysis.

MIL-STD-1338-1A provides a detailed description of a set of forms that can be used to record logistics support analysis data for input to a computerized system. The blocks of Figure 3.9-21 are correlated with these forms by the letters at the upper left corners of the blocks. These forms are completed while the analyst develops the system maintenance concept using process diagrams as the window to the planned reality.

Inputs are part of the total for his or her item. As the program moves into the design phase and the design concept matures into drawings, the logistics analyst must examine these design concepts in more detail and evaluate whether they are consistent with the planned logistics concept. Where inconsistencies are uncovered they must be fed back to the designer or used as a basis for changing the logistics concept. The logistics engineers needs a model of this process of maintenance and operations in order to identify logistics requirements before the design has been consummated in formal engineering drawings.

3.9.6.5.2 LSA Example

Figure 3.9-22 illustrates a top level process diagram for a high altitude, high speed target (called the HAST in its Demonstration-Validation Phase). During that prior phase the USAF customer became very concerned because the target had to be returned to the manufacturer after each flight for refurbishment and the Air Force intended that the vehicle would be turned around in the field. Teledyne Ryan Aeronautical won the competition for the full scale development phase and renamed the flight vehicle the Firebolt. The vehicle was propelled by a hybrid rocket motor using inhibited red fuming nitric acid (IRFNA) as oxidizer and a solid rubber-like material for fuel. Figure 3.9-23 shows a

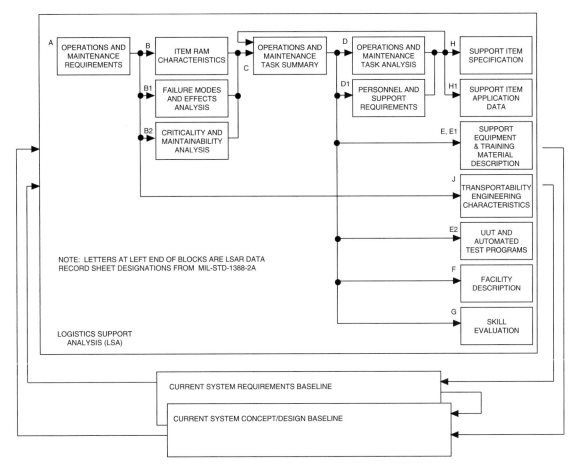

Figure 3.9-21 *Logistics support analysis process flow*

small fragment of the expansion of the QC000 block titled Post Flight Maintenance (Wet) shown in Figure 3.9-22. Using a model like this, the logistics support analyst seeks to break down the maintenance actions in increasing detail until arriving at a point where very elementary steps are identified, the time for which can easily be estimated, and it is possible to imagine all of the detailed steps necessary to complete the task and identify all of the resources needed.

The logistics engineer must simultaneously lay out the logistics concept and requirements for the product for consistency with that concept. Concurrently, the product design engineers should be developing the product design requirements consistent with the logistics concept using LSA.

Rather than cover the many forms covered in military standards for this purpose, we will use a simple text-oriented form to capture the information in an example shown in Figure 3.9-24 which reports upon task QCR00, Install Burst Disk. Then we will pick out the requirements and associate them with items in the system. Where there is nothing in the current system architecture to accomplish a task, we will add something to the system. The purpose behind using rigid forms is to force the analyst to think

through every aspect of each task analyzed and to ensure capture of computer data needed to print customer-required reports. The relatively loose form used here may not stimulate the analyst to identify all aspects of the task or transform the results into meaningful consequences for the system but it is representative of the intent in use of a set of standard forms for that purpose.

Given that we have done a good job of analyzing this task, we now have to draw conclusions from our analysis about the impact on the product system and implement the actions suggested by these conclusions. We have to influence system requirements and design concepts based on the results of our analysis. To do this work and then not benefit from it is complete stupidity. Product system requirements that flow out of this analysis:

a. There is a new interface requirement identified between the added Burst Disk Alignment Tool and the forward flange of the tank assembly. The tool is attached by two screws and the nut plates for these two attach points have to be accurately located by tooling rather than simply match drilled with mating nose assembly during the manufacturing process.

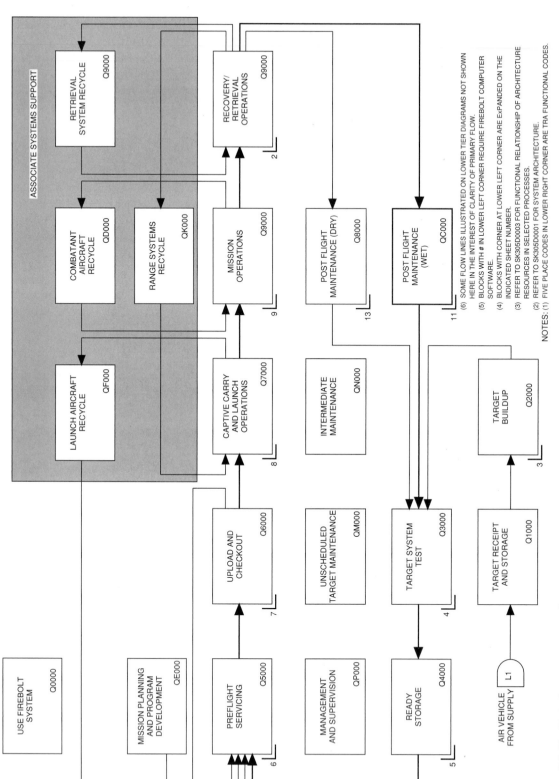

Figure 3.9-22 *Typical system process diagram*

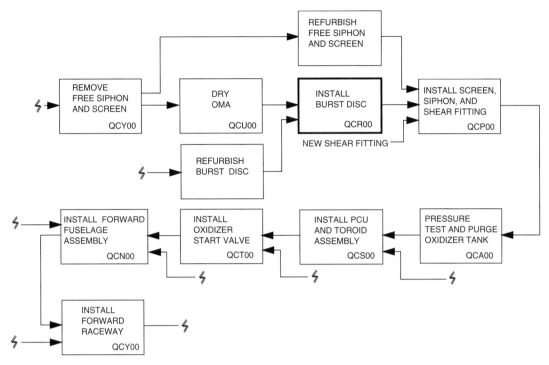

Figure 3.9-23   *Postflight maintenance process flow*

b. We have to add a new piece of support equipment to the baseline called a Burst Disk Alignment Tool that will be locally manufactured in accordance with instructions in the technical data.

c. Engineering drawings must identify use of KRYTOZ 240AB lubricant on threads and seals. This material was selected for compatibility with IRFNA.

Note also in this analysis that it was necessary to iterate the analysis after it was found in another analysis that the tank pressure test concept was invalidated by prior conclusions. The added tool was driven by a decision to remain with the bubble pressure test method. A trade study might have been done to check this alternative against using a pressure decay method, but even if that had been used, it would still be necessary to find a leak detected by a pressure decay approach.

This change reflected into another task that would have to be changed as well. The data used in this example was reported using typewriter technology in the early 1980s. You can see that it would be useful to capture this data in a computer database such that it can be easily updated and applied to many other purposes, like technical data development, efficiently.

3.9.6.5.3 Product Operation Analysis

Process analysis can be used to understand needed operator actions in essentially the same way as described for maintenance purposes. But, often the operator is confronted by a much more complicated situation than the maintenance person due to real time concerns. The operator is therefore

more closely wedded to the machine or system he or she is trying to control than is the maintenance person.

This more complex situation can be defined and analyzed using a special diagram that is essentially a multi-channel process diagram allowing the analyst to model the required human operator actions and the actions and responses of the system controlled in a coordinated way. Figure 3.9-25 illustrates a fragment of such a diagram. Some selected set of system elements, and possibly some elements in the system environment, are modeled such that each element has its own channel on the diagram. Symbols are introduced into those channels to indicate specific stimuli and responses. The diagram can be drawn with the channels horizontally or vertically arranged with time being the along channel dimension.

Commonly, the human operator initiates a string of interaction which is finally terminated by human observation of the desired response but the system (machine) may initiate as well as terminate strings of behavior.

3.9.6.5.4 Modification Development

Modification of a system is similar to the process used to create a system except that the principal material input is in a more organized state. The development of the modification is simply a system development process that should follow the same process as that used to develop the system originally. We should understand the customer needs, translate them into clear requirements as they apply to product and process elements of the existing system, design the changes defined by the requirements, and finally verify that the design satisfies those requirements.

TASK CODE	TASK NAME	SH
QCR00	INSTALL BURST DISK	1

TASK DESCRIPTION

The QCR00 step named "INSTALL BURST DISK" involves installing a device that seals a high pressure nitrogen supply from the IRFNA tank prior to flight. When starting the target engine, this burst disk must be pierced to pressurize the tank and feed oxidizer into solid propellant cast into the motor casing. After flight, the burst disk must be decontaminated, removed, refurbished, and re-installed in preparation for the next flight. QCR00 is the step that re-installs the device.

ANALYSIS NOTES

1. Opened analysis 30 June 1981.
2. The burst disk assembly arrives at this task from QM800 (Burst Disk Shop Work) where the disk and seals were replaced and the assembly was cleaned. The high and low pressure port unions are not installed until after the Retainer is torqued to the pressure supply port of the IRFNA tank. The unions and their associated packings arrive at the vehicle from the shop with the burst disk. The cap is not torqued to the housing at task entry, that is accomplished in this task. Burst disk assembly sketch below.

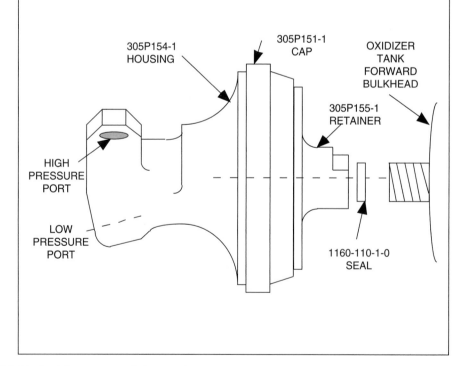

Figure 3.9-24   *Logistic support analysis example*

(Continued)

TASK CODE		TASK NAME	SH 2
	QCR00	INSTALL BURST DISK	

3. Wrench flats are machined into the retainer where it attaches to the pressurization port.
4. The ports should be kept capped until the burst disk is connected to its associated plumbing to avoid foreign object entry.
5. Closed 1 July 1981.
6. Re-opened 22 Sept 1981 and closed. Added burst disk cap torque to this task from task QCS00 (see analysis IIA-48). In order to properly position the burst disk housing so ports will line up with rigid lines in task QCS00, it is necessary to use a new special tool to hold the housing firmly in place while torque is applied to the cap. See item 9 for sketch showing special tool, a Burst Disk Alignment Tool.
7. The problem with the previous method of torqueing the cap during installation of the PCU and Toroid was that it would have defeated the plan to pressure test the interface between the burst disk retainer and the oxidizer tank port. After the Toroid and PCU are installed, one cannot see this interface even with a mirror and flashlight so the bubble leak detection method could not be used. Added step to Burst Disk Refurbishment (QM800, Table IIA-102) to apply KRYTOZ 240AB to the cap threads and mating surfaces between cap and housing to decrease friction during torqueing of cap in this task.
9. The figure below illustrates the Burst Disk Alignment Tool addressed in procedure

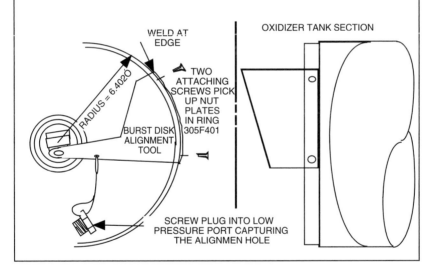

WELD AT EDGE

OXIDIZER TANK SECTION

TWO ATTACHING SCREWS PICK UP NUT PLATES IN RING 305F401

RADIUS = 6.4020

BURST DISK ALIGNMENT TOOL

SCREW PLUG INTO LOW PRESSURE PORT CAPTURING THE ALIGNMEN HOLE

Figure 3.9-24, cont'd Logistic support analysis example

(Continued)

The content of this part of the book is based on development of an unprecedented system following the form follows function approach in which one first determines needed functionality and then determines through function allocation what physical resources will be used to accomplished the needed functionality. When developing modifications of existing systems it is most often more effective to approach the solution first from the perspective of the architecture of the existing system. In the Define Modification block of Figure 3.9-26, one should evaluate the existing architecture for compliance with the new system needs.

Some of the architecture will be found to be acceptable as is. Other elements will be ineffective and have to be deleted from the system or modified. Other new elements may have to be added to the system. The new and modified elements will be the subject of new design work.

From this point forward, the process may be more complex than the original development process, however. The reason for this is that there is already a system in operation out in the world and it may be necessary to preserve normal operation of that system while the modified system is deployed and placed in service.

TASK CODE QCR00		TASK NAME INSTALL BURST DISK				SH 3	
TASK ANALYSIS		PERSONNEL		SUBTASK		TIME(MIN)	
STEP	TITLE	NO	SPEC	DESCRIPTION			
1	VERIFY CAP LOOSE	1	MECH	Verify that cap is loose on the retainer threads.		0.3	0.3
2	INSERT SEAL	1	MECH	Lightly coat seal with KRYTOZ 240AB and insert seal into outlet port		0.7	0.7
2A	LUBRICATE THREADS	1	MECH	Lightly coat threads with KRYTOZ 240AB		1.0	1.0
3	INSTALL DISK ASSY	1	MECH	Screw disc assy onto tank pressurization port hand tight.		1.0	1.0
4	TIGHTEN DISK ASSY	1	MECH	Tighten until it bottoms. Do not overtighten.		1.0	1.0
5	INSTALL BURST DISK ALIGNMENT TOOL	1	MECH	Install tool as shown in sketch with 2 screws and one plug.		2.0	2.0
6	HAND TIGHTEN CAP	1	MECH	Tighten burst disk hand tight.		0.5	0.5
7	TORQUE CAP	1	MECH	Couple torque wrench to cap with spanner adapter. Correct offset.		2.0	2.0
8	REMOVE TOOLS	1	MECH	Disengage torque wrench and spanner and remove alignment tool.		2.0	2.0
9	INSTALL UNIONS	1	MECH	Install O-rings and install two unions into high and low ports.		2.0	2.0
10	CAP PORTS	1	MECH	Install plastic caps on ports.		0.5	0.5
	TOTALS					13.0	13.0
	MANHOURS						0.22

Figure 3.9-24, cont'd *Logistic support analysis example*

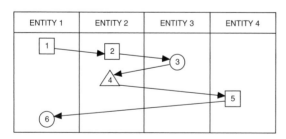

Figure 3.9-25 *Operational sequence diagram*

3.9.6.6 Disposal analysis

It is very difficult to gain acceptance of the need to define the disposal process while the development of the product is being planned and requirements defined. But, it is essential that the complete life cycle be defined at one time prior to the design of the product and processes. And this includes disposal as well as manufacturing, verification, logistics support, and operation. This can be done in exactly the same way as all other process analyses with analysis of each block in the exposed steps to define resource needs.

In the normal life cycle, the product system will be manufactured and delivered, placed in service for an extended period of time, and eventually, possibly after a

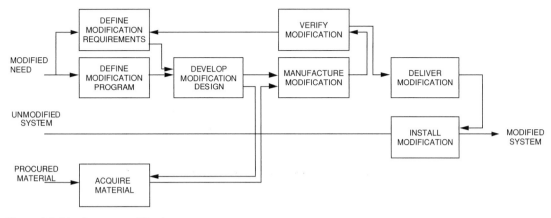

Figure 3.9-26 *System modification process*

series of life-extending modifications, be retired from service. The disposal process should have been planned as part of the development activity and updated as any modifications are implemented. This is especially true for systems that use hazardous materials as so many military products do, for example.

3.9.7 Quality Function Deployment

3.9.7.1 Introduction to quality function deployment (QFD)

The name QFD was selected as one of several possible English names for an analytical technique developed in Japan that provides for exploring the relationships between two sets of data in a quantitative way within the context of a group effort and consensus management. The meaning of the original Japanese words looses something in translation so most practitioners are quite content using only the acronym. The technique can be accomplished on one pair of data in a single matrix structure or several matrix structures coupled together in a string to study the relationships between several variables in a pair-wise sequence. It has become popular in many organizations as an element of an overall product development method. Its strength is in providing a team of several people a means to cooperate very intensely in the development of the relationships between two sets of objects. Therefore, it is a good tool for cross functional or concurrent development teams. The QFD approach discussed follows the techniques primarily popularized in the United States by the American Supplier Institute in Dearborn, Michigan, and GOAL/QPC in Methuen, Massachusetts.

Figure 3.9-27 illustrates a typical QFD matrix composed of six specific areas. This one is commonly called the house of quality because of the roof-like structure on top and is the most commonly employed matrix of the several possible ones. In some QFD applications involving use of up to 18 matrices, this matrix is called the A1 matrix. The left axis of the matrix (area 1 of Figure 3.9-27) is used to record the customer requirements, perhaps in a hierarchical listing. In a more general application of QFD, the data on this axis is referred to as the "whats." Most QFD descriptions call for identification of the requirements on the left axis of the top

level matrix through customer interviews and questioning or market analysis.

Development team members are encouraged to list the customer requirements using a technique called an affinity diagram. This diagramming technique entails a group of people contributing their ideas of the customer needs on file cards laid out on a table or sticky-back slips, like post-its, stuck to the wall. The whole group or team can see all of the inputs and cooperate in structuring them into a hierarchical arrangement built from the bottom up. An alternative tree diagram contains the same information in a hierarchy but is built from the top down. Either of these techniques can be used to organize the requirements inputs.

Ideally, this list of "whats" would evolve from a functional decomposition of the customer need used as the basis of detailed discussions with the customer/user to identify an accurate and complete picture of needed functionality. In some organizations, one will find an independent functional analysis done in parallel with QFD work with no conscious correlation between the two. These two analyses should have the same result if independent in that in both cases we are trying to find out what the system must do. An organization may also use marketing research data as the top level QFD input but, once again, this should be essentially the same story as the results of a functional analysis or QFD voice of the customer exercise.

Area 2 of Figure 3.9-27 is used to record what are called, in the generic situation, the "hows." Different experts in this field refer to these as the voice of the manufacturer, producer, or developer; design requirements; or design features. Where the latter is the case, the QFD house of quality becomes an effective means to make the transform between customer wants or needs (requirements) into a conforming design concept within the context of a cross-functional team. Where the content of area 2 is thought of as requirements, you can think of these items as detailed requirements or derived requirements. In the context of MIL-STD-961D Appendix A, which describes performance and detailed program peculiar specifications for Department of Defense programs, the area 1 requirements could be thought of as the performance requirements and area 2 requirements as the detailed requirements. In the generic situation, the content of area 1 and area 2 could be anything where you

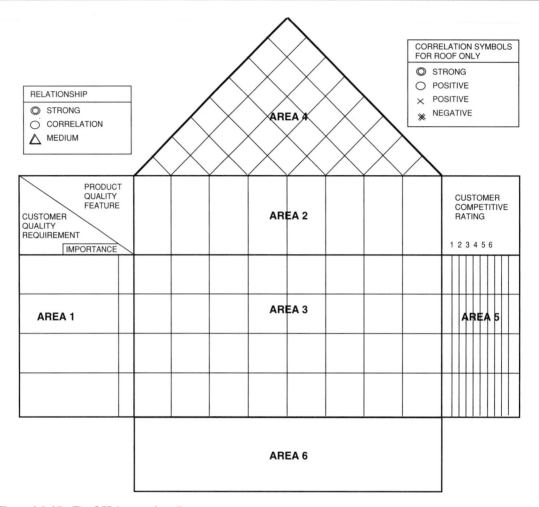

Figure 3.9-27   *The QFD house of quality*

are trying to establish a degree of correlation which is accomplished in areas 3 and 4 of the diagram.

The content of area 2 can be determined through brainstorming by the team members, appeal to historical precedent, or through trade studies evaluating alternative implementations.

Area 3 provides an organized space within which to relate the "whats" and the "hows." Different symbols are commonly used to denote the strength of the relationship. One would be very interested in ensuring that all "whats" have at least one "how" supporting it.

Area 4, often called the roof for obvious reasons, is used to identify relationships between the "hows" using different symbols to denote the strength of the relationship.

Area 5 offers the analyst a place to chart the relative merits of two or more candidate designs, possibly including a competitor's product, against the requirements in area 1. This is often done by placing a scale of some kind across the top of area 5 and for each requirement in area 1 a symbol unique to each of the competing designs or concepts is entered on that line corresponding to its value on the scale.

Then, the like symbols are joined by lines showing a kind of candidate profile relative to the others.

Area 6 offers a place to capture the relative goodness of the several candidates covered in area 5 relative to the area 2 requirements. Also, specific numerical targets are entered for each design requirement and these are the values that would go into the specification. One could also make the case that the area 1 requirements are design independent and belong in a performance specification while the area 2 requirements are design dependent and belong in a detail specification.

3.9.7.2 Physical implementation

The size of the QFD house of quality matrix is a function of the number of customer and design requirements, obviously, but the user might choose to build a set of standard matrices where this work is done with felt marker pen and paper. For any particular analysis, one would pick the smallest matrix thought to be large enough. There is no real problem using a matrix that is too large in that it does not change the use of any of the areas of the matrix so we could simply print a single size. These could be printed on easel paper tablets or taped to the wall. In this configuration, the voice of the customer can be built using post-it notes to the left of the matrix on the wall by someone facilitating the voice of the customer development.

Computer applications are available for QFD and this capability can also be implemented in a concurrent development team environment. The team space should include one computer that is equipped with projection capability and oriented to project the screen image onto a white board or wall space. Post-it notes can still be used as the computer operator responds to verbal inputs from the team and post-it notes or comments written on the white board. The content of the computer-projected data is progressively improved until the team reaches a point of diminishing returns.

3.9.7.3 A Problem with QFD

The strength of QFD in providing a creative environment within which several people can cooperate to understand a common problem is also its weakness. The user is encouraged to use market research, brainstorming, customer questioning, and careful thought to uncover the customer requirements on the left axis of the matrix and to interact in a group environment to list and organize requirements from these experiences. These are all valuable techniques but they are very close to the free style strategy discredited earlier. Many people favoring structured analysis, including the author, feel that most explanations of this process describe a touchy-feely approach that can miss important requirements. The value of structure in requirements analysis is that it tends to encourage the identification of the necessary characteristics and avoid the unnecessary ones when preparing a specification and QFD does not have as a part of its normally described or applied technique to do this.

The premise of this chapter is that it is possible to merge QFD and the structured methods described in earlier chapters. The offered solution involves attaching structured models to the left hand axis as a means of gaining insight into "customer requirements."

3.9.7.4 Linking QFD with structured analysis

Figure 3.9-28 illustrates how QFD can be integrated into the stream of work termed structured analysis. The structured approach, described in detail earlier in this chapter, follows the method attributed to Sullivan, an architect of the late 19th and early 20th centuries, who believed that a building form should follow from its function, that is, form follows function. The ultimate function is the customer need and we apply some form of functional analysis to decompose that need into lower tier functions and assign or allocate these functions to things that will satisfy the functionality. This process identifies the physical system structure and encourages the identification of one or more performance requirements for the item to which the functionality was allocated based on the function allocated. Three other structured approaches augment this process to identify design constraints: interface requirements analysis, environmental requirements analysis, and specialty engineering requirements analysis. The results from all of these activities, not necessarily accomplished by more than one person, but this is commonly the case, are collected into specifications for the items identified in the system architecture.

This traditional process can be augmented by QFD as shown in Figure 3.9-28. The structured process defines what the things are in the system and QFD can be applied selectively to these elements, perhaps only for the new items or those that have to be modified in a case where the project

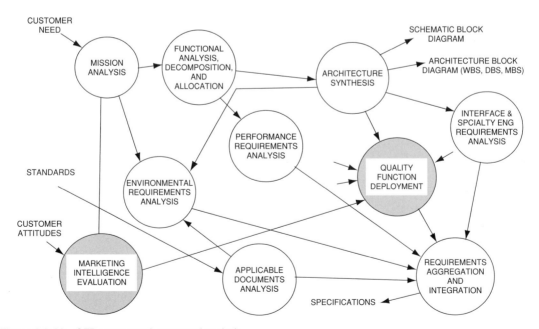

Figure 3.9-28 *QFD augmented structured analysis*

Table 3.9-3 *Process cycle*

Stage 1 enterprise	*Stage 2 proposal*	*Stage 3 program*	*Stage 4 sustainment*
Generic Process	Create Program Proposal Plan Program	Implement Program Refine Program Process Design Use Process	Implement Use Process

involves re-engineering a prior product. One could apply QFD by itself as the medium for item requirements analysis or could complement the traditional process to explore relationships between requirements identified by the structured approaches and through this process additional requirements may be uncovered as well as a clearer understanding of the requirements thus identified as a prerequisite to design work to satisfy these requirements.

All of the clear bubbles in Figure 3.9-28 relate to the normal structured analysis process feeding requirements to the specification being created. The QFD augmentation entails the two gray bubbles. While the marketing intelligence activity should be active whether using QFD or not, in a commercial situation this is a particularly important input to the QFD process, especially where QFD is being used alone without the structured analysis apparatus surrounding it.

In the method exposed in Figure 3.9-28, the intent is that all of these activities are in use together with QFD applied to observe relationships between requirement identified through the structured approach and to explore for requirements that might otherwise be missed. The organized team participation energy that QFD harnesses so well can be very effective in exposing holes in the current understanding.

3.9.7.5 Derived requirements generator

In Part 2 we discussed briefly what many engineers call derived requirements. The author's opinion is that all requirements except the need are actually derived, but, none the less, this is a popular term that is not going to go away. In an organization where this term is forever embedded in the lexicon, QFD can be used as a tool to uncover derived requirements. The customer requirements are placed on the vertical axis and the requirements derived from them through brainstorming, structured analysis, or careful individual thought and analysis are placed on the horizontal axis as they come to the surface. The matrix is used to assess the relationships between these sets. Surviving derived requirements are included in the specification. This approach is not recommended as a requirements analysis model but it may have merits over no method at all.

The house of quality concept (Matrix A1) of QFD has a particularly useful application in structuring the process for acquiring sure knowledge of the customer's needs and wants. This is especially true where the customer is composed of multiple offices and organizations. It is not uncommon in the DoD procurement environment for the immediate customer to be a different organization from the ultimate user and the two to have very different expectations. Some examples of this relationship are: (1) U.S. Air Force Systems Command as an acquisition agent and an operational aircraft wing as user of an aircraft system or (2) U.S. Air Force Space Systems Division as the acquisition agent and U.S. Air Force Space Command as the operator of a satellite system. These two commands have since been joined to form an odd case where the user and acquisition agent are one. In each of these cases, at a major program review, you would find ten or more other agencies represented and each of these agencies forms a part of the voice of the customer.

A CDR in the late 70s for an unmanned aircraft ground launcher offers a case in point that different elements of the customer do not always speak with the same voice. The launcher included a rotating red beacon light activated whenever the vehicle solid rocket motor was armed in preparation for launch. This seemed like a good idea to the contractor and procurement safety communities. A USAF Lt. Colonel from NATO (a user) viewed the beacon only as a homing beacon for hostile file at a critical point in the use of the system – the opening shots of war.

The traditional contractor approach to gaining insight into customer requirements is for one or more contractor system engineers to work with one or more members of the customer, often in an unstructured way, to define needed system requirements using models interrelating MOE, functional analysis, and mission analysis to gain insight into needed functionality and requirements. The customer eventually formally reviews and approves the requirements and analytical basis for them. This method is very dependent on the skill and experience of the system engineer and his or her ability as a good listener. There is also a tendency for this process to focus only on the immediate customer or the user while not bringing other customer agencies, particularly safety, into the initial work. The QFD house of quality concept forces a conscious consideration of competing alternatives and a selection based on demonstrable evidence agreed upon by all parties.

The house of quality approach begins with a clear definition of the voice of the customer, a process for gathering information from customers and identifying input categories. The voice of the customer is translated into a tree diagram that helps organize the input for the vertical axis of the house of quality matrix. This matrix features a list of necessary characteristics, or customer quality requirements, on the vertical axis and a list of product quality features on the horizontal axis that support satisfying the requirements.

3.10

RAS-Complete and RAS-Centered Analysis

Contents

3.10.1 A System Defined

A man-made system is a collection of entities that are meant to interact in predictable ways with an environment and with each other via relations between them to achieve a useful function identified and articulated by a customer as a system need statement. Therefore, systems are composed of entities and relations between the entities. The system is intended to satisfy the system need statement, the system's ultimate function, depicted on system diagrams as a rectangular block titled System Need and identified with a functional identifier F. The need is allocated to the system depicted on system models by a rectangular block named "system" (or a particular system) and identified with an architecture identifier A.

A system interacts within an environment as shown in Figure 3.10-1. The environment for every system is everything in the Universe less the entities that are part of the system architecture (E = U – A). One can reduce the scope of the environment to those entities that will have some influence on the system. The line joining the system and environment in Figure 3.10-1 (I2) indicates the relations between the two (external interfaces). The line joining the system on both terminals (I1) indicates the internal relations between system entities (internal interfaces) yet to be defined within the system. The line labeled I3 relates to relations between environmental elements that are of interest to the system.

3.10.2 Descriptors of Interest

A system, then, is defined by identifying its functionality starting with the need (F), allocating that functionality to entities that become part of the system architecture (A). These entities that form the system architecture are selected by determining the performance requirements that the system must satisfy to meet the top-level customer's need. The pairs of functions and architecture allocations pre-determine how the entities will have to relate to each other through interfaces (I) between the architecture entities. The environmental elements (E) are defined at the system level in terms of the spaces within which the system must function and the corresponding characteristics of those spaces drawn from appropriate environmental standards covering those spaces. As depicted in Figure 3.10-2, the traditional structured analysis effort attempts to define the most cost effective solution such that in N cycles of the process axis of the physical system (generally cyclical in the interest of the economy that reuse of system elements provides) the relation P (process) maps the cross product of the power sets of architecture (A^*), interface (I^*), and environment (Q^*) to the function set F such that F is covered. For every process Pi, there exists a combination of architecture, interface relations, and environmental stresses such that some subset of the function set is covered or accomplished. The power sets of these entities include all of the possible subsets of these entities within their own set thus the power set of A includes every useful combination of architectural entities

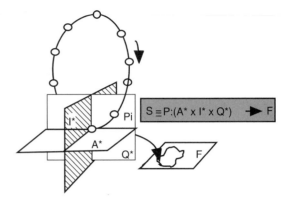

Figure 3.10-2   *The system relationship*

relative to every process step. Useless subsets are also included in the power set as well, of course. It is important that the functions be covered in the correct order determined by the sequence of the processes linked together in the process axis. If all of the functions are satisfied in N revolutions of the process axis as planned, then we may say that the system is consistent relative to the use of its architecture, interfaces, and environmental stresses. If there are architecture elements that are not used in the process or some that are needed but not available we may not have the optimum architecture. This whole process happens in practice somewhat backwards in that for an unprecedented system, one begins the development process only knowing the ultimate function, the need, and must expand everything from that one perspective. We will add one additional descriptor to this mix to control the inclusion of desired features that tend to add customer value to the system in the form of specialty engineering features.

3.10.3 System Functionality

A function is a necessary activity for a system to perform. It may be static, dynamic, or both. It should be named using an action verb followed by a noun phrase. A function is depicted in modeling the system as a rectangular block identified by an action verb name centered in the box and a function identifier (ID) in the lower right corner. The ultimate function for any system is the customer need the name of which is the need statement possibly paraphrased to fit into the space provided and with a function identifier F.

Two or more functions can be linked together using directed line segments to show a sequence of functions. In Figure 3.10-3 the understanding is that function F1 must be accomplished before function F2. Combinatorial symbols may be added to permit more complex sequential relationships. The combinatorial symbols encouraged are AND, inclusive OR (IOR), and exclusive OR (XOR) with the common logical and language meanings. Diagrams so constructed are called

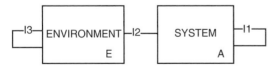

Figure 3.10-1   *Ultimate system diagram*

Figure 3.10-3   *Function sequence*

functional flow diagrams. These diagrams may be oriented on the page with their primary flow axis arranged horizontally or vertically with the flow in either direction.

For any function with identifier F@ (where @ is a string of length n (including n=0) composed of characters from the set {(A through Z less O)U(a through z less l)U(0 through 9)} there may exist one or more functions F@# (where # is a single character from the same set identified above which differentiates other functions at that level from one another). This is illustrated in Figure 3.10-4. Every function need not have an expansion. There is no need to assign function identifiers in alpha numeric sequence on a page of the diagram but it helps the human to use the diagram if they are assigned initially coordinated as much as possible with the function sequence. If a function is deleted subsequent to a release of the diagram, that identifier should not be used again. If the number of functions on any one diagram exceeds the maximum number of symbols available, 60, change the diagram to reduce the number to less than 60.

Ideally, who ever accomplishes the initial analysis of the need, would do so using the functional analysis process described here where the first decomposition is the system life cycle as shown in Figure 3.10-5 and the second is an

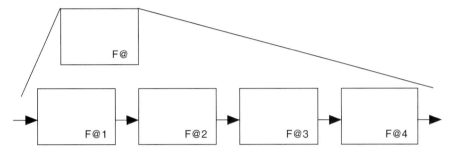

Figure 3.10-4 *Function decomposition*

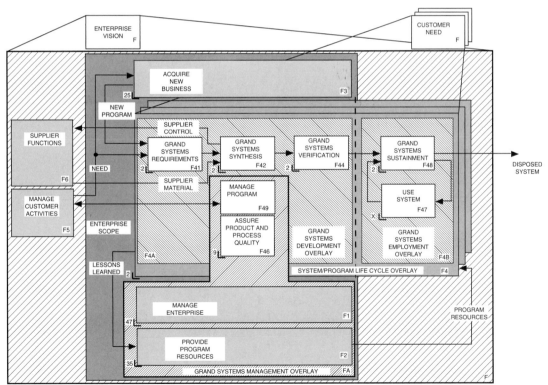

X: REFER TO PROGRAM SYSTEM DEFINITION DOCUMENT FOR EXPANSION

Figure 3.10-5 *System life cycle*

expansion of the life-cycle function "Use System" (F47) to expose the top level operational intent and initial content of the user requirements documentation or preliminary system specification. If the customer or other initial analysis agent applies an unstructured or ad hoc approach, then the development organization may have to accomplish a functional analysis and try to map the requirements identified by the customer (user and acquisition agent) to the functionality exposed when they do accomplish this work.

Ideally, the development organization would extend the functional analysis into the remainder of the system life-cycle functions as well as the Use System function determining appropriate resources for the process steps of the development program and the product system being developed such that the physical product delivered will be jointly optimized relative to its product and process. Commonly, process functions do or should influence product functions and the corresponding architecture as well as the opposite case.

3.10.4 Performance Requirements Derivation and Allocation

The functions identified in the functional flow diagram must be translated into performance requirements that tell what the system and its parts must do and how well it must do them as shown in Figure 3.10-6. These statements can be first developed as primitive statements—for example phrased as "Velocity ≥ 600 knots"—without complete sentence structures and subsequently transformed into complete sentences in the chosen language. Traditionally a requirements analysis sheet (RAS) has been used to capture the function identification, the primitive performance requirements statements, and the allocation of these performance requirements to architecture. One could allocate the function names directly to architecture but often one finds a one-to-many allocation result this way and allocation of performance requirements tends to follow a one-to-one pattern. To fully characterize a function it may require identification of multiple performance requirements and these several performance requirements may be allocated to different architectural entities.

3.10.5 Conventional RAS Limitations

Most system engineers are familiar with the requirements analysis sheet (RAS) employed in traditional structured analysis though which one associates functions and derived performance requirements with architecture entities thus identifying architectural entities that must be part of the system under development and the performance characteristics those entities must possess as a result of design synthesis of the requirements. This was described in Chapter 3.3. The problem with the conventional RAS is that it is incomplete. It only displays the relationship between functions and architecture when it could be used to display and correlate all of the requirements related artifacts relative to entities and relations needed for the complete traditional structured analysis process. In the RAS-Complete, the traditional RAS becomes only the function-architecture matrix which is connected to several other matrices and which collectively provide insight into the complete requirements analysis process.

3.10.6 The Beginning of the Complete RAS

Figure 3.10-7 illustrates the function-architecture matrix. The need is the ultimate function and appears on the function axis at the origin. The need is instantly allocated to the architecture entity system which also appears on the architecture axis at the origin. The need and the system coincide at the origin of the function-architecture matrix. The need is decomposed into a life-cycle model, an example of which is shown in Figure 3.10-5, and those functions, particularly function F47 covering operational use of the system, further

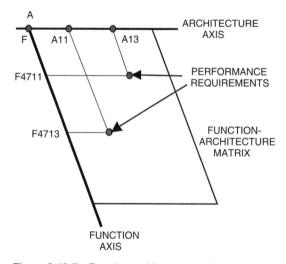

Figure 3.10-7   *Function-architecture matrix*

FUNCTION		REQUIREMENTS	ARCHITECTURE	
ID	NAME		ID	NAME
F4711	Provide Thrust	Thrust > 10,000 pounds at sea level	A11	Engine

Figure 3.10-6   *Traditional requirements analysis sheet*

decomposed as needed to define all needed system functionality. These functions and their derived performance requirements are allocated to architecture entities, either ones that already exist or ones which are added when no existing architecture is satisfactory as the implementing entity for the function. Each intersection on the function-architecture matrix represents a performance requirement, which should eventually be found in the specification for the architecture entity to which it was allocated. Thus is completed the function-architecture matrix commonly referred to as the RAS but it is only a single component part of the RAS-Complete.

3.10.7 System Architecture

System functionality is accomplished by physical entities that form part of the physical system. The entities in the system, in the aggregate, comprise the system architecture. The function F is accomplished by architecture A, the system, by definition. Lower-tier entities in the architecture should be exposed following Sullivan's encouragement of form follows function. Trade studies may be appropriate to make hard decisions on the best implementation of a particular function in early program phases. The physical entities that accomplish functionality can be partitioned into five classes: (1) hardware, (2) computer software, (3) facilities and improvements to real property, (4) procedural definition, and (5) humans following procedures or acting autonomously. We could merge the last two into one.

The relationship between functions and physical entities in the architecture in addition to the corresponding performance requirements is depicted on a requirements analysis sheet as shown in Figure 3.10-7. The requirements provide a selection criterion that assures that the correct architecture is selected for procurement or designed in-house to satisfy the function.

The aggregate architecture for a system is illustrated in a hierarchical model connecting the architecture entities arranged into a breakdown block diagram as illustrated in Figure 3.10-8. The architecture identifiers follow the same

pattern as the function identifiers explained earlier. The architecture entities stream out of the RAS available for structuring into the architecture block diagram. Ideally, this work would be accomplished by a team of people representing hardware and software engineering, manufacturing, procurement, and verification with strong system/program leadership. Initial functional analysis and allocation must concentrate on understanding user mission needs. This will generally require intense interaction with the user, ideally using system models to encourage mutual understanding. Two or more top level alternative architectures should be evaluated with a trade study to determine and select a preferred concept.

3.10.8 Allocation Pacing Alternatives

The conduct of the functional analysis and allocation work can be paced in one of four different ways:

1. *Instant:* As soon as functions and/or their corresponding performance requirements are identified, they must be immediately allocated to the expanding architecture.
2. *Terminal:* All of the functional analysis must be complete before any functions and/or their corresponding performance requirements can be allocated to architecture.
3. *Layered:* The analyst completes one layer of the expansion of a function and must allocate all of the exposed functionality and/or the corresponding performance requirements to architecture before further expanding that function. This works best if all artifacts related to a layer are developed before pursuing the next functional layer and its allocation.
4. *Progressive:* The analyst completes several layers of the functional analysis without allocating any of it to architecture. At a layer guided by experience the analyst begins allocating higher tier functionality to architecture. Design concepts are defined for the architecture entities at the higher levels and these act to both guide

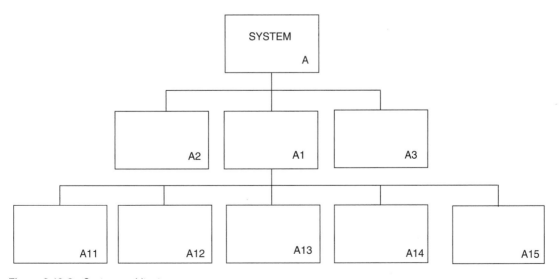

Figure 3.10-8 System architecture

and constrain lower-tier functionality progressively. Allocation is delayed throughout the analysis such that higher-tier design concepts help to steer lower-tier functional analysis tending to isolate iteration to one structure (functionality, architecture, or allocation) at a time. The two extreme cases (1 and 2) are flawed due to a need to iterate F, A, and allocations excessively in the case of 1 and a common need to significantly change lower-tier functionality after the higher-tier allocations begin in the case of 2. The progressive approaches either by layer or guided by experience generally produce better results with less modeling hysteresis.

There exists a downward limit in decomposition or expansion of the functional portrayal. This limit for any one branch in the expansion is best determined on the architecture plane based on the analyst's understanding of the problems related to the lowest-tier architecture entities. Where all of the lowest-tier entities are fully characterized supporting procurement or in-house design, the functional analysis work can be reduced to maintenance of the models and related data.

3.10.9 System Relations

As the architecture is formed, the analyst can begin to identify the needed internal relations between the system entities by using an N-Square diagram where the architectural entities are identified down the diagonal at some level of indenture. For a given analysis where the N is the number of entities being studied for interface relations in an N-Square diagram, the largest possible number of relations is N (N–1) counting each direction as one possibility between each pair of entities. Functionally driven internal relations between these entities is pre determined by the way that functionality is allocated to the entities. Therefore, one may explore the list of function-architecture pairs associated with each architectural entity in the N-Square diagram such as that shown in Figure 3.10-9. This is referred to as a pair-wise analysis of the function-architecture pairs and we can use the intersections of an N-Square diagram such as that shown in Figure 3.10-9 to capture the results of that analysis. We use both sides of the square to permit us to identify

A11		X		X	
	A12	X			X
X	X	A13	X	X	X
	X	X	A14		
X		X		A15	X
	X	X			A16

Figure 3.10-9   *Traditional isolated N-Square diagram*

direction of a relation as well as its existence. In this example, the author has chosen to use clockwise flow as indicated by the arrow at the top right corner of the square.

Figure 3.10-9 unfortunately disconnects the relationship between function-architecture pairs and interfaces that is the key to the identification of interfaces. Figure 3.10-10 juxtapositions the diagonal of the n-square diagram on top of the architecture axis of the function-architecture matrix where it belongs. The process for marking the intersections of the function driven relations matrix (N-Square plane) entails a pair-wise analysis of the function-architecture pairing relationships marked on this matrix. Interface Ix is encouraged by the conclusion that if F4711 maps to A12 and F4712 maps to A13 then there is a demand for an interface between A12 and A13 to implement those functional relationships. So, we map functions to architecture but we map F-A pairs to interfaces and those interfaces are predetermined by the way we allocate functions to architecture.

If the system is a modern war ship or the system that will take man to Mars, a pair-wise analysis of the function driven relations matrix would be beyond human comprehension if attempted all at one time, We can, however, partition the

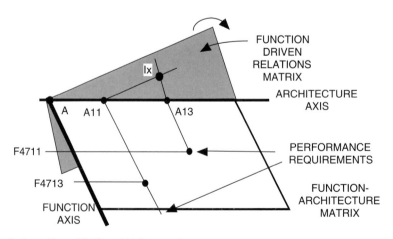

Figure 3.10-10   *Juxtaposition of RAS and N-Square diagrams*

analysis work to one interface expansion at a time and it is not so overwhelming. At any one level of architectural granularity, we can explore one layer of architecture expansion for internal interfaces within the parent item. If there are five subordinate entities then the number of possible interfaces to be examined in a pair-wise fashion would be 5×4 or 20. Granted, there will commonly be cases where more than one function is allocated to an architecture entity so the number of F-A pairs will commonly be greater than N (N − 1). Similarly, external interfaces can be analyzed individually. Of course, it will still be necessary to accomplish considerable interface integration work because of the partitioned process. This process will go very fast if the analyst is very familiar with the problem space and the evolving solution concepts. The complete algorithm is extended in subsequent paragraphs.

The requirements analysis sheet (RAS) identifies every pairing of functions and architecture entities. We may sort this listing so that all of the functions (or performance requirements derived from those functions) allocated to each entity are grouped by entity. Then we can pile up the allocations onto the architecture entity squares on the diagonal of a physical N-Square diagram as suggested in Figure 3.10-10. At the higher tiers we do not need the full N-Square diagram as we are not concerned with directionality only the binary question of whether or not there is a needed relation between particular pairs of entities and a semi N-Square diagram is adequate at the higher tier. As the analysis presses downward, we will have to use the full N-Square diagram showing directionality. In Figure 3.10-10 we have tilted the N-Square diagram in order to permit us to superimpose additional matrices in the process of assembling a part of the RAS-Complete.

So far our discussion of interface has only dealt with internal interfaces, so we need to extend the N-Square diagram to include external entities as well as internal system entities. In order to do that, we must explore the environment of the system.

3.10.10 The System Environment

Systems function in an environment and the environment is composed of the subsets illustrated in Figure 3.10-11. While all environmental effects on the system are relations, they may be partitioned between those that are commonly considered environmental stresses and the cooperative environmental elements which are treated as external interfaces commonly developed by a pair of teams or contractors responsible for the terminal architecture entities.

A context diagram, such as that shown in Figure 3.10-12, even though similar in nature to Figure 3.10-1, offers a useful simple model for focusing attention on identifying all external relations. Some of these terminators will be natural, non-cooperative, or induced environmental stresses. Others include hostile stresses determined through a system threat analysis as well as both stresses and useful relations with cooperative systems.

The system natural environment is determined by defining all of the spaces within which the system will be employed based on an analysis of the intended mission and basing concept. The spaces are coordinated with a set of environmental standards. Each standard is studied for necessary content and the remainder tailored out. Each selected parameter is then studied for an appropriate range. The system natural environment is then the union of the selected parameters from the selected standards.

The non-cooperative environment is defined by determining what stresses will be applied to the system from manmade systems which are neither hostile nor cooperative.

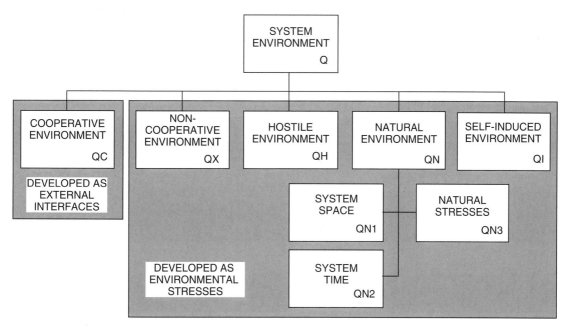

Figure 3.10-11   *System environment*

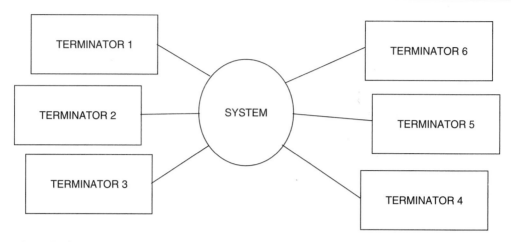

Figure 3.10-12   *System context diagram*

An example of non-cooperative stress is electromagnetic energy. Self-induced environmental stresses are not easily determined at the system level because one needs to understand energy sources and other stressors within the system determined as part of the design of end items.

System cooperative environmental relations are defined by determining how the system to be developed will associate with other friendly systems already in existence or being developed. These associations may be coupled into or out of the system in terms of information, physical association, materials, or energy.

3.10.11 Environmental Relation Algorithm

3.10.11.1 System environmental relations

The system environment consists of all entities in the Universe less those which are in the system. That is, $Q = (U - A)$ where Q is the environment, U is the Universe, and A is the architecture of the system being developed. The system environment as depicted in Figure 3.10-1 is illustrated on Figure 3.10-13 at the diagram "origin" on the environmental axis that happens to coincide with the architecture "origin" which corresponds to the whole system. That is the system is exposed to the system environment. It is convenient to partition all system level environmental relations into the sets illustrated in Figure 3.10-11. The system cooperative environment (QC) can actually be treated as an external interface and can be developed using the algorithm covered in Paragraph 3.10.9 very effectively. External cooperative systems are simply located on an extension of the architecture axis forming the cooperative environment axis as illustrated in Figure 3.10-13. The hostile environment (QH) can best be understood through analysis of threats posed by hostile forces. The non-cooperative environment will yield to the same thought process applied in the threat analysis except that the stresses applied to the system are not applied for hostile purposes, rather simply because the system being developed is sharing a common operating space with other systems. Electromagnetic interference is an example of the stresses applied in this set.

System time (QN2) is studied using time lines oriented about the functions that the system must satisfy. When we allocate those functions (or their corresponding performance requirements) to architectural entities the timing requirements corresponding to the functions are applied to the entity as timing requirements.

System space (QN1) is defined through mission analysis such that it is determined in what volumetric spaces of the Earth (surface, subsurface, and aerodynamic) and/or surrounding space and/or distant bodies the system shall function within, on, or around. For each space, that space is teamed up with one or more natural environmental standards which define that space. Each of these standards is then studied to determine which natural environmental (QN3) parameters included in the standard shall apply to the system being developed. Those that do not apply are tailored out of the standard. The selected parameters are then studied to ensure that the range of values is appropriate for the system. If the range is too broad it is tailored to narrow the range of values to that for which the system shall be designed. This process is repeated for each standard linked to a system operating space.

The earlier discussion of interface dealt only with internal interface identification, all defined on the function driven relations matrix of Figure 3.10-10. To cover external interfaces we will add the larger n-square diagram on Figure 3.10-13. The diagonal in this case includes all of the architecture entities plus all of the external entities in the cooperative environment. We can identify relations between these external and internal entities in the same way we did the internal ones. Interface Iy represents an interface between external entity QC11 and system entity A12.

3.10.11.2 End item service use profile

An end item is a major element of a system which generally retains its physical configuration throughout its mission performance and has an end use function. The end items may remain fixed in place during use or move over great distances and maneuver within the system spaces as a function of the system mission and the end item's application in the system. Each end item should be designed to endure only those natural environmental stresses anticipated so it is necessary to determine what subset of the system natural environment each end item shall be exposed to. To accomplish

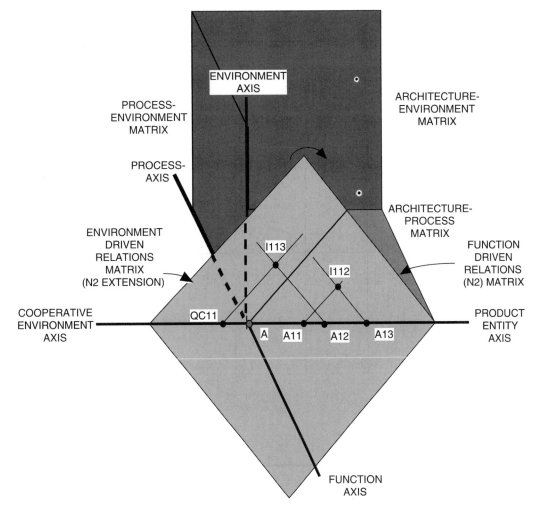

Figure 3.10-13   *Environmental requirements RAS addition*

this, one must create a physical process flow diagram. This is not the same as the functional flow diagram used to identify system architecture and performance requirements. The blocks on a functional flow diagram represent things that have to happen whereas the block of a process diagram represent a real world analogy. You cannot profitably consider system entities flowing through the functional flow diagram, indeed we are using the diagram to determine what those entities should be. However, we can imagine the system architectural entities flowing in the process diagram where each process acts as a transformation on the system entities. The first step in defining end item environmental relations is to map the system environmental parameters to the process steps at some level of indenture, generally at a level where the environmental map does not change significantly below that process level. This work defines an environment for each process. The next step is to map the architectural entities to the process blocks. If an entity only maps to a single process step, it simply inherits the process

environmental set. If, as is so often the case, the architectural entity maps to two or more process blocks, then it will be necessary to apply some kind of integrating process to any differences in environmental stresses and their values observed between the two or more process blocks. The rule most often selected initially is to pick the worst case range for each parameter across the process values being evaluated. If this rule does not adversely influence system cost, then it is an adequate solution. If this approach either results in an adversely narrowed system solution space, then an alternative to "worst case" must be derived. Often time lines will show that the problem environmental stress will be applied over such a short time as to be insignificant. In other cases, one may find that the problem can be narrowed to some particular combinations of values which are very unlikely to occur. If the problem is intractable, it may be necessary to restrict one or more system environmental parameters more severely than is currently being done. Generally, this will have an adverse effect on system performance.

The self-induced environment (QI) can best be studied and defined at the end item level since it is end items which contain the sources of energy and other stresses of interest which will reflect commonly through a natural environmental parameter right back into the system. Since the self-induced stresses are commonly greater in magnitude than the corresponding natural stresses for that same parameter, these induced relation values have the effect of extending the range defined through the application of the end item service use profile algorithm discussed above.

3.10.11.3 Component environmental relations
The environmental relations appropriate for components installed within end items are simply the end item environ-mental stresses if the end item has no altering effect on those stresses and all spaces within the end item offer the same environmental stresses to components installed within them. Where an end item does have a modifying effect on end item stresses but all spaces within and upon the end item offer the same stresses, it is necessary to determine the end item design effect on end item environmental stresses and the result is the set of installed component environmental stresses. The most complex case occurs when the end item must be partitioned into two or more spaces more often called zones of common environmental stresses. The value of each end item parameter must be determined for each zone thus defining the environment for each zone. Then one maps the components to the zones and they inherit the zone environments.

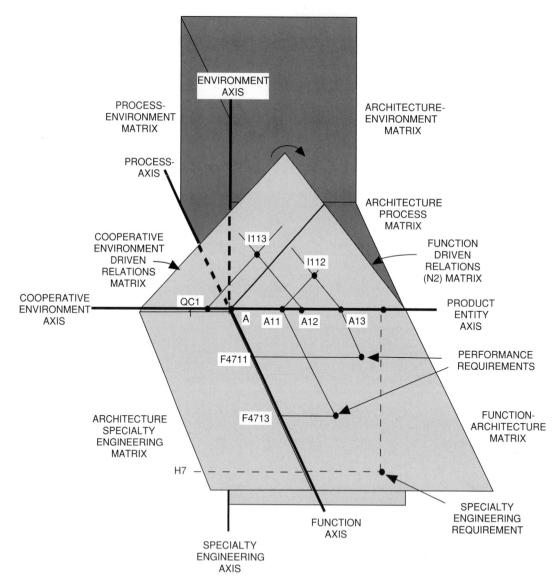

Figure 3.10-14 *RAS-complete in graphical form*

Commonly, the job is not complete at this point because it is found that there is no zone within which one or more components can be installed in a particular end item which will cause their environmental stress limits to be satisfied. When this happens, it is necessary to either change the component environmental specification values or include an environmental control system as an added entity into which the components with the environmental range shortfall problem are placed.

3.10.12 Specialty Engineering and RAS Complete

The system engineering agent for the system must build a list of all of the specialty engineering disciplines that will be applied in the development of the system. A map should be prepared between specialty engineering disciplines that may be required in development of the system and the architectural entities. This will help to determine team staffing needs in that area and connect people in those disciplines with a need to do specialty engineering requirements analysis for

MODEL ENTITY		REQUIREMENT ENTITY				PRODUCT ENTITY	
MID	ENTITY/RQMT TITLE	REL	VALUE	TOL	UNITS	PID	TITLE
F4713	Maximum Flight Speed	$\geq$	760		Knots	A1	
H11	Reliability	$\geq$	0.93			A1	
H12	Maintainability	$\leq$	28		Minutes	A11	
I2311	Electrical Power Voltage	=	28	$\pm$ 0.3	VDC	A12	
I2311	Electrical Power Voltage	=	28	$\pm$ 0.3	VDC	A13	
QH2	Explosion Pressure	$\leq$	—				
QI3	Acoustic Vibration	$\leq$	—				
QN12	Rain Volume	$\leq$	4		Inches/Hour	A1	
QX2	EMI	$\leq$	—				

Figure 3.10-15   *RAS-complete in tabular form*

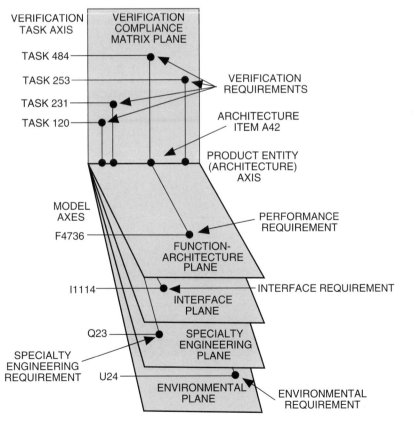

Figure 3.10-16   *Verification extension*

the indicated items. Figure 3.10-14 adds one more plane to the construct previously illustrated in Figure 3.10-13 providing for allocation of specialty engineering disciplines listed to architecture. The function-architecture plane extension to drive allocation of needed functionality to cooperative environmental elements is not shown in the interest of exposing some of the specialty engineering plane.

Specialty discipline H7 is shown mapped to architecture item A11. This must be followed by analyst definition of one or more discipline H7 requirements which will flow into the specification corresponding to the architecture entity. The structure exposed in Figure 3.10-14 is a complete RAS showing all of the important requirements related relationships supporting the requirements analysis process leading to the identification of every kind of requirement appropriate to the system and hardware specifications.

We have discussed how all four models used in traditional structured analysis (functional, interface, environmental, and specialty engineering) can be coordinated with a single three-dimensional construct. The question remains about whether or not we can apply the result to a real requirements analysis sheet. Figure 3.10-15 offers a partial RAS that does exactly that. The left hand column lists analytical identifications (IDs) for the following analysis areas: (H) specialty engineering discipline, (F) function, (I) interface, and (Q) environment. Some representative records are included to illustrate typical content. Note that an interface must be listed twice recognizing the two terminals, one for each of the two architectural entities. The reader may have applied some more obvious letters to these classes but the author was constrained by prior choices for modeling artifacts used in enterprise and program planning.

3.10.13 Verification Extension

The ideas expressed in this chapter can also be extended to verification as suggested by Figure 3.10-16. The functional-architecture plane includes the architecture axis to which all of the requirements derived from the four models indicated are allocated. Each product requirement is mapped to a verification requirement on the verification compliance matrix plane which is in turn allocated to a verification task of one of the four types (analysis, test, demonstration, or examination). Figure 3.10.16 reorganizes the planes from Figure 3.10.14 for easy communication with the verification compliance matrix plane. Granted, this is an oversimplification of a verification model, but it offers the general ideas of interest. This concept is explored more fully in Chapter 6.6 connecting up the plans, procedures, and reports developed in verification work.

3.10.14 Conclusions

The complete RAS provides a graphical model for identification of all four kinds of requirements appropriate for systems and hardware, performance requirements, and the three kinds of constraints (interface, environment, and specialty engineering). It is especially useful in suggesting an effective interface algorithm to replace the method most often used by experienced system engineers—experience. The complete RAS encourages the treatment of all environmental stresses, not just cooperative systems ones, as interfaces. Finally, the RAS Complete suggests that it is not only possible to allocate function derived performance requirements to architecture entities but also F-A pair driven interface requirements, every kind of environmental parameter, and specialty engineering disciplines as well.

Chapter 3.11 offers the tools to implement the RAS Complete idea on programs through a system definition document (SDD) that includes the machinery for full allocation of requirements from the several model sets exposed in this chapter. It also offers a more comprehensive view of the requirements analysis sheet like the one we have used through out the chapters in this part.

3.11

Traditional Structured Analysis Documentation

Contents

3.11.1 The Common Failure

It is not common for a development program to document the results of the structured analysis process by capturing the resultant work products. The reason for this is two-fold: (1) a lack of vision before and during the work and (2) a loss of interest after the fact. There is no generally accepted document format for capturing the results of the traditional structured analysis and few companies have practices which define one, call for its development, control its application, encourage its retention, or provide for its maintenance. The author experimented at one employer with using released engineering drawings but concluded that the information would be more useful if concentrated into one work product. This began a 20 year long search for the right content and format. This chapter provides the results of that search in a form that captures the results of the system requirements analysis work so as to provide a foundation for lower-tier requirements analysis that simply extends the scope of the analysis downward into the system definition.

The two final steps in finalizing the author's views about the ideal content and format for such a document were the development of the RAS Complete exposed in Chapter 3.10 and establishing a paragraph level lateral traceability map between the traditional structured analysis process and a specification template provided by MIL-STD-961E.

3.11.2 SDD Content and Format

When used to support the application of traditional structured analysis on a program, the preferred SDD format consists of a main body and seven appendices, each providing a capture point for the work products of one of the several fundamental analytical system requirements analysis process areas. Figure 3.11-1 shows how the document is structured. A series of seven interactive system analysis activities feed the development of the appended data explained in subordinate paragraphs. The appended data then becomes the basis for lower-tier analysis which produces content for the lower-tier specifications and adds to the appended data.

3.11.2.1 Document main body

The main body simply contains a table of contents, list of illustrations, and list of tables for the document plus it should provide text explaining the capture of work products in the seven work areas during system and lower-tier analyses. The body should also explain that the SDD couples the structured analysis work and its work products to specification content as guided by the selected specification standard templates.

3.11.2.2 Appendix A, functional analysis

This appendix captures the functional flow diagram starting with the identification of the system need and the life-cycle flow diagram. The Use System Function is initially decomposed progressively to expose more details about the user need. For each block in the functional flow diagram, there should be one line in the function dictionary also contained in Appendix A.

3.11.2.3 Appendix B, system environment analysis

The environment consists of several subsets of stresses that are applied to the system. This appendix identifies and characterizes them as described in Chapter 3.8. Timelines capture critical timing requirements. The spaces within which elements of the system must function are identified and the corresponding environmental stresses defined in terms of standards which describe those spaces. Service use profile analysis is applied to uncover end item environmental requirements. Finally, zoning of end items exposes component environmental requirements.

3.11.2.4 Appendix C, system architecture analysis

The system architecture results from the allocation of functionality to things. As these pairs are defined on the function-architecture matrix, they must be entered into the architecture block diagram, This work should be accomplished by a team of people knowledgeable in system, hardware, and software engineering, manufacturing engineering, verification engineering, logistics, material and procurement, and logistics in order to evolve an optimum architecture which will be universally respected on the program. This architecture is also the basis for the specification tree. Each item on the tree must have a responsible agent identified, a template selected, and a release date established. Refer to Chapter 3.5 for supporting text.

3.11.2.5 Appendix D, system interface analysis

Interfaces are identified by pair-wise evaluation of function allocations to architecture entities using an N-Square diagram. This appendix identifies all interface needs internal to the system as well as externally to the cooperative systems identified in Appendix B. The work is covered in Chapter 3.6

3.11.2.6 Appendix E, specialty engineering definition analysis

Appendix E provides a space in which system engineers can capture their work directed at identifying the specialty engineering disciplines that will have to accomplish work on the various entities in the system architecture to define the appropriate requirements and subsequently the needed analyses to confirm that those requirements are being satisfied. A specialty engineering scoping matrix is used to report the results of that analysis. Refer to Chapter 3.7 for details of the process.

3.11.2.7 Appendix F, system process analysis

Appendix F captures the results of a physical process analysis in the form of a process flow diagram. This is used by logistics engineers to drive out requirements related to training, support equipment, maintenance procedures (tech data content), and spares consumption. It is also needed to complete the environmental use profile study reported upon in Appendix B which drives environmental requirements for end items.

3.11.2.8 Appendix G, requirements analysis sheet

The exposed functions are listed in the Requirements Analysis Sheet (RAS) contained in this appendix. Related performance requirements are defined and allocated to an architecture entity. These performance requirements have to have a paragraph number assigned, title identified, and they can be outputted into a specification following a particular template. That part of the work can be done inside a requirements database system. Ideally, all of this work would take place within a requirements database tool but some organizations may find it preferable for their purposes to do the traditional structured analysis work using pencil and paper followed by capture of the resulting requirements

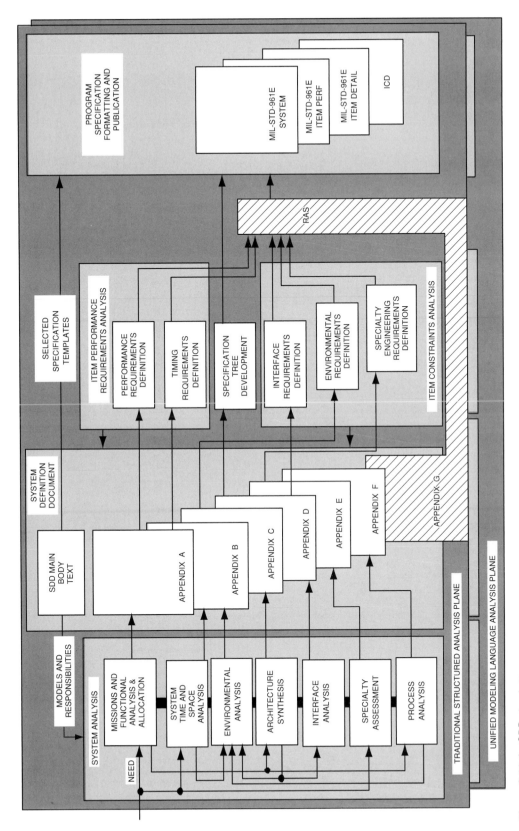

Figure 3.11-1 SDD structure

Para Nbr	Paragraph Title	Preferred Model	Discipline Specialist
1	Scope		System Engineer
2	Applicable Documents		System Engineer
3	Requirements	Traditional Structured Analysis	System Engineer
3.1	Functional and Performance Rqmts.	Performance Rqmts Analysis	System Engineer
3.1.1	Missions	Mission Analysis	System Engineer
3.1.2	Threat	Threat Analysis	System Engineer
3.1.3	Required States and Modes	Functional Analysis	System Engineer
3.1.4	Entity Capability Requirements	Functional Analysis	System Engineer
3.1.5	Reliability	Reliability math Model	Reliability Engineer
3.1.6	Maintainability	Maintainability Math Model	Maintainability Engineer
3.1.7	Deployability	Logistics Support Analysis	Logistics Engineer
3.1.8	Availability	Availability Math Model	Logistics Engineer
3.1.9	Environmental Conditions	Three Layer Model	System Engineer
3.1.10	Transportability	Logistics Support Analysis	Logistics Engineer
3.1.11	Materials and Processes	M&P Controls	M&P Engineer
3.1.12	Electromagnetic Radiation	EMI Analysis	EMI Engineer
3.1.13	Nameplates and Product Markings	Boilerplate	Quality Engineer
3.1.14	Producibility	Manufacturing Engineering	Manufacturing Engineer
3.1.15	Interchangeability	Logistics Support Analysis	Logistics Engineer
3.1.16	Safety	System Safety Analysis	Safety Engineer
3.1.17	Human Factors Engineering	Human Factors Analysis	Human Factors Engineer
3.1.18	Security and Privacy	Threat Analysis	Security Engineer
3.1.19	Computer Resource Requirements		Computer Hardware Engineer
3.1.20	Logistics	Logistics Support Analysis	Logistics Engineer
3.1.21	Personnel and Training	Logistics Support Analysis	Logistics Engineer
3.1.22	Requirements Traceability		System Engineer
3.2	Interface Requirements	N-Square Diagrams	System Engineer
3.2.1	GFP Interfaces	N-Square Diagrams	System Engineer
3.2.2	External Interface Requirements	N-Square Diagrams	System Engineer
3.3	Design and Construction	Engineering Domain Techniques	Engineering Domain Engineer
3.3.1	Production Drawings		Engineering Domain Engineer
3.3.2	Software Design		Computer Software Engineer
3.3.3	Workmanship		Quality Engineer
3.3.4	Standards of Manufacture		Quality Engineer
3.3.5	Process Definition	PMP Controls	M&P Engineer
3.3.6	Material Definition	PMP Controls	M&P Engineer
3.4	Precedence and Criticality of Rqmts.		System Engineer/Project Engineer
4	Verification		Verification Engineer (T&E)
4.1	Methods of Verification		Verification Engineer (T&E)
4.2	Classes of Verification		Verification Engineer (T&E)
4.3	Inspections		Verification Engineer (T&E)
5	Packaging		Logistics Engineer
6	Notes		All

Figure 3.11-2 *Specification management matrix*

in a word processor or a computer database tool from which specifications can be generated. Refer to Chapters 3.3 and 3.10 for details on the performance of functional analysis.

In keeping with the RAS-Complete idea advanced in Chapter 3.10, the RAS is not restricted to the performance requirements. Specialty engineering, interface, and environmental requirements can also be included so that every requirement appearing in every specification on a program will transition from the analytical model from which it was derived into a specification via the requirements analysis sheet.

3.11.3 Recommended Responsibility Pattern

In the author's view, a program should staff a program integration team (PIT) that should begin the requirements analysis process at the system level and develop the top level diagrams in the SDD. This work should continue as necessary to develop the content of the system specification and the specifications corresponding to the top level teams. The structured analysis for each of these teams should be taken over by the corresponding IPPTs in each case until they have completed the content of the specifications that define the problem for any subordinate teams. If no subordinate teams have been identified then they would have to complete the analysis needed to develop all of the specifications subordinate to their top level specification. This same pattern carries down to the lowest level.

Each team should act as the system agent for all of its lower tier teams and principal engineers. This starts at the PIT for the system and works its way down through the lower-tier teams. The Program Manager and Chief Engineer/PIT Manager should review and approve the system specification and all top level specifications. PIT should establish rules for review and approval of lower-tier specifications created by the teams.

With different parties doing the structured analysis, it is necessary to apply process integration and the PIT should do that accepting data into the several appendices, numbering the figures, and cross-checking the data submitted. At least one team will be involved in software development and if traditional structured analysis has been applied for the system level, then that or those teams responsible for software will want to switch to some form of software modeling such as UML. Refer to Part 4 for the suggested SDD treatment for parts of the system applying software models.

Figure 3.11-2 illustrates a way to tie together a MIL-STD-961E template for a hardware item specification, direction for use of particular methods/models, and the disciplines from which a person should be selected to accomplish the requirements analysis work indicated.

4

Computer Software Structured Analysis

Contents

4.1

Introduction

Contents

4.1.1 Computer Software Development Environment

Computer software came into existence with the development of digital computers using stored programs, but there were precedents from prior machines. Analog computers previously used to solve engineering problems had to be programmed in terms of gain settings and patch panel configurations that could be characterized in an instruction set. Early electromechanical computers relied on a stack of punch cards initially developed to control textile machines. The earliest forms of computer programs were prepared in the detailed language of the machine, driven by the way the arithmetic unit and registers had to be controlled. The instructions were prepared as a sequence of words of a length corresponding to the size of the machine, written in the ones and zeros of binary arithmetic, indistinguishable from the ones and zeros of the data processed by the machine when not associated with their intended function.

Binary digital computers have come to be the principal means of accomplishing numerical and logical operations in machines because they can be made very reliable. This is accomplished by driving binary electrical circuits between the two states of cutoff and saturation, where one condition represents the number 1 and the other the number 0. If the machines employed base 10 arithmetic, they would forever be in maintenance while technicians tweaked and aligned the electrical levels for each circuit. Any numerical computation can be accomplished in binary arithmetic, and industry has found ways to produce very dense, fast, and economical electronic circuitry to accomplish these arithmetical functions that is also very stable and reliable relative to the two levels.

Computer software is a preplanned series of actions accomplished on data by a machine. The data represented in binary numbers within digital computers can correspond to any physical phenomena, such as the current aircraft or automobile speed, a fuel tank pressure reading on a spaceship to Mars, or the number of dollars in your bank account. But, within the computer, these numbers are simply numbers that can be manipulated by instructions that compare one to another, add, subtract, multiply, or divide them to yield results that may be further combined in other operations to solve very complex mathematical relationships.

Software is a product, and there is little to distinguish it from hardware in the system development process except that there is no physical reality for the software. It is an intellectual entity that has a physical representation external to the machine, true, in paper tape, magnetic tape, or computer disk, but these are only containers for the real product, which is intellectual in nature.

In software development, as in hardware development, we should seek to understand the problem through requirements analysis before we attempt to solve the problem through design and manufacture, the latter by writing lines of code for computer software. Throughout the development process we should test the evolving product to ensure that the product satisfies the predefined requirements.

Software is unique in that it is implemented as an intellectual entity, not a physical one. This makes it very difficult for the people involved in its development to communicate effectively about it even among themselves, much less with those not familiar with software. Over a period of several decades, people working in this field have developed very specialized methods for analyzing and describing problems to be solved in software so that many people can interact in the development process. Throughout this period it has been common practice for hardware people and system engineers to ignore the details of the development of software, leaving it totally in the hands of software specialists. The result has been less than fully satisfactory.

It is impossible for a pure software product to function because it needs to operate in some kind of hardware device. Therefore, computer software should be developed as an integrated part of the whole. In order for this to be a reality, it is necessary that hardware, software, and grand system people communicate about their evolving product elements throughout the development process. This is a bidirectional experience across the hardware–software divide, but the principal responsibility for improving the development of systems containing software falls on those who often do not understand software development. It is a current and future reality that more and more of the difficult system functionality will fall to software, with hardware providing a physical reality within which the software operates to connect sensor inputs and action-oriented loads. Hardware-oriented domain engineers must accept this reality and make an effort to improve their ability to communicate with software people. The alternative future for a system engineer is unemployment. It is commonplace for a system engineer, who had to come into the field from some prior specialized discipline or educational experience, to depend on others to complete his or her knowledge base; but, if you are able to understand only the hardware world, the future will be very difficult, because an increasing percentage of difficult system functionality resides in software. If you have knowledge of, or access to knowledge about (through people whose knowledge you trust), only 20 to 30 percent of the system functionality, you will find it very difficult to be an effective system engineer.

Computer software engineers have developed many effective ways to enable this communication process through development models. They may possibly have developed too many different models, making it difficult for a novice to focus on one effective development model that he or she might come to understand. Unfortunately, software development was, in the 1990s, still a new and changing industry. No one had created the ultimate development modeling technique that we human developers could standardize. Very bright people continued to come up with new and often better ways to develop software with fewer errors, and these methods flowed into the mix of available ideas about this very difficult work.

The whole software development process requires sound management throughout and good skills in analysis, design, coding, and testing. The software engineer and manager should be familiar with at least the fundamentals of this whole sweep of development activity, rather than just the difficult task of writing computer software code in particular languages. Yet, if you look into the available software engineering education opportunities in many communities, you find many courses and programs for writing code in this or that language but few full software engineering programs that provide good skills in management, task estimation and planning, requirements analysis (problem definition), system design, and testing to complement skills in writing lines of code in particular languages.

4.1.2 Software Development Models for Analysis

Some very gifted programmers have been able to simply write code that generally worked without the benefit of a consciously applied method; that is, they employed an artistic or ad hoc approach. This has been and continues to be the exception. The inexpressible work of great software heroes, however meritorious the final results, is not a method worth emulating. Nonetheless, it survives as an existing method and should be recognized if not defended. It is the lack of method, a simple appeal to the creativity of the individual. The complete opposite of this problem could be realized by constructing such a rigid analytical framework that there remained no room for creativity, resulting in mechanically constructed software within a context of static technology.

There is a middle ground between these two unacceptable extremes. The tremendous complexity of the problems we seek to solve today in developing new systems or modifying existing ones demands a degree of order, as discussed earlier in this book. The successful methods for introducing order developed to date appeal to the powerful connection between vision and the mind in the form of simple graphical models. These models use very simple constructs, including bubbles or blocks to denote entities connected in some form by directed lines representing flow or transition.

These methods respect only two fundamental characteristics of computer software. Computers process data translating input data into output data. The two key words here are *processing* and *data*. These are the two focuses of software requirements analysis. I will discuss several examples of each of these approaches, followed by several methods that have been created in attempts to join these two views more closely together. The serious student of software models is encouraged to read Edward Yourdon's *Modern Structured Analysis,* which gives an excellent accounting of how many of these methods evolved. History, as is so often the case, is a good investment because many of these structures return in the latest modeling methods. All of these methods apply the simplest possible symbols with a powerful visual connection to represent ideas and artifacts. There are only so many simple graphical symbols, such as bubbles, rectangles, and the like. They, therefore, keep returning in the latest models but, perhaps, with slightly different meanings.

All of the computer software development models and methods can be collected into the three sets used in this part to discuss all of them: (1) computer processing–oriented development (discussed in Chapter 4.2), (2) computer data development (discussed in Chapter 4.3), and object-oriented development (discussed in Chapter 4.4). Chapter 4.5 covers a Department of Defense model for developing computer software for information systems common to weapons systems called DOD AF. Chapter 4.6 addresses where I believe system development is heading, with a merger of traditional structured analysis and the most recent version of object-oriented analysis called unified modeling language (UML). This will reunite hardware and software development, which parted company when software people departed from the use of flowcharts and migrated to modern structured analysis as a first step on a road paved with several modeling techniques.

4.1.3 Model Comparisons

While reading the several chapters in this part, you are encouraged to consider and contrast three characteristics of these models: (1) their common use of very simple artifacts, (2) the application of two-dimensional artifacts versus one-dimensional artifacts, and (3) the relative visual complexity of the diagrams included.

All of these models apply very simple artifacts such as blocks, bubbles, and directed line segments. The diagrams are intended to be relatively simple to encourage the import of ideas expressed on them into the human mind via vision, the most powerful way to move ideas that reside outside the human mind into the human mind. As *Life* magazine said many years ago, "A picture is worth a thousand words."

While the models all use simple artifacts, they are not all characterized by visual simplicity. The analyst may choose from simple models that do not model the problem space very richly, but their content passes into the mind of the analyst easily. At the other extreme are models that very richly model the problem space while the ideas expressed make their way into the human mind with difficulty.

The simple artifacts employed on these diagrams are of two varieties. Two-dimensional artifacts may represent data, functionality, product entities, or states, among other things. Directed lines may represent the flow of data or sequence of events, for example. Especially in software models, the two-dimensional objects coordinate with the strength of the diagram.

4.1.4 Design and Manufacturing Differences

Software models share a unique capability quite different from models used to develop hardware and systems. In the latter case, the model is used to try to gain an understanding of the problem to be solved. Model artifacts end up being associated with essential characteristics of the design of the objects that will form the system or system hardware item. The design team interprets the requirements included in the system and item hardware specifications in context with the available technology, resulting in engineering drawings and physical parts lists that manufacturing people can understand and convert into manufacturing processes.

Some software models, by contrast, are used to understand the requirements and then adapt to support development of the design solution complying with the requirements. Some modeling tools can actually generate code based on the modeling work accomplished. Some might argue that the hardware development process is not that far removed from the software process, however, in that the design accomplished on CAD can be moved through CAM applications to drive machine tools. The degree of automation is not as advanced for hardware as for software, however, as it requires considerable human interaction to produce the hardware item from the machine tool than is the case of converting the software requirements into a working software program. Very likely, the degree of difference between software and hardware design and manufacturing in terms of human involvement will continue to narrow, but for the time being it remains significant.

4.1.5 Software Deficit Disorder

Many system engineers currently working in industry have a hardware background, and many of them are in denial about the software content of systems on which they work. There was a time when there was no computer software in any system, of course. But today, all of the hard functionality goes into computer software. System engineers who are still in denial are in danger of becoming obsolete because there is not enough of a system outside the software on which to base a useful career. It is not necessary to advance to the point where you can write the code, but it is essential for a system engineer to be able to communicate with software people and integrate across the hardware–software boundary. I am an old hardware-dominated system engineer, but I finally came to understand the need to learn how to talk to software engineers. It is not a hopeless task. There are excellent books on the various models applied by software people, and these models express how software engineers describe the problem and solution space. It is possible to understand how these models work and use the knowledge obtained from them as a basis for conversation.

Linguists say that after one has learned several languages it is not so difficult to master a new one. This is also probably the case with software models. By learning another model we do not necessarily forget something else. Rather, this learning encourages a broader knowledge about modeling in general. One also begins to realize that there are only so many artifacts of interest in modeling a problem space and that the models that have been used and continue to be developed simply recycle similar sets of artifacts, sometimes with different representations of the same artifact.

4.2

Computer Processing–Oriented Analysis

Contents

4.2.1 A Little History

Computer software is composed of two entities, computer program instructions that direct the processing of data and the data to be processed. The models developed to analyze problem spaces to be solved through software all address these two entities. The first models that appeared were process oriented, focusing on the stream of computer instructions that were needed to guide computer operation. In the earliest days of the computer revolution, these models dealt very directly with the computer hardware configuration in terms of its registers and storage locations, using something called "machine language." The analysts had to write the program lines of code in binary numbers composed of ones and zeros. This was not a great problem because the only people who could write the code were people who understood the machine. For computers to become a great success, however, it was going to be necessary for people to be able to write the code without having detailed knowledge of the machine. Two solutions evolved, involving what came to be called "higher-order languages." Some languages, like Basic, must be interpreted, because they execute in the machine to translate the lines of code into machine language. Other higher-order languages require compiling before they can be introduced into the machine. The compiler translates the higher-order language into machine language for that particular machine.

4.2.2 Flowcharts and Other Things

The earliest organized approaches for software requirements analysis were process oriented, and the most popular of these was flowcharting. The flowchart, an example of which is shown in Figure 4.2-1, is built from three simple symbols: process blocks, directed flow lines, and decision symbols (other symbols were also commonly used to address the complete system). A flowchart looks like a functional flow diagram running vertically rather than horizontally. This pattern may have evolved because it was easier to program a mainframe computer to output graphics using ASCII characters on a line printer that way.

The blocks of a flowchart reflect required computer processing. The directed lines join these processing blocks together into a sequence telling which process must be accomplished first, second, and so forth. Decision symbols are used to indicate places where the flow may branch based on certain specified conditions between two or more pathways. Figure 4.2-1 illustrates a typical chart for discussion. Figure 4.2-1a requires, in one block of a flowchart, the calculation of a relative velocity between two vessels moving on the sea surface. In Figure 4.2-1b, this problem is expanded, or decomposed, to define in more detail how this must be done. Note that the diagram gives a good understanding about how the problem must be solved but does not give good insight into the data, where it comes from, and how it must be treated prior to entry into the illustrated processes.

Figure 4.2-2 provides a higher-level flowchart for this problem, showing process requirements in the context of a decision-making process that would have to be implemented in the software developed. By some means we compute range and make a decision on what source data to use to calculate target position based on that range figure. It is not sufficient, of course, to calculate a static range figure in a changing situation, so this process must continue at some rate until our weapon is ready for a data dump and launch.

The position calculation process continues until we need to use that data.

You can imagine a very elaborate flowchart for a very complex problem with many branches and many processes leading to development of many lines of code to implement this required processing. The principal problem with flowcharting, and functional flow diagramming, is that data is not well modeled. It is consumed and created by the blocks in the flowchart, but there are no elements of the modeling environment to address this data; what is it, where must it come from, and where must it go?

4.2.3 Modern Structured Analysis

The shortcomings of flowcharting were felt by many software developers, and improved models were documented by a number of people. Edward Yourdon, Tom DeMarco, and Larry Constantine were three of the leaders in developing, documenting, and teaching what has been called modern structured analysis (MSA). It is always hazardous to call anything "modern," including architecture, auto styling, and software modeling. There are newer models available now; but, at the time, this approach was a great advance on earlier methods or the lack of them, and many people continue to use this approach today.

The ultimate model of the software is called a "context diagram" and is illustrated in Figure 4.2-3. At this time, we have no detailed knowledge of the system, which is represented by a single bubble. But we know the system must interact with its environment in ways identified through terminals or terminators, the rectangles shown. This diagram is similar to the ultimate system diagram, composed of a system block interacting with a system environment block. Now, we must decompose the system bubble to expose more detail of the problem space, just as we would in a functional analysis of the ultimate functionality expressed by a need statement.

Modern structured analysis uses data flow diagrams (DFD) to expose needed functionality. The context diagram is the top-level DFD. The bubble represents all the internal workings of the system, and the terminators reflect external connections between the system and its environment. We can associate the system need with the single bubble of the context diagram. This bubble is decomposed into a set of lower-tier bubbles connected by the data that must flow between them. The DFD, illustrated in Figure 4.2-4, shows a context diagram for a weapons system and the decomposition of one level. This diagram includes only three different objects: bubbles representing computer processing that must be accomplished, directed line segments representing data that must flow between the bubbles, and temporary stores for data symbolized by two parallel lines.

The remainder of the model is linked to DFD constructs. The internal workings intended for the bubbles are expressed in what are called "P-specs," which can consist of normal descriptive text, structured English following some predefined pattern, a state diagram, or tabular data reflecting input–output constructs. A complex problem will become represented by an extensive set of dataflow diagrams. There could be many layers in the expanding diagram. Only the bubbles on the lowest-tier diagrams, however, require P-specs. It is important to number the bubbles in some way so they can be coordinated with the corresponding P-specs contained in different appendices, perhaps of the analysis report.

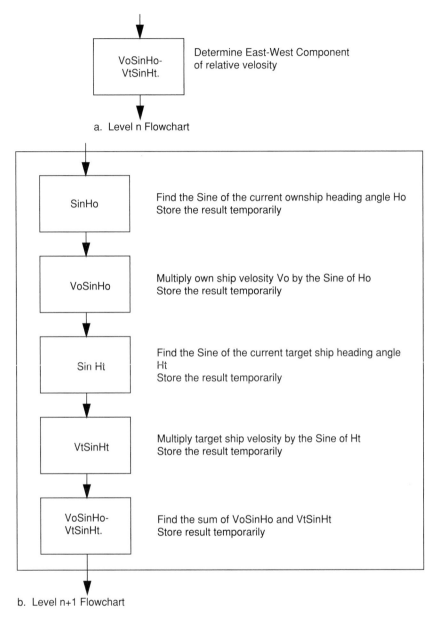

a. Level n Flowchart

b. Level n+1 Flowchart

Figure 4.2-1   *Flowchart example*

The strength in models tends to be in the two-dimensional symbols used to represent entities rather than in those represented by the one-dimensional symbols, lines. In this model, the two-dimensional bubbles represent computer processing. What gets processed in these bubbles is data. The model identifies data in three ways: a data dictionary, dataflow lines and stores on the DFD, and data entities called out in P-specs.

A data dictionary lists every data line and data store on every DFD and provides a clear definition of those data elements. If you use a computer tool, such as "Software through Pictures," to develop the model, you will be obligated to employ a prescribed syntax, but there are several benefits for that small surrender. The tool will apply discipline by insuring that the data items are consistent across the DFD, data dictionary, and P-specs. A tool might, for example, require you to use all caps, leading letter caps, or all lowercase for all references to data names, and the tool will insist that the analyst comply. If data items that are not in the data dictionary have been identified on a DFD, the

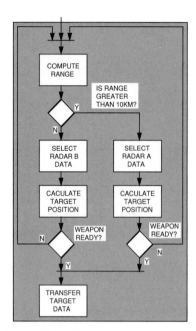

Figure 4.2-2   *Higher-tier flowchart*

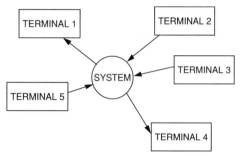

Figure 4.2-3   *Context diagram*

same data item is listed in the fragment of the corresponding data dictionary shown in Figure 4.5 and appears in the statements of the P-spec for bubble 5, as shown in Figure 4.2-6.

A P-spec is prepared for each lowest-tier bubble on the aggregate DFD set. Theoretically, this represents all computer processing that must be accomplished by the system. That is, it is not necessary to write P-specs for every bubble on every diagram. This is a different technique than that employed in functional flow diagramming. Every block throughout the hierarchy of functional flow diagrams must be allocated to something in the system and have performance requirements derived for the thing(s) to which the function is allocated. Figure 4.2-6 illustrates a P-spec for bubble 5 of Figure 4.2-4. Note that the model used by this book applies all caps for data names.

We model to gain an insight into the appropriate requirements for the software to be and its structure. As discussed under the models more appropriate to hardware, there

tool will alert the analyst to the inconsistency. Some tools will even generate lines of code in a selected language based on the model developed.

Figure 4.2-4 illustrates the use of data flow lines on the DFD. The line named X flows from bubble 5 to bubble 7 conveying a data result from bubble 5 needed in bubble 7. This

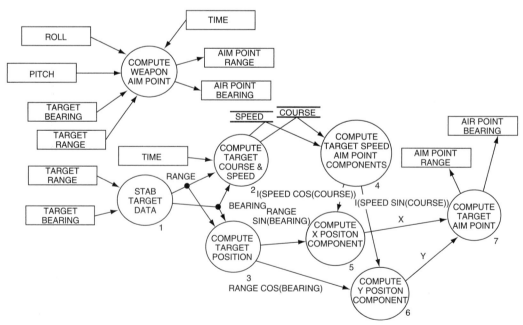

Figure 4.2-4   *Dataflow diagram*

DATA ITEM NAME	DATA ITEM DEFINITION
BEARING	Target bearing in stabilized horizontal plane measured clockwise from North in degrees
COURSE	Target course measured clockwise from North in degrees
RANGE	Target range in stabilized horizontal plane measure in nautical miles
RANGE Cos(BEARING)	Distance from own ship to present target position in the North-South orientation measured in nautical miles
RANGE Sin(BEARING)	Distance from own ship to present target position in the East-West orientation measured in nautical miles
SPEED	Target speed in nautical miles per hour
X	RANGE Sin(BEARING) + Integral (SPEED Cos(COURSE)) the East-West component of weapon aim point
	RANGE Cos(BEARING) + Integral (SPEED Sin(COURSE))
Y	the North-South component of weapon aim point

Figure 4.2-5 *Data dictionary*

BUBBLE 5 P-SPEC

Add RANGE Sin(BEARING) to I(SPEED Cos(COURSE)) where I is the integral of the X component velocity of target motion

Figure 4.2-6 *Processing specification (P-spec)*

should be a clear connection between entities in the models and content of the specification. Figure 4.2-7 illustrates the software requirements specification (SRS) format encouraged by MIL-STD-498, which was retired by DoD when EIA produced EIA J STD 016, which followed 498 very closely. The states from the model P-specs flow right into paragraph 3.1 subparagraphs. The performance requirements corresponding to these states flow right into paragraph 3.2 subparagraphs. The data requirements suggested by the data lines and stores (data dictionary content) flow right into paragraph 3.5. Wherever we decide to put the product partitions, we identify content for paragraphs 3.3 and 3.4 for external and internal interfaces, respectively.

In the example used to illustrate MSA, states do not logically come into play, because it is a fairly simple straightforward mathematical problem. Thus we might choose to substitute the computer-processing bubbles in place of states for a definition of the performance (entity capability) requirements in the corresponding specification.

At one time, standards differentiated between states and modes in a very specific but ambiguous way, and the dictionary meanings of these words are not that easy to differentiate. MIL-STD-961E, MIL-STD-498, and EIA J STD 016 use the word *states* in a very broad sense, saying that

1	Scope
1.1	Identification
1.2	System Overview
1.3	Document Overview
2	Referenced Documents
3	Requirements
3.1	Required States and Modes
3.2	CSCI Capability Requirements
3.2.x	(CSCI Capability)
3.3	CSCI External Interface Requirements
3.3.1	Interface Identification and Diagram
3.3.x	(Project Unique Interface Identifier)
3.4	CSCI Internal Interface Requirements
3.5	CSCI Internal Data Requirements
3.6	Adaptation Requirements
3.7	Safety Requirements
3.8	Security and Privacy Requirements
3.9	CSCI Environment Requirements
3.10	Computer Resource Requirements
3.10.1	Computer Hardware Requirements
3.10.2	Computer Hardware Resource Utilization Requirements
3.10.3	Computer Software Requirements
3.10.4	Computer Communications Requirements
3.11	Software Quality Factors
3.12	Design and Implementation Constraints
3.13	Personnel-Related Requirements
3.14	Training-Related Requirements
3.15	Logistics-Related Requirements
3.16	Other Requirements
3.17	Packaging Requirements
3.18	Precedence and Criticality of Requirements
4	Quality Provisions
5	Requirements Traceability
6	Notes
A, B,...	Appendixes

Figure 4.2-7 *MIL-STD-498 SRS format*

these may be states, modes, functions, or objects and that modes may be substates as one alternative meaning. These standards encourage the use of a structured analysis method but do not prescribe any particular one.

4.2.4 Hatley–Pirbhai Real-Time Extension

Mr. Derek Hatley and Mr. Imtiaz Pirbhai concluded, from a great deal of work they had accomplished using modern structured analysis, that it was adequate for mainframe batch processing problems such as the payroll problem but unsuited for developing solutions to real-time control problems like air traffic control. They developed an extension of the methods described above that added some degree of complexity while providing a richer modeling environment. We see this pattern throughout all the models discussed in this book—the choice between rich but visually complex model graphics versus less rich, simple graphics that are easily taken into the human mind visually. The engineer or team performing the analysis must pick from models that span the continuum between two extremes. Most mere mortals will not be able to make this decision solely based on a reading of the books explaining these models. It will be necessary to try them in certain situations to find out how well the models work for them in specific problem situations.

The Hatley–Pirbhai (HP) model uses the foundation provided by MSA, composed of a problem statement, context diagram, and scenario description (event list or icon flow, for example). The context diagram is expanded to develop a second-tier dataflow diagram (DFD), which may have to be expanded, as in MSA, for many more tiers. A P-spec is developed for every lowest-tier bubble on the DFDs. They referred to the data dictionary as a "requirements dictionary," even though this is not the only indicator of requirements. This dictionary will have a line item for every line and store on every DFD. In addition, it will have entries for every line and store on all of the control flow diagrams (CFD). The CFD is the significant difference between the MSA of Yourdon and Demarco and HP. For every sheet of the DFD series for which the analyst concludes there are control influences, he or she creates a CFD that uses the same bubbles as the corresponding DFD; but the bubbles are connected by directed data line segments as a function of control influences, while the DFD bubbles are connected by directed line segments indicating dataflow related to computer processing needs. Special control-oriented temporary stores may also appear on the CFD. The final difference from MSA is that the analyst must create a C-spec for every sheet of the CFD, defining the control requirements for that sheet using a state diagram, structured English, or tabular input–output data, for example, to explain the control problem space.

4.2.5 Transform from Models to Software Entities and Their Requirements

Particular portions of the DFD set can be coordinated with, assigned to, or allocated to particular computer software entities—programs, components, modules by whatever names. The total DFD diagram set is essentially partitioned into collections that correspond with particular software entities. This is a part of the system design because it is in the use of traditional structured analysis where performance requirements derived from functions are allocated to architectural entities. The required states and modes paragraph, 3.1 of MIL-STD-498/EIA J STD 016 shown in Figure

2.4-7, will employ modern structured analysis as the structured analysis model identifying computer processing requirements derived from the P-spec content. Each sheet of the DFD associated with a particular software entity can be assigned a different subparagraph of paragraph 3.1. Then the paragraphs under paragraph 3.1 can be mapped to the 3.2 subparagraphs. For example, for a given software entity, perhaps sheets 23 through 42 and 56 have been assigned to the entity. These can be related to paragraphs 3.1.1 (sheet 23) through 3.1.21 (sheet 56) and coordinated with paragraphs 3.2.1 through 3.2.21, with subparagraphs of the latter as required in each case to capture processing requirements derived from the P-specs and corresponding data requirements derived from the data dictionary content.

Paragraph 3.3 of the previously referred to standard deals with external interfaces for the software entity in question. The external interfaces are clear at the top software level from the context diagram. The context diagram terminators flow down to the collection of DFDs associated with the lower-tier specification level. We also need to determine the inter software entity interfaces, based on where the software entity boundaries appear. These boundaries are determined by how we partition the complete DFD set into software entities. This can be done by simply drawing rough lines around the parts of the diagram that belong to those entities. Data lines crossing those boundaries are the lower-tier item interfaces. These can be defined in a separate interface requirements specification for the whole or retained in the lower-tier specifications pair-wise.

Figure 4.2-8 illustrates a software model for discussion of transformation of the model into specification content that may or may not represent a practical software system. The system (dark gray background) has been partitioned into four software entities about which we will write lower-tier specifications. A software system specification would have dealt with the aggregate software functionality and the external interfaces shown, piercing the system boundary corresponding to the terminators of the context diagram. The specification for software entity D1 will have to recognize bubbles D11 through D13 in paragraph 3.1.1 through 3.1.3 and derived requirements from the corresponding P-specs. Software D1 external interfaces discussed in paragraph 3.3.x include I21, I22, and I23, which are system-external interfaces, as well as entity-external interfaces I1161, I1162, I15, and I17. All of the internal interfaces are also internal data entities, so we could cover their related requirements in paragraph 3.4 or 3.5. The data requirements would also have to address the need for temporary storage associated with store S11. The data dictionary provides the details that must go into those data-related requirements. The remaining paragraphs deal with quality factors or what I would call specialty engineering requirements determined by experts in the several related fields.

4.2.6 Are These Models Appropriate Only for Software?

The modern structured analysis approach does work well for software, and some practitioners claim that it can be used to develop insight into a system that will come to consist of a combination of hardware and software or even strictly hardware. The author believes that none of the methods discussed in this chapter has a chance of providing the universal solution to the system modeling and design problem. In Chapter 4.4 we will look into a variation of object-oriented

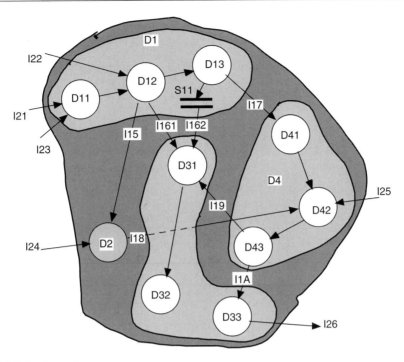

Figure 4.2-8 *DFD for discussion*

analysis called UML that does have a chance of satisfying all analysts' needs.

The author is not a software engineer, and his training and early experience was in fact in hardware and mathematics. When confronted by a problem space, the author first tries to apply functional flow diagramming to that problem space. If that fails to result in useful model artifacts, he will try state diagramming. One of these two simple models will almost always produce useful results. Once the author has identified a need for software and the functionality associated with this software, he would in the past encourage the analyst to switch to an appropriate software modeling approach for those parts of the system that will be implemented in software. This situation is in the process of changing; over the first decade of the 21st century, a variation on unified modeling language (UML) will emerge as a truly universal modeling language useful in cases where the problem will be solved through hardware and/or software synthesis. As this book is being written, something called SysML Version 1.0 in combination with UML Version 2.0 offered the great hope.

4.3   Data-Oriented Analysis

Contents

4.3.1 Data Augmentation of Modern Structured Analysis

4.3.1.1 Data lines, stores, and dictionaries

In Chapter 4.2 we saw several cases where data played a role in modern structured analysis (MSA), but that modeling environment was primarily focused on processing data. The student and user of models should notice what is being depicted by the symbols and what is being depicted by the lines in a particular model. In MSA, the bubbles represent processing and the lines represent data. In traditional structured analysis (TSA) using functional flow diagramming, IDEF 0, or computer flowcharting, the blocks represent functionality, and the lines represent sequence. The strength in a model is generally with the two-dimensional symbols (blocks or bubbles) and the weakness with the lines. In the models noted above, the strength is with processing or functionality.

In the MSA model, we saw data represented by lines flowing between the bubbles and by temporary store symbols. Further, a data dictionary focused our attention on the definition of all data needs. So, MSA is not without some aspect of data coverage. But, all graphically expressed modeling approaches offer the greatest strength in the entities represented by the two-dimensional plane figures and less strength in the entities represented by one-dimensional lines connecting the figures. Any computer software model that fails to represent both data and processing is, of course, incomplete, because computer software must entail both computer processing of data and routing of data between processing nodes, stores, and I/O.

4.3.1.2 Entity relationship diagrams

Many software engineers have employed entity relationship diagrams (ERD) to augment MSA to gain a better understanding of data relationships than they possible with MSA alone. In Figure 4.3-1 the blocks represent data, and the lines with included diamonds represent relationships. We read the relationship between the data items illustrated as REQUIREMENT CONTAINED IN SPECIFICATION. Some analysts use directed arrows to make the direction of the relationship clear, while others feel it can generally be easily determined from the context.

4.3.2 Relational Database Development

There are kinds of database structures other than relational, such as hierarchical; but relational databases are so pervasive in their appeal that I have selected that kind for discussion in this book. The simple rules for a relational database are:

1. The database is composed of tables.
2. A table is composed of columns and rows.
3. Entries in columns are single valued.
4. Entries in columns are from the same domain.

5. The order of the rows (also called records) and order of the columns (also called fields) in a table is insignificant.
6. Each row in a table must be unique—no duplicate rows.

4.3.2.1 Relational database development using table normalization

While accumulating a lot of experience in relational database development, engineers have gained a great deal of insight into some rules of behavior in building them that encourage success. This process is called "checking a table for normal structure." It entails a series of tests that are applied to the structure in a particular order to verify certain conditions. A table that survives the tests or is modified to pass the tests will result in a database that is easy to maintain with integrity.

The first test is called "testing for first normal form," which looks for repeating groups in the columns. There should be only a single value in a column–row intersection. When confronted by this problem, the designer must separate the data into two or more columns or restrict the entry of multiple elements into any one column–row intersection.

The second normal form test should not be made until the table is in first normal form. All of the columns in a table may be separated into two kinds. There will be one or more columns (fields) that establish uniqueness, and this field or these fields are called key field(s). The other columns are called nonkey fields. All nonkey fields must be functionally dependent on only key fields in a table. So, a table is composed of one or more key fields to which all nonkey fields are related. If the table has multiple combinations of key fields and their corresponding nonkey fields, it must be partitioned into two or more tables such that all tables conform to this rule.

A third normal form test depends on the table being in second normal form. It checks the structure of a table for existence of any transitive relationships between nonkey fields. The nonkey fields must be functionally related only to their key field. To satisfy this test, we may have to partition the table again into multiple tables. We may have started with one large table and ended up with many small tables that can now be relationally linked through the key fields.

Additional tests that can be made include Boyce–Codd, forth normal form, and fifth normal form; information on these tests is available in any book on relational database development. The end result of performing these tests correctly and responding to the anomalies detected is that the database will be characterized by integrity, and, in the normal process of using the database, needed data will not be lost or damaged.

4.3.2.2 Relational database development using IDEF 1X

A whole series of IDEF models continue to be developed. In Chapter 3.9 we discussed IDEF 0 for process/functional analysis. IDEF 1 was assigned for data modeling, but only IDEF-1X, dealing with relational database development, has been documented in Department of Commerce Federal Information Processing Standard (FIPS) Publication 184. The book *Designing Quality Databases with IDEF1X Information Models* by Thomas A. Bruce also offers a good story for business models.

IDEF1X offers a graphical modeling environment where blocks represent tables in a relational database and the blocks are connected together with lines representing the table

CONTAINED IN

Figure 4.3-1 *Entity relationship diagram (ERD)*

relations. For some reason not entirely understood, when the author uses IDEF1X to define a database structure, the table structures commonly come out normalized; but table normalization can be used to double check the structure. Figure 4.3-2 illustrates the graphical structures used in this model, and they are explained below.

This is a rudimentary requirements database structure, consisting of specification and requirements tables. Each table captures information related to a particular entity identified by the name at the top of the table rectangle. In the rectangle is a list of attributes that name fields needed in that table. Some of these fields appear above a horizontal line and others below it. Those above the line are key fields through which uniqueness is established for each record in the table data. Those listed below the line are nonkey fields.

The line joining the specification and requirement tables is identified with a cardinality symbol explained in the legend on the figure, in this case meaning that for each specification listed in the specification table there will be 0, 1, or more records in the requirements table. We have to cover the zero case because, in using the system, we may list several specifications in the database before we get around to writing any requirements in them.

Below the requirements table is a categorization symbol, meaning that there are several kinds of requirements, each of which requires different fields. This database structure is arranged to permit the analyst to enter the requirements in primitive form, which generally consists of the characteristic that must be controlled, a value and units for that characteristic, and the relation between characteristic and value. A typical requirement statement might be, "weight ≤ to 320 pounds." A requirement table attribute named "type" identifies the structure of the requirement, which permits the database system to connect the requirement table content by a categorization symbol to the appropriate categorization table below. In that the data requirements related to five of the requirement types (L, G, S, B, and E) are the same, there are actually only five different data types, with all of those contained in the gray outline being one. The data contained in the gray outlined data type would be used differently as a function of the type code in assembling the sentence generated from the data.

This kind of database structure encourages the analyst to focus on capturing the requirements in the simplest pos-sible form; when the requirements are printed for publication, software code creates a complete sentence for each requirement based on different cases for each requirement type. For example, the following case statement could be used to assemble the specification content corresponding to a type S requirement for weight:

CASE Type="S"

[ParaNbr]+" "+[Title]+". "+Item "+[Attribute]+" shall be less than or equal to "+[Value]+" "+[Units]+"."

In this case, it might result in a specification statement: "3.7.2 Weight. Item weight shall be less than or equal to 320 pounds."

In addition to the several kinds of numerical requirements statements, the system includes a type that references an applicable document and a header-only paragraph such as "3. Requirements." In all cases, the requirement could include added text material. Also, a text type requirement is allowed, consisting of just a block of text.

4.3.3 Transition to Specification

An IDEF-1X diagram pretty clearly identifies the database performance requirements in the form of the entities defining the tables needed, the data elements listed in each table, the data requirements, and the cardinality relationships the intertable relationships must respect. These requirements could be fashioned into a paragraph structure if necessary but could be used directly. Additional requirements would have to be developed to form product quality factors. The IDEF-1X model actually bridges the gap between database requirements analysis and database design, as so often happens in software requirements analysis.

4.3.4 DoD Architecture Framework

The Department of Defense has developed an extensive information system model for the development of architectures appropriate for the development of DoD systems for fighting wars and business applications. This framework is covered in Chapter 4.5.

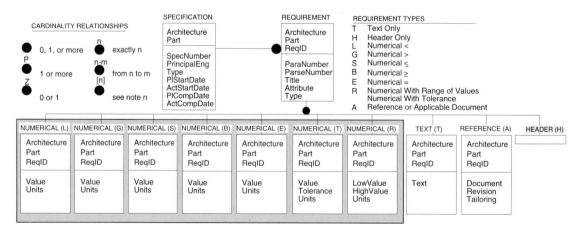

Figure 4.3-2 *IDEF 1X diagram*

4.4

Object-Oriented Analysis

Contents

4.4.1 The Early Combined Analysis Techniques

We have seen in Chapters 4.2 and 4.3 that separate efforts to understand the two principal entities in a software problem space have been applied in the past. On many development programs, this was seen as an impediment to the development of good software. Several attempts have been made to overcome this separate and unequal development capability culminating today in object-oriented methods.

4.4.1.1 Input-process-output (IPO) analysis

This method was developed when 80-column Holerith cards were popular as input/output devices for mainframe computers. The Holerith card was one example of punched paper media common in the early days of computer applications that had a long heritage, starting in the 1800s as weaving machine control media. In IPO, it was understood that the cards provided data to a mainframe computer that accomplished a process on the data and generated an output that might be stored in memory for subsequent use in other calculations or output to cards or other media such as magnetic tape. Engineers with design and analysis experience in the 1960s will recall that they would carry their card stack box (containing their FORTRAN program and variable cards) to the computer room and come back in a few hours or days to pick up the results, only to find in some cases that it did not run due to flawed code or variables out of range.

It is hard today to imagine an engineer peacefully queuing up like this to have some critical problem solved. This is one of the many ways that system development time has collapsed, removing much of the slack from the process and compacting the remaining work tightly about the work that can be done only by the thinking human being. This is one way the work speedup has occurred and, in combination with downsizing, has placed the engineer in an often uncomfortable position of appearing to be the only thing in the way of a solution and an end to a costly task. Our computers have succeeded in extracting all the slack from the process without adequately relieving the engineer of the many mundane work elements that must be performed, forcing long hours under pressure to get the job done. IPO was also related to some early attempts to computerize the requirements analysis process. Computer I/O was punched card oriented. requiring the specialized skill of a keypunch operator. The analyst would commonly fill out a paper form based on the corresponding 80-column card data demands. The form would go to keypunch and get punched. The resultant card stack would then be run into the computer via a card reader. We will return to this story in Part 7 of the book and see how the evolution to client server architecture from mainframes radically improved the use of requirements tools and collapsed this chain to analyst thinks/analyst enters.

To return to the topic at hand, however: In a less pressurized world, an 80-column card contained spaces for 80 punches horizontally in 8 rows. These cards survive today in U.S. government checks, airline tickets, and checks paid out to many system engineers working in industry. Figure 4.4-1 illustrates an example of IPO diagramming. Data is input (presumably from a particular card in a stack) on the left side of the first block. The computer acts on this data in accordance with a computer program producing output data that may be stored temporarily for use in a subsequent computer processing action or output to a card.

So, this technique consists of our old friend the flowchart running down the page, superimposed with data connec-

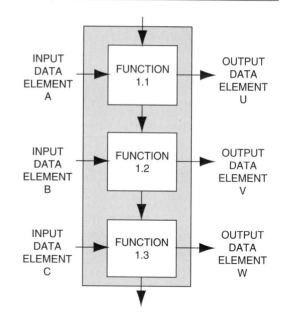

Figure 4.4-1   *IPO diagram*

tions leading left to right across the page. As you might imagine, these diagrams can become very busy, leading to a conflict between their utility for describing a problem that one or more humans must solve and the efficiency of humans in taking this problem perspective into mind visually. In all problem modeling media, this is a critical relationship. Most models are very simple visually to encourage efficient paper-to-mind transfer of displayed information; as a result, they cannot contain a rich description of the problem. This is the reason we often augment the diagrams with some form of dictionary, providing details that we can look up. If this data description were all included on the diagram, it would clutter the visual view. IPO is an example of a rich diagrammatic treatment that borders on excessive complexity, and you will see others in this book.

4.4.1.2 SADT and IDEF-0

The SADT diagramming technique was developed at Softech in the 1970s as a software modeling technique. These ideas were picked up by the U.S. Air Force and renamed IDEF, meaning Integrated Computer Aided Manufacturing (I) Definition (DEF) language, with the intent that contractors would use this common language to define their manufacturing processes. The Department of Commerce published a standard on IDEF-0 (more precisely identified with a zero because in the meantime other IDEFs had been developed) in the form of Federal Information Processing Standard (FIPS) Publication 183. SADT is a variation on a dataflow diagram as shown in Figure 4.4-2. In addition to dataflow entering a box on the left edge and leaving the box on the right edge, one draws connections related to control flows into the box on the top edge and supporting mechanisms into the box on the lower edge. Note that this early attempt to differentiate control and processing functionality was later implemented in the Hatley–Pirbhai modeling approach.

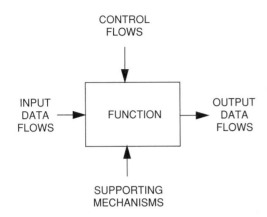

CONTROL
FLOWS

INPUT
DATA
FLOWS

FUNCTION

OUTPUT
DATA
FLOWS

SUPPORTING
MECHANISMS

Figure 4.4-2   *SADT diagramming*

The author made the point earlier that one has a choice of modeling methods using very simple diagrams that are not very rich in exposing the problem space but that communicate powerfully with the human mind through vision, on one end of the spectrum, to very rich modeling structures that are so complex that they do not communicate very well with the human mind through vision. SADT compared to functional flow diagramming, as covered in Chapter 4.2, is very intense and does impose problems in drawing the diagrams because of the directionality requirements related to the sides of the blocks. The added richness includes development of controlling flows and supporting mechanisms by earlier blocks in the model that are fed to subsequent function blocks. Some people choose to create IDEF-0 diagrams with very bold lines for control and supporting mechanism lines, which makes it very difficult for the user to move the exposed information through vision into the mind. Using the same size lines for all lines reduces this problem somewhat.

The IDEF-0 standard calls for the line entering the top of the block to denote controlling influences and the line entering the lower side for supporting mechanisms as befitting its application initially for manufacturing process modeling. The IDEF-0 standard provides an opportunity to add a call line at the lower right edge of a block that is actually a downward-pointing detail arrow that can be used to enable a sharing of detail between models. The standard encourages drawing the function blocks in a kind of stair-step fashion as a way to ameliorate some of the topological diagramming problems caused by the directionality constraints in two dimensions. Inputs to top and bottom sides can come from previous block outputs.

4.4.2 Early Object-Oriented Analysis

4.4.2.1 A dynamic beginning
There are two distinct OOA versions, the many early attempts and unified modeling language (UML). The former has many different expressions in the form of dozens of books written on the subject. The four that author has been most impressed with are *Object-Oriented Analysis* by Coad and Yourdon; *Object-Oriented Analysis and Design* by Booch; *Object-Oriented Modeling and Design* by Rumbaugh, Blaha, Premerlani, and Lorensen; and *Object-Oriented Systems Analysis* by Shlaer and Mellor.

Each of these authors and many more had their own opinions on how the object-oriented approach should work. Each had his or her own diagrammatic treatment and process explanation. So, object-oriented modeling became an extension of the model wars already in effect.

Something all the early authors and practitioners agreed on was that the model should cover both computer processing, or functionality, and data. Earlier modeling had generally focused on one or the other, when both are essential in any system applying computers. Looking back on the early development of OOA from the perspective of the year 2006, one can say that it was worth it to explore the many variations on the concept because it forced to the surface a generally preferred approach on which many of the early authors and practitioners could agree.

4.4.2.2 Misplaced beginnings
In all of the early books on object-oriented analysis, the early chapters dealt with finding those pesky classes and objects in the problem space as a beginning for the analysis. We should ask, "What do the objects represent?" They represent the static structure of the software to be. They are akin to the architecture of the hardware system discussed earlier under traditional structured analysis. The fact that we would attempt to identify the objects (physical architecture) before we discovered what the system must do (functionality) appears to reverse the idea that Sullivan expressed so succinctly as "Form follows function."

Early OOA appeared to encourage function following form. Frank Lloyd Wright, a protegé of Sullivan, was once asked which came first in developing a design, and he responded that they were both important. The author has fairly recently heard a software manager exclaim that all that analysis work is wasted and that one should dive right into the design process in a spiral fashion. This manager apparently did not realize that the spiral approach encourages analysis as the entry step for each revolution of the spiral. One should enter the problem space via the functionality facet for an unprecedented problem and the architecture facet for a fully precedented problem space (one for which there is an existing physical reality). In between, one should progressively favor one or the other as a function of the degree of solution precedent. The author has concluded that Sullivan was right for unprecedented systems and that Wright was right in the grander sense, depending on the nature of the problem space.

It appears that early OOA as practiced was OK as a means to transform a current hardware and people solution into a software solution. But, where the problem space is new, Sullivan's ideas seem appropriate, and early OOA faulted. Regardless, there are many people who continue to use the early OOA models, and the system engineer should be able to communicate with them as they apply it. Most of the early OOA books have many redeeming qualities, but the author has chosen Rumbaugh and his several companion authors as the model of choice to discuss in subsequent paragraphs.

4.4.2.3 The class and object model
A class is a collection of objects that have some common characteristics. Each object in a class is an instance of the class. The early Rumbaugh book illustrates a class or object with a rectangle divided into three spaces, as shown in Figure 4.4-3. The top space includes the class or object name. The middle space includes all the data elements or

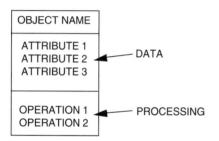

Figure 4.4-3   *Class and object artifact according to Rumbaugh*

attributes with which the object must deal. The lower space lists all operations or processing that the object must provide. Thus, object-oriented analysis does marry data and processing orientations. Other authors and analysts prefer different shapes for class and object artifacts (Booch preferred clouds in his early work, and some other authors use rounded-corner boxes), so be careful to understand the model elements and how they are illustrated on diagrams.

A class or object can be used to represent anything in the problem space. It can represent a machine, a person, an organization, or almost anything else that has some relationship to your view of the problem space. These are things about which one is interested in their corresponding data and the services they provide.

These objects are joined together in several ways to form class and object diagrams. They can be hierarchically joined to signify generalization or inheritance as shown in Figure 4.4-4a, to show aggregation as in Figure 4.4-4b, and associations with cardinality of the relation defined as shown in Figure 4.4-4c. You may write the name of the association by the line representing it.

4.4.2.4 The dynamic model

Within any one class or object, we must define its dynamic behavior, and that can be done with an old friend called a state diagram or state chart. Rumbaugh's book used a flattened bubble to represent a state and connected these bubbles using directed line segments. An object is said to be in a particular state reflected by one of the flattened bubbles on the state diagram. It can be in only one state at a time, thus excluding simultaneity, a very important characteristic in some problem spaces.

The object can change state, as defined by the directed line segments on the diagram in Figure 4.4-5, excited by a specific event. The diagram shows to what state the object moves when that event occurs. One should identify an initial state that unfolds from some initial condition stimulated by an initial event. The state diagram can reach a conclusion or go on essentially indefinitely (though that may or may not be very practical in a real situation).

Once in the first intermediate state on this diagram, different events can stimulate different responses, In one case, the event may stimulate a transition to a second intermediate state, while other situations stimulate a terminal behavior. These diagrams allow us to explore how we wish the object to behave and to communicate this behavior to others involved in the software development work. It is a good idea to supplement this model with a state dictionary and transition dictionary to provide a place to define with precision what the states mean and the exact events that trigger a transition without cluttering the diagram excessively with notes.

4.4.2.5 The functional model

We must also describe the functionality of the objects in the object diagrams. This is done using dataflow diagrams explained under modern structured analysis. Rumbaugh used oval bubbles on these diagrams, connected by directed

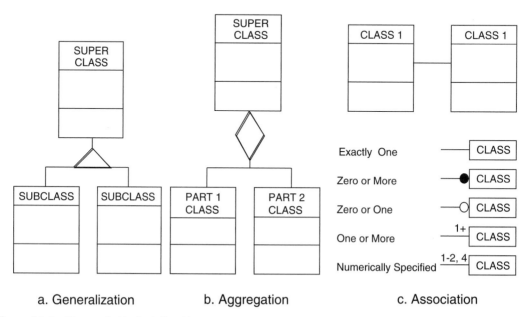

a. Generalization   b. Aggregation   c. Association

Figure 4.4-4   *Class and object relationships*

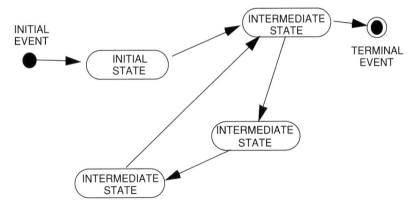

Figure 4.4-5   *State diagram notation*

line segments to show data flowing between the bubbles. A pair of parallel lines was used to show a data store. The bubbles are named and numbered. Other information keyed to their numbers provides specifics on what functions (processing) are required. Figure 4.4-6 illustrates such a diagram, once again assumed to exist inside an object explaining the operations called for in that object.

Rumbaugh coined the term *actor* for what was often called a *terminator* in modern structured analysis. These are the external influences on the system providing source data and data sinks (output data destinations). The term *actor* survives in UML.

4.4.3 Function-Driven Early OOA

Some professional system engineers whose experience in requirements analysis has been based on Sullivan's idea of form following function, as explained under the traditional structured analysis model, were unhappy about the initial OOA approach because it appeared to feature a "function follows form" sequence. The authors of several books on OOA make a point about how difficult it is to identify objects, which is the first order of business in early OOA.

This search of the problem space is difficult because it is done in an ad hoc fashion. It is like performing traditional structured analysis by looking for physical elements of the architecture and then trying to figure out what the system does. Painfully, this describes how some people do develop hardware systems, but it is not an optimum approach. Is it possible that we can form a bridge between traditional analysis and OOA and in the process link up Sullivan's ideas with OOA, provide some structure for the hunt for objects, and extend OOA utility more obviously to grand systems of hardware, software, and people in unprecedented problem spaces? The development of the unified modeling language was a step in the right direction in answering this question.

A Sullivanlike approach that one could apply is to first develop the system functional view using the modern structured analysis provided by the dataflow diagram (DFD), which is applied in early OOA to describe the internal functional characteristics of objects. Once the DFD is exposed, one can study it for best selection of object boundaries, drawing a boundary line around the DFD bubbles to be "allocated" to a particular object. Then the behavioral facet can be added relative to the objects and their enclosed functionality.

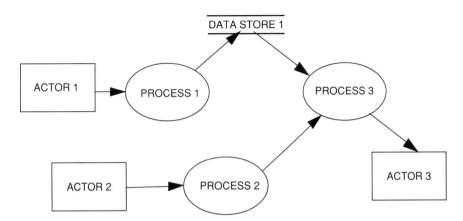

Figure 4.4-6   *Functional model notation example*

The author asked one system engineer of great renown how one should handle this inversion of the normal Sullivan approach. He related a conversation he had with another system engineer, also of considerable renown, regarding this matter. Engineer one asked engineer two how he handled the difficulty of understanding the needed functionality using early OOA. Engineer two told him that he first modeled the problem the way the books called for (objects first); then, when he understood the needed functionality, he threw away the work he had done and started over. Engineer one asked, "What program manager would ever approve that approach?" To which engineer two replied, "Oh, I would never tell my program manager what I was doing."

4.4.4 Unified Modeling Language (UML)

The word *language* in the title of this modeling approach is instructive. There is more to UML than a collection of diagrams. The authors of *The Unified Modeling Language User Guide,* Grady Booch, James Rumbaugh, and Ivar Jacobson, summarized an introduction to UML by saying, "The UML is a graphical language for visualizing, specifying, constructing, and documenting the artifacts of a software intensive system." So, UML, like some of the other modeling structures, goes beyond analysis, with which the UML book noted above is primarily concerned, extending to the design and "manufacturing" of the product as well.

UML is said to encompass three kinds of building blocks: things, relationships, and diagrams. The things are modeled in the diagrams that include exposition of relationships between those things. In this chapter, we will focus on the diagrams, of which there are nine. The first five diagrams illustrate the dynamic nature of the system being developed and reflect the functional and behavioral facets of the sys-

tem being created. Generally, one will begin modeling the dynamic behavior of an unprecedented system in UML rather than the static one as in early OOA; but one could reverse this orientation, beginning with the deployment diagram and working from the top of the physical view down through the components and classes, then determining the corresponding behavioral needs of the classes and objects. The latter would be more appropriate for a precedented development program.

The last four diagrams discussed are for modeling the static nature of the system from its microstructure in terms of classes and objects up through components and nodes. These diagrams describe the physical facet of the system to be. There is a hierarchical relationship between objects, classes, components, and nodes. A class commonly consists of several specific object expressions of it. A component consists of one or more classes, and a node consists of one or more components. Therefore, there is a top-down hierarchical relationship among the content of deployment, component, class, and object diagrams, as suggested in Figure 4.4-7.

4.4.4.1 Problem space entry and continuation

We could take our pick on the direction of development of the structures illustrated in Figure 4.4-7. One could work from the top down, bottom up, middle out, or outside in as a function of the available information about the application and the strength of the constraints applied by existing elements embraced in a partially precedented development. This would be an application of UML as described under the earlier OOA approaches, where one first identifies the objects and then tries to examine their functionality and behavior using statecharts and DFDs. There is no DFD in UML because it is replaced by the activity diagram. Alternatively, we could choose to enter the problem space

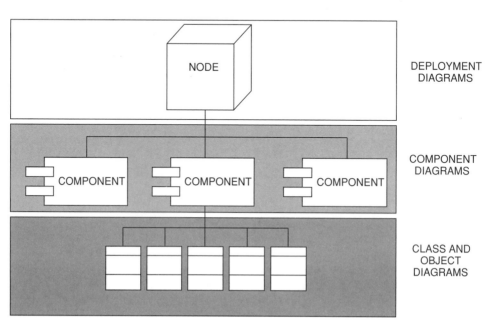

Figure 4.4-7 *Hierarchical static structure relationships*

via the dynamic aspects of the problem using some combination of five diagrammatic treatments. This alternative is not available to you when using the earlier OOA models because the state diagrams and DFDs are used to understand and explain the previously defined objects.

As mentioned earlier in this chapter and covered in more detail in Chapter 3.1, it is wise to follow Sullivan's idea of form following function and enter the problem space through the functionality facet where the problem is new to the analyst/team, to enter through the physical facet where a solution already exists that is going to be subjected to a modification, and to enter with some combination in between these extremes. Within the context of UML, this means to enter the unprecedented problem space using the dynamic models and the precedented problem space using the static models. In either case, we should also build or modify the other set of constructs based on the results of the entry analysis.

In the unprecedented case, use cases can be used to expose the analyst to statechart and activity expressions that can be then expressed in the two interaction diagram treatments, collaboration and sequence diagrams, reflecting a kind of preliminary design of the system that exposes classes and objects. The class and object diagrams follow and are grouped into component and node diagrams to reflect the physical organization of the evolving system expressing the detailed design of the system being created. At this point, the developer should be ready to attack the detailed design problem of writing lines of code responsive to the requirements exposed in the analysis work.

4.4.4.2 Dynamic model elements

Five dynamic models are included in UML: a use case diagram that permits us to visualize the relationships between the system to be created and the outside environment, a statechart diagram used to explore behavior, an activity diagram to express needed functionality, a sequence diagram through which we explore timing or sequence relationships, and a collaboration diagram through which we look into structural relationships between entities and messages that must flow.

4.4.4.2.1 Use Case Diagram   The use case diagram identifies activities that the system must address. It is similar to the context diagram of modern structured analysis in that it exposes the relationship between the system and actors external to the system referred to as "terminators" in modern structured analysis. These external entities, called "actors," are drawn as human stick figures, as shown in Figure 4.4-8, and they are intended to interact in specific ways to derive benefit from the system. These actors need not, of course, be humans at all. They can be other cooperative systems, and very complex in their own right, but the concept is the same, providing and receiving benefit relative to the system under development. The actors are connected to the related use cases by actor–use case communication lines. The use cases may be interconnected within the system boundary by lines representing association, extension, use case generalization, or inclusion.

4.4.4.2.2 Statechart Diagram   A statechart, state machine diagram, or state diagram explores the behavior of a use case or an object. The understanding is that an object can exist in only one configuration at a time and that it can switch from one configuration to another. In defining these

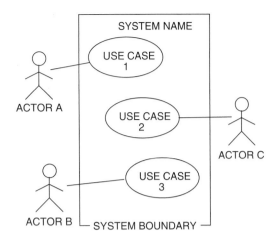

Figure 4.4-8   *Use case diagram*

configurations or states and the sequences through which the states can change, one becomes familiar with the behavior of the object. It is also understood that the object may exist in only one of several states at any one time for some period of time until a particular triggering event occurs that instantaneously switches the object to another state. A statechart diagram defines the states within which the object may exist, the permissible transitions between these states, and the triggering events that stimulate transitions. Figure 4.4-9 illustrates such a diagram. In UML, rounded-corner boxes represent states joined by directed line segments annotated with triggering events. Initial and final states are identified by the indicated symbols.

State diagrams are built from two kinds of entities interconnected by directed line segments. Action states are atomic in nature and cannot be further decomposed. A use case or object is thought of as existing in an action state only for an insignificantly short time span. Activity states can be further decomposed into combinations of activity and action states and are considered to require some time duration to occur.

4.4.4.2.3 Activity Diagram   A lot of software engineers have, for reasons unknown, become estranged from expressions of functionality in relationship to the problems they seek to understand and solve. Perhaps this is traceable to the flowcharts used in early software development

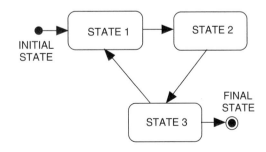

Figure 4.4-9   *Statechart diagram*

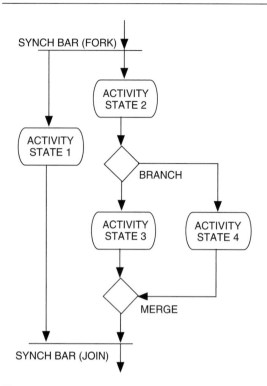

Figure 4.4-10 *Activity diagram*

work, which are essentially the same as the functional flow diagrams that were the center point of Chapter 3.3. As discussed in that chapter, these diagrams are insufficient by themselves to encourage understanding of a difficult problem space, but they are very effective in exposing what an object must accomplish—its functionality.

These diagrams are essentially linear state diagrams with a beginning and end, whereas statecharts can cover endlessly cycling behavior. One could say an activity diagram covers a string or scenario that is capable of supporting a use case. Figure 4.4-10 illustrates an activity diagram. The activity state strings between a pair of synchronization bars are intended to be accomplished concurrently, while the strings of states between a pair of branch symbols are alternative flows.

It may be necessary to use two or more activity diagrams to model different use case scenarios sometimes referred to as "strings of behavior." The flow from an initial state to a stop state can also be partitioned into two or more lanes laterally, with each lane correlated to specific resources or set of objects. Some analysts refer to these as "swim lanes" because the diagrams seem to be laid out like a swimming pool set up for competition. This same technique is used in what is called by some an "operational sequence diagram," as discussed in Chapter 3.9.

The system engineer will see a strong resemblance between the activity diagrams, the old flowchart reborn, and functional flow diagrams with the branch symbol replacing the OR symbol and the synchronization bar replacing the AND symbol. There is nothing to prevent a system engineer steeped in traditional structured analysis

from using activity charts to model system functionality. One would have to make the switch from horizontal flow diagrams to vertical flow diagrams, but if it encouraged more effective communications between systems, hardware, and software engineers, it could not be all bad. I will discuss this possibility further later.

4.4.4.2.4 Collaboration Diagram   A collaboration diagram is one of two kinds of interaction diagrams used in UML. These interaction diagrams provide a transition between the requirements analysis work performed in association with use case, state, and activity diagrams and the detailed design of the software system embodied in the state and class, component, and node diagrams. Thus, the interaction diagrams offer what could be called a "preliminary design environment." The collaboration diagram illustrates the structural relationships among objects and the messages that must interconnect them to accomplish some activity. Figure 4.4-11 illustrates the diagram.

Blocks represent objects conceived to implement functionality and behavior exposed in use case, state, and activity diagrams. These blocks are interconnected by links that correspond to messages among the objects. The system engineer will see the similarity between the collaboration diagram and the schematic block diagram to illustrate interface relationships between architecture (static) elements in a hardware system.

4.4.4.2.5 Sequence Diagram   A sequence diagram is one of two kinds of interaction diagrams used in UML. It is used to describe a time/sequence relationship between actions taken by objects identified to implement a use case diagram. We can use these diagrams to illustrate and describe particular sequences, strings of activities, or scenarios that will lead to a desirable result. Figure 4.4-12 illustrates this diagram. The system engineer will notice the similarity to a time line diagram with time increasing top to bottom.

4.4.4.3 Static model elements
Developing use case diagrams corresponds to the requirements analysis stage of hardware. Developing the statecharts and activity diagrams corresponds to preliminary

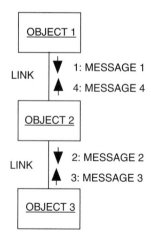

Figure 4.4-11 *Collaboration diagram*

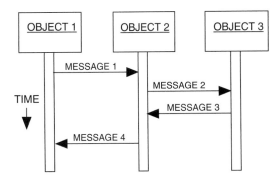

Figure 4.4-12 Sequence diagram

design, and developing the static elements corresponds to the detailed design of the software entity. Actually, writing the code based on the requirements that are exposed in the modeling process could be correlated with hardware manufacturing.

4.4.4.3.1 Class and Object Diagrams The class diagram illustrates a collection of classes representing static entities in the problem space through which we choose to view that space in a physical sense. Figure 4.4-4 illustrates the essential diagrammatic elements exposed in UML models as well as early OOA models. An object is a specific static instance of a class. Object diagrams are, therefore, essentially the same as class diagrams, reflecting specific objects rather than classes of objects. If we modeled a development program on a class diagram, we might identify a class called "integrated product and process team (IPPT)" and on a corresponding object diagram would identify the specific IPPT 3 that provides certain services or operations relative to a particular system entity that are similar to the services performed by IPPT 2 on a different part of the system undergoing development.

4.4.4.3.2 Component and Deployment Diagrams A component diagram illustrates the organization and interrelationships among a set of components, each of which covers

a major portion of the physical system in terms of a set of class and object diagrams. A deployment diagram collects components together into nodes and shows how they are related. Figure 4.4-13 illustrates both of these diagrams in one. The component diagram does not include the nodes identified. These diagrams are both part of the solution rather than part of the description of the problem space. They reflect the intended design solution. They represent particular deliverable software applications.

4.4.4.4 Unprecedented Application UML can be effectively applied to unprecedented problems, unlike early OOA. The customer need statement can be represented by the ultimate use case diagram, similar to a context diagram with one oval interacting with all known actors (external interfaces) connected to it. We know that the system shall accomplish this ultimate functionality, and the challenge is to apply UML to bridge the gap between the ultimate use case and the system expressed as components and nodes of known capabilities. The first step would be to analyze the relationships between the system to be created and the actors needing service. For each service element, we may draw one use case diagram, and we may find it necessary to craft many use case diagrams. Some people are concerned that this process can run as an open loop, yielding a proliferation of use cases, because there is insufficient structure in this creative part of the analysis. Some analysts think in terms of each use case being a string connecting an actor needing service and the provision of that service. No matter what our vision is at this point, we must come to an understanding of the necessary system functionality, behavior, and timing. The activity, statechart, collaboration, and sequence diagrams provide a means to do that. Not all of these diagrams are necessarily needed for every analysis, but it is a strange and complex problem that cannot be taken apart with the whole set of tools.

The statechart exposes system behavior. The activity diagram exposes system functionality. The collaboration diagram exposes needed relationships between system entities. The sequence diagram permits consideration of the timing needs. We apply these models to the use case expressions collectively and iteratively to evolve patterns of functionality

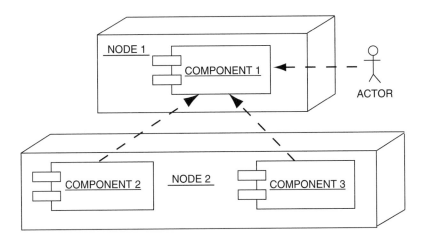

Figure 4.4-13 Component and deployment diagrams

and behavior that can be associated with elements representing the static structure of the system being created (classes, objects, components, and nodes). The precise way this is done includes some art as well as ULM discipline.

4.4.4.5 Precedented Application In a precedented case, one may begin with the existing software product represented by node and component diagrams. If the system was developed using UML and the model data is still available, the class and object, component, and deployment diagrams will be of value in representing the current solution. A modification of this existing system will be phrased in the context of what the system must accomplish that is different from current capabilities. This different functionality and behavior may be traced back to the appropriate interaction diagrams and analytical diagrams (use case, state, and activity diagrams) resulting in alterations that reflect back into the static modeling components (state and class, component, and node diagrams).

Unfortunately, it is not so easy to reverse the application of early object-oriented modeling. In my view, it is useful only in the precedented case because one first has to model the static representation as an entry into the functional and behavioral models, which is backward from all of the other modeling techniques I know. However, once the system has been created and both the static and dynamic diagrams are available, it could be possible to apply them in much the same way as discussed for UML above in modifying the system.

4.4.5 Moving to Specification

MIL-STD-498 data item description for a software requirements specification very interestingly notes that it is not necessary in all cases to prepare a paragraph-structured document to specify the requirements for the software entity to be developed. It permitted the use of models, diagrams, or tables to define software requirements. Alternatively, one could use the paragraph style, referencing the diagrammatic treatments in paragraph 3.1 on states and modes and using paragraph 3.2 to cover entity capability requirements.

When using one of the early OOA models, the states and modes paragraph can focus on the objects, with each one broken down under the corresponding entity capability requirements paragraph covering the needed behavioral requirements extracted from the state diagram included in the object and the functional and data requirements in the case of each object extracted from the embedded DFD. The remainder of the specification largely addresses quality factors and constraints.

4.5

System Modeling Using the DoD Architecture Framework

Contents

4.5.1 Background

For many years, the Department of Defense (DoD) permitted its component services to independently develop their systems, but as the weapon systems became more complex and the amount of funding available dropped off after the fall of the Soviet Union, coordination across the services became a vital need. The interface presenting the most serious difficulty was, of course, of an information nature. Tremendous advantages became obvious if various weapons systems could interact—if, for example, one system could detect the presence of a threat and another system with a more effective capability against that particular kind of threat could attack it using information derived from the first system.

This interest occurred in coincidence with a tremendous period of change in computer science leading to rapid advancements in computer capabilities, software implementation languages, networking technologies and standards, and system modeling methodologies. The proliferation of solutions that came about were not in the general interest, though experimentation by the several services did unintentionally advance the state of the art by offering several alternatives from which advantages and disadvantages could be observed.

The common interest among the services came to be referred to as C4ISR, which means "Command, Control, Communications, Computers, Intelligence, Surveillance, and Reconnaissance," the essence of the focuses of interest in military information, why we need it, how we process it and move it around from place to place, and how we make it available for human decision making. It took several years to focus clearly across the Department of Defense, and general success occurred after many independent advancements in the several services and government components.

The movement to cooperative efforts began with direction from the Deputy Secretary of Defense in 1995 that a DoD-wide effort be undertaken "...to define and develop better means and processes for ensuring that C4I capabilities meet the needs of warfighters...." A C4ISR Integration Task Force (ITF) was formed under the direction of the Assistant Secretary of Defense for Command, Control, Communications, and Intelligence. Representatives were assigned from the Joint Chiefs of Staff, the military services, and DoD agencies. The Integrated Architectures Panel (IAP) of the ITF released the first effective architecture information system model based on a study of the successes realized in the following initiatives:

1. Joint Chiefs of Staff/Commander and Chiefs (JCS/CINCs) standardized joint warfighting based on UJTL
2. Office of Secretary of Defense (OSD)/CISA joint integration and analysis methods as used in the integrated broadcast service (IBS)
3. U.S. Air Force node-to-node data exchanges, as in Horizon-Link
4. U.S. Marine Corps information flows, as in MAGTF C4I
5. DISA joint technical reference models, as in TAFIM
6. Defense Information Agency (DIA) joint intelligence system architecture, as in DODISS/SIM
7. U.S. Navy warfighting focus, as in Copernicus
8. U.S. Army standardized data elements, as in Enterprise Strategy

The results of this study were released in "The C4ISR Architecture Framework, Version 1.0," dated 7 June 1996; this was intended as a baseline from which the services could work collectively to develop a mature architecture, eventually leading to promulgation of directives and guidance instructions to bring about the desired condition, systems that cooperated in providing warfighting effectiveness. The discussion of the model offered in this chapter is based on and draws from the second generation of this DoD effort released by the C4ISR Architecture Working Group, a follow-on organizational entity to the prior effort, as "C4ISR Architecture Framework Version 2.0," dated 18 December 1997.

"DoD Architecture Framework Version 1.0" was issued on 15 August 2003 in two volumes with a supplementary deskbook. This document evolved from those noted above under the auspices of the DoDAF Working Group, and it supercedes them and was the basis for this chapter. A formal DoD letter called for all architectures developed and approved after December 1, 2004, to be in compliance with this framework. It should be noted that the Department of Defense was pursuing this matter on many fronts at the time this was being written, some of which are noted in paragraph 4.5.4. No organization compared to DoD has such a complex information system challenge nor more dire consequences of failure, so it is not odd that DoD leads the way on massively coordinated information systems. Currently, DoD is developing a model for a global information grid for warfighting that may very well be copied by commercial interests for more mundane purposes as it reaches some success. This grid for military purposes will have to accept data related to targets in real time from many sources and make it available as targeting information for many possible shooters. This has been successfully accomplished on a particular relatively confined battlefield already, but our military now fights on a global stage.

4.5.2 Overview

The DoD Architecture Framework embraces the definition of architecture given in IEEE STD 1471, 2000: "An architecture is the fundamental organization of a system embodied in its components, their relationships to each other, and to the environment, and the principles guiding its design and evolution." The framework identifies the architecture across its development time frame, starting with the operational description and ending with the implementation of a particular physical configuration. The framework does not describe how to make this transition on a real program but references policies that are useful in that process.

The analyst may use a set of work products encouraged in the DoD Architecture Framework to define a problem space related to an information system or information aspects of a system with a broader purpose under development as a means of defining its requirements and modeling its capabilities. This is the same advantage observed in using any of the other models described in prior chapters of this part and Part 3. The framework encourages the use of 26 particular architectural views organized into four view sets, making it the richest model covered in this book. Traditional structured analysis (Part 3) employed functional flow diagrams, requirements analysis sheets, architecture diagrams, schematic block or N-Square diagrams, and additional models for environmental requirements and specialty engineering, but nowhere near 26 in total. Unified modeling language (UML) requires eight (or nine, depending on how you count objects and classes) different modeling constructs to fully characterize a problem space.

The careful reader will find that he or she is already very familiar with many of the 26 constructs in this model because, as we discussed in Chapter 3.2, there are only so many facets of interest in problem and solution spaces, and we can represent those facets in only so many effective ways that take advantage of simple graphical constructs that enter the human mind efficiently through human vision. The authors of the DoD Architecture Framework have been similarly constrained to select from among some of those preexisting useful artifacts with only a few additions. Also, some of the framework products are repeated from one view to another.

4.5.3 Framework Products

The architecture is described using framework products, or work products. They are partitioned into essential and supporting roles in the development effort. The intent is that all of the essential framework products must be developed, and the analyst is free to develop the others to the extent that they are helpful in modeling the system. This is the same encouragement found in applying UML. Some of the diagrammatic treatments are useful in the beginning, and others come into play later in the development effort.

The framework products are also partitioned into three different views. In Chapter 3.2 we similarly partitioned the problem space into three facets of interest, recognizing functional, behavioral, and architecture (physical or object) facets or views. The rationale for that was that the problem spaces we have to deal with today in the development of systems are simply too complex to unveil themselves in terms of a single view. The views chosen by the authors of the DoD Architecture Framework are operational, systems, and technical standards with an additional one (called "all views") that stretches across all of the other three views. Figure 4.5-1 shows how the 26 framework products map

into these two partitioning sets using the product reference designators that are included parenthetically after the title of the framework products in subordinate paragraphs.

The descriptions offered in some cases in this chapter come verbatim from the DoD Architecture Framework Version 1.0, dated 15 August 2003, because they are well stated there and efforts to recast these descriptions may cause confusion among those who already understand the framework. In some cases, comments are added to draw attention to specific facets of the model artifacts or to correlate them with models exposed in prior models covered in this book.

4.5.3.1 All views
Two very general framework products are included in the architecture that stretch across all three views. These are high-level system products that are useful in placing the rest of the views into context.

4.5.3.1.1 Overview and Summary Information (AV-1)
This product is similar to a system need statement providing a user perspective of the architecture in text format. It identifies the system and describes the purpose and scope of the architecture. It provides system context, dealing with mission and geographical related information as well as rules, criteria, and conventions to be followed. Findings are included that identify the results of related analyses that might include identification of shortfalls, recommended systems implementations, and opportunities for technology insertion. Finally, this product includes identification of the file names, file formats, file locations for each element of the overview, and summary information.

4.5.3.1.2 Integrated Dictionary (AV-2)   The integrated dictionary provides a single authoritative source for definitions of terms and meta data (that is, data about an item)

ROLES	VIEWS			
	ALL VIEWS	OPERATIONAL	SYSTEMS	TECHNICAL
ESSENTIAL	AV-1 AV-2	OV-1 OV-2 OV-3 OV-4 OV-5	SV-1	TV-1
SUPPORTING		OV-6a OV-6b OV-6c OV-7	SV-2 SV-3 SV-4 SV-5 SV-6 SV-7 SV-8 SV-9 SV-10a SV-10b SV-10c SV-11	TV-2

Figure 4.5-1   *Framework product partitioning*

used in all framework products. Every term that has special meaning relative to the architecture being analyzed should appear in this dictionary. Wherever possible, the system architects should use standard terms from existing, approved dictionaries and lexicons. Where it is necessary to fashion new terms, they should be offered to those responsible for the related dictionary or lexicon.

4.5.3.2 Operational Architecture Views

The operational architecture view is a description of the tasks and activities, operational elements, and information exchanges required to accomplish DoD missions, which includes both warfighting and business processes. The user could develop these views.

4.5.3.2.1 High-Level Operational Concept Graphic (OV-1)

These diagrams are often called "lightning diagrams" because they are commonly drawn with icons joined by things looking like lightning bolts signifying communication of some form. The diagram is the highest-level operational view of a system provided for in the architecture. It has a lot in common with a system interface description, except that this diagram tends to be crafted from a user orientation while the interface description is an attempt to portray an engineering view of the architecture. While these diagrams do not provide much engineering content, it is

amazing how much valuable information they do contain for the population involved in the program. Discussions on programs about what is depicted on one of these diagrams can be very intense and useful because they are generally understood by program managers and almost everyone else on the program. Figure 4.5-2 shows an example.

4.5.3.2.2 Operational Node Connectivity Description (OV-2)

A node is defined as a representation of an element of architecture that produces, consumes, or processes data. An operational node performs a role or mission. The operational nodes in effective systems will have to exchange information. The nodes in operational node connectivity descriptions may be virtual where a clear decision has not been made at the time on precisely how the system will be organized, or they may represent real physical assets.

Figure 4.5-3 shows an example of a very generic form of this description. These diagrams can be drawn with specific icons representing collection and targeting nodes or be very generically drawn to reflect a future decision that will be made to associate particular generic nodes with specific assets. Each information exchange must be defined as indicated and the activities associated with each node of interest should be identified especially if an activity diagram is not created. Note that this diagram is a kind of

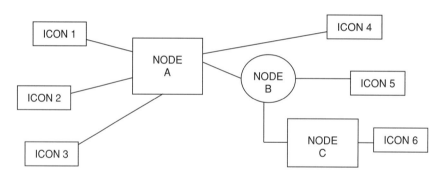

Figure 4.5-2   High-level operational concept graphic example

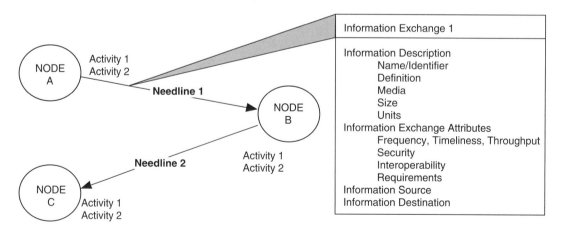

Figure 4.5-3   Operational node connectivity description

reflection of an activity diagram. In the operational node connectivity description focuses first order on the nodes and second order on the activities while the activity diagram reverses this relationship.

4.5.3.2.3 Operational Information Exchange Matrix (OV-3) This matrix provides a means to offer details of an information exchange identifying *who* exchanges *what* information with *whom, why* the information is necessary, and *how* the information exchange must occur. There may not be a one-to-one correspondence between OV-3 information exchanges and OV-2 needlines. Many information exchanges may be associated with one needline. Figure 4.5-4 shows an example of the matrix.

4.5.3.2.4 Organizational Relationships Chart (OV-4) This diagram is a very familiar organization diagram, generally hierarchically structured as in Figure 4.5-5. It clarifies the various relationships that can exist between organizations that have a role in a particular architecture.

4.5.3.2.5 Activity Model (OV-5) The dictionary defines the word *activity* as a specific deed, function, or sphere of action. People interested in computer software seem to prefer the word *activity* to *function*, whereas system engineers like to use the word *function*. No matter which word is applied, these are very important expressions of one view of an architecture. Required system activities can be depicted hierarchically or in a sequence-oriented fashion. If using a sequence-oriented diagrammatic treatment, the analyst could employ IDEF-0, functional flow diagramming, enhanced functional flow block diagramming, behavioral diagramming, or a flowcharting technique used years ago in the early days of software development. The blocks on these diagrams do not represent the physical entities the system consists of. Rather, the blocks represent activities or functions that the system must achieve. Figure 4.5-6 shows a sample activity model using IDEF-0, which is covered in Department of Commerce FIPS Publication 183.

The sequence of activities follows the generally lateral directed line segments from left to right. The activity blocks are drawn in echelon formation because earlier blocks often produce outputs that become controlling influences on later blocks. The lines entering the top edge of the block are for controlling influences, while the lines entering the lower side are supporting resources.

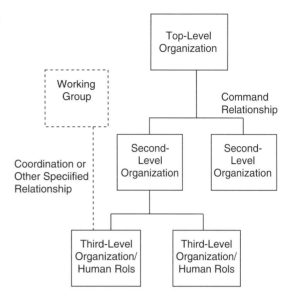

Figure 4.5.5 *Organizational relationships chart*

4.5.3.2.6 Operational Activity Sequence and Timing Descriptions (OV-6) The framework uses three different diagrams to express the behavior of the architecture being developed. The first one is a way of recording a set of rules. The second one is a state diagram, and the third one is a way to explore and report the timing relationships between events.

4.5.3.2.6.1 Operational Rules Model (OV-6a) The operational rules model is a text document that specifics operational or business rules acting as constraints on an enterprise, a mission, an operation, a business, or an architecture.

4.5.3.2.6.2 Operational State Transition Description (OV-6b) State diagrams, such as the one depicted in Figure 4.5-7, capture and communicate the several states within which a system may exist and the conditions under which they should transition from one state to another. Three artifacts are required to describe the situation illustrated by a state

Needline Identifier	Information Exchange Identifier	Information Element Description					Producer			Consumer			Nature of Transaction				Performance Attributes				Information Assurance					Security			
		Information Element Name and Identifier	Content	Scope	Accuracy	Language	Sending Op Activity Name and Identifier	Sending Op Node Name and Identifier	Receiving Op Activity Name and Identifier	Receiving Op Node Name and Identifier	Mission/Scenario UJTL or METL	Transaction Type	Triggering Event	Interoperability Level Required	Criticality	Periodicity	Timelines	Access Control	Availability	Confidentiality	Dissemination Control	Integrity	Accountability	Protection(Type, Name, Duration, Date)	Classification	Classification Caveat

Figure 4.5-4 *Operational information exchange matrix*

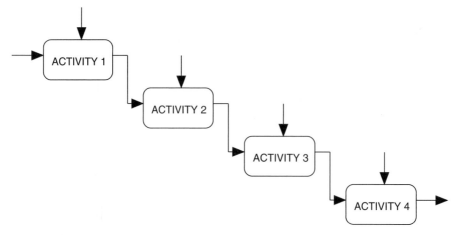

Figure 4.5-6 *Activity model example*

transition diagram: (1) the state diagram, which illustrates the states shown by bubbles or rounded-corner boxes and relations shown by directed line segments between the states; (2) a state dictionary, which precisely defines what it means to be in each state; and (3) a transition dictionary, which precisely defines the events that cause a transition between one state and another.

4.5.3.2.6.3 Operational Event/Trace Description (OV-6c) This diagram, shown in Figure 2.4.5-8 is referred to as a sequence diagram in unified modeling language. It identifies nodes of interest and links these nodes by events of interest that transpire. A timeline is included down the left side of the diagram, as noted in Figure 2.4.5-9. The inter-

section between the directed line segments and the node lines reflect the nodes becoming aware of the event in question. If the event is essentially instantaneous relative to both source and destination node, then the directed line segment for that event will be horizontal. If there is some delay at work in the passage between nodes the event line will be at some angle. An example of the latter would be a signal that must pass from a terrestrial source via a satellite at geosynchronous orbit and back to a terrestrial destination.

4.5.3.2.7 Logical Data Model (OV-7) The logical data model employed in the C4ISR Architecture Framework is IDEF 1X, described in Department of Commerce FIPS publication 184. As shown in Figure 4.5-9, IDEF-1X can be used to model a relational database. The blocks represent the tables the database must consist of, and the lines joining them show relationships between these tables. A table name describing the kind of information the table contains is included above each block. The text in a block lists specific attributes contained in the table. A horizontal line partitions the attributes into those above the line, which are key fields, and those below the line, which are nonkey fields. A key

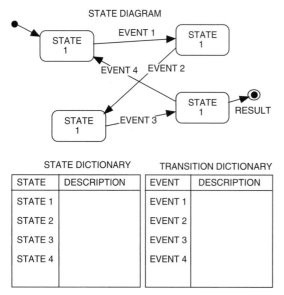

Figure 4.5-7   *Operational state transition description example*

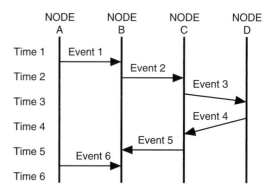

Figure 4.5-8   *Operational event/trace description example*

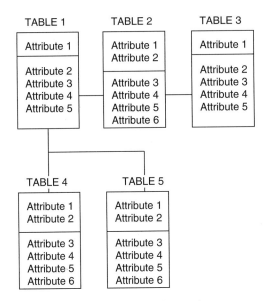

TABLE 1	TABLE 2	TABLE 3
Attribute 1	Attribute 1 Attribute 2	Attribute 1
Attribute 2 Attribute 3 Attribute 4 Attribute 5	Attribute 3 Attribute 4 Attribute 5 Attribute 6	Attribute 2 Attribute 3 Attribute 4 Attribute 5

TABLE 4	TABLE 5
Attribute 1 Attribute 2	Attribute 1 Attribute 2
Attribute 3 Attribute 4 Attribute 5 Attribute 6	Attribute 3 Attribute 4 Attribute 5 Attribute 6

Figure 4.5-9 Logical data model example

field is an attribute through which each record achieves uniqueness. It is possible that a particular table may require two or more fields concatenated together to identify record uniqueness.

4.5.3.3 Systems view
The systems view is a set of graphical and textual products that describes systems and interconnections providing for,

or supporting, DoD functions, both warfighting and business. The systems view relates system resources to the operational view. Physical nodes are identified for these resources, and information exchange between these nodes is facilitated in the system views.

4.5.3.3.1 System Interface Description (SV-1)   The system interface view illustrates the interface relationships between the representations of elements of architecture that produce, consume or process data, that is, nodes. Figure 4.5-10 shows an example of one of these diagrams, though they may be drawn abstractly using blocks or bubbles, as shown here, or icons to represent the entities (airplanes, ships, computers, or satellites, for example) of interest. Lines, directed or not, define interface needs between the entities illustrated. The interfaces may be implemented through wires, radio in space, or optics, for example, as a function of the particular system and its characteristics. A node represents an element of an architecture that produces, consumes, or processes data. A node is a general entity relative to the concept of an architecture level and may represent a system of systems, a system, or some element of a system.

The interface relationships can be illustrated on one of these diagrams at whatever architecture level is desired and given a particular representation; the interfaces shown on one diagram can be expanded upon on a more detailed diagram forming a family of interface description diagrams.

4.5.3.3.2 Systems Communications Description (SV-2)   The systems communications description models the specific communications systems pathways or networks and the details of their configurations through which the physical nodes and systems interface. This view is a design document calling for specific physical solutions for the connecting media. Figure 4.5-11 shows several internodal

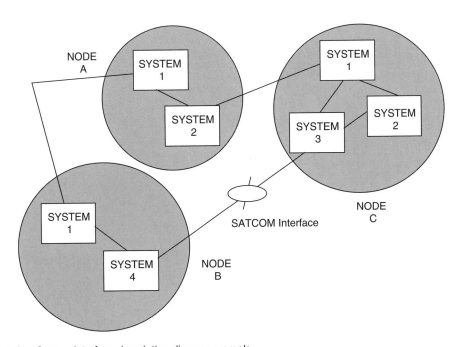

Figure 4.5-10   System interface description diagram example

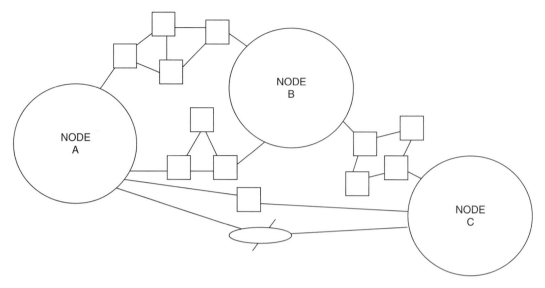

Figure 4.5-11   *System communications description*

connections, but the same treatment can be extended to the intranodal interfaces.

4.5.3.3.3 Systems–Systems Matrix (SV-3) The systems–systems matrix is similar to the N-Square matrix employed by system engineers to depict interface relationships among system elements. There are many ways this matrix can be used because it is useful in exposing the relationship among n things. A square matrix, such as that shown in Figure 4.5-12, can be employed if the analyst is interested in directionality of the relationships; or a triangular shape can be used if one is only interested in a binary fashion, whether or not a relationship must exist between two entities. The entities analyzed for relationships can be entered into the diagonal boxes because they represent the internal interfaces of these entities. The intersections of the matrix are marked to indicate a needed interface.

4.5.3.3.4 Systems Functionality Description (SV-4) In the DoD Architecture Framework, the notion of functionality is connected to the dataflow diagramming construct. The analyst might initially construct a hierarchical depiction of functionality and then express the exposed functions in a dataflow diagram format, as shown in Figure 4.5-13. The bubbles represent computer processing that must occur, and the directed line segments show dataflow needs between the computer processing bubbles and data stores. The inside–outside relationship is shown by connections between blocks representing external data sources and sinks and the computer processing that must be accomplished.

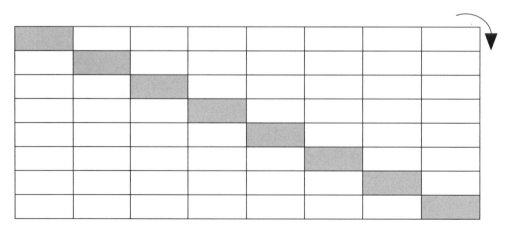

Figure 4.5-12   *System–systems matrix*

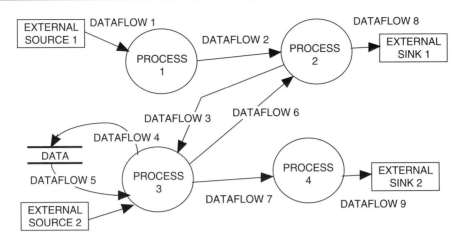

Figure 4.5-13 *System functionality description*

4.5.3.3.5 Operational Activity to System Function Traceability Matrix (SV-5) This matrix provides a transition between the operational activities and system architecture functions. The former are identified from a user perspective, while the latter are an engineering view of the system to be created to satisfy the user needs. This is similar to the principal area of the quality function deployment (QFD) matrix, which relates the voice of the customer to the design requirements. Figure 4.5-14 shows how this matrix makes the connection. The matrix intersections are marked to show relationships. There may be a one-to-one correspondence between these entities, or the relationship could be more complex all the way to many-to-many.

4.5.3.3.6 Systems Data Exchange Matrix (SV-6) This matrix is identical to the one developed under OV-3, discussed in paragraph 4.5.3.2.3, but the content relates to system data to be exchanged between systems, and it contains

only automated information exchanges. It replaces the OV-3 operationally stated content with corresponding system data characteristics. The OV-3 matrix also contains exchanges completed through verbal orders, which the SV-6 matrix excludes.

4.5.3.3.7 Systems Performance Parameters Matrix (SV-7) This matrix is essentially a set of primitively stated performance requirements discussed in Chapter 2.1. Requirements are quantitatively stated for the characteristics of systems and system hardware and software items and their interfaces and functions. The matrix can have several columns to capture the values for the listed parameters at different points in time through the development period, which is similar to the way one might capture the requirements for several prototypes in a tabular fashion.

4.5.3.3.8 Systems Evolution Description (SV-8) The systems evolution description provides a graphic that depicts the evolutionary development of the system in time. A totally unprecedented system that evolved through a waterfall development program would call for a very dull systems evolutionary description, but one that followed a more realistic evolution today would involve some preexistence and spiral evolutionary characteristics, with capabilities added over time through prototyping activities, trials with COTS followed by some development to clean up problems observed, merging efforts between to or more projects, and linking the projects of two or more services when coincidence is observed and due to other causes. Figure 4.5-15 illustrates the form of this kind of diagram. The different contributing efforts merge with a main line of development at different points in time, coordinated with version configuration identifiers shown.

4.5.3.3.9 Systems Technology Forecast (SV-9) This forecast is critical on projects that are heavily dependent on new technology to achieve very difficult operational capabilities. It provides a means to capture expectations correlated with the SV-8 product, either enabling the planned evolution or inhibiting it. Clearly, these two products have to be coordinated. It should evolve from the TV-2 forecast made early in the project. The product can be depicted as a simple matrix with columns for the periods of time required for evolution.

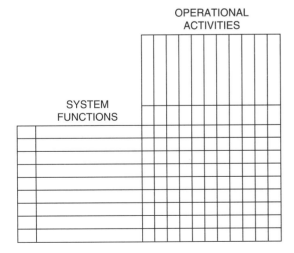

Figure 4.5-14 *Operational activity to function matrix*

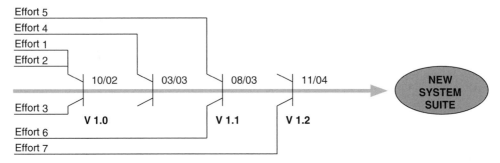

Figure 4.5-15 *Systems evolution description*

4.5.3.3.10 System Activity Sequence and Timing Descriptions (SV-10) The SV-10 products are physical expressions of the operationally characterized OV-6 products, and the diagrammatic treatments will not be repeated from paragraph 4.5.3.2.6. They include a systems rules model, systems state transition description, and systems event/trace description, which all explore the dynamic behavior of the system.

4.5.3.2.11 Physical Schema (SV-11) This product should evolve from the OV-7 logical data model showing how the information requirements are actually to be implemented. The systems model product can be represented in a number of ways including, for example, a physical database design.

4.5.3.4 Technical standards view
The technical view identifies a set of rules that governs system implementation and operation. Generally, these rules are contained in standards that will have to be respected in the development of the system.

4.5.3.4.1 Technical Architecture Profile (TV-1) The technical architecture profile can take the form of a simple list of standards by service area. Each standard listed should be studied for a possible need to tailor it to ensure that unnecessary characteristics are not introduced into the system that could affect cost adversely without adding system value. Generally, standards are general in their content, and not all of the content of any one standard necessarily applies in every instance of application.

4.5.3.4.2 Standards Technology Forecast (TV-2) Standards represent plateaus of human knowledge that are useful over finite time spans. In a time of rapid advance in human knowledge about the technical aspects of solutions to complex problems, those plateaus may come and go over a disturbingly brief time span. In developing systems that may require several years, it is possible that particular standards respected early in the development period will have passed from the scene or have been significantly updated over the development span. The standards technology forecast is a formal way to provide a project with insight into the volatility of the standards with which they are dealing. The forecast lists the service area as in the TV-1 product but also provides an evolutionary profile of the standards listed with columns for different time frames. One set of standards might be listed for the

beginning period of time and other columns for later periods of time, perhaps keyed to major project phasing points or baselines.

4.5.4 Other Related Efforts
The Department of Defense has put a great deal of work into the many facets of information system development because of the tremendous complexity of the systems needed to detect targets, communicate location and motion parameters, and hit those targets, as well as the need to be successful in joint operational situations. Many of these efforts have produced a surviving document of value to continuing development of information systems, and they are referred to in the DoD Architecture Framework Version 1 as universal reference resources.

4.5.5 Architecture Product Interrelationships
The reader will already have discovered several relationships among the several products included in the architecture. The intent, of course, is that the operational views would be developed early in the project based on user perspectives telling what is needed. These should evolve over time, respecting the content of the TV products' maturing into SV products illustrating real-world solution-oriented descriptions of the system under development. We will see relationships between the several problem space elements depicted in the OV products, between the several solution space elements in the SV models, and between OV products and SV products.

4.5.5.1 Operational view relationships
Figure 4.5-16 illustrates the relationships between the several operational view products. Organizations in OV-1 should match those in OV-2 and annotations identifying responsible operational nodes in OV-5. Operational nodes in OV-2 match the lifelines of OV-6c. Operations in OV-4 map to the operational nodes in OV-2. Information elements in OV-3 map to the inputs and outputs that belong to OV-5 conducted across two more operational nodes of OV-2. Events in OV-6b map to events in OV-6c. State transitions in OV6b and events in OV-6c should be consistent with activities in OV-5. Dynamic rules in OV-6a may constrain state transitions in OV-6b decision points in OV-5. Events in OV-6b and OV-6c map to triggering events in OV-3.

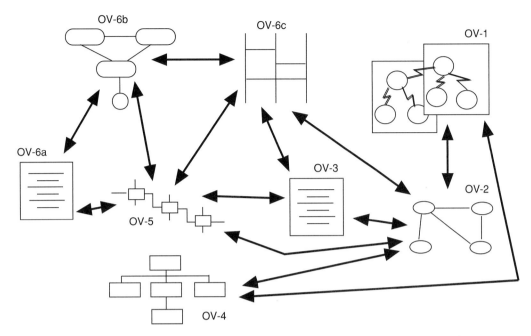

Figure 4.5-16 *Operational view relationships*

4.5.5.2 Systems view relationships

Figure 4.5-17 shows the relationships among system view products. Interfaces in SV-1 map to system data exchanges in SV-6. Systems data input or output by system functions in SV-4 map to system data elements of SV-6. Interfaces in SV-3, communications links in SV-2, and some performance requirements in SV-7 map to interfaces in SV-1. SV-8 phases interface requirements and implementation in SV-1. SV-8 phases the applicability of performance requirements in SV-7 and availability of new products and technologies in SV-9. Rules in SV-10a constrain transitions in SV-10b, events in SV-10c, and functions in SV-4. Events in SV-10b map to events in SV-10c. Lifelines in SV-10c map to systems in SV-1 and systems functions in SV-4.

4.5.5.3 Operations to systems view traceabilities

There are some operations views that translate directly into the systems views, as shown in Figure 4.5-18. SV-1 is a physical version of the operations scenario chart OV-1. SV2 translates the nodes of OV-2 into the intended physical systems view. OV-6a through c move right into SV-10a through c. The logical model exposed in OV-7 matures into the physical schema of SV-11. OV-5 is a significant input to developing SV4, where the former is exposing needed activities and the latter is exposing the intended functionality of the solution systems. SV-6 shows the mapping between these to maps appearing like the primary matrix of a QFD matrix and serving the same purpose.

4.5.6 The Six-Step Architecture Description Process

The first three steps of the architecture description process recommended by the authors of DoD Architecture Framework, Version 1, are focused on understanding the problem to be solved. Step 4 is a planning step to determine which

of the views and which framework products within those views will be needed to define the architecture. Step 5 entails building the products, and the final step involves using the results of the description to make decisions on the actual physical product development that will solve the original problem. As a result of having accomplished the analysis called for in the six-step process, the development team should be able to craft a set of requirements appropriate for the development of the system to actually accomplish the needed capability.

4.5.6.1 Determine intended use of the architecture

Descriptions of an architecture should be built with a specific purpose. Determine what the purpose is from an operational perspective.

4.5.6.2 Determine architecture scope, context, environment, and assumptions

Determine the scope, the appropriate level of detail to be captured; the architecture's context within the bigger picture; operational scenarios, situations, and geographical areas to be considered; and the projected availability and capabilities of specific technologies during the time frame to be considered.

4.5.6.3 Determine what information the architecture description needs to capture

Determine what information is needed to adequately understand the problem. Care must be exercised to ensure that all needed information will be captured and that no unnecessary information is identified.

4.5.6.4 Determine views and products to be built

The analyst or team must decide which of the 26 framework products must be developed, based on an understanding of the problem space acquired in steps 1 through 3 and

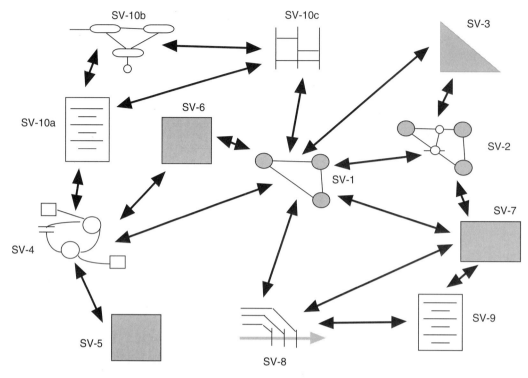

Figure 4.5-17   *Systems views relationships*

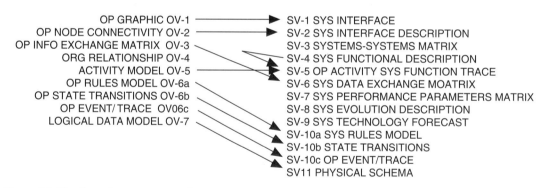

Figure 4.5-18   *Operations to systems view traceability*

planned project work phasing and scheduling. It is not necessary to build every product for every system and may be an unnecessary cost to do so. It may also be helpful to selectively build the products in time. Some of these products can be easily built early in a project, when the available information is thin; in the process of building those products, valuable insights are driven out that enable work on other products. So, the job in this task is to determine which products to build and how their availability should be distributed in time. This planning work will have to be accomplished in context with the amount of available fund-

ing, or it may be what drives the decision on how much funding is going to be necessary to accomplish the project.

4.5.6.5 Build the requisite products

The analyst or team simply follows the plan to build the framework products covered in the plan in accordance with the schedule. This work should not be accomplished as a totally disconnected activity; rather, interaction between the building team and outside persons who may be able to contribute understanding about the application should be encouraged.

4.5.6.6 Use the Architecture for Its Intended Purpose
Given the availability of the framework products, it should be recognized that nothing has yet happened that puts operational resources into the hands of the user. The physical assets necessary to provide the services defined by the products must be designed, purchased, assembled, and tested before their delivery and the user implementation of the system. Throughout this implementation phase, the framework products should provide guidance as well as having been the basis for the identification of the requirements that the physical assets must comply with in the design process and which will be used as the basis for testing. These products may have to be maintained to some extent during the implementation period if significant changes are called for.

4.6

Structured Analysis Fusion and Reunification

Contents

4.6.1 Functional Flow or Die!

There was a time when some DoD procurement offices apparently felt that there was only one way to do structured analysis. Contractors working in compliance with AFSCM 375 series standards had to create and submit data prepared in accordance with those standards. The standard contained a series of forms. In doing structured analysis work, one would fill in the blanks on the forms. These forms evolved into mainframe computer data-entry forms based on 80-column Holerith card keypunch data entry. The punched cards were entered into the mainframe computer and operated on to produce various reports (including specifications), requiring a great deal of tolerance on the part of the reader to glean information from them. Many kinds of problems would surrender to this structured approach, but some kinds of problems were resistant. Many software engineers at the time were using flowcharting, so there was not a great mismatch, flowcharting being very closely related to the functional flow diagramming that was the model required by the standard.

Functional flow diagramming is pretty hard to beat for most large problems that will entail product entities moving dynamically in space and time and where it is yet undecided how to implement the exposed functionality (an unprecedented problem). It is a very general and very simple model, wherein lie its strengths and weaknesses. It communicates well with the human mind visually, but it cannot capture a very rich problem context. It is ill suited as the sole basis for software analysis in particular because there are many much better models available for software. Where you have decided that some functionality shall be accomplished in software, functional flow diagramming is not the model of choice for continuing that analysis. So, if one starts a system analysis using functional flow diagramming and encounters a need for some software in the system, there will be at least two modeling approaches in use on the program, functional flow diagramming and whatever software model is selected.

There are other reasons why a program may unfold, applying more than one structured analysis modeling approach. The author does not happen to think that is a problem but rather a sign of good structured analysis sense, applying the most effective modeling capability to the parts of the problem space. The problem appears when you are required to show some kind of traceability across the different model gaps. The tools commonly available at the time this book was written were all built as if they were islands separated from each other by a sea of competition and proprietary data. In some cases, it was possible to link these tools up, but the result was seldom perfect. The ideal solution from a user's perspective would be a general agreement on the database structure that all tools would be capable of feeding requirements into. Within this "big dumb database," traceability could be consummated, specifications printed, and all other specification work accomplished. The tools could focus on gaining insight into what requirements should be identified and accepted, applying various modeling schemes to do so. Under this regime, a dozen different tools could be applied on one program as a basis for modeling, but all of the results of the modeling would be coupled into the "big dumb database," where traceability would be maintained.

Figure 4.6-1 illustrates such a structure. The human analysts draw information from the program problem space in the context of a particular modeling tool (A, B, or C) and the results of their work in the form of fully formed requirements statements flows into the big dumb.

With this kind of structure, one could perform the modeling work inside an application best suited to the selected structured model and, regardless of application, apply the

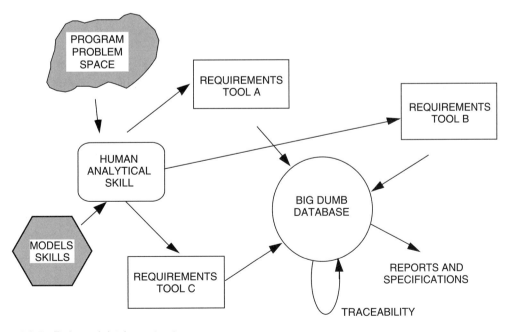

Figure 4.6-1   *Federated database structures*

results channeled to the common database. Traceability to the source would be automatic and other forms of traceability established and maintained within the "big dumb database." In the context of Chapter 3.10, the big dumb database performs the function of the requirements analysis sheet (RAS) into which requirements flow from the several tools with traceability to the tool providing the insight into the need for the requirement.

4.6.2 Structured Analysis Boundaries

All structured analysis models and methods do not necessarily include the same problem space facets, as discussed in Chapter 3.1, but whichever one or ones selected, they would all, presumably, approach the same problem space. That problem space is generally thought of as being chaotic or unorganized initially; through our application of a structured analysis model, we organize it on the way to understanding it. Could one say that the problem space is characterized by some form of organization that we are tapping into through structured analysis? If this were the case, then different modeling facets could be said to connect or trace to specific problem subspaces, and model features could be correlated across the different models that might be applied.

A problem is defined in the dictionary as "any question or matter involving doubt, uncertainty, or difficulty or a question proposed for solution or discussion." These definitions imply that human awareness is a necessary condition for problem existence. The author therefore accepts that unprecedented problem spaces are not preorganized. It is through human thought that the organization takes place in context, with an evolving appreciation of the problem and potential solutions.

We should establish constraining boundaries on the scope of the problem space we shall attempt to understand and solve through system development work. It does no harm to evaluate a space larger than current resources and immediate interest prescribe, but it is essential that we clearly establish the boundary of the space for which a given program is responsible.

4.6.2.1 Multiple paths
4.6.2.1.1 Decomposition Methodology Flexibility If a program team subscribes to a distributed system engineering process such that every engineer is responsible for applying the system engineering approach to his or her work, then the team should be more flexible than to require everyone to use either functional analysis or any other single method to gain insight into needed functionality. One notable exception to a rigid prescription is the case where system functionality has been allocated to an element of computer software. In this case, the principal engineer and team will likely prefer to apply a structured software approach such as the Yourdan–Demarco methods built into many commercial computer-aided software analysis tools or a variation of OOA/UML.

An avionics engineer, who has to determine how to package the required functionality into a radar altimeter, may prefer to use a more physically oriented circuit partitioning logic scheme foreign to either functional, process, or state transition analysis.

The point here is that we should not force fit everyone into the functional flow analysis process throughout the depth and span of the system architecture across the program development time span. Once the responsibility for developing a branch of architecture is established, that agent should be free to choose whatever reasonable tool he or she finds effective so long as it produces a traceable decision path from the jump-off point down through the analysis.

4.6.2.2 Functional traceability
Given that the reader finds this freedom of expression acceptable, how would we link up the traceability between these different methods, assuming we do not have access to the big dumb database center surrounded by cooperative tools? We may even first ask how we would do so using just functional flow analysis. The functional flow analysis approach has the traceability of functions ingrained in the very structure of the diagram. A lower-tier flow diagram is always expanded from a parent diagram, so the method of coding the flow diagram boxes carries the traceability information visually and analytically with no need for additional features. The traceability between the functions and the architecture is defined in the functional allocation process, as is the traceability of performance requirements associated with these functional allocations.

Where we permit a principal engineer to use some other method of functional decomposition, we need a way to connect the product of that method with the work performed at a higher level. The assumption here is that functional flow analysis will always be performed at the system level as the mechanism for gaining insight into system level functionality. This may not be true, but the logic of this argument will fit whatever the top-level methodology used.

At the lower fringe of the functional flow diagram, where a particular function is unambiguously allocated to a particular system element, the principal engineer for that element may choose to use some other approach for further decomposition. At these interfaces you need to show traceability to the function from which the analysis is derived. This can be done in a purely manual approach by referencing the block or bubble on the functional flow diagram, process diagram, functional hierarchy diagram, state transition diagram, packaging partitioning analysis diagram, or text write-up. In a real situation, there may be a more complex relationship between functions and architecture, of course, such that a bundle of functions has been allocated to a particular "black box."

For example, we may use functional flow analysis at the system level and identify the functions with the base-60 system described previously. A particular function F47135, Navigate to Target, may be allocated to an avionics box and contained computer software. The box principal engineer may choose to use state diagramming as a means to expand item functionality during his or her requirements analysis process. The top state could be called "state S," with subordinate states S1, S2, S3, S4, and S5. The state S maps to system function F47135 while the lower-tier states map to state S. We could extend the state analysis to lower-tier states such as states S31, S32, S33, and S34 subordinate to state S3. We would decide how to implement each state in terms of things that comprise the avionics box and thus expand the architecture of the box. For each of these allocations, we would force ourselves to define one or more performance requirements.

Lower-tier diagrams or analysis sheets in the new media then need to reference the higher-level analysis item for that analytical island. This pattern can be implemented in a computer database, probably with the same structure needed for pure functional analysis. If the analysis methods include information types that are sufficiently different, it may not be possible to use the same database structure for all approaches, but the data can be related to the source function with a memory variable or key database field.

Traceability in the special case of progressive conversion of a system functional flow diagram into a system process diagram, suggested earlier as a natural system description maturation progression, can be provided for within a database used to capture the function/process dictionary. A field for Function ID and another field for Process ID will allow the analyst to grow the analysis into a process diagram with good traceability back to the past rationale for system composition and performance requirements heritage. Alternatively, a separate, matrixing database can be used to interrelate the functions and processes.

In Part 3 we used something called a model ID or MID to identify modeling artifacts that requirements were derived from. This field could be the basis for traceability from the big dumb database to the several modeling databases. In the case of traditional structured analysis, the following MIDs were used: (F) function, (A) architecture of product entity, (I) interface, (Q) environmental factor, and (H) specialty engineering characteristic. This alphabet can be extended to

cover the artifacts in alternative models more appropriate to software development. Table 4.6-1 lists the IDs the author was experimenting with at the time this book was written. The letters are obviously not well selected for some of these artifacts and will probably be changed, as a particular universal set appears to be forming.

4.6.3 Expanding Zigzag

The analyst or program must decide what requirements strategy and methods to apply in determining appropriate requirements for the system under development and the items within it. This part and Part 3 of the book have covered many methods. There is another component to the requirements analysis process, however, no matter what analytical methods are applied. As the requirements analysts are defining requirements for system and items, design engineers must start the synthesis of those requirements into concepts in step with the top-down requirements flow. The timewise relationship between the requirements and synthesis work for the same item should be requirements before design concept. But the concept for an item should be used as one of the inputs in determining the requirements for items immediately subordinate to that item. The parent concept will properly influence child requirements.

Table 4.6-1 *Modeling artifact IDs*

MID	Model entity name	Description
A	Architecture	Traditional structured analysis product entity MID
B	Class	UML artifact
C	Component	UML static artifact
D	Department Number	Functional department number responsible for enterprise resources provided to program organizations
E	Event	Program planning event which closes out a program time axis period. Appears in an IMP planning string
F	Function	Traditional structured analysis function MID. In modern structured analysis could be used for the DFD bubbles
G	Process	Traditional structured analysis process MID meaning process task
H	Specialty Engineering	Traditional structured analysis specialty discipline MID which has to be linked to the functional department number
I	Interface	Traditional structured analysis interface relationship between two product entity items
J	Data	Modern Structured Analysis data line or data store
K	Classifier	UML term related to a physical artifact
L	State	State Diagramming artifact. Also used in UML. May be appropriate for state diagrams included in p-specs of modern structured analysis
M	Manual	Functional department manual
N	Activity	UML and flow charting artifact
O	Object	UML artifact, an instance of a class
P	Program	Program planning program number, the first ID in the P-string that is part of an IMP string
Q	Environment	Traditional structured analysis environmental MID
R	Sequence	UML dynamic artifact useful to illustrate a time-ordered relationship
S	Stage	Program planning phase number. Appears in an IMP string
T	Transition	State Diagramming artifact permitting transitions between states. Also used in UML
U	Use Case	UML artifact
V	Ad Hoc	The requirement in question has no structured analysis source reference (an undesirable but real possibility)
W	Work Product	Enterprise Planning artifact for task work product
X	Collaboration	UML static artifact
Y	Activity	UML activity
Z	Zone	UML deployment diagram artifact

When considering the whole system architecture on an unprecedented program, this process takes on a kind of expanding zigzag pattern between requirements and concept development. The requirements for an item drive the design concept for the item, followed by the design concept for the item influencing the child requirements immediately subordinate, as the structured analysis continues to produce insights into child requirements. So, we zig from requirements to concepts at level X branch Y and then zag to requirements for each of the subordinate requirements sets on that branch followed by series of zigs from parent concept to each subordinate requirements set.

4.6.4 Evolution of the Ultimate Method

In the 1960s, system, hardware, and software engineers all used the same modeling method, flowcharting. Functional flow diagramming described in Part 3 is a direct descendent of that model, as is UML activity charting covered in Chapter 4.4. Over a period of many years, computer software developers moved away from flowcharting for what appear to have been good reasons, while systems and hardware engineers remained with flowcharting, preferring a horizontal depiction rather than the vertical form preferred by software people. Figure 4.6-2 illustrates this progression, picking up with traditional structured analysis as a more complete version of flowcharting. Software moved from flowcharting to many different models, including modern structured analysis and extensions, many versions of early OOA, and most recently unified modeling language. The reader will note that traditional structured analysis is useful in developing the requirements and concepts associated with hardware and procedural elements of a system. The software modeling approaches are useful only in developing the software elements of a system. What is needed is a single method useful in developing all three elements of a new system.

As this book was being written, work was underway to extend UML to the more general system modeling application on the road to the Holy Grail of system modeling, the universal model. Unified Modeling Language (UML) certainly has the modeling structures to encourage this. Table 4.6-2 lists several necessary constructs in the traditional structured analysis approach, followed by a UML construct that is similar in purpose. The activity diagram is essentially the software flowchart of years gone by, which was a vertical cousin to the horizontal functional flow diagram preferred by those applying functional analysis for systems implemented through hardware. The UML sequence diagram is a timeline with a little more richness. Hierarchical architecture relationships called for in traditional structured analysis are akin to the object/class, component, and deployment diagram aggregations in UML when considered from their hierarchical relationships. Finally, interface relationships are expressed in several ways in UML but perhaps more nearly parallel with TSA in the relationships expressed between objects/classes, components, and nodes on object, component, and deployment diagrams. The interrelationships between these simplistic pairings are not perfect, but the UML is such a rich modeling environment that it will be possible to respect traditional structured analysis as a subset and approach a truly universal unified modeling language.

There is a significant difference between hardware and software design, of course. Of the five UML dynamic models, three are useful in defining the system: use case, activity, and state diagrams. The two remaining dynamic models, collaboration and sequence diagrams, are useful in making a transition from dynamic modeling to static modeling, which is documented in three additional diagrams: object/class, component, and deployment diagrams. These three could be considered the design of the "physical" product. UML object association can be used to hierarchically represent system architecture, whether hardware or software. The other two static objects should be replaced by conventional engineering drawings to capture hardware design, however.

Over the first decade of the 21st century, a restoration has been occurring between hardware and software problem space modeling, focused around the unified modeling language, as a result of the work of the International Council On Systems Engineering (INCOSE) and the Object Modeling Group (OMG). The first work product of this evolution has been created in the form of SysML Version 1.0. While I would accept that the traditional structured analysis subset approach would work today, the people working on this are a great deal more thoughtful and wish for the result to be logically correct as well as practically useful. The result of these good works will be a tremendously expanded capacity for communication between hardware and software people that should improve the development of the HW/SW interfaces. In order to make this happen, however, hardware-dominated system engineers will have to accept that the world has changed, that most of the hard functionality has moved to software, that they must learn to talk to people in the software domain, and that their preferred modeling process is but a subset of a grander model that has evolved through the work of software people. It is a philosophically pleasing outcome that we can now see approaching, where a process that began as a unified approach became separated for 30 years and eventually returned to a common perspective.

Many hardware-dominated system engineers will have great difficulty accepting this reunification outcome based on a software model. Many feel that the software people have already had too much influence on system analysis and specification formats by influencing the format of the system specification and causing alignment between system and hardware specifications and software specifications relative to states and modes and entity capability capture. I believe that my hardware-dominated colleagues should get over it and embrace the coming unification. If current system engineers cannot make this transition, then the transform may take a little longer awaiting the retirement or some more aggressive form of the separation of those system engineers from the workplace.

Organizations that are able to overcome the resistance of their own hardware-dominated system engineers to master common structured analysis will, of course, pass through a valley of inefficiency on the road to realizing significant benefits because a great deal of the difficulty in HW/SW interface development will disappear.

4.6.5 Model-Driven Development

In addition to movement to common structured analysis models, the future will bring into the development process a dramatically improved efficiency in the form of model-driven development. We have passed through a document-driven development process over the past 75

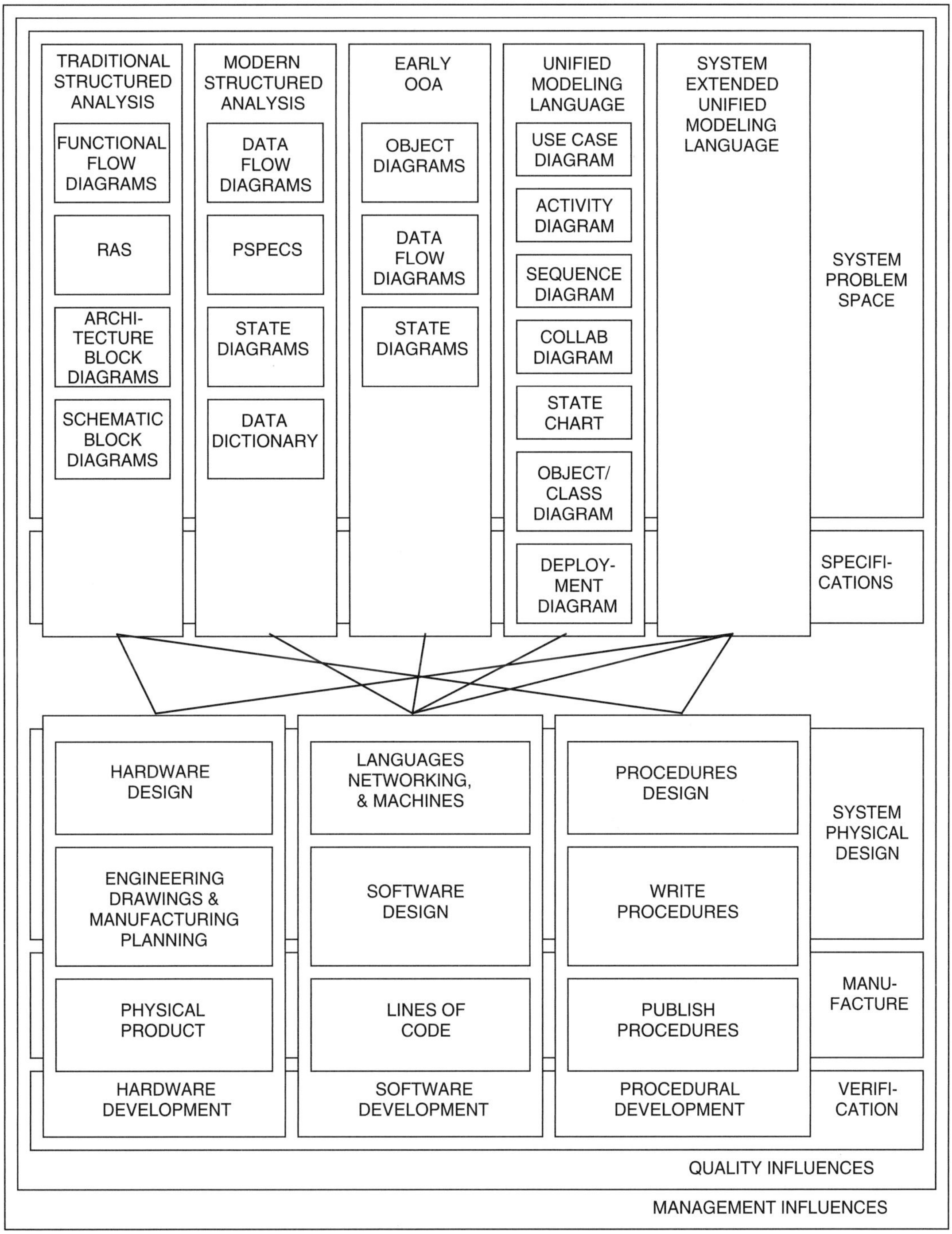

Figure 4.6-2 *Closure toward a common method*

Table 4.6-2   *Universal model coupling*

Traditional structured analysis construct	Comparable UML construct
Functional Flow Diagram	Use Case Diagram and Activity Diagram
Time Line Diagram	Sequence Diagram
Architecture Diagram	Object/Class, Component, and Deployment Diagrams, Hierarchical Perspective
Schematic Block Diagram	Object/Class, Component, and Deployment Diagrams, Lateral or Relationship Perspective and Collaboration Diagram
State Transition Diagram	Statechart Diagram

years, and some organizations have moved on to database-driven development relative to specifications. In the latter case, the developer captures the requirements in a database, and people use the content of the database directly rather than using the database to print paper documents. In these development models, whatever tools are used, they are isolated; and interaction between their content must be accomplished by human beings interacting via human communication (publishing and reading paper documents, conversation, and meetings), resulting in other engineers reacting to the work products of one engineer by changing their own work products.

In the model-driven development case, the many computer tools are directly linked so that changes in one are communicated to those that are related. If the mass properties engineer has to change the weight of one element of the product, the resultant changes in gains and other influences will be accomplished in an automated fashion or be offered to the responsible team for approval. The computer applications are not now available for purchase but will be evolving over a period of several years, starting as experimental applications joining off-the-shelf applications on small projects in large aerospace companies.

As this capability matures, the advantages are obvious in that the inefficiency of human interpretation and response will be overcome. Essentially, the computer system serving the program will become a team member accomplishing a great deal of work previously accomplished by human team members. Initially, of course, the humans and their managers will not trust the systems to interact directly and will require that the applications produce changes that other applications must be human tended to accept. Even as trust matures, engineers and managers may prefer human-assisted connections for configuration and baseline control purposes.

In this new environment, we will be forced to rethink the current process of preparing specifications. The software development process applied today may be instructive. When using some software development applications, the problem space is modeled in the tool, and the team can transition from the problem space model to a solution space model within the tool. Based on the modeling work accomplished, the tool can generate code. Note that there is no published specification in this case; all of the requirements have been modeled within a computer application and have been the basis for software code. MIL-STD-961E actually recognizes today that the specification need not be a narrative paper document. It can be a model.

It is true that computer software is a narrowly characterized product consisting of data and processing instructions. Hardware is a much more physically complex entity, implemented in hydraulic devices and plumbing, mechanical connections, electrical wiring and circuits, and artifacts from many other design domains. However, today hardware design is implemented in CAD, where the results of the creative capture of the engineer's understanding and synthesis of the driving requirements takes the form of lines of code behind the pictorial screen. These lines of code can be coupled into machine tools, resulting in machining of physical parts faithful to the design engineer's intent. The big difference between these two cases is that the SW tool is much more involved in the requirements-to-solution transform whereas, in the hardware case, the human mind is making the transformation almost entirely. It is unlikely that this condition will change much in the foreseeable future, though CAD will continue to be augmented in support of design for particular disciplines. Computer software has no physical representation, so it can be configured to make essentially intellectual transformations that are extremely difficult to implement where a physical reality must materialize from the process.

Is it possible that paper specifications will stop being developed as system development moves to a common modeling environment and model-driven development? This outcome is entirely possible, but it is not now obvious what the outlines of this process will look like. The author believes that it will still be important for engineers to clearly identify requirements as a prerequisite to design development but that these requirements will be captured in a requirements database, providing the analyst with effective models for exposing the problem space.

The principal inhibiting influences on movement into this world today are the tremendous complexity of the modeling connections for the hardware case and the very opaque nature of the contractual and acquisition management influences. As a consequence, this infrastructure may evolve first in commercial development programs rather than large customer acquisition programs, under the Department of Defense.

5 Specification Content Standards

5.1

Specification Development Fundamentals

Contents

5.1.1 Overview

5.1.1.1 What is a specification?

A specification is a document that contains requirements. It is written to cover the requirements for a specific product item or process. Requirements are necessary characteristics for an item or process, identified prior to crafting a design responsive to those requirements. A specification, therefore, provides a definition of the design problem. Synthesis is the solution to a problem defined in a specification.

A specification is a published list of requirements for the design or the fabrication, assembly, and manufacture, or procurement of an object in a prescribed format. It is a detailed statement of particulars about an object, setting forth characteristics or attributes that must be controlled. A specification captures in one place the requirements identified for an object through the process we have called "requirements analysis."

5.1.1.2 Specification format control

There are many kinds of specifications, but they fit into two broad classifications at the top of the hierarchy, program-peculiar and general. When the latter are called or referenced in a program-peculiar specification, they are called "applicable documents," and they are relevant to that application. First we will explore the program-peculiar specification variety. Later we will explore applicable documents that contain general requirements that can be made to apply to the products developed on any program by a referencing mechanism in program-peculiar specifications, essentially a bibliography.

Large customers involved in procurement of very complex systems, like DoD and NASA, require their contractors to create several kinds of program-peculiar specification types. Contractors serving these kinds of customers will also find it necessary to prepare in-house and procurement specifications beyond the customer-defined types. An enterprise, whether or not it deals with large customers like these, would find it useful to craft internal specification standards or comply with external standards. The formats prescribed for these documents provide an important controlling influence on one's requirements analysis process. The requirements analysis process should be structured purposely to feed appropriate input into the structure of these document types. The requirements analysis strategies and methods covered in this book do precisely that.

5.1.1.3 Document controls

A specification must be formally released and controlled because it is the basis for a design that in turn must be configuration controlled. An enterprise, therefore, needs to place released specifications into a library, providing absolute control over their content while making them available for use by everyone.

5.1.1.4 The case for uniformity

It is a sound practice to insist on specification standards that are used to guide development of the specifications. Standards encourage completeness of the needed content. They provide for easy access to content of interest because we know where to find particular kinds of information.

We can make two errors in our requirements analysis efforts that result in poor specification content. First, we may fail to identify all of the needed requirements for an item. In this case, we will run the risk of the designer solving the small-scale problem in such a way that the solution does not contribute well in some way to the larger system solution. Alternatively, we could identify an excessive number of requirements. Is that possible, you ask? Yes, it is possible, and the result is that the design engineer is overly constrained and perhaps precluded from selecting the best possible overall design concept.

All requirements act as constraints on the design engineer. That is their very nature. In requirements analysis we seek to define the minimum necessary set of requirements for an item so as to provide the design engineer with a safe envelope within which to use his or her creative genius to convert today's technology and science into a practical solution. We wish to capture this minimum necessary set in our specifications uniformly to ensure the inclusion of essential requirements and to aid in the use and analysis of specification content. This goal also encourages the maximum possible solution space.

The argument for format uniformity, or format compliance, in specifications from the user perspective is that it allows us to expend minimum energy and time to find the information we need in any specification. Format and content controls also help the team building the specification to ensure they have thought of all of the needed requirements.

5.1.2 DoD Specifications under MIL-STD-490A

MIL-STD-490A was the military standard for specifications for many years. It was discontinued and replaced by Appendix A of MIL-STD-961D in the 1990s. MIL-STD-490A identified DoD specification types A through E as noted in Table 5.1-1. This section is largely extracted from MIL-STD-490A and augmented with additional information on internal requirements documents. Table 5.1-1 also cross-references the NASA type names given in MM 8040.42. The NASA CEI acronym means "contract end item," roughly equivalent to the DoD configuration item term. DoD requirements documentation may also employ the two-part arrangement noted for NASA documents. While MIL-STD-490A is no longer being identified for compliance on new contracts, there are many contracts still being worked on where it applies. Another reason for including this content is that there are many engineers and managers who spent 20 years or more responding to MIL-STD-490, and many will not rid their vocabulary of its outdated content for the remainder of their career. This means that younger engineers may have to understand some of the history behind current standards.

5.1.2.1 MIL-STD-490A specification types

5.1.2.1.1 Type A System/Segment Specification   This type of specification stated the technical and mission requirements for a system or segment as an entity, allocated requirements to functional areas, documented design constraints, and defined the interfaces between or among the functional areas. Normally, the initial version of a system or segment specification was based on parameters developed during the DoD concept exploration phase.

This specification (initial version) was used to establish the general nature of the system that was to be further defined and finalized during the demonstration and validation phase. The system or segment specification was maintained current during the demonstration and validation phase, culminating in a revision that formed the future performance base for the development and production of the prime items and configuration items. The system or seg-

Table 5.1-1 *Specification types*

DoD Code	DoD type names from MIL-STD-490A	Nasa type names from MM 8040.42
A	System/Segment	Program, System, or Project
B1	Prime Item Development	CEI Prime Equipment Detail, Part I
B2	Critical Item Development	Critical Component Detail, Part I
B3	Noncomplex Item Development	CEI Identification Item Detail, Part I
B4	Facility or Ship	CEI Facility Detail, Part I
B5	Software Development	CEI Computer Program Detail, Part I
C1a	Prime Item Product Function	CEI Prime Equipment Detail, Part II
C1b	Prime Item Product Fabrication	CEI Prime Equipment Detail, Part II
C2a	Critical Item Product Function	Critical Component Detail, Part II
C2b	Critical Item Product Fabrication	Critical Component Detail, Part II
C3	Noncomplex Item Product Fabrication	CEI Identification Item Detail, Part II
C4	Inventory Item	CEI Requirements Item Detail
C5	Software Product	CEI Computer Program Detail, Part II
D	Process Specification	NA
E	Material Specification	NA

ment specification would normally be prepared by the contractor for the customer in accordance with the format and content of the system or segment specification data item description (DID), defined by the customer.

This specification type was often preceded by a system or segment requirements document (SRD). The SRD normally followed a simpler and less standard format than that prescribed for a system specification, reflecting the relatively small amount of money available to the customer in the early phases. These early program documents were generally drafted by the user community and most often did not include clearly or precisely defined requirements suitable as the basis for a development contract.

It was not uncommon for DoD to award multiple contracts for concept and demonstration or validation phases, where all competing contractors had to create and maintain SRD or preliminary system specifications corresponding to their selected concept. In these cases, the contractors might have to maintain traceability from their version of the requirements to those received from the customer.

One of the fundamentals that should be included in the SRD or system specification was the identification of the major elements of the system and how their requirements would be captured in specifications. DoD, NASA, and most large development agents evolved an organized way to manage a development program through a subset of the system architecture. DoD called these items "configuration items," and NASA called them "contract end items," as noted in Table 5.1-1. Table 5.1-1 also identifies several different configuration item types referred to as Type B and C specifications. Type B captured the development requirements and type C the production requirements. These two sets could also be handled through a two-part scheme, which will be covered later.

5.1.2.1.2 Type B Development Specifications Development specifications state the requirements for the design or engineering development of a product during the development period. Each development specification must be in sufficient detail to describe effectively the performance characteristics that each configuration item is to achieve when a developed configuration item is to evolve into a detail design for production. The development specification should be maintained current during production when it is desired to retain a complete statement of performance requirements.

Because the breakdown of a system into its elements involves configuration items of various degrees of complexity, which are subject to different engineering disciplines or specification content, it is desirable to classify development specifications by subtypes. The characteristics and some general statements regarding each subtype are given in the following paragraphs.

5.1.2.1.2.1 Type B1 Prime Item Development Specification
A prime item development specification is applicable to a complex item such as an aircraft, missile, launcher equipment, fire control equipment, radar set, training equipment, or the like. A prime item development specification may be used as the functional baseline for a single configuration item development program or as part of the allocated baseline, where the configuration item is part of a larger system development program. Normally configuration items requiring a Type B1 specification meet the following criteria:

1. The prime item will be received or formally accepted by the contracting agency (on a DD Form 250 within the U.S. Government), sometimes subject to limitations prescribed on delivery documentation.
2. Provisioning action will be required.
3. Technical manuals or other instructional material covering operation and maintenance of the prime item may be required where the item is operated by the customer.
4. Quality conformance inspection of each prime item, as opposed to sampling, will be required.

5.1.2.1.2.2 Type B2 Critical Item Development Specification
A Type B2 specification is applicable to a configuration item that is below the level of complexity of a prime item but that is engineering critical or logistics critical:

1. A critical item is engineering critical where one or more of the following applies:
   i. The technical complexity warrants an individual specification.

ii. Reliability of the critical item significantly affects the ability of the system or prime item to perform its overall function, or safety is a consideration.

iii. The prime item cannot be adequately evaluated without separate evaluation and application suitability testing of the critical item.

2. A critical item is logistics critical where the following apply:

i. Repair parts will be provisioned for the item.

ii. The contracting agency has designated the item for multiple source reprocurement.

5.1.2.1.2.3 Type B3 Noncomplex Item Development Specification This type of specification is applicable to configuration items of relatively simple design that meet all of the following criteria:

1. During development of the system or configuration item, the noncomplex item can be shown to be suitable for its intended application by inspection or demonstration.

2. Acceptance testing is not required to verify performance.

3. Acceptance can be based on verification that the item, as fabricated, conforms to the drawings.

4. The end product is not software.

Examples of configuration items that normally meet the noncomplex criteria are special tools, work stands, fixtures, dollies, and brackets. Many such simple configuration items can be defined adequately during the development phase by a sketch and during production by a drawing or set of drawings. If drawings will suffice to cover all requirements, and unless the Government contracting agency requires a specification, a specification for a particular noncomplex item need not be prepared.

However, when it is necessary to formally specify several performance requirements to ensure development of a satisfactory configuration item or when it is desirable to specify detailed verification procedures, the use of a specification of this type is appropriate.

5.1.2.1.2.4 Type B4 Facility or Ship Development Specification A facility or ship development specification is applicable to each HWCI that is both a fixed (or floating) installation and an integral part of a system. Examples of facility or ship requirements are basic structural, architectural, or operational features designed specifically to accommodate the requirements unique to the system and that must be developed in close coordination with the system; the facility or ship services that form complex interfaces with the system; facility or ship hardening to decrease the total system's vulnerability; and ship speed, maneuverability, and the like. A development specification for a facility or ship establishes the requirements and basic restraints or constraints imposed on the development of an architectural and engineering design for such facility or ship.

5.1.2.1.2.5 Type B5 Software Development Specification Software development specifications are applicable to the development of computer software and consist of a software requirements specification and interface requirements specification(s). DoD has written a special military standard, MIL-STD-2167A, and a special series of software specification data item descriptions to cover software development, reflecting their concern for adequate documentation of software requirements. A contractor may also be required to develop a software development plan or manual defining exactly how all software will be developed and maintained:

1. *Software Requirements Specification.* This type of specification describes in detail the functional, interface, quality factor, special, and qualification requirements necessary to design, develop, test, evaluate, and deliver the required computer software configuration item (CSCI).

2. *Interface Requirements Specification.* This type of specification describes in detail the requirements for one or more CSCI interfaces in the system, segment, or prime item. The specified requirements are those necessary to design, develop, test, evaluate, and deliver the required CSCI.

The interface requirements may be included in the associated software requirements specifications under the following conditions: (1) There are few interfaces; (2) few development groups are involved in implementing the interface requirements; (3) the interfaces are simple; or (4) there is one contractor developing the software.

5.1.2.1.3 Type C Product Specifications Product specifications are applicable to any configuration item below the system level and may be oriented toward procurement of a product through specification of primarily functional (performance) requirements or primarily fabrication (detailed design) requirements. Subtypes of product specifications to cover equipments of various complexities or requiring different outlines of form are covered in the following paragraphs:

1. A product function specification states: (1) the complete performance requirements of the product for the intended use and (2) necessary interface and interchangeability characteristics. It covers form, fit, and function. Complete performance requirements include all essential functional requirements under service environmental conditions or under conditions simulating the service environment. Quality assurance provisions for hardware include one or more of the following inspections: qualification evaluation, preproduction, periodic production, and quality conformance.

2. A product fabrication specification will normally be prepared when both development and production of the HWCI are procured. When a development specification (Type B) has been prepared, specific reference to the document containing the performance requirements for the HWCI shall be made in the product fabrication specification. These specifications shall state: (1) a detailed description of the parts and assemblies of the product, usually by prescribing compliance with a set of drawings, and (2) those performance requirements and corresponding tests and inspections necessary to assure proper fabrication, adjustment, and assembly techniques. Tests normally are limited to acceptance tests in the shop environment. Selected performance requirements in the normal shop or test area environment and verifying tests therefore may be included. Preproduction or periodic tests to be performed on a sampling basis and requiring service, or other, environment may reference the associated development specification. Preproduction or periodic tests to be performed on a sampling basis and

requiring service, or other, environment may reference the associated development specification. Product fabrication specifications may be prepared as Part II of a two-part specification when the contracting agency desires close relationships between the performance and fabrication requirements.

5.1.2.1.3.1 Type C1 Prime Item Product Specifications
Prime item product specifications are applicable to configuration items meeting the criteria for prime item development specifications (type B1). They may be prepared as function or fabrication specifications as determined by the procurement conditions.

5.1.2.1.3.2 Type C1a Prime Item Product Function Specification   A type C1a specification is applicable to the procurement of prime items when a "form, fit, and function" description is acceptable. Normally, this type of specification would be prepared only when a single procurement is anticipated and training and logistic considerations are unimportant.

5.1.2.1.3.3 Type C1b Prime Item Product Fabrication Specification   Type C1b specifications are normally prepared for procurement of prime items when: A detailed design disclosure package needs to be made available; it is desired to control the interchangeability of lower-level components and parts; and service maintenance and training are significant factors.

5.1.2.1.3.4 Type C2 Critical Item Product Specifications
Type C2 specifications are applicable to engineering or logistic critical items as described for type B2 above and may be prepared as function or fabrication specifications.

5.1.2.1.3.5 Type C2a Critical Item Product Function Specification   A type C2a specification is applicable to a critical item where the critical item performance characteristics are of greater concern than part interchangeability or control over the details of design and a "form, fit, and function" description is adequate.

5.1.2.1.3.6 Type C2b Critical Item Product Fabrication Specification   A C2b specification is applicable to a critical item when a detailed design disclosure needs to be made available or where it is considered that adequate performance can be achieved by adherence to a set of detail drawings and required processes.

5.1.2.1.3.7 Type C3 Noncomplex Item Product Fabrication Specification   A noncomplex item product fabrication specification is applicable to noncomplex items as described above. Where acquisition of a noncomplex item to a detailed design is desired, a set of detail drawings may be prepared in lieu of a specification.

5.1.2.1.3.8 Type C4 Inventory Item Specification   This type of specification identifies applicable inventory items (including their pertinent characteristics) that exist in the DOD inventory and that will be incorporated in a prime item or in a system being developed. The purpose of the inventory specification is to stabilize the configuration of inventory items in the DOD inventory, on the basis of both current capabilities of each inventory item and the requirements of the specific application, or to achieve equipment/component item standardization between or within a system or prime item. This puts the government on notice as to the performance and interface characteristics that are required, so that, when engineering change proposals (ECP) for an inventory item are evaluated, the needs of the various applications may be kept in mind. If this is not done, design changes implemented in response to needs identified in another pro-

gram also using that item may make an inventory item unsuitable for the system. A separate inventory item specification should be prepared, as required, for each system, subsystem, prime item, or critical item in which inventory items are to be installed or that require the support of inventory items. Appendixes, one for each unique inventory item, cover requirements for the items aggregated to a particular inventory item specification.

5.1.2.1.3.9 Type C5 Software Product Specification   The software product specification is applicable to the delivered CSCI and is sometimes referred to as the "as-built" software specification. This specification consists of the final updated version of the software top-level design document, the software detailed design document, the database design document(s), the interface design document(s), and the source and object listings of the software:

1. *Software Top-Level Design Document.* The software top-level design document describes how the top-level computer software components (TLCSCs) implement requirements allocated from the software requirements specification and, if applicable, interface requirements specification(s).
2. *Software Detailed Design Document.* The software detailed design document describes the detailed decomposition of TLCSCs to lower-level computer software components (LLCSCs) and units. The software detailed design document shall be prepared by the contractor and shall be in accordance with the format and content of the software detailed design document data item description.
3. *Database Design Document.* The database design document describes one or more database(s) used by the CSCI. If there is more than one database, each may be described in a separate database design document.
4. *Interface Design Document.* The interface design document provides the detailed design of one or more CSCI interfaces. When interface requirements specifications have been prepared, associated interface design documents shall be prepared as well.

5.1.2.1.4 Type D Process Specifications   This type of specification is applicable to a service that is performed on a product or material. Examples of processes are heat treatment, welding, plating, packing, microfilming, marking, and so on. Process specifications cover manufacturing techniques that require a specific procedure so a satisfactory result may be achieved. Where specific processes are essential to fabrication or procurement of a product or material, a process specification is the means of defining such specific processes. Normally, a process specification applies to production but may be prepared to control the development of a process.

5.1.2.1.5 Type E Material Specifications   This type of specification is applicable to a raw material (chemical compound), mixtures (cleaning agents, paints), or semifabricated material (electrical cable, copper tubing) that are used in the fabrication of a product. Normally, a material specification applies to production but may be prepared to control the development of a material.

5.1.2.2 DoD specification forms under MIL-S-83490
Military specification MIL-S-83490 defined four specification forms as noted below. This specification was dropped

at the same time MIL-STD-490A was discontinued but may remain in effect on some old programs that are still active;

1. Form 1a specifications with maximum format control to MIL-STD-490A
2. Form 1b specifications with limited format control
3. Form 2 specifications prepared to commercial practices with supplemental military requirements
4. Form 3 specifications prepared to commercial practices

From this forms list, it is apparent that it would be entirely possible for a contractor to prepare a configuration item specification for a military customer simply by placing an appropriate cover sheet on an existing commercial specification.

The action that would make this possible would be an agreement in the contract that one or more specifications produced on the contract may be prepared to form 3 of MIL-S-83490. This can be done by tailoring the instructions in the contract data requirements list (CDRL) and referenced data item descriptions (DID). This may be a good solution for a given contract where the military customer is acquiring a product that has been produced for commercial purposes in the past and provided good service under conditions that are not vastly different from those it would experience in the military application.

The military customer could protect the configuration of their procurement for future purchases by identifying the item as a configuration item that must satisfy the requirements in a product function specification prepared by the contractor through the simple expedient of placing a MIL-

STD-490A inspired cover sheet on the existing commercial specification the contractor has in hand.

Generally, military customers will need material designed to specifically satisfy requirements driven by military situations not common in commercial practices. For example, a missile launcher may have to withstand the tremendous forces and stresses applied to it by a tactical nuclear weapon explosion within one mile and still function in order to continue to provide military capability in a battle obviously underway. Military requirements are driven by use of an article in a military situation, potentially involving very hostile actions by adversaries. However tough the competition becomes in the commercial world, these kinds of events are most uncommon in a commercial environment. As a result, military customers will commonly need specifications that comply with form 1 or 2.

5.1.2.3 Coordinated MIL-STD-490A references
Table 5.1-2 lists some documents that provide additional information about specifications for DoD and NASA applications when MIL-STD-490A was effective on DoD contracts. Many of these documents have also been discontinued or revised from the status shown.

5.1.2.4 MIL-STD-490A specification baselines
A baseline is a list of documents, each in a particular version under configuration control. On a program. the current aggregate configuration identification is established by a set of baseline configuration identification documents and all effected changes. This set of configuration identification documents includes all those necessary to provide a full

Table 5.1-2 *Specification references*

Number	Title
DOD-D-1000	Drawings, Engineering, and Associated Lists
MIL-S-83490	Specifications, Types, and Forms
FED-STD-102	Preservation, Packaging, and Packing Levels
MIL-STD-12	Abbreviations for Use on Drawings, Specifications, Standards, and in Technical Documents
DOD-STD-100	Engineering Drawing Practices
MIL-STD-109	Quality Assurance Terms and Definitions
MIL-STD-129	Marking for Shipment and Storage
MIL-STD-130	Identification Marking of U.S. Military Property
MIL-STD-1472	Human Engineering Design Criteria for Military Systems, Equipment, and Facilities
DOD-STD-480A	Configuration Control—Engineering Changes, Deviations, and Waivers
MIL-STD-483A	Configuration Management Practices for Systems, Equipment, Munitions, and Computer Programs
MIL-STD-490A	Specification Practices
MIL-STD-499A	Engineering Management
MIL-STD-961B	Military Specification and Associated Documents, Preparation of
DOD-STD-2167A	Defense System Software Development
Handbook H2	Federal Supply Classification
Handbook H4	Federal Supply, Code for Manufacturers
Handbook H6	Federal Item Identification Guides for Supply Cataloging
DOD 4120.3-M	Standardization Policies Procedures and Instructions
DOD-5220.22-M	Industrial Security Manual for Safeguarding Classified Information
MM 8040.12A	NASA Standard Contractor Configuration Management Requirements
None	GPO Style Manual
None	*Merriam-Webster's New International Dictionary*
DI-E-1126	Notice of Revision or Specification Change Notice
DI-E-3128	Engineering Change Proposals
DI-E-21430	Specification Revision Pages

technical description of the characteristics of the configuration item that require control at the time that the baseline is established. DoD has established three major baselines, and NASA had an equivalent baseline approach.

5.1.2.4.1 Functional Configuration Identification Functional configuration identification (functional baseline and approved changes) will normally include a Type A specification or a Type B specification supplemented by other specification types as necessary to specify: (1) all essential system functional characteristics; (2) necessary interface characteristics; (3) specific designation of the functional characteristics of key configuration items; and (4) all of the tests required to demonstrate achievement of each specified characteristic. This baseline should be established prior to entry into full-scale development.

5.1.2.4.2 Allocated Configuration Identification Allocated configuration identification (allocated baseline and approved changes) normally consists of a series of Part I Type B and C specifications defining the requirements, including functional, for each major configuration item. These may be supplemented by other types of specifications, engineering drawings, and related data, as necessary, to specify: (1) all of the essential configuration item characteristics, including delineation of interfaces; (2) physical characteristics necessary to assure compatibility with associated systems, configuration items, and inventory items; and (3) all of the tests required to demonstrate achievement of each specified functional characteristic. The allocated baseline should be established by the preliminary design review in full-scale development. The allocated baseline is the basis for system and item qualification and is audited at the FCA.

5.1.2.4.3 Product Configuration Identification The product baseline is defined by product specifications and engineering drawings and should be nearly complete at critical design review (CDR). The product baseline is the basis for acceptance and is audited at the PCA.

5.1.3 MIL-STD-961D Specification Standard

An attempt was made to release MIL-STD-490B, but it was never approved. MIL-STD-490A was discontinued in the early 1990s, as DoD attempted to move toward performance specifications and commercial standards. MIL-STD-961 had been in the inventory for years as a military standard for general military specifications, while 490A focused on program-peculiar specifications. An altered form of 490A content was introduced via a change notice into MIL-STD-961D Appendix A, providing continuing program-unique specification coverage at the time this book was first published.

5.1.3.1 Specification types

A significant change was made in the types of specifications with the introduction of the new standard. All program-unique specifications fell into one of two high-level types as before but with two different names. What were called "development" or "Part I" specifications under 490A were now called "performance specifications," in keeping with DoD policy to eliminate, where possible, specifications that contained the solution in them. The intent was to move to specifications that defined the performance expected and allow qualified engineering organizations to develop products that satisfied those requirements. What were called "product" or "Part II" specifications now were called "detail

specifications." DoD, Office of Secretary of Defense, published SD-15, Performance Specification Guide, in June 1995 in an attempt to communicate guidance on preparation of performance specifications and to distinguish between appropriate content for the two kinds of specifications.

The number of different types within these two broad categories was reduced significantly to the following five: system, item, software, material, and process. The item specification is presumably nonsoftware. In a complex system involving many levels of product definition, one would obviously have to associate these types with more than one level. A system may be composed of two or more systems, some of which in turn have three or more layers of items, some of which might be hardware and some of which might be software.

5.1.3.2 Structure and content

The structure of these specifications will be covered in another chapter, but they do contain six sections, like MIL-STD-490, called for with a couple of variations. Section 4 is now called "verification" rather than "quality assurance provisions"; and Section 5, involving packaging, was discouraged because it was said to be better expressed as a contractual matter. Minor changes in the internal structure seemed to me to be generally an improvement. In the effort to still respect many kinds of requirements in the new "performance" specification format, it was necessary to place a lot of requirements under that heading in the specification that the author would argue do not belong under that heading. As a result, there is a conflict between MIL-STD-961D content under performance requirements and the two major requirements categories covered in Part 3 of this book. The author partitioned all requirements into performance requirements and constraints. In the MIL-STD-961D format, all of the constraints except interface are included under the "functional and performance" heading.

5.1.4 MIL-STD-961E

The program-peculiar specification content was inserted into MIL-STD-961D via a change notice because MIL-STD-490A was withdrawn. The E revision of 961 completely rewrote the standard to cover all forms of military specifications, integrating the content added in a change notice into the document proper.

5.1.5 Other Requirements Document Types

In addition to the types of specifications identified by large customers like DoD and NASA, enterprises find it necessary to prepare four other types internally:

1. *Procurement Specifications.* Procurement specifications satisfy the same function between the contractor and their suppliers that a configuration item specification does between the ultimate customer and the contractor. They clearly define the essential characteristics of a product that they expect delivered from the supplier.
2. *In-House Requirements Documents.* The contractor may find that it is useful to identify requirements for intermediate levels of architecture within a system between the system and configuration item levels. An example of this is a requirements document for the fluids subsystem of a space launch vehicle (segment) core vehicle (prime item). This fluids subsystem itself would not

normally be identified as a configuration item but will be composed of many component parts, some of which may be configuration items and others procurement items.

The contractor may find it necessary to identify the requirements for the subsystem as a precursor to the identification of the requirements for these many components. A good solution to this problem is to prepare an in-house fluid subsystem requirements document that acts as a conduit for the core vehicle prime item requirements to the component level requirements by providing a place to capture the results of the requirements analysis process conducted at the subsystem level. This is also useful because this subsystem is probably the highest level in that branch of the system that has been assigned in total to a specific design department in the contractor's organization.

This same internal requirements document approach can also be used for component-level items fabricated by the contractor. Why should you do this? Why is it not adequate to simply place any requirements of note on the face of the engineering drawing? The reason is that the drawing captures the solution to the problem. The question is, "What is the problem?" This question should be clearly answered before the solution is defined.

The capture of these requirements need not necessarily impede design schedule progress. The internal requirements document can be a very simple document with a cover page and a numbered list of requirements (as described in Chapter 2.1) reviewed and approved by engineering management. Generally this step and the associated discipline necessary to create the item requirements as a prerequisite design will save a lot of money, schedule time, and grief.

3. *Parts Specifications.* Parts specifications are the third additional type of specification not defined in the documents noted in Table 5.1-1. These specifications are generally considered book drawings rather than specifications and, as a result, are often handled differently than specifications. They, along with materials and processes specifications, form the underpinnings of the system architecture. They are the lower fringe of which every item in the architecture is composed. This book, however, is focused on requirements analysis from the system level through the component level and does not do justice to the parts, materials, and processes levels, though the same techniques can be applied.

4. *Interface Specifications and Control Documents.* The final type of requirements document of note is the interface control document, or ICD. There are few good standards for the content of these documents. Specifications are prepared to define the requirements of the things in the system. ICDs are prepared to define the characteristics required in the interrelationships (interfaces) between two things of a system or between two systems. Chapter 6.5 covers interface documents, including a suggested template.

5.1.6 Coverage of Specifications

MIL-STD-490A identified two levels of specification coverage. A specification may be prepared to cover a group of products, services, or materials or a single product, service, or material, and are general or detail specifications, respec-

tively; either may be prepared as any of the types listed above. MIL-STD-961D Appendix A uses the word *detail* as a replacement for the use of the word *product* for a specification that includes design-dependent content, so one should be careful how these words are used on new programs.

5.1.6.1 General specification

A general specification covers requirements common to two or more types, classes, grades, or styles of products, services, or materials; this avoids repetition of common requirements in detail specifications. It also permits changes to common requirements to be readily effected. General specifications may also be used to cover common requirements for weapon systems and subsystems. Note that the general specification is a program-peculiar specification. Applicable documents that we will cover later apply to multiple programs.

5.1.6.2 Detail specification

A detail specification covers all requirements for one or more types of configuration items or services so as not to require preparation and reference to a general specification for the common requirements. A detail specification may also take the form of a specification sheet, which is incomplete without reference to a general specification. The detail and referenced general specification (which contains the requirements common to the family of configuration items) together then constitute the total requirements.

5.1.7 One- and Two-Part Specifications

Specifications commonly contain two kinds of information, and this information may be included in a single specification, separated into two independent specifications having two different identifying numbers or collected into coordinated two-part specifications. These two components relate to the development requirements that properly define the technical problem that must be solved. These requirements are focused on the problem and not on the solution. They are solution independent. These are the requirements that properly drive the qualification verification process. The second kind of content is design dependent, captured as a means for determining the requirements for the acceptance process.

Two-part specifications, which collectively combine both development (performance) and product fabrication (detail) requirements under a single specification number, are a sound option for large systems acquired by customers like DoD and NASA. Under this practice, the development specification remains alive during the life of the article as the complete statement of performance requirements. Proposed design changes must be evaluated against both the development and the product parts of the specification. To emphasize the fact that two parts exist, both parts are identified by the same specification number, and each part is further identified as Part I or Part II, respectively. The alternative, offering separate specifications under different specification numbers, is equivalent; but it is more difficult to keep track of the relationship between these parts.

The single specification containing both kinds of requirements is a viable option for smaller programs, procurement specifications, and in-house requirements documents. This specification should be developed in two steps, however. The first stage includes the development requirements that are design independent. Once the design is determined, based on that version of the specification, the detail requirements should be added that are design dependent. There is a great danger that engineers preparing these kinds of specifications will develop the attitude that it is acceptable to release them only after the design is determined in that part of their content (the detail requirements) should follow this pattern. A system specification is, by its very nature, a performance specification, while materials and processes specifications are detail specifications.

5.1.8 A Strange Specification Format

MIL-STD-961E permits specifications that are not paper documents at all. It permits the use of models as specifica-tions. This is a radical departure from decades of insistence on paper specifications. Perhaps a software model would be more commonly called as a specification than a model for a hardware item, but this is an area that will surely grow over time. The slow movement to model-driven development will encourage a drift to model-based specifications. Eventually we may actually reach the old paperless office goal for spec-ifications. In this new world, analysts will accomplish the work covered in this book, clearly defining problem space model entities that will represent physical entities to be pro-duced. Those problem space models will be transformed into solution space models from which the hardware com-ponents can be manufactured through a CAD/CAM trans-form and software code generated directly by the model, to some extent.

5.2 General Specification Style Guide

Contents

5.2.1 Style, Format, and Identification of Military Specifications

This section covers style, format, and general instructions for preparing a specification, including material arrangement, paragraphing, numbering, heading, and concluding material. It is essentially a copy of the corresponding section of MIL-STD-490A with some changes to recognize a more recent standard and include some author preferences.

5.2.1.1 Sectional arrangement of specifications

According to both MIL-STD-490A and MIL-STD-961D/E, military specifications should contain six numbered sections as identified in Table 5.2-1. There are obviously many ways that one could organize the information captured in a specification, but if we are to have a standard approach we must accept one standard. The structure defined in MIL-STD-490A and MIL-STD-961D/E benefits from many years of experience and should be used unless there is a compelling reason to deviate from it. Many companies do find reasons that appear to make sense to them for adding additional sections for their procurement specifications. Often this added material is more related to contractual matters than requirements and could be better covered in the supplier contract or statement of work.

Subject matter should be kept within the scope of the sections so that the same kind of requirements or information will always appear in the same section of every specification. Except for appendixes, if there is no information pertinent to a section, the following should appear below the section heading, "This section is not applicable to this specification."

5.2.1.2 Language style

The paramount consideration in a specification is its technical essence, and this should be presented in language free of vague and ambiguous terms and using the simplest words and phrases that will convey the intended meaning. Inclusion of essential information must be complete, whether by direct statements or reference to other documents. Consistency in terminology and organization of material will contribute to the specification's clarity and usefulness. Sentences should be as short and concise as possible. Punctuation should aid in reading and prevent misreading. Well-planned word order requires a minimum of punctuation. When extensive punctuation is necessary for clarity, the sentence(s) should be rewritten. Sentences with compound clauses should be converted into one or more short and concise sentences.

5.2.1.3 Primitive requirement statement

The requirements statements should contain three fundamental entities surrounded by a minimum of language to ensure understanding in the form of one or more complete English sentence structures: (1) the attribute that must be controlled, (2) the value of the attribute that must be satisfied (with a possible tolerance) and corresponding units of measure, and (3) a relation term that gives the relationship between the attribute and value. In numerical requirements, the value will be a number with units coupled with one of the following relations: less than, greater than, equal to, less than or equal to, greater than or equal to, or between a range of values. In requirements stated without numerical content, the relation statement will be words like *is, be,* or *conform to.*

5.2.1.4 Capitalization and spelling

Except where DOD requirements differ, the *United States Government Printing Office Style Manual* should be used as the preferred guide for capitalization, spelling, punctuation, syllabification, and so on. *Merriam-Webster's New International Dictionary* (latest revision) is an accepted guide when the *Style Manual* does not provide the guidance needed. Alternative dictionary sources embedded within computer word processors may also be used.

5.2.1.5 Abbreviations

The applicable standard abbreviations listed in MIL-STD-12 offer a guide for DoD specifications. The only other abbreviations employed should be those in common usage and not subject to misinterpretation. The first time an abbreviation is used in text, it should be placed in parentheses following the word or term spelled out in full; e.g., "pounds per square inch (psi)." This rule does not apply to abbreviations used for the first time in tables and equations. Uncommon abbreviations so used should also be explained in the text or footnotes.

5.2.1.6 Symbols

Symbols should be avoided in text but may be used in equations and tables. Graphic symbols, when used in figures, should conform to customer usage. (Any symbol formed by a single character should be avoided if practicable, because an error destroys the intended meaning.) Avoid the use of any text content that cannot be easily accommodated by the word processing capabilities used on the program. If, for example, the hardware and software in use will not permit the use of a square root symbol, you may have to spell it out: "the square root of x."

Table 5.2-1 *Specification section titles*

Section number	MIL-STD-490A Section title	MIL-STD-961D/E Section title
1	SCOPE	SCOPE
2	APPLICABLE DOCUMENTS	APPLICABLE DOCUMENTS
3	REQUIREMENTS	REQUIREMENTS
4	QUALITY ASSURANCE	VERIFICATION
5	PREPARATION FOR DELIVERY	PREPARATION FOR DELIVERY
6	NOTES	NOTES
APPENDIX	APPENDIX I, II, III, ...	APPENDIX A, B, C, ...

5.2.1.7 Proprietary names

Trade names, copyrighted names, or other proprietary names applying exclusively to the product of one company should not be used unless the item(s) require source control or cannot be adequately described because of the technicality involved, construction, or composition. In such instances, one or even, if all pertinent requirements are specified, several commercial products may be included by inserting the words *or equal* after the trade name to assure wider competition and that bidding will not be limited to a particular make specified. The same applies to manufacturer's part numbers or drawing numbers for minor parts when it is impractical to specify all detail requirements in the specification. In all instances where "or equal" is permitted, the particular characteristics required must be included to define "or equal."

5.2.1.8 Commonly used words and phrasing

Certain words and phrases are frequently used in a specification. The following rules for these words and phrases are offered:

1. Referenced documents are cited thus: (1) "conforming to …," "as specified in …," or (3) "in accordance with …."
2. "Unless otherwise specified" is used to indicate an alternative course of action. The phrase shall always come at the beginning of the sentence and, if possible, at the beginning of the paragraph. This phrase is used only when it is possible to clarify its meaning by providing a reference such as to Section 6 of the specification for further clarification in the contract or order or otherwise.
3. When making reference to a requirement in the specification and the requirement referenced is rather obvious or not difficult to locate, the simple phrase "as specified herein" is sufficient and should be used.
4. The phrase "… to determine compliance with …" or "… to determine conformance to …" should be used in place of "… to determine compliance to …." In any case, use the same wording throughout.
5. In stating positive limitations, the phrase shall be stated thus: "The diameter shall be no greater than …."
6. The emphatic form of verb should be used throughout the specification; that is, state in the requirements section that, "The indicator shall be designated to indicate …," and, in the section containing test provisions, "The indicator shall be turned to zero and 230 volts alternating current applied." For specific test procedures, the imperative form may be used, provided that the entire method is preceded by "the following tests shall be performed," or related wording. Thus, "Turn the indicator to zero and apply 230 volts alternating current."
7. Capitalize the words "drawing," "bulletin," and the like only when they are used immediately preceding the number of a document. However, federal and military standards and handbooks should be identified in the text only by their symbol and number; thus "MIL-E-000," and not "specification MIL-E-000."

5.2.1.9 Use of "Shall," "Will," "Should," and "May"

Use "shall" whenever a specification expresses a provision that is binding. Use "should" and "may" wherever it is necessary to express nonmandatory provisions. "Will" may be used to express a declaration of purpose on the part of the contracting agency. It may be necessary to use "will" in cases where the simple future tense is required, for example, "Power for the motor will be supplied by the ship."

5.2.1.10 Use of "Flammable" and "Nonflammable"

The terms *flammable* and *nonflammable* should be used in specifications in lieu of the terms *inflammable, uninflammable,* and *noninflammable.*

5.2.2 Paragraph Numbering and Identification

Each paragraph and subparagraph should be numbered consecutively within each section of the specification, using a period to separate the number representing each breakdown. There follows an example in the context of Section 3 of a specification:

3.   REQUIREMENTS
    3.1 First Paragraph
        3.1.1 First Subparagraph
    3.2 Second Paragraph
        3.2.1 First Subparagraph
        3.2.2 Second Subparagraph

Itemization within a paragraph or subparagraph should be identified by lowercase letters to avoid confusion with paragraph numerals. For clarity of text, paragraph numbering should be limited to seven levels. The use of letters or bullets was found necessary in any voluminous specification because of the limitation of seven levels. The author encourages numbering every paragraph, regardless of the number of indentures used, to support simple traceability and verification referencing.

5.2.2.1 Paragraph identification

If practicable, each paragraph and subparagraph should be given a subject identification. MIL-STD-490A and MIL-STD-961E require that only the first letter of the first word in the paragraph identification be capitalized. The author believes that each word in a paragraph title should be capitalized, but you need to follow the practices required by your customer. Paragraph identifications in any one section should not be duplicated. MIL-STD-490A called for primary paragraph identifications in boldfaced type and subparagraph identifications italicized when typeset and underlined when typewritten. The author encourages plain text without underlining in all cases. The standards also require that a paragraph title end with a period that the author believes to be unnecessary.

5.2.2.2 Underlining

Do not underline any portion of a paragraph or capitalize phrases or words for the sake of emphasis with the exceptions noted in paragraph titles. All of the requirements are important in obtaining the desired product or service.

5.2.2.3 Cross-references

Cross-references, that is, references to parts within the specification, should be held to a minimum. Cross-references are appropriate only to clarify the relationship of requirements within the specification and to avoid inconsistencies and unnecessary repetition. When the cross-reference is to a paragraph, subparagraph, or the like within the specification, the cross-reference should be only to the specific paragraph number. It is unnecessary to include the word *paragraph.*

5.2.2.4 Figures, tables, and foldouts

A figure is a picture or graph and constitutes an integral part of the specification. It must be clearly related to, and consistent with, the text of the associated paragraph. Figures should not be confused with numbered and dated drawings referenced in the text, which should be listed in Section 2 and not physically incorporated in the specification.

5.2.2.5 Location of figures in specification
According to MIL-STD-490A, each figure should be placed following, or within, the paragraph containing a reference to it. If figures are numerous and their location, as indicated above, would interfere with correct sequencing of paragraphs and cause difficulty in understanding or interpretation, they may be placed in numerical sequence at the end of the specification before any appendix or index. Depending on the word processor or desktop publisher used, there may be advantages to placing the figures at the end as well.

5.2.2.6 Preparation of figures
All figures should be titled and numbered consecutively with Arabic numerals in the order in which they are initially referenced in the specification. A table is an arrangement of data in lines and columns. The tabular format is appropriate when data can thus be presented more clearly than in text. Elaborate or complicated tables should be avoided. References in the text should be sufficiently detailed to make the purpose of the table clear, and the table should be restricted to data pertinent to the associated text.

5.2.2.7 Location of tables in specifications
A table should be placed following, or within, the paragraph containing a reference to it. If space does not permit, a table may be placed at the beginning of the succeeding page or, if extensive, on a separate page. If tables are numerous and their location, as indicated above, would interfere with correct sequencing of paragraphs and cause difficulty in understanding or interpretation, they may be placed in numerical sequence at the end of the specification before any appendix or index.

5.2.2.8 Preparation of tables
Any tables should be numbered consecutively with Roman numerals in the order in which they are initially referenced in the specification. The number and title are placed above the table. The contents of a table should be organized and arranged to show clearly the significance and relationship of the data. Data included in the text should not be repeated in the table. Tables should be boxed in and ruled. When a table is of such width as to make it impracticable to place it in normal position on the page, it should be rotated such that the lower edge is outboard on the page.

5.2.2.9 Foldouts
Foldouts should be avoided except where required for legibility. Large tables or figures may be broken out so that they may be printed on facing pages. Where foldouts are required, they should be grouped in one place, preferably at the end of the specification (in the same location as noted for figures above) and suitable reference to their location included in the text.

5.2.3 Footnotes

5.2.3.1 Footnotes to text
Footnotes to the text should be avoided if possible. Their purpose is to convey additional information that is not properly a part of the text. A footnote to the text, when absolutely necessary, should be placed at the bottom of the page containing the reference to it.

5.2.3.2 Footnotes to tables and figures
Footnotes to a table or figure are placed below the table or figure. The footnotes may contain mandatory information that cannot be presented as data within a table. Footnotes should be numbered separately for each table. Where numerals will lead to ambiguity (for example, in connection with a chemical formula), superior letters, asterisks, daggers, and other symbols may be used.

5.2.4 Contractual and Administrative Requirements

Specifications should not include contractual requirements that are properly a part of the contract. MIL-STD-490A specifically warned against including cost figures in the specification, suggesting it was a programmatic or contractual factor and not a technical one. I would encourage including cost figures where design to cost or life-cycle cost is specifically defined as a valid design constraint. Some other contractual factors include: time of delivery, instructions on reworking or resubmitting rejected items or lots, method of payment, liquidated damages, and provision for configuration items damaged or destroyed in tests.

Contractual, administrative, and warranty provisions, such as those covered in general provisions of contracts, should not be made part of the requirements in the specification. Contractual and administrative provisions not covered in the general provisions, but considered essential for procurement, may be indicated as "ordering data" or "features to be included in bids or in contract" in Section 6. This provision will be exercised with caution and limited to essential matters.

5.2.4.1 Definitions in specifications
The inclusion of a definition can be avoided if requirements are properly stated. When the meaning of one or more terms must be established in the specification, definitions should be placed in the text. However, it is often clearer to list one or more definitions in Section 6, especially where the terms are used in many places throughout the specification. When this is done, a parenthetical reference to the applicable paragraph in Section 6 should follow to indicate the existence of a definition.

5.2.4.2 References to other documents
Referencing is the approved method for including requirements in specifications where this eliminates the repetition of requirements and tests that are adequately set forth elsewhere. Refer to Chapter 5.4 for more information about applicable documents. Chain referencing should be avoided. References should be restricted to documents that are specifically and clearly applicable to the specification and are current and available. Care must be taken in writing the specification to indicate in a positive manner the extent to which a referenced document is applicable. The specification should also include any special details called for by the referenced document. Reference to paragraph numbers in other documents should not be made. The reference is made to a title, method number, specifically identified requirement, or other definitive designation.

SPECIFICATION NUMBER 12345B
CODE IDENT XXXXX
PART I OF TWO PARTS
(Date)

PRIME ITEM DEVELOPMENT SPECIFICATION
FOR
(APPROVED TITLE)
(TYPE DESIGNATOR, CONFIGURATION ITEM NUMBER, ETC.)

a. Example of Identification for Part I

SPECIFICATION NUMBER 12345B
CODE IDENT XXXXX
PART II OF TWO PARTS
(Date)

PRIME ITEM FABRICATION SPECIFICATION
FOR
(APPROVED TITLE)
(TYPE DESIGNATOR, CONFIGURATION ITEM NUMBER, ETC.)

b. Example of Identification for Part II

Figure 5.2-1 *Method of identifying two-part specifications*

5.2.4.3 Limitation on references

A specification should not contain anything in conflict with provisions in referenced documents unless it is desirable to make special exceptions to such provisions, in which case the specific provision to which exception is made shall be stipulated or the application of a specific portion of the referenced document shall be clearly defined. It is not intended that other documents be made a part of a specification by reference unless the items, materials, tests, or other services in the referenced documents are required in the quality and detail that these documents are designed to produce. The applicability of all referenced documents listed in Section 2 of a specification should be defined in Section 3, 4, or 5, as appropriate. The extent of applicability of referenced documents should also be indicated. The whole of a referenced document need not be made applicable by reference unless all of its provisions are clearly required.

5.2.4.4 Security marking of specifications

Specifications containing U. S. Government classified information must be marked and handled in accordance with customer-defined security regulations. Security of classified information is very important, but we must understand that it is a two-edged sword. Security provisions that prevent adversaries from acquiring access to sensitive information have a way of preventing easy access to that same information by people who need to know. There is no good solution to this problem, but one that is often used is to place the classified information in a classified appendix, leaving the basic document unclassified.

5.2.4.5 Identification of specifications

Each specification must be numbered and dated on each page. The identification number, with the date below it, should always appear at the top of the page opposite the binding edge. U.S. Government specifications must be identified by the code identification of the U.S. Government design agency as listed in Cataloging Handbook H4 and by a number assigned by the government design agency. This may be either a number or a combination of letters, numbers, and dashes. The number must not contain more than 15 characters, excluding dashes and revision letter. Specifications for HWCIs, materials, or processes intended for multiple application may be identified by a military specification number rather than a program-specific number.

Specifications will require change after they are initially released, and the different versions of the specifications must be differentiated one from another. The common method of doing this is to assign revision letters to the basic document number. These revision letters should start with

"A" for the first revision, and be assigned alphabetically for each succeeding revision. Letters such as I, O, or Z that can be confused with numerals should not be used. The revision after Y (assuming the letter Z is not used) would be AA. The second letter is advanced through AY as needed. If a third alphabet cycle is needed, the next revision letter would be BA. This pattern is repeated as needed. When a two-part specification concept is used, the parts are identified on the title page and both parts are assigned the same specification number (see Figure 5.2-1). Revision status of each part is separately maintained.

A specification sheet, when used, is identified by the same number and code identification as the associated applicable general specification followed by a virgule (slash) and a sequentially assigned Arabic numeral for the sheet. Example: Code Ident 10001 WS 1967B/1A designates revision A of sheet 1 issued for the B revision of general specification, numbered WS 1967.

5.2.4.6 Titling the specification

The approved basic name of the material, product, or service covered by the specification should be the first part of the title. The customer may have a preference for naming items, as in the case of DoD, which prefers the guidance in Cataloging Handbook H6. The basic noun in the title should be in the singular form if the specification covers only one product and in the plural form if the specification covers more than one product, that is, various types, grades, classes, sizes or capacities, and so on, except where the only form is plural or where the nature of the product unavoidably requires the plural form. Where there is no approved configuration item name, DoD prefers that the title be developed in accordance with DOD-STD-100. For general specifications the words *General Specifications for* should be the closing phrase of the title.

The title of the specification should include, where appropriate, and in addition to the approved basic name, the minimum number of modifiers, including type designators, that are necessary for distinction and ready identification of the coverage of the specification. Modifiers should be arranged in reverse order and separated from each other and the noun name by punctuation.

The type of specification should be included above the specification title. As a minimum, the main specification type should be specified as "SYSTEM/SEGMENT," "DEVELOPMENT," "PRODUCT," "PROCESS," or "MATERIAL." A subtype may be specified when desired by the contracting agency.

5.3

Specification Content Guidance

Contents

5.3.1 MIL-STD-490A Content Standard

We will now look into the proper content of specifications. This section has been essentially copied from MIL-STD-490A, with some adjustments to soften the harsh edge of government language. The reader should realize that we will address the content issue in a generic way and that particular kinds of specifications will require some differences in content needs within the major six sections generally called for in all specification types. In general, the actual specification paragraph number would be the same as the paragraph number in this chapter if "5.3.1" is deleted from it.

5.3.1.1 Section 1: Scope

General information pertaining to the extent of applicability of a configuration item, material, or process covered by a given specification and, when necessary, specific detailed classification thereof, are placed in the appropriate subdivision of Section 1. This section should not contain requirements properly part of other sections of the specifications.

5.3.1.1.1 Scope   A statement of the scope consists of a clear, concise abstract of the coverage of the specification and may include, where necessary, information as to the use of the configuration item other than specific detailed applications covered under "intended use" in Section 6 of the specification. This brief statement should be sufficiently complete and comprehensive to describe generally the configuration item, material, or process covered by the specification in terms that may be easily interpreted by manufacturers, contractors, suppliers, or others familiar with applicable terminology and trade practices.

5.3.1.1.2 Classification   Where a specification covers more than one category of a configuration item, designations of classification such as types, grades, classes, and the like are listed under this heading in accordance with accepted industry practice. The same designation should be used throughout the specification. The name of the configuration item covered by the specification is followed by the words "shall be of the following types, grades, classes, etc., as specified," listing only the applicable designations. When more than one type, grade, class, and the like is listed, each must be briefly defined. When only one (type, grade, or other) is covered, a statement to this effect should be included in the scope paragraph, and the classification paragraph omitted. The types, grades, classes, and so on should not change when the specification is changed or revised except when industry practice changes or for other good reason a change is required. Where the characteristics of a configuration item change enough to affect interchangeability, the original designation is deleted and a new type, grade, class, or the like is added. Whenever it becomes necessary to change the designation without changing the characteristics of the configuration item, a cross-reference should be included in Section 6 indicating the relationship between the new and old designations. Because such changes may require cataloging and other record changes, they should be kept to a minimum:

1. *Type*. This term implies differences in like configuration items or processes as to design, model, shape, and the like and generally will be designated by Roman numerals, thus "type I," and so on. Note the different use of the word *type* here from that used to define the level of a specification and kind of product, as in type A, B1, C2, and so forth.

2. *Class*. This term provides additional categorization of differences in characteristics other than those afforded by type classification that do not constitute a difference in quality or grade but are for specific, equally important uses. These generally will be designated by Arabic numerals; thus, "class 1," "class 2," and so on.

3. *Grade*. This term implies differences in quality of a configuration item and generally will be designated by capital letters; thus "grade A," "grade B," and so on.

4. *Composition*. This term is used in classifying configuration items that are differentiated strictly by their respective chemical composition, and it generally will be designated in accordance with accepted trade practice when satisfactory to the government design agency.

5. *Style*. This term is used to denote differences in design or appearance.

6. *Other*. If the terms *types, grades,* and *classes* do not serve accurately to classify the differences as indicated above, other terms such as *color, form, weight, size, power supply, temperature rating, condition, unit, enclosure, rating, duty, insulation, kind, variety,* and the like, suitable for reference, may be used. When a specification contains a multilevel reliability requirement, Section 1 of the specification shall identify the levels covered.

5.3.1.2 Section 2: Applicable Documents

All and only those documents referenced in Section 3, 4, 5 and Appendixes of the specification must be listed in Section 2 of the specification. If the documents listed are numerous, Section 2 may reference an appendix or other appropriate document containing a complete listing. References should be confined to documents available at the time the current revision of the specification is issued. Figures bound integrally with the specification should not be listed in Section 2. In general, the content of these documents referenced will be of a more general nature than the specific application covered by the specification; and, unless they are tailored to correct this excess of requirements, the design team will be required to create a design solution that covers the full content of those documents, resulting in unnecessary capability and cost coordinated with small benefit for the buyer. Refer to Chapter 5.4 for tailoring coverage.

5.3.1.2.1 Kinds of Documents Referenced

5.3.1.2.1.1 Government Documents   Federal and military specifications (as well as government design agency specifications), standards, drawings, and other government publications may be referenced in specifications. Government regulations or codes that are mandatory on the military services (such as Federal Insecticide, Fungicide, and Rodenticide Act; Drug and Cosmetic Act; Federal Hazardous Substances Labeling Act; Atomic Energy Act; Department of Transportation Regulations; and Screw-Thread Standards for Federal Services) are referenced in specifications, where applicable.

5.3.1.2.1.2 Nongovernment Documents   Reference may be made to nongovernment specifications, standards, and publications promulgated by commercial organizations, technical societies, and other nongovernmental agencies when such documents are accepted by the using governmental agency. Care must be taken in referencing nongovernmental publications so as to assure the availability of copies and prior approval of the copyright owner.

5.3.1.2.2 Listing of References

References are listed by document numbers and titles and may include specific issue or revision where necessary to rigidly control the configuration or implementation of the configuration item, material, or process. The title of each document should be that appearing on the document itself rather than the title shown in an index.

5.3.1.2.2.1 Government Documents

Government specifications, standards, drawings, and other publications intended to be made available to bidders are listed under the appropriate preceding headings and in alphabetical-numerical order in individual groups, such as federal, military, and departmental agency (such as Weapons Command or the like). These listings are included under a paragraph similar to one of the following:

Example 1:
2.1 *Government Documents.* The following documents of the issue in effect on date of invitation for bids or request for proposal, form a part of the specification to the extent specified herein.

Example 2:
2.1 *Government Documents.* The following documents of the exact issue shown form a part of this specification to the extent specified herein. In the event of conflict between the documents referenced herein and the contents of this specification, the contents of this specification shall be considered a superseding requirement.

Government documents should be listed under the paragraph 2.1 heading in the following order:

Specifications
  Federal
  Military
  Other government agency
Standards
  Federal
  Military
  Other government agency
Drawings
  (Where detailed drawings referred to in a specification
    are listed on an assembly drawing, it is necessary to
    list only the assembly drawing.)
Other Publications
  Manuals
  Regulations
  Handbooks
  Bulletins
  etc.

Copies of specifications, standards, drawings, and publications required by suppliers in connection with specified procurement functions should be obtained from the contracting agency or as directed by the contracting officer.

5.3.1.2.2.2 Nongovernment Documents

Nongovernment documents should be listed in appropriate order under a paragraph similar to one of the following subparagraphs:

Example 1:
2.2 *Nongovernment Documents.* The following documents form a part of this specification to the extent specified herein. Unless otherwise indicated the issue in effect on date of invitation for bids or request for proposal shall apply.

Example 2:
2.2 *Nongovernment Documents.* The following documents of the exact issue shown form a part of this specification to the extent specified herein. In the event of conflict between the documents referenced herein and the contents of this specification, the contents of the specification shall be considered a superseding requirement.

Nongovernment documents are listed in the following order:

SPECIFICATIONS:
STANDARDS:
DRAWINGS:
OTHER PUBLICATIONS:
  (List source for all documents not available through normal government stocking activities.)

The following source paragraph should be placed at the bottom of the list when applicable.

Technical society and technical association specifications and standards are generally available for reference from libraries. They are also distributed among technical groups and using Federal agencies.

5.3.1.3 Section 3: Requirements

The essential requirements and descriptions that apply to performance, design, reliability, personnel subsystems, and so on of the configuration item, material, or process covered by the specification are stated in this section. These requirements and descriptions define as applicable the character or quality of the materials; formula; design; construction; performance; reliability; transportability and product characteristics; chemical, electrical, and physical requirements; dimensions; weight; color; nameplates; product marking; and so on. This section is intended to indicate, as definitively as practicable, the minimum requirements that a configuration item, material, or process must meet to be acceptable. The requirements section must be so written that compliance with all requirements will assure the suitability of the configuration item, material, or process for its intended purpose, and noncompliance with any requirement will indicate unsuitability for the intended purpose. Only those requirements that are necessary and practicably attainable should be listed.

Section 3 of a general specification contains all requirements that are common to a family of systems, configuration items, materials, or processes. When detail specifications are to be prepared to supplement the general specification to fully define an individual configuration item, or the like, the following paragraph should be included in Section 3 of the general specification:

3.x.x *Detail Specification.* Requirements for individual (insert the proper term from among the following) parts, configuration items, materials, process, systems shall be as specified herein and in accordance with the applicable detail specification.

Section 3 of a detail specification should contain the requirements only for the particular system, configuration item, material, or process covered by that specification. However, if the specification does cover more than one type, class, grade, or the like, it should first specify the general requirements for all types, classes, grades, and so on. The differentiating requirements may then be specified for the individual types, classes, grades, or the like in the proper sequence. In general, each requirement is covered in a separate paragraph; and where one requirement differs for the various types, classes, grades, and the like, a separate paragraph immediately following the general requirements should be devoted to each type, class, grade, or the like. The various detailed requirements are then contained in appropriate subparagraphs. Where it is necessary to include additional data, descriptive and appropriate headings should be used and assigned in logical order.

Section 3 of system or development specifications (Type A or B) sets forth requirements in terms of performance, reliability, design constraints, functional interfaces, and so on, that are necessary to assure a practical and reasonable development effort. Development specifications may include design goals in addition to minimum requirements; but, in such case, goals and requirements must be clearly identified to avoid confusion. Only essential design constraints should be included as requirements, such as restriction of use of certain materials due to toxicity, dimensional or functional restrictions to assure compatibility with associated equipments, and so on.

Section 3 of a product, process, or material specification (Type C,D,E respectively) should contain all requirements necessary to assure delivery of an acceptable end product. Requirements in product function specifications should include both physical (dimensional and interface characteristics) and performance requirements in sufficient detail to assure procurement of interchangeable, but not necessarily identical, HWCIs. Requirements in a product fabrication specification should include all requirements necessary to assure delivery of identical HWCIs from suppliers. This is normally accomplished by invoking a set of drawings (DOD-D-1000 Level 3) as a primary requirement. Product requirements should be included for performance, reliability, and the like when such features or characteristics are not completely controlled by detail drawings.

5.3.1.3.1 Definition (Paragraph 3.1)

Where applicable, a definition of the system or configuration item should be provided in the form of a brief description. It should identify major physical parts, functional areas, and functional and physical interfaces. System logic diagrams, block diagrams, schematic diagrams, and pertinent operational, organizational, and logistic considerations and concepts may be included.

5.3.1.3.2 Characteristics (Paragraph 3.2)

Development, product, and material specifications specify all required performance characteristics, physical characteristics, and requirements for reliability, maintainability, environmental consideration, and, as appropriate, relative priority of design disciplines or characteristics.

5.3.1.3.2.1 Performance Characteristics These characteristics include general and detail requirements, under appropriate subheadings, for all performance requirements, that is, what is expected of the system, configuration item, or material. This is clearly the most difficult part of the specification to prepare, and it is the principal target of the requirements analysis process this book is focused on.

5.3.1.3.2.2 Physical Characteristics These characteristics in a development, product, or material specification sets forth requirements such as weight limits, dimensional limits, and the like necessary to assure physical compatibility with other elements and not determined by other design and construction features or referenced drawings. They shall also include considerations such as transportation and storage requirements, security criteria, durability factors, health and safety criteria, command control requirements, and vulnerability factors. Where applicable, protective coating requirements should be specified under this heading to assure protection from corrosion, abrasion, or other deleterious action. Where feasible, color and protective coating should be combined.

5.3.1.3.2.2.1 Reliability Reliability requirements should be stated numerically with confidence levels, if appropriate, in terms of mission success or hardware mean time between failures. Initially, reliability may be stated as a goal and a lower minimum acceptable requirement. During contract definition, or an equivalent period, realistic requirements must be determined and incorporated in the specification with requirements for demonstration. Never state reliability requirements in terms of a goal in Type C (product) specifications.

5.3.1.3.2.2.2 Maintainability Numerical maintainability requirements should be stated in such terms as mean-time-to-repair (MTTR) or maintenance person-hours per flight or operational hour. Realistic maintainability requirements should be identified as discussed above for reliability. Qualitative requirements for accessibility, modular construction, test points, and other design requirements may be specified as required.

5.3.1.3.2.2.3 Environmental Conditions Environments that the system or equipment is expected to experience in shipment, storage, service, and use must be specified. Where applicable, it must be specified whether the equipment will be required to meet or be protected against specified environmental conditions. Subparagraphs may be included as necessary to cover environmental conditions such as climate, shock, vibration, noise, noxious gases, and the like.

5.3.1.3.2.2.4 Environmental Conditions Transportability Any special requirements for transportability and materials handling are specified under this heading.

5.3.1.3.2.3 Design and Construction (Paragraph 3.3) Minimum or essential requirements that are not controlled by performance characteristics, interface requirements, or referenced documents are included here. They include appropriate design standards, requirements governing the use or selection of materials, parts and processes, interchangeability requirements, safety requirements, and the like.

5.3.1.3.2.3.1 Materials Requirements for materials to be used in the item or service covered by the specification are stated under this heading, except where it is more practicable to include the information in other paragraphs. Requirements of a general nature should be first, followed by specific requirements for the material. Definitive documents shall be referenced for the material when such documents cover materials of the required quality.

5.3.1.3.2.3.2 Materials: Toxic Products and Formulations Specifications requiring or permitting toxic products and formulations must demand compliance with the requirements of the applicable regulations promulgated by the appropriate federal regulatory agency or the official compendia governing such products and formulations.

5.3.1.3.2.3.3 Electromagnetic Radiation Where applicable, requirements pertaining to electromagnetic radiation are stated in terms of the environment that the item must accept and the environment that it generates.

5.3.1.3.2.3.4 Nameplates or Product Markings The nameplate or markings in some cases may be the only way to identify a product after delivery. Such identification is important from the standpoint of stock, replacements, and repair parts. All requirements having to do with nameplates or markings should be placed under this or another appropriate heading, referencing applicable specifications (for example, MIL-STD-130), drawings, or standards.

5.3.1.3.2.3.5 Workmanship Where applicable, reference to workmanship is stated and includes the necessary requirements relating to the standard of workmanship desired, uniformity, freedom from defects, and general appearance of the finished product. This paragraph is intended to indicate as definitively as practicable the standard of workmanship quality that the product must meet to be acceptable. The requirements must be so worded as to provide a logical basis for rejection in cases where workmanship is such that the item is unsuitable for the purpose intended. Generally, no definite tests other than visual examination of workmanship will be applicable to the requirements of this paragraph.

5.3.1.3.2.3.6 Interchangeability This paragraph specifies the requirements for the level at which components shall be interchangeable or replaceable. Entries in this paragraph are for the purpose of establishing a condition of design and are not to define the conditions of interchangeability that are required by the assignment of a part number.

5.3.1.3.2.3.7 Safety This paragraph specifies requirements to preclude or limit hazard to personnel, equipment, or both. To the extent practicable, these requirements are imposed by citing established and recognized standards. Limiting safety characteristics peculiar to the item due to hazards in assembly, disassembly, test, transport, storage, operation, or maintenance shall be stated when covered neither by standard industrial or service practices nor the system specification. "Fail-safe" and emergency operating restrictions should be included when applicable. These include interlocks and emergency and standby circuits required to either prevent injury or provide for recovery of the item in the event of failure.

5.3.1.3.2.3.8 Human Engineering Human engineering requirements for the system or configuration item are specified herein, and applicable documents (for example, MIL-STD-1472) are included by reference. This paragraph should also specify any special or unique requirements, for example, constraints on allocation of functions to personnel and communications and personnel or equipment interactions. Specified areas, stations, or equipment that require concentrated human engineering attention due to the sensitivity of the operation or criticality of the task, such as those areas where the effects of human error would be particularly serious, should be included.

5.3.1.3.2.4 Documentation (Paragraph 3.4) Where applicable, requirements for documenting the design are specified in general terms in development specifications under this heading. Requirements identify types of documents required for design review and approval, manufacture or procurement, testing, inspection installation, operation, maintenance, and logistic support as appropriate. This paragraph is not intended as a requirement for procurement or delivery of data, which is accomplished by use of DD Form 1423 for DoD contracts.

5.3.1.3.2.5 Logistics (Paragraph 3.5) Where applicable, logistic considerations and conditions that will apply to the system or configuration item should be specified in development specifications and, if applicable, in product specifications. Logistic conditions such as maintenance considerations, modes of transportation, supply system requirements, and impact on existing facilities and equipments are considered.

5.3.1.3.2.6 Personnel and Training (Paragraph 3.6) Where applicable, requirements imposed by or limited by personnel or training considerations are specified in development specifications. Training considerations include existing facilities, equipment, special or emergency procedures (associated with hazardous tasks), and training simulators, as well as the need for additional facilities, equipment, and simulators.

5.3.1.3.2.7 Characteristics of Subordinate Elements (Paragraph 3.7) Subsequent paragraphs may be added as necessary to system, development, or product specifications to specify requirements for subordinate elements of the subject system or configuration item. Requirements for each selected subordinate element are grouped under a major heading titled with the name of the subordinate element and shall include all of the pertinent types of requirements discussed in previous paragraphs for the parent system or configuration item. Requirements imposed directly on the subelement by a requirement on the parent system or configuration item must not be repeated. Allocation or apportionment of a parent system (or configuration item) requirement may be specified for the subelement. Subelements may be functionally or physically integrated portions of the parent system (or configuration item) but would not usually be both in a single specification.

We must be very careful when implementing the subordinate element paragraph that we not include the same requirements in two different documents. It is a cardinal rule of good specification management that each unique requirement appear in only one place in the specification structure.

5.3.1.3.2.8 Precedence (Paragraph 3.8) This paragraph specifies the order or precedence of requirements, such as specification over drawings, functional requirements over physical requirements, adherence to specified processes over other requirements, and the like. The paragraph may also require that the contractor notify the contracting agency of each instance of conflicting, or apparently conflicting, requirements. Alternatively, this paragraph may specify that the requirements of the specification shall take precedence over referenced documents. In system or development specifications, this paragraph should specify the relative importance of requirements (or goals) to be achieved by the design.

5.3.1.3.2.9 Qualification (Paragraph 3.9) Qualification, as used in MIL-STD-490A, refers to the verification or validation of item performance in a specific application. This qualification results from design review, test data review, and configuration audits. Where performance qualification of a design or an end configuration item (including its components) is required, on either a one-time basis or a periodic basis, to achieve design approval, proof of producibility, assessment of production, or other reason, provisions for such qualification testing should be stated in this paragraph. Requirements are included that state the conditions for testing, the time (program phase) of testing, period of testing, number of units to be tested, and other requirements relating to qualification or requalification.

Qualification, as used in Defense Standardization Manual 4120.3-M, refers to the testing or review of test data to judge configuration items from various sources as being suitable for general application and is intended to lead to the establishment of a Qualified Products List (QPL). Therefore, this type of qualification is subject to the provisions of Manual 4120.3-M and is not part of the specification problem covered in this book.

A standard sample is one considered essential to supplement or illustrate certain requirements of the specification. Use of standard samples should be kept to a minimum, because their use can create problems in determining the acceptability of HWCIs subsequently produced. Adequate inspection requires that all requirements be made available, such as the approved tolerances of dimensions, performance, and so on. A standard sample does not provide all this information but must be supported by specification requirements and drawings. The use of the standard sample shall be limited to the illustration of qualities and characteristics that cannot be readily described because detailed test procedures or design data are not available or because certain qualities and characteristics cannot be definitely expressed, such as the texture of fur, the color of cloth, or the grain of wood. Further, the specification should state the specific characteristics and the degree to which these characteristics are to be observed in the standard sample. When a standard sample is to be furnished, it should be so stated in Section 3. Means of obtaining or viewing standard samples would be specified in Section 6.

Where it is essential that a preproduction or periodic production sample, a pilot model, or a pilot lot be tested for design approval prior to or during regular production on a contract or order, the requirements are specified in this section under the appropriate paragraph identification.

5.3.1.4 Section 4: Quality Assurance Provisions

For software, this section is titled "Qualification Requirements" and specifies the qualification requirements, including methods, levels of testing, tools, facilities, test formulas, algorithms, and acceptance tolerance limits required to show that the requirements stated in Sections 3 and 5 have been met. The software requirements and interface requirements specification data item descriptions contain further information for specifying qualification requirements. For software embedded in firmware devices, the application of quality assurance provisions or qualification requirements depends on whether the software is designated as a CSCI or part of an HWCI. When the software is designated as a CSCI, qualification requirements apply, but when it is designated as part of an HWCI, quality assurance provisions apply.

For hardware, this section includes all of the examinations and tests (by reference where applicable) to be performed in order to ascertain that the product, material, or process to be developed or offered for acceptance conforms to the requirements in Sections 3 and 5 of the specification. Section 4 should be arranged in an orderly sequence that will indicate clearly which inspections (examinations and tests)

apply directly to the process, material, HWCIs, or lots of HWCIs that were developed or produced and which apply to requirements such as evaluation, qualification, preproduction sample, pilot model, or pilot lot. The order of presentation of Section 4 material should, insofar as practicable, follow the order of requirements as presented in Section 3 of the specification or, alternatively, in the most logical order of conducting the examinations and tests listed.

5.3.1.4.1 General Where applicable, the general test and inspection philosophy should be described with a statement of responsibility for inspection, classification of examinations and tests, sampling, lot formation, and other information pertinent to the quality assurance provisions but not directly associated with a specific test or examination.

5.3.1.4.2 Responsibility for Inspection The DOD concept of quality assurance places primary responsibility for quality assurance of delivered products, materials, or services on the supplier who is responsible for offering to the contracting agency only those products, materials, or services that conform to all specified requirements. In system specifications, however, where assembly of the system or segment is at a government facility or on a government-owned vessel involving government-furnished property and personnel, responsibility for the conduct of tests will probably be split between the contracting agency and the contractor. Accordingly, the supplier's responsibility for inspection must be clearly stated and the contracting agency's role, either as a partner or monitor, must also be clearly specified. A typical statement of responsibility is:

4.1.1 *Responsibility for inspection.* Unless otherwise specified in the contract or order, the supplier is responsible for the performance of all inspection requirements as specified herein. Except as otherwise specified, suppliers may utilize their own facilities or any commercial laboratory acceptable to the contracting agency. The contracting agency reserves the right to perform any of the inspections set forth in the specification where such inspections are deemed necessary to assure that supplies and services conform to prescribed requirements.

5.3.1.4.3 Special Tests and Examinations Any special tests and examinations or associated actions required for sampling, lot formation, qualification evaluation, and the like are covered under an appropriate heading. Two examples follow:

4.1.2 *Preproduction sample, pilot model, or pilot run.* When Section 3 specifies a requirement for preproduction sample, pilot model, or pilot run, Section 4 shall include, under an appropriate identification, a description of the testing routine, sequence of tests, number of units to be tested, data required, and the criteria for determining conformance to specified requirements.

or

4.1.2 *Qualification provisions.* When the requirements for HWCIs covered in Section 3 contain a qualification provision, the applicable examinations and tests shall be listed under appropriate headings in Section 4.

These inspections should be specified for initial and higher levels (reliability levels) of qualification, including the test methods for continuous testing, and periodic qualification reevaluation, as covered in Section 3 of the specification.

When a tabular form of presentation will provide a better understanding of the correlation between tests of Section 4 and requirements of Section 3, or would clarify the test requirements for acceptance, performance, qualification, preproduction, and so on, a tabular presentation similar to that below can be made.

Test Procedure

Requirement	Pre-prod	Acceptance	Periodic prod
3.3.1	4.2.1	4.2.1	4.2.1
3.3.2.1	4.2.2.1		
3.3.2.2	4.2.2.2	4.2.2.1	4.2.2.1
3.3.2.3	4.2.2.3		
3.3.2.4	4.2.2.3	4.2.2.3	4.2.2.3
3.3.3.1	4.2.3.1	4.2.3.1	4.2.3.1
3.3.3.2	4.2.3.2	4.2.3.2	

5.3.1.4.4 Quality Conformance Inspections This section lists all examinations and tests required to verify that all requirements of Section 3 and 5 have been achieved in the HWCI, material, or process offered for acceptance. These examinations and tests include or reference the following, as appropriate:

1. Tests and checks of the performance and reliability requirements
2. A measurement of comparison of specified physical characteristics
3. Verification, with specific criteria, for quality of the work
4. Test and inspection methods for assuring compliance, including environmental conditions for performance
5. Classification of characteristics as critical, major or minor, as defined in MIL-STD-109. When required for reference purposes in reporting inspection results, the characteristics may be numbered. When numbered, numbers should be in accordance with the following:

1 through 99	Critical Characteristics
101 through 199	Major Characteristics
201 through 299	Minor Characteristics

5.3.1.5 Section 5: Preparation for delivery

5.3.1.5.1 General This section is generally applicable to product specifications only, and it should include applicable requirements for preservation, packaging, and packing the configuration item, as well as markings of packages and containers.

This section states the general requirements for preservation, packaging, packing, and package marking. If more than one level of preservation and packaging is included, the conditions for selection of levels should be explained. See FED-STD-102 for DoD contracts. The specific requirements for materials to be used in preservation, packaging, and packing a product are covered here either directly or by reference to other specifications, publications, or drawings.

5.3.1.5.2 Detailed Preparation Requirements may be included by reference to other specifications and applicable standards or, where these do not exist or are not applicable, by detailed instructions. The requirements are included with appropriate headings, as required, for disassembly, cleaning, drying, preservation, packaging, packing, and shipment marking. These requirements must be specifically related to each required level of preparation in a manner that will leave no doubt regarding requirements applicable to such level. Detailed preparation for delivery requirements should be covered as far as practicable in three basic categories.

5.3.1.5.2.1 Preservation and Packaging The requirements for preservation and packaging cover cleaning, drying, and preservation methods adequate to prevent deterioration; appropriate protective wrapping; package cushioning; interior containers; and package identification-marking, up to, but not including, the shipping container. Where no suitable reference is available, step-by-step procedures for preservation and packaging shall be included.

5.3.1.5.2.2 Packing The requirements for packing cover the exterior shipping container, the assembly of configuration items or packages therein, and any necessary blocking, bracing, cushioning, and weatherproofing.

5.3.1.5.2.3 Marking for Shipment Normally, marking requirements are established by reference to MIL-STD-129. Markings essential to safety and to the protection or identification of the configuration item that are not required by MIL-STD-129, or are required on a "when specified" basis by that standard, are specified in detail under this heading. In any instance where reference to MIL-STD-129 is not applicable, requirements in detail or by reference to recognized documents should include: appropriate identification of the product, both on packages and shipping containers; all markings necessary for delivery and for storage, if applicable; all markings required by regulations, statutes, and common carriers; and all markings necessary for safety and safe delivery.

5.3.1.6 Section 6: Notes

Section 6 of specifications contains information of a general or explanatory nature, and no requirements should appear therein. It contains information, not contractually binding, designed to assist in determining the applicability of the specification and the selection of appropriate type, grade, or class of the configuration item, such as additional supersession data, changes in product designations (grades, class, and the like), standard sample (if required), and so on. This section should include the following, as applicable, in the order listed:

1. Intended use
2. Ordering data
3. Preproduction sample, pilot model, or pilot lot, if any
4. Standard sample, if any
5. Definitions, if any
6. Qualification provisions
7. Cross reference of classifications
8. Miscellaneous notes

5.3.1.6.1 Intended Use Information about the use of the configuration item covered by the specification should be included under this heading. The differences among types, grades, and classes in the specification shall be explained herein. If particular applications exist for which the material is not well adapted, this information also may be included.

5.3.1.6.2 Ordering Data Detailed information to be incorporated in invitations for bids, contracts, or other purchasing documents is stated in this paragraph. Reference shall be made to all parts of the specification where it is required that options be exercised, such as requirements for a preproduction sample for qualification; selection of grade, type, class, level of preservation and packaging; and so on.

5.3.1.6.3 Instructions for Models and Samples If Section 3 specifies a preproduction sample, a pilot model, or a pilot lot, the necessary instructions for the arranging for its examination, test, and approval are stated in this section under an appropriate paragraph identification.

If Section 3 specifies a standard sample, information for obtaining or examining the standard sample (source and address) is stated under this paragraph identification.

5.3.1.6.4 Qualification Provisions Where provisions for qualification of a product are a requirement of the specification, information concerning such qualification is stated in this section.

5.3.1.6.5 Cross-Reference of Classifications A cross-reference of old to new classification (types, grades, classes, and the like) of the configuration item, material, or service should be included if such changes are made by specification revision. If new classes, grades, or types of configuration items or materials are being added to, and others are being removed from the coverage of the specification, a cross-reference showing substitutability relationships should be included.

5.3.1.6.6 Miscellaneous Notes In Chapter 2.2 we covered some requirements about interrelationship information (rationale and sourcing) that can be included in Section 6 of the specification under miscellaneous notes. This section can also be used for a temporary paragraph and table listing immature requirements during the initial preparation of a specification. This is accomplished by placing a TBD-N remark in the text where information is missing and including a table of TBDs in Section 6 correlated to responsible engineers and planned closure dates.

A traceability matrix could be included that lists all of the requirements in Sections 3 and 5 along with the requirements for other items to which they trace. It is more satisfactory to report traceability to a customer in a separate requirements analysis report if required. A verification matrix may be included here, though it would be more appropriate in Section 4.

5.3.1.7 Appendix and Index

Where required, appendixes and an index may be included as integral parts of a specification. An appendix, identified by the heading "Appendix," is a section of provisions added at the end of a specification. An appendix may be used to append large (multipage) data tables, plans pertinent to the submittal of the configuration item, management plans pertinent to the subject of the specification, classified information or other information or requirements related to the subject configuration item, or materials or processes that would normally be invoked by the specification but would, by its bulk or content, tend to degrade the usefulness of the

specification. In all cases where an appendix is used, reference to the appendix shall be included in the body of the specification.

5.3.1.7.1 Appendix Numbering Appendixes to a specification are numbered as Sections 10, 20, and so on, in multiples of 10 for each succeeding appendix. Divisions and paragraphs within an appendix are similarly numbered, such as 10.1, 10.1.1, and so on. Page numbers for the appendixes normally will be consecutive and in sequential order, with the page numbers used throughout the specification. Each page of the appendix should be identified with the specification number as in the specification.

5.3.1.7.2 Scope An appendix may include a statement of scope to indicate the limitations of the appendix and to insure its proper application and use.

5.3.1.7.3 Headings Headings should be used as necessary but need not duplicate the structure of the specification of which the appendix is a part.

5.3.1.7.4 References References that may be required and that relate to the appendix are listed in Section 2 of the basic specification and may also be listed in a section of applicable documents in the appendix itself.

5.3.1.7.5 Index An alphabetical index may be placed at the end of a specification to permit ready reference to contents. It should only be used in lengthy specifications.

5.3.2 MIL-STD-961D Content Standard

When MIL-STD-490A was discontinued, MIL-STD-961D was changed by Notice 1 to include Appendix A, continuing a modified version of MIL-STD-490A content preserving DoD coverage for program-unique military specification content. This change was coordinated with the DoD movement to what it called "performance specifications." As part of several changes driven by reduced DoD resources following the Soviet Union's removal from the global stage, the Department concluded that its specifications and standards unnecessarily constrained its suppliers and contractors, telling them how to design or build the product rather than what DoD wanted or needed.

The structure is similar to 490A, with a few notable exceptions. Many kinds of requirements were moved under the performance requirements paragraph heading to encourage the capture of all of the kinds of requirements DoD felt should be included in a performance specification. Section 3 also included a paragraph for traceability not included in 490A, reflecting increased interest in this concept on DoD programs. Specification Section 4 was renamed "Verification" from "Quality Assurance Provisions."

Significant differences appeared in the names of the kinds of specifications. Instead of development and product (or Part I and Part II) specifications, 961D with Notice 1 referred to performance and detail specifications. It identified only five kinds of specifications related to the level and application: system, item, software, material, and process.

MIL-STD-490A included the caution against including cost requirements in a specification with the rationale that it was a contractual matter. This was probably appropriate during the days of cost as a dependent variable, dependent on whatever it took to satisfy the performance requirements.

But now DoD, strapped for funding, had to recognize that cost had to be treated as an independent variable and that suppliers would have to trade cost right along with performance parameters. MIL-STD-961D did not include this restriction, but neither did it include a paragraph heading for cost, even though affordability had become a critical criterion in the acquisition process. The new standard (961 had been in existence for years; but, with Notice 1 to the D Revision, it applied to program-peculiar specifications for the first time) did recommend that Section 5, packaging and shipping, be a void (no content). The rationale was the same as the prior restriction on cost—that it was a contractual matter. The author does not like this restriction because often the packaging of equipment shipped from suppliers was inadequate to preclude damage. There is less chance of this occurring if these matters are treated as item requirements during the development of the item specification. If there is no Section 5, one might ask why DoD retained a six-section specification format. Why not go to a five-section specification? The tradition of six-section specifications was probably too strong.

5.3.3 MIL-STD-961E Content Standard Delta

Notice 1 to Revision D was rushed into publication in August 1995 to cover the change to performance specifications—perhaps too fast. Revision E cleaned up some of the loose ends and made available a more cleanly integrated standard, covering all military specifications, including program-peculiar ones that this book is primarily focused on rather than general specifications that MIL-STD-961 had always covered when MIL-STD-490A applied to program-peculiar ones.

To understand the MIL-STD-961E preferred paragraph structure for program-peculiar performance specifications, one must consult data item description DI-SDMP-81493A, which calls for compliance with paragraphs 4.1 through 5.14 of MIL-STD-961 and DI-SDMP-81565A on performance specifications, which reference essentially the same standard paragraphs. This is supplemented by an example of a performance specification format in Figure 6 of the standard.

5.3.4 Software Specification Standards

5.3.4.1 Military Standards

Technically, MIL-STD-490A and MIL-STD-961D/E covered both hardware and software specifications but never very satisfactorily. Several other standards evolved as software became an increasingly important aspect of military products. Some of these, including DOD-STD-2167A, in my opinion, were not very satisfactory either. MIL-STD-498 evolved as an exception to this trend. It was approved for an interim period until an adequate commercial standard could be created, reviewed, and issued. It replaced three standards of questionable value: DOD-STD-2167A, DOD-STD-7935A, and DOD-STD-1703. Data item descriptions (DID), coordinated with 498 content, defined a series of specification documents described below largely using quotes from those documents. EIA J-STD-016 was prepared as the commercial standard and is almost word-for-word from MIL-STD-498, no longer maintained by DoD, except that some of the harsh-sounding terms, such as *computer software configuration item (CSCI)*, are softened somewhat or eliminated.

5.3.4.1.1 Software System Specification While it is difficult to understand how one could realize a software system specification in that software must run on a computer that is a hardware device, if one places the boundary at the software extent, then perhaps it would be possible to characterize that entity as a system. The military structure of system specification, no matter whether hardware or software, is essentially the same.

5.3.4.1.2 Software Requirements Specification (SRS) A software requirements specification specifies the requirements for a computer software configuration item (CSCI) and the methods to be used to ensure that each requirement has been met. Requirements pertaining to the CSCI external interfaces may be presented in this specification or in one or more interface requirements specifications referenced in this specification. The DID for this document encourages a six-part specification, as outlined below:

1. Scope
2. Referenced document
3. Requirements
4. Qualification provisions
5. Requirements traceability
6. Notes

The requirements section encourages use of a modeling environment for identifying required states and modes from which performance requirements may be derived. This represents a tremendous improvement in specification structures and is reflected also in the MIL-STD-961E structure. Some hardware-dominated system engineers complain about software domination of the hardware and system specification structure, but the author believes we should get over it because the changes were primarily positive.

5.3.4.1.3 Software Product Specification (SPS) The software product specification contains or references the executable software, source files, and product support information, including "as-built" design information and compilation, build, and modification procedures, for a computer software configuration item (CSCI). The SRS is the performance or development specification (Part I) for a software entity, and this document provides the detail or product specification (Part II) of the pair.

5.3.4.1.4 Interface Requirements Specification (IRS) The interface requirements specification specifies the requirements imposed on one or more systems, subsystems, hardware configuration items (HWCI), computer software configuration items (CSCI), manual operations, or other system components to achieve one or more interfaces among these entities. The document is structured like the SRS in six sections. Section 3 of the IRS identifies all interfaces for the software entity and clearly defines all requirements corresponding to each interface.

5.3.4.2 Commercial Standards

At the time this book was being written, sound commercial standards were emerging from several societies joined together to craft American and international standards organization (ISO) standards. DoD now shows a preference for commercial standards and, indeed, does not have a standard for computer software development, depending on EIA J STD 016.

5.3.5 A Standard for the Ultimate Simplicity

Chapter 2.1 of this book introduced a specification structure embodying the ultimate simplicity, called a "concept requirements list." The list is formed of primitive requirements statements in no particular order but numbered. Each statement consists of four fundamental components: attribute or characteristics to be controlled; a relationship between the characteristic and the value, such as "equals" or "greater than" or "equal to"; a numerical value; and units appropriate to the characteristic and value. Some characteristics will be difficult to phrase in this fashion, but even some qualitative requirements can be placed in this format using nonmathematical relationships. An example of a requirement stated in a primitive format is, "weight ≤ 132 pounds." It is not necessary to use capital letters, punctuation, or complete sentence structures with this form of requirements statement.

This form of documentation is encouraged in the earliest program phases populated by very creative engineers and analysts who commonly do not respond well to a need for reading or writing voluminous specifications rigorously responsive to an enforced standard. These same people need to know what these very simple requirements statements communicate so powerfully, however, and can generally be relied upon to cooperate in their identification. At some point in the program evolution, these simple statements will probably have to be fashioned into complete sentences, easily done, and dropped into a formal document structure.

5.3.6 An Updated Content Standard

The specification format definition contained in MIL-STD-490A evolved over a period of decades, with little relationship to the evolving and improving requirements analysis process accomplished to provide the content. When writing *System Requirements Analysis*, the author was surprised to find that the map between the MIL-STD-490A paragraphing structure and the contents of a sound requirements analysis toolbox of techniques for identifying good specification content was a crazy quilt. The MIL-STD-490A specification format predates efforts to organize the process of writing specification content. It was crafted at a time when specifications were typeset or typewritten. The author tried to realign the analysis process components to preserve the 490A structure with a resulting undesirable analysis process complexity.

Figure 5.3-1 offers the author conclusions about an optimum organization of section 3 paragraphing for a revised standard encouraging coordination with the requirements analysis components covered in this volume. Figure 5.3-2

PARA	PARAGRAPH TITLE	MODELING METHOD	
3.1	Performance	STRUCTURED ANALYSIS	
3.2	Interface	INTERFACE ANALYSIS	PRODUCT DESIGN CONSTRAINTS ANALYSIS
3.3	Environmental	ENVIRONMENTAL ANALYSIS	
3.4	Specialty Engineering		
3.4.1	Physical Characteristics		
3.4.2	Reliability	SPECIALTY ENGINEERING ANALYSIS	
3.4.3	Maintainability		
3.4.4	Interchangeability		
3.4.5	Portability and Transportability		
3.4.6	Human Engineering		
3.4.7	Safety		
3.4.8	Producibility		
3.4.9	Parts		
3.4.10	Materials and Processes		
3.4.11	Identification and Marking		
3.4.12	EMI/EMC		
3.5	Product Requirements		
3.5.1	Cost	DTC/LCC ANALYSIS	PROCESS DESIGN CONSTRAINTS ANALYSIS
3.5.2	Product Schedule	SCHEDULING ANALYSIS	
3.6.3	Manufacturing Requirements	MANUFACTURING ANALYSIS	
3.6.4	Test and Evaluation	TEST PLANNING	
3.6.5	Procurement Process	MATERIAL ANALYSIS	
3.6.6	Quality Process Requirements	QUALITY ANALYSIS	
3.6.7	Item Operation and Logistics Requirements	LOGISTICS SUPPORT ANALYSIS	

Figure 5.3-1 *Updated specification paragraphing coordination*

provides a partial example of this structure for a hypothetical item called a Framsat Widget. You may, of course, add or subtract from the specialty engineering characteristics as a function of the product line. You will note that all of the requirements corresponding to a particular modeling method or tool set are contiguous in the specification content, not scattered about in the structure.

Paragraphs 3.1 through 3.4 are intended to be focused on development or performance requirements in the context of MIL-STD-961D/E. Paragraph 3.5 is provided for

---

1.   IDENTIFICATION

1.1   Scope

This document defines the requirements for the design, production, test, and logistics support of the Framsat Widget.

1.2   Item Function

The function of the Framsat Widget is to fill the emptying fuel tank during mission performance with inert pressurant media.

1.3   Architecture

Item architecture in accordance with Figure 1-1.

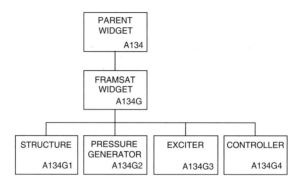

Figure 1-1 Item Architecture Block Diagram

2.   APPLICABLE DOCUMENTS

Documents listed in subordinate paragraphs form a part of this requirements document to the extent tailored (if applicable) and appropriate to the specific application of the document within the context of its reference in this document. Documents referenced in listed documents provide guidance only unless they are also listed herein. Paragraphs of this document where each applicable document is called out (first instance) are included in parentheses.

2.1   Government Documents

2.1.1 WIMSICAL 23-203

Christmas Island Launch Site Safety Criteria (5.4.4.1). Tailor the document as follows: (1) delete paragraphs 3.4.5 through 3.7 and (2) delete Appendices A and D.

2.1.2 CDD-10-19000 Cleanroom Requirements (3.6.3.3)

2.2   Non-Government Documents

2.2.1 133-109 Cleanroom Parts Entry Preparation (3.6.3.3)

2.2.2 CC-45-9090 Wamsat Program Verification Plan (4.1)

Figure 5.3-2   *Sample updated specification, sheet 1 of 7*

augmentation of this structure to include product requirements (detail requirements in MIL-STD-961D/E).

Paragraph 3.6, "Programmatic Requirements," is an attempt to give the development team a place to capture all item requirements, including those that actually define interfaces with the development process. These requirements could be allowed to flow out to other program documentation before printing the product item specification for final release; or they could be kept in the specification throughout its life cycle.

---

3.   REQUIREMENTS

3.1 Performance

3.1.1 Timeline

Framsat functional sequence and period of performance during mission as shown in Figure 3-1.

TBD-1

Figure 3-1 Framsat Widget Timeline

3.1.2 Pressurant Rate

Item shall generate TBD-2 pounds of pressurant in a 5 minute period.

3.2 Interface

Item interfaces are as defined in Figure 3-2. Interface 4 is the item internal interface defined in lower tier documents, if applicable.

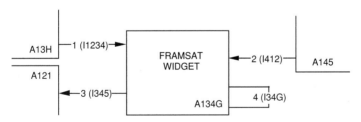

Figure 3-2 Item Schematic Block Diagram.

3.2.1 Interface 1, Framsat Widget Electrical Power Input

3.2.1.1 Input Voltage

Electrical power input voltage shall be equal to 28 VDC plus or minus 0.5 VDC.

3.2.1.2 Redundancy

Redundant power supply connectors shall be used to guard against connector or human failure.

3.2.1.3 Electrical Load

Framsat electrical load shall be less than 530 watts.

3.2.2 Interface 2, Framsat Widget Exciter Signal.

Figure 5.3-2   *Sample updated specification, sheet 2 of 7*

(Continued)

3.2.2.1 Exciter On Command

Exciter on signal corresponds to a +5 VDC +/- 0.5 VDC voltage applied to Framsat Widget input connector terminal.

3.2.2.2 Exciter Off Command

Exciter off signal corresponds to open circuit applied to Framsat Widget.

3.2.3 Interface 3, Framsat Widget Output Pressurant

3.2.3.1 Steady State Pressurant Flow Rate

Pressurant flow shall be greater than 100 pounds per minute steady state.

3.2.3.2 Time to Flow Stability

Pressurant flow shall reach a steady state level within 5 seconds of exciter signal application.

3.2.3.3 Flow Stability Duration

Pressurant flow shall continue at the steady state level for a time greater than 5 minutes.

3.3 Environmental

Item shall satisfy performance characteristics subsequent to and while subjected to environmental conditions defined in subordinate paragraphs.

3.3.1 Atmospheric Pressure

Item shall operate in an ambient pressure range of 0.1 through 1.5 of normal sea level atmospheric pressure.

3.3.2 Vibration

3.4 Specialty Engineering

3.4.1 Physical Characteristics

3.4.1.1 Mass Properties

3.4.1.1.1 Space

Item shall fill a space less than 125 cubic inches.

3.4.1.1.2 Form Factor

Item form factor shall be between 1:1:1 through 1:3:2 (H:W:D, H in vertical axis of item while in launch orientation).

3.4.1.1.3 Maximum Height

Item maximum height shall be less than or equal to 8 inches.

3.4.1.1.4 Weight

Item weight shall be less than or equal to 56 pounds.

3.4.2 Reliability

Item reliability shall be greater than or equal to 0.988.

3.4.3 Maintainability

Figure 5.3-2   *Sample updated specification, sheet 3 of 7*

3.4.3.1 Refurbish Time

Item refurbish time shall be less than 6 hours with one technician.

3.4.3.2 MTTR

Item MTTR shall be less than or equal to 30 minutes.

3.4.4 Interchangeability

3.4.5 Portability and Transportability

3.4.6 Human Engineering

3.4.7 Safety

Item design shall comply with WIMSICAL 23- 203 as tailored.

3.4.8 Producibility

NOTE

This paragraph is concerned with product requirements driven by ease of manufacturing concerns. See paragraph 3.6.3 for manufacturing requirements driven by product features.

3.4.9 Parts

3.4.10 Materials and Processes

3.4.11 Identification and Marking

3.4.12 EMI/EMC

3.5 Product Requirements

To be supplied in accordance with requirements schedule in IMS.

3.6 Programmatic Requirements

3.6.1 Cost

3.6.1.1 Design To Cost (DTC)

Item DTC shall be less than or equal to $ 12,500.

3.6.2 Schedule

Figure 5.3-2 *Sample updated specification, sheet 4 of 7*

(Continued)

3.6.2.1 Development Schedule

Figure 3-2 provides the item development schedule. Note the need for completion of DET by M Day 1200.

TBD-3

Figure 3.2 Item Development Schedule

3.6.3 Manufacturing Requirements

This paragraph defines manufacturing processes requirements that are driven by the design that have to be accounted for in manufacturing facilities development, planning cards, and tooling. These requirements may be deleted from this document as they transition to other documents more appropriate to their capture.

3.6.3.1 Manufacturing Facility

3.6.3.1.1 Plant Facilities

Plant 13 line 2 will be the manufacturer of the item. Addition of 32 square feet of floor space to station 5 required to support filter test bench.

3.6.3.1.2 Cleanroom

Final assembly shall require a cleanroom satisfying CDD-10-19000 requirements. All parts entering final assembly shall be cleaned and bagged in accordance with company procedure 133-109.

3.6.3.1.3 Process Effects

Planning cards must cover coating threads of cover plate screws with anti-seize compound.

3.6.3.2 Tooling

The flaxen spacer grove will require use of a conical shaper machine in two inverse runs.

3.6.4 Test and Evaluation

3.6.5 Procurement Process

3.6.5.1 Long Lead Items

The following item components have been identified as long lead: electrical connectors (3 each), high precision resistors (3 values) and capacitors (2 values), and a special lock screw required by the customer for compatibility with technician tool sets in use.

3.6.6 Quality Process Requirements

This paragraph defines quality engineering process requirements driven by product characteristics. Content may be deleted from this paragraph if and when it is moved or integrated into quality documentation.

Figure 5.3-2 *Sample updated specification, sheet 5 of 7*

3.6.7   Item Operation and Logistics Requirements

This paragraph defines user process requirements driven by product characteristics. Content may be deleted from this paragraph if and when it is moved or integrated into logistics documentation.

3.6.7.1   Preventive Maintenance Requirement

Replace item filter after 120 hours of operation.

3.6.8   Documentation

4.   QUALITY ASSURANCE

4.1   Compliance Matrix

Table 4-1 defines the methods to be used to verify design compliance with listed requirements. Refer to the Wamsat Program Verification Plan for compliance matrix and references to supporting data.

Table 4-1 Requirements Verification Methodology Matrix

PARA NBR	REQUIREMENT TITLE	VER EVNT NBR	VERIFICATION METHODS			
			A	T	I	D
3.4.2.1	MTBF	1	X			
		2				X

4.2   Verification Events

4.2.1   MTBF

Predict failure rates of item components, combine to yield item MTBF, and convert to a reliability figure to verify compliance with paragraph 3.4.2.1.

5.   PREPARATION FOR DELIVERY

Not Applicable

Figure 5.3-2   *Sample updated specification, sheet 6 of 7*

(Continued)

6 NOTES

6.1 Requirements Maturation

Requirements with immature values denoted by "TBD-n", where "n" is a number, are listed below followed by planning data on resolution of TBD status by a particular person by a given date. The paragraph shall be replaced with the word "Reserved" and the table deleted when all requirements have matured.

Table 6-1 TBD Resolution Control Table

TBD NBR	TBD TITLE	PARA NBR	PRINCIPAL ENGINEER	DUE DATE
1	Timeline	3.1.1	Jones	10-25-91
2	Pressurant Rate	3.1.2	Burns	11-02-91
3	Item Dev Schedule	3.6.2	Adams	11-15-91

6.2 Requirements Traceability, Source, and Rationale

Table 6-2 links requirements to sources, rationale and parent item requirements.

Table 6-2 Requirements Traceability Matrix

PARA NBR	ITEM REF	PARA REF	RATIONALE	SOURCE REFERENCE
3.4.2.1	A134	3.6.1	Allocation	R&M AAA Report

Figure 5.3-2   *Sample updated specification, sheet 7 of 7*

5.4

Applicable Document Analysis

Contents

5.4.1 Introduction to Applicable Documents

5.4.1.1 Applicable documents defined

There are many ways to categorize specifications, as we have seen. The most fundamental category depends on the program applicability of a document. A specific requirements document may apply only to the products and services of a single program or system, in which case it is a program-unique document. On the other hand, a document may contain requirements of a generic nature so as to apply to a broad class of things that could be acquired under a number of different programs for the specific needs of those programs. It is common practice to refer to these generic requirements documents in program-unique documents as a way of applying these generic requirements to a program without having to include them all in one or more program-unique documents. The program-unique document simply references the generic document in a statement such as, "The design of pressurized storage vessels shall comply with MIL-X-XXXX." This simple statement causes all of the content of this document to apply to the design of any pressure vessels needed on the program.

The utility of these generic requirements documents is pretty obvious. It is far cheaper to prepare a program-unique specification by referring to a standard set of requirements than to perform a requirements analysis for each program to determine appropriate values independently. If each program under DoD acquisition authority developed the requirements now contained in military standards and specifications independently, the cost of all of the independent requirements analysis work would be repeated on each program and DoD would not realize a standard approach to solving some of these requirements problems. At the same time, the unrestricted use of generic documents called out in program-unique specifications can add tremendously to the cost of an overall system development effort. In this chapter, we will explore a solution to this dilemma by tailoring of these generic requirements documents to alter the content to that which is really required to satisfy the particular customer's needs.

Applicable documents are documents referenced in a specification Section 2 (titled "Applicable Documents") or the statement of work so as to apply to the contractor's performance on a contract. These documents may apply in total, apply only to the extent called out in the program peculiar document, or be tailored by editing to change the content.

Before a program begins to generate specifications, someone must be assigned to conduct a structured applicable document assessment resulting in the: (1) identification of a preferred set of applicable documents where the customer does not provide a list, or (2) identification of tailoring or streamlining actions that will permit the contractor to satisfy customer needs while pursuing standard contractor procedures. This section offers a standard, structured, systematic process for accomplishing the assessment of applicable documents, reporting results, and maintenance of an applicable document baseline.

5.4.1.2 Bidirectional tailoring

The objective in applying a systematic approach to applicable document assessment is a clear, mutual understanding between the contractor and the customer about required compliance with cited reference documentation so that the contractor proposal for accomplishing the contract task is in concert with a cost estimate that is responsive to customer needs, fair, and competitive.

The desired result of this analysis is a condition of consistency between customer requirements and the company approach to satisfying them. This may be achieved by: (1) tailoring the customer-applicable documents; (2) tailoring or modifying the way you do business, as defined in company procedures manuals; or (3) some combination of these actions. The latter is the normal condition. The tailoring of your procedures is implied by your acceptance of applicable documents that are in conflict with your procedures.

It is important to recognize this effect, even though we may not formally go through the motions to physically tailor the internal procedures. A sound method of tailoring company practices for a particular program application is through preparation of a program-peculiar procedure contained in a program directives manual or a program-specific plan. It is especially important to communicate the effects of this tailoring to personnel working on the program who may otherwise follow normal internal procedures as closely as possible, thinking that is the correct thing to do.

Every company should develop a set of internal procedures consistent with their product line, facilities, and workforce. This set of procedures results in a quality product at a minimum cost. Where possible, a company should wish to follow these procedures to permit them to offer the most competitive price for their product. To the extent that they can reach agreements with the customers to permit use of internal procedures to satisfy tailored applicable documents, the contractor will succeed in the goal of satisfying customer needs at a competitive price and frequently be selected to do so. In the process, the company personnel will be able to follow internal practices on all programs, encouraging continuous improvement of capabilities through common practice repetition.

Every contractor organization should have in place a continuous process improvement program, actively supported by their top management, that results in a continuous review of lessons learned from programs and improvement of their internal methods. The continuous process improvement activity should work hand-in-glove with the program-applicable document tailoring efforts. Programs provide a company with the opportunity to experiment with different practices and determine, based on experience and information, whether or not to implement these changes.

5.4.1.3 Document tailoring

Applicable document assessment and tailoring is appropriate for all phases of the DOD and NASA development cycle and may also apply to commercial sales. While all program phases will benefit from the methods covered here, the most critical application is in preparation for full-scale development. Normally, the assessment process will be accomplished in association with a proposal, but at least a part of the job may continue into the contract period of performance. Engineering and contract change proposals (ECP and CCP) may provide reasons for identification of additional applicable documents, and they should be subjected to the same assessment process as that described here.

Applicable documents cited in the statement of work generally apply against the development process employed by the contractor to design and manufacture a product covered by the contract. Applicable documents cited in the system specification (or other top-level specification) generally apply to the product developed. Some of these documents apply to both the development process and the product and, as a result, may be listed in both the statement of work and the system specification. As a rule, question the appropriateness of listing any document in both places.

There are three classes of applicable documents: (1) compliance, (2) guidance, and (3) self-imposed. The contractor must fully comply (to the extent tailored) with applicable documents identified in a statement of work or top-level specification as compliance documents. The contractor should make every reasonable effort to comply with documents identified as "for guidance" in statements of work or top-level specifications. Self-imposed applicable documents are applicable documents other than those identified by the customer in the statement of work or top-level specification that the contractor chooses to impose on themselves on lower-tier system elements in the interest of good engineering practices that can be shown to be cost effective.

In all cases, it is desirable to clearly tailor applicable documents for the precise meaning intended for a given contract. But, it is imperative that compliance and self-imposed documents imposed on the contractor by the customer and imposed by the contractor on the subcontractors be carefully analyzed for tailoring alternatives. The general term *applicable documents* will be used throughout this section, but the procedures really apply most forcefully to these two classes of applicable documents.

5.4.1.4 Applicable document levels

First-tier applicable documents are those that are called out directly in the top-level specification or statement of work. Lower-tier applicable documents are those that are identified in first-tier applicable documents that are not also first-tier applicable documents in their own right. At one time, the government required contractors to comply with all applicable documents referenced in other applicable documents, no matter how convoluted the interreferencing became. The current government policy, driven by a clear understanding of the unnecessary cost that can be added to a program by unrestricted applicability, is covered in DoD Directive 5000.43, "Acquisition Streamlining." MIL-HDBK-248A, "Guide for Application of Tailoring of Requirements for Defense Material Acquisitions," provides government guidance on the streamlining of applicable documents.

DoD contractors should fashion their policy for consistency with current government policy, which states that second-tier and lower-tier applicable documents apply as guides. This means that one should make every reasonable effort to follow them as long as doing so will not result in a program cost or schedule risk. It is unnecessary to prepare tailoring statements for the ways in which we fail to satisfy guidance documents. But each of these cases should be discussed openly with the appropriate customer representative and, in some cases, be formally agreed upon through contract letters.

5.4.1.5 DoD policy changes

In the 1990s, many changes were implemented in DoD to streamline acquisition, and one of the big ones was to dramatically reduce dependence on military specifications and standards. The government eliminated many documents that had not been called on a program in prior years or were involved in manufacturing or management activities. With these went many of importance in the system engineering field. The intent was to move toward commercial specifications in these areas and avoid the cost of maintaining these documents.

This policy was extended to many other areas as well. Military specification parts, important when it required special attention to produce electronic parts of the quality needed for military systems, were found to be unnecessary in many applications because commercial parts production had improved so significantly. Also, the use of electronic parts had undergone a tremendous change with only a small fraction of the total going into military equipment, leading to little DoD leverage over parts production.

5.4.1.6 Definitions

Specification terms used in this volume are subject to interpretation. Some of these terms are defined in Table 5.4-1.

5.4.2 Initiation of the Program-Applicable Documents List

5.4.2.1 An enterprise-applicable documents list (EADL)

Your company should consider developing and maintaining an enterprise-applicable documents list for use on programs. This list provides a streamlined set of government and industry standards and specifications that are consistent with internal company procedures. If a customer accepted this list in total, your company would be able to completely satisfy all customer requirements while operating in response to its own internal procedures. This document may also be used as a basis for early procurement action with suppliers.

The enterprise-applicable documents list should be maintained by the program integration team (PIT) and made available to programs in both hard copy and electronic media. Someone should be picked from the functional department responsible for requirements and specifications to lead the performance of applicable document assessment. This person has been called the assessment coordinator in the remainder of this section, or simply the coordinator.

When a program is initiated without a customer-supplied applicable documents list, the program team should begin with the enterprise-applicable documents list as a basis for estimating the contract task. Simply use an electronic copy of the report containing the list as a boilerplate and edit the cover page and text matter into the program-applicable documents list. When a program is initiated with a customer-supplied list of applicable documents in a statement of work and/or a top-level specification, those documents should be copied over into the boilerplate program-applicable documents list (PADL) format. In either case, the coordinator will end up with a list of candidate-applicable documents with which to initiate the assessment process.

5.4.2.2 Applicable document assessment sources

The input to the assessment process may be from one of two sources: (1) the enterprise-applicable documents list, or (2) a list supplied by the customer in a statement of work and top-level specification Section 2, as noted above. In either case, the same process described in this section is applied to the resultant program-applicable documents list, which initially contains only candidate-applicable documents.

The normal mode of organization for applicable document assessment entails coordination of the effort by an assessment coordinator, or coordinator. The coordinator will develop a list of all documents requiring assessment and assign (with the cooperation of supervision) a principal

Table 5.4-1   *Definitions*

Term	Explanation
Applicable Document	MIL-STD-490A and MIL-STD-961D Appendix A both define specification section 2 as "Applicable Documents." Documents listed in section 2 of a specification are referenced somewhere in the document as applying to the item covered by that specification. These documents are referenced in the statements of work and system specification to avoid repetitive definitions of standard requirements common to many programs and systems. There are three kinds of applicable documents: (1) compliance, (2) guidance, and (3) self-imposed.
Compliance Document	An applicable document identified by the customer in the statement of work or top-level specification that the contractor must comply with in the performance of a contract. This same relationship holds between the contractor as the customer for a supplier.
Contract Documents	The complete collection of documents that collectively define the product and work we shall perform to produce it. The contract, customer program-specific specifications (including the system or other top-level specification), and applicable documents cited in the top-level specification and statement of work are included. Interface control documents to which the contractor is a party as an associate contractor may or may not be, in a legal sense, contract documents.
Guidance Document	An applicable document to be used as a guide in the performance of a contract, tailoring the process of evaluating individual potential requirements to determine the pertinence and cost effectiveness for a specific system or equipment acquisition, and modifying these requirements (sections, paragraphs, or sentences of the selected specifications and standards) to ensure that each contributes to an optimal balance between need and cost.
Top Level	Generally this is the system specification but may be a lower-level specification where the contractor is in an associate or subcontract relationship. This is the specification imposed by the immediate customer.
Acquisition	Any action that results in more efficient streamlining and effective use of resources to develop, produce, and deploy quality systems and products. This includes ensuring that only cost-effective requirements are included, at the most appropriate time, in system and equipment solicitations and contracts.

engineer to the program from the department most clearly associated with the document.

The coordinator should create and maintain a program-applicable documents status report giving: (1) the name and number of each document requiring assessment, (2) the precise revision and date of the document, (3) the name of the principal engineer assigned, (3) the current assessment class (from Table 5.4-2), and (4) current status of the tailoring effort for each document. The coordinator should also maintain a summary status briefing viewgraph with the data presented in Figure 5.4-1. These viewgraphs will be used at the chief engineer's periodic staff meetings while the assessment process is active to update the chief engineer and staff on status and to encourage management support for completion of assessment assignments.

Each assessment action will follow a common pattern of events depicted in Figure 5.4-2 and explained in paragraph 5.4.3. The coordinator is responsible for managing the overall process of accomplishing this same pattern of work for each document on the program-applicable documents list, requiring assessment so that each document has associated with it: (1) a compliance class (see Table 5.4-2), (2) an assessment completion status, and (3), for those requiring tailoring, a set of tailoring statements and a reference to the internal documents that implement the document requirements.

Six possible top-level assessment conclusions may be reached that correspond to six different compliance situations for the contractor. Table 5.4-2 lists these classes described from the contractor's perspective, and Figure 5.4-2 illustrates how they relate to the assessment process.

The process has two principal inputs: (1) the customer's top-level specifications (wherein are referenced the contractual technical requirements) and statement of work, and (2) the contractor's corporate knowledge described in the current company standard practices in terms of manuals, design and manufacturing specifications, the specification callouts on current production drawings (where existing designs are used), and the collective experience embodied in the functional organizations.

As a program matures, it is also possible that the design organizations may choose to select other cost-effective applicable documents not required by the customer in the interest of following sound design standards. In the absence of a set of customer-required applicable documents, the contractor should substitute their enterprise-applicable documents list (EADL). This list is mutually consistent with the contractor's standard practices. In cases where the contractor is proposing an existing product defined in released drawings and specifications to a different customer's requirements, the contractor assessment process input may also include an existing system specification compatible with current practices.

The process has three principal outputs: (1) a list of applicable documents based on either the enterprise-applicable documents list or the combination of the customer top-level specification and statement of work lists with tailoring statements, (2) a set of tailored applicable documents in the form of red-lined applicable documents, and (3) possible adjustments to the proposal cost estimates to cover nonstandard approaches (it being assumed that the contractor's current standard practices, embodied in their EADL, offer the least-cost approach).

Table 5.4-2 *Compliance classes*

Term		Explanation
1	MATCH	A perfect match between the customer's requirements contained in an applicable document and our standard practices
2	ACCEPT	Those customer requirements that are different from our standard practices that we can accept and easily accommodate with no significant impact on cost, schedule, or performance risk
3	TAILOR	Those customer requirements that are different from our standard practices that we believe can be tailored for essential agreement with our standard practices with no adverse affect on our ability to fully satisfy the customer's expectations
4	IMPACT	Those customer requirements that are different from our standard practices without compromising the customer expectations; we must adjust our practices to comply with the applicable document and adjust our cost estimate into alignment with those changes
5	REJECT	An applicable document that should, for good reasons, be rejected completely
6	REPLACE	An applicable document that is inappropriate and should be replaced with another

```
PROGRAM XYZAPPLICABLE  DOCUMENT STATUS

NUMBER OF APPLICABLE DOCUMENTS       _____
ASSESSMENT COMPLETE                  _____
MINOR PROBLEMS REMAIN                _____
SERIOUS CONFLICTS NOT RESOLVED       _____
PLANNED COMPLETION DATE              _____
ESTIMATED COMPLETION DATE            _____
```

Figure 5.4-1 *Typical summary status briefing viewgraph*

If the customer supplied a list of required applicable documents, the first output identified above must include a redlined system specification and statement of work reflecting tailoring consistent with the cost estimate.

5.4.3 Detailed Process Description

Figure 5.4-2 illustrates the process described in subordinate paragraphs. Refer to Table 5.4-2 for a definition of the six classes of applicable-document assessment responses that relate to the five obvious paths through Figure 5.4-2. Table 5.4-3 defines the organizations or persons responsible for leading the tasks defined below for each task block on Figure 5.4-2. The task numbering is coordinated with presentation material used in a system-engineering certificate program offered by the author's company, JOG System Engineering.

5.4.3.1 Create and maintain program-applicable document list, F3131

The coordinator creates the initial program-applicable documents list (PADL) from customer-supplied documents or the enterprise-applicable document list (EADL). Throughout the assessment process, the coordinator maintains the list, noting program approval status of all listed documents. Approval should be obtained through review by the chief engineer and program director or proposal manager, as appropriate.

Each document on the list should be assigned to one principal engineer, who accomplishes or coordinates all assessment actions against that document. Each document assessment action accomplished in accordance with the process illustrated on Figure 5.4-2 should be scheduled for assessment completion on a specific date mutually agreed to by the principal and the coordinator.

In a paper-driven system, the coordinator can initiate the assessment process by tasking the assigned principal engineer with an applicable document assessment tasking, containing a partially completed applicable-document assessment form, a set of instructions for the assessment, a copy of the applicable document, and the mutually agreed due date. The tasking letter also contains a form letter for use by the principal in reporting assessment results to the coordinator. The principal's response refers to this as an applicable-document assessment report.

In a fully computerized assessment system, the assignments may be made to a principal engineer's account on a computer network. The principal completes the data requested directly on a network terminal. The coordinator reviews the data on another network terminal before including it in the growing database.

Years ago it would have been possible for a computer database system supporting the applicable documents work to publish and email a form letter to principal engineers assigned to the documents, soliciting tailoring information that, upon receipt, could then be entered into the database.

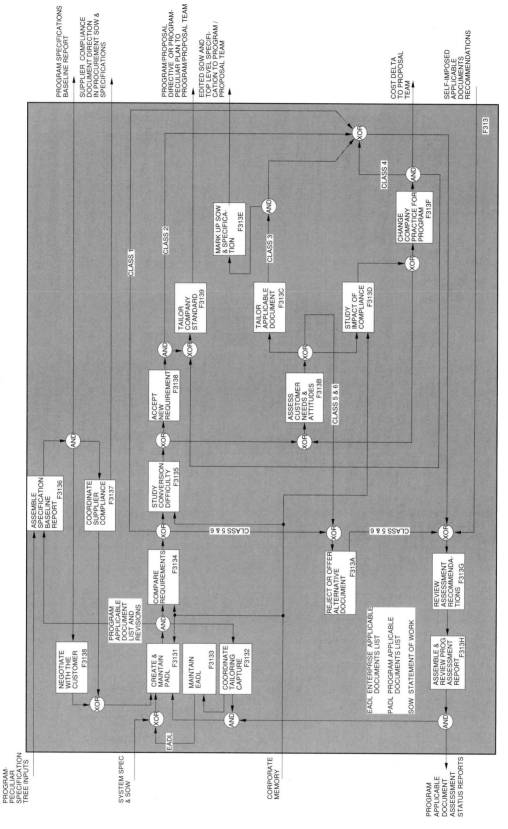

Figure 5.4-2 *Applicable document assessment workflow*

Table 5.4-3   *Principal organizational responsibilities*

Task NBR	Task Title	Responsible Department/Persom
F3131	Create and Maintain Program Applicable Applicable Document List	Coordinator
F3132	Coordinate Tailoring Capture	Coordinator
F3133	Maintain EADL	PIT
F3134	Compare Requirements	Coordinator
F3135	Study Conversion Difficulty	Assigned Principal/Subcontractor Team
F3138	Accept New Requirement	Assigned Principal
F3139	Tailor Company Standard	Assigned Principal
F313A	Reject or Offer Alternative Document	Assigned Principal
F313B	Assess Customer Needs and Attitudes	Assigned Principal
F313C	Tailor Applicable Document	Assigned Principal
F313D	Study Impact of Compliance	Assigned Principal
F313E	Mark Up SOW and Specification	Assigned Principal
F212F	Change Company Practice for Program	Assigned Principal
F313G	Assemble and Review Assessment Recommendations	Coordinator
F313H	Review Assessment Report	Proposal/Program Manager
F313I	Negotiate With the Customer	Proposal/Program Manager

Technology available today can speed this process up by giving principle engineers direct access to document data for which they are responsible.

5.4.3.2 Coordinate tailoring capture, F3132

The coordinator will capture all tailoring statements resulting from task F313C and include them in the program-applicable documents list, tailoring statements coordinated with the document to which they are related. The principal should enter these statements on the assessment form.

5.4.3.3 Maintain EADL

The PIT should monitor program-applicable documents work to minimize the program deviation from enterprise-standard practices and to evaluate improvements considered while assessing customer needs.

5.4.3.4 Compare requirements, F3134

The assigned principal engineer must carefully compare the assigned applicable document with current company practices covered in company manuals, standard practices, and department instructions. The assigned principal must be someone very familiar with current company practices corresponding to the applicable document and should the person responsible for implementing or responding to those requirements during the program. As a minimum, he or she should be someone whom the appropriate manager or director will accept as an authority.

The coordinator should assemble an initial assessment and provide it to the proposal manager or other management representative in the form of an initial applicable-document assessment report for review and initial steering of the assessment process. In the most extreme case, this could result in a decision not to bid a program because of extreme differences between requirements and our capabilities.

Some listed applicable documents may offer little reason for in-depth analysis by a specialized group because of known equivalence with company standard practices. In these cases, the coordinator may accomplish the assessment, being careful to note his or her own name as the prin-

cipal. These decisions should be reviewed by management and by the organization most closely concerned with the document. This approach may be elected for the complete assessment process where funds and time are in very short supply. The problem is that, unless the coordinator is a very experienced system engineer who is also very familiar with company practices, the outcome may include considerable hidden risk.

The output from this task will flow to one of three paths, as noted on Figure 5.4-2. If there is no significant difference between the customer requirement and company practices, the document is assigned to assessment class 1, with no further work required. If there are significant differences between customer requirements and company practices, the principal engineer should consider two courses of action: (1) offer an alternative pursued in Task F313A or (2) study conversion difficulty in Task F313D.

5.4.3.5 Study conversion difficulty, F3135

Given a perceived significant difference between the listed applicable document and company standard practice, the assigned principal engineer must determine how difficult it would be to convert company standard practice to full compliance with the applicable document. The goal in all cases should be to fully comply with customer requirements at a competitive price. The conclusion will be a recommendation either that we should accept the applicable document requirement in full and alter our own standards for the program (Task F3138) or that we should further evaluate other alternatives (Task F313B).

5.4.3.6 Assemble specifications baseline report, F3136

The program team should create and maintain a specifications baseline report that includes both listings for the program-peculiar (specification tree) specifications and applicable documents. The program-applicable documents list can form Appendix A of the baseline report. Appendix B could provide a comprehensive listing of all applicable documents data, including detailed tailoring statements.

5.4.3.7 Coordinate supplier compliance, F3137

Suppliers have exactly the same relationship to the prime contractor that the contractor has to their customer. They may have good reasons to tailor the applicable documents imposed on them in statements of work and procurement specifications, just as the contractor wishes to tailor its customer's requirements. Any such tailoring must be included within the SOW or procurement specification provided to the supplier. Either the responsible IPPT or a subcontract management team could be assigned the responsibility for coordinating subcontract tailoring and compliance with applicable documents.

5.4.3.8 Accept new requirement, F3138

The coordinator must review a recommendation for accepting a condition of full compliance with a compliance document that is in conflict with company standard practice (assessment class 2). The coordinator may ask for a peer review by another member of the responsible discipline, a review by someone in functional management for that discipline, or a program or proposal review board. Assessment class 2 should be used only for very minor differences between applicable documents and company procedures, where the cost for converting to the applicable document approach is very small (final judgment to be made by the program manager or proposal manager) or zero.

5.4.3.9 Tailor company standards, F3139

The work is not commonly expended to actually physically tailor the company practices. In cases where our proposal has accepted applicable documents that require practices different from our standards, the proposal takes precedence. Where time and funds are available, one should prepare a proposal or program directive that essentially tailors the standard practice to comply with the applicable document in question. A sound alternative is to include the different practices in a program-peculiar plan with a clear identification of the differences.

5.4.3.10 Reject or offer alternative document, F313A

This task provides a path for two alternatives: (1) a recommended replacement document or (2) the rejection of a listed document. If the assigned principal finds that the customer-supplied list contains applicable documents that are less favorable to company participation than others would be (those on the company enterprise-applicable document list, for example), has a sound rationale for the change based on the customer's view, and has some reasonable expectation that the customer might accept a replacement, and if the program environment and timing are amenable to changes, an alternative document may be proposed for inclusion in the program-applicable documents list, subject to review by the chief engineer and program director or proposal manager, as appropriate. If the alternative is accepted, it must be assigned a principal (normally the same principal as the replaced document) and passed back through the assessment process.

Rejection of a listed document, especially if the document appears on a customer-supplied list, should be reviewed carefully. It may happen that a customer or a program assessment coordinator has assembled a shopping list from which the final list will be screened. The principal should supply a sound rationale for deletion of a document. The best rationale is one based on cost.

5.4.3.11 Assess customer needs and attitudes, F313B

Before rushing to tailor, it can be useful to first try to understand customer needs and attitudes about the applicable document. A particular document may have been added to the SOW or specification list from the list used on a previous program or out of habit. It may also be considered a keystone by customer management or a customer specialty group. This is simply a case of knowing your customers and working hard to satisfy their needs as they perceive them. Working group meetings with the customer can help define hard requirements with respect to applicable documents.

5.4.3.12 Tailor applicable document, F313C

This whole procedure is built around this one task, illustrated on Figure 5.4-2 as block F313C. The goal of tailoring is to edit an applicable document so that it maximally reflects company practices. This is done by adding, deleting, and changing figures, tables, and text to read in a different way. Each of these changes is called a "tailoring statement." The principal engineer may be allowed to simply red-line the document, or the coordinator may require all tailoring statements to be entered on the data collection form provided with the tasking letter, depending on the anticipated workload and distribution of available personnel resources.

5.4.3.13 Study impact of compliance, F313D

Where the customer is adamant about using a particular applicable document that is characterized by a significant conflict with company practices (assessment class 4), the principal must study how difficult and costly full compliance would be. This is done by understanding the consequences of compliance in terms of staffing, plant, facilities, equipment, capital expenditures, and schedules. In the case of a major conflict involving multimillion-dollar expenditures, the impact analysis may involve a large team of specialists headed by a principal engineer selected from upper management with follow-on review and approval at an internal review board.

5.4.3.14 Mark up SOW and top-level specification, F313E

The lead specification representative must collect all approved changes to the system specification, and the person assigned SOW responsibility must collect all changes to the statement of work and put them in a format suitable for communication to the customer.

5.4.3.15 Change company practice for program use, F313F

Determine the changes that must be made to company standard practices to bring about coincidence between our practices and the applicable document. This task covers only the identification of the changes required. The changes are reported on the assessment form. They are fully documented in task F3138.

5.4.3.16 Assemble and review assessment recommendations, F313G

The coordinator should initially assemble the program-applicable documents status report using the entries included in the initial program-applicable documents list derived from one of the two sources addressed earlier. The report will include: (1) the name and number of each document requiring assessment, (2) the precise revision and date of the document, (3) the name of the principal engineer assigned, (4) the current assessment class (from Table 5.4-3), and (5) current status of the tailoring effort for each document. This report will be updated as the assessment process continues.

The coordinator should also maintain a summary status briefing viewgraph with the data presented in Figure 5.4-1. These viewgraphs will be used at the chief engineer's or proposal manager's periodic staff meetings while the assessment process is active to update management on status and encourage management support for completion of assessment assignments.

5.4.3.17 Review assessment report, F313H

All recommendations for assessment action (classes 1, 2, 3, 4, 5, and 6) should be formally reviewed by the chief engineer and program or proposal manager. This may be done all at one time where the assessment process does not stretch over a long time span or accomplished on a weekly or preplanned subset basis.

The coordinator should arrange the meeting, and each principal should review the action recommended with alternatives and recommendations. The coordinator captures the decisions made at the meeting and incorporates changes directed in the assessment status report.

5.4.3.18 Negotiate with customer, F313I

Our final conclusions about applicable documents must be coordinated with the customer. During the proposal phase, this may not be possible because of rules put in place to protect the competitive environment. During the contract period of performance, it is extremely important to gain customer agreement on applicable documents issues as early as possible. The coordinator must work closely with the program office to coordinate a schedule for customer review and approval of the final list. In the case of the top-level specification, this can be accomplished in association with specification authentication. It can be more troublesome to get customer agreement on applicable documents referred to only in the statement of work, but this must not be neglected.

5.4.4 Team Tailoring

It is possible to apply the approach covered in this section in a team environment with subcontractors, associates, the customer, academic institutions, and commercial groups. It may be particularly useful to team up with customer representatives during DoD demonstration validation phase to accomplish assessment actions so as to enter full development with an excellent mutual understanding of the tailoring situation.

The U.S. Air Force even applied team tailoring to a competitive situation involving several large defense contractor teams awarded study contracts on the Advanced Launch System in 1989. The Air Force program office called periodic meetings with attendance by representatives from each contractor team.

Together, under the guidance of an Air Force officer, the team worked toward a mutual agreement on tailoring a set of documents that would be included in the next program phase request for proposal against which they would all bid. The fact that this team was able to agree in large part on a set of documents and tailoring was no small achievement for the Air Force program office because each con-

tractor looks at tailoring in the context of its internal development process, production plant, and procedures. The Air Force had as much or more difficulty gaining agreement among their own internal experts in the several fields affected by the documents as they did among the contractors.

Team assessment can best be implemented in a working group format. All potentially useful applicable documents (possibly using the contractor's generic applicable documents list as a beginning) are listed and partitioned into related subsets. Each subset is assigned to one of several teams established to review the documents. The teams are structured to permit assignment of appropriate technical personnel to review related documents. The exact relationship between the teams and the document subsets may be at least partially driven interactively and may be different on different contracts. The tailoring process defined in DOD-HDBK-248A offers one insight into a possible subject-related teaming structure.

Each assessment working group will pursue the review and assessment of its assigned list of documents and consider recommended additions and deletions. From time to time, each working group should report to an executive board formed by the joint management represented on the working groups. The executive board makes assessment decisions and enters them into the growing program-applicable documents list with appropriate tailoring. At the terminal point of the team assessment process, a complete list of tailored applicable documents should be available and jointly approved.

Experience has shown that completion of the tailoring job is often held up by only 5 percent of the applicable documents. Therefore, it may be helpful to also prioritize documents to recognize the degree of difficulty in reaching agreement. This is especially important where tailoring is accomplished in parallel with ongoing preliminary design. By rapidly releasing the tailoring for 95 percent of the documents, it is possible to minimize the impact on preliminary design schedules and related schedule risk-management problems.

5.4.5 System Engineering Standards Relating to Requirements Analysis

Many older standards that included coverage of the system requirements analysis process have passed from the scene and are no longer applied on programs, such as MIL-STD-490A, AFSCM 375 series, U.S. Army Field Manual 770-88, and MIL-STD-499A. Two very good commercial standards appeared on the scene in 1999 in the form of IEEE 1220 and ANSI/EIA 632. These standards cover the whole system development process but do embrace the requirements work. Some of the earlier standards cover, in the author's opinion, an incorrect sequence of work that should not be followed. In some cases, process flow diagrams are included, suggesting that requirements analysis precedes functional analysis. This book makes it clear that in order for the requirements process to have continuity throughout, one should decompose the ultimate function and performance requirement, the need. Performance requirements derive from functions and are allocated to product entities, so performance requirements work cannot precede functional analysis; rather, it is intertwined with it. The design constraints cannot be identified until you know what the items are in the product physical architecture, so this part

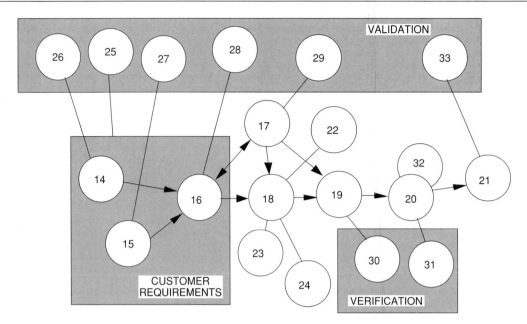

Figure 5.4-3 *ANSI/EIA 632 requirements work sequence*

of the requirements analysis process also cannot lead functional analysis.

The author believe that this flawed sequence is based on the attitude that requirements analysis is accomplished by the customer in the process of completing a system specification to use as the basis of a procurement. When the contract has been initiated, the contractor must perform functional analysis to allocate these system requirements downward through the architecture. This is a flawed view. The system requirements should be derived through structured analysis as well as the lower-tier requirements, no matter who or what organization is doing the work. It is true that the initial system specification crafted by a customer seldom is created as a result of a structured analysis process, but that does not detract from the good sense of applying a process characterized by logical continuity.

ANSI/EIA-632 breaks the system development process down into 33 requirements. The notion is that if a system development process is designed to satisfy these 33 requirements it should produce good results if managed and implemented well. Figure 5.4-3 illustrates a proper sequencing of the work correlated with the 632 requirements. The numbers correspond to the requirements numbers included in the standard. All 33 are not included because all of them do not relate to requirements. The author initially perceived a problem with the sequence of requirements 16 (system technical requirements), 17 (logical solution representations), 18 (physical solution representations), and 19 (specified requirements) but has since concluded that requirement 19 can be interpreted to mean an item detail or Part II specification, which should contain design-dependent content as a basis for acceptance.

5.5

Part II
Specifications

Contents

5.5.1 The Part Situation

Specifications may be developed in a one- or two-part format. The one-part format includes all requirements, whereas the content of a two-part format is partitioned into that which drives design and qualification and that which drives acceptance. Commonly, contractors will write one-part specifications for things that they will develop themselves and for things they procure (procurement specifications). Customers may require that a contractor develop two-part specifications for any items through which they wish to manage the program, sometimes called "configuration items." System specifications are, by their very nature, performance specifications in the current DoD phraseology, and materials and process specifications are detail specifications.

The content of this section applies primarily to what are called Part I, development, or performance specifications. The content of these specifications is intended to drive the design process as well as the qualification verification work. The content of the Part II, production, or detail specification drives the acceptance process for a particular article manufactured. You can see why it has been popular to prepare these specifications in two parts. The first part is prepared without detailed knowledge of the design solution—it is a definition of the problem to be solved in the design process. Note that the content of these Part I specifications should be design independent, which is what DoD means by the term *performance specification* applied to these documents. These requirements are what some people refer to as "design-to requirements."

The content of the Part II specification is all design dependent, based on a selection of the design features that the product article must have to be acceptable to the customer for delivery. You must know the design to prepare the Part II specification, and some time passes on a program before the design has progressed far enough to prepare the Part II specifications.

In preparing one-part specifications, some engineering organizations commit a development sin of no small proportions. The engineer responsible for preparing a one-part specification, knowing that the specification will not be whole until the design is fairly well known, decides that he or she should save the company some money and release the specification only once, which will have to be after the design effort. The specification may very well be released after the engineering drawings. This is not a good pattern and can be prevented by insisting that a one-part specification be released twice, first as a "Part I" version and secondly as a "Part II" version.

5.5.2 Specification Timing

Figure 5.5-1 shows a typical two-part item specification development schedule coordinated with design and manufacturing activities. Note that the Part I specification should be prepared before the design work, though some preliminary design work may have to be done to evaluate and mitigate potential risks. This pattern, requirements before design, is practically a mantra for system engineers; but, as

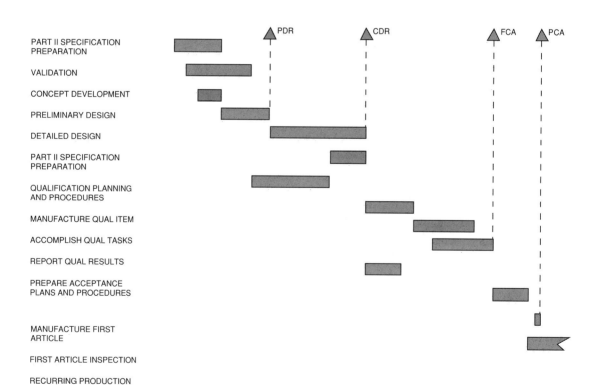

Figure 5.5-1   *Specification development timing*

noted above, Part II specifications have to be prepared after the design is known and the content is design dependent. However, if we recognize the intent of preparing the Part II specification, this will resolve this apparent conflict.

A Part II specification provides requirements that are key to the manufacturing process and are the basis for the design of the acceptance process, just as the Part I specification content is used as the basis for the design of the product and the item design of the qualification process.

5.5.3 Military Standards

MIL-STD-490A called for two generations of program-peculiar specifications, a development specification and a product specification, also referred to as Part I and Part II specifications, respectively. The former contained design-to requirements intended to be design solution independent, while the latter were design specific, based on the design solution, and often called "build-to requirements." The former was the basis for qualification testing and the functional configuration audit, where the evidence from qualification testing and analysis is offered to the customer to prove that the design satisfies the development requirements. The Part II specification was the basis for acceptance testing and the physical configuration audit conducted on the first article acceptance test results.

The Department of Defense canceled 490A in 1995 and replaced it with Appendix A (Change 1) of MIL-STD-961D, thereby changing the names of some old friends. The two-step process was retained, but the first part is now called a "performance specification" and the second part a "detail specification." Subsequently, the standard was completely revised to MIL-STD-961E, and the appended data was absorbed into the main body of the document.

The motivation for these changes was that the Department of Defense recognized that they had gradually evolved into a habit pattern of telling contractors what they wanted developed in terms too specific, with the effect of unnecessarily constraining the solution possibilities. To emphasize this new attitude, they restructured specifications to focus on performance requirements that tell what is needed and minimize the design-specific or detailed content in the development phase. In so doing, they included things in the performance specification that were previously not listed subordinate to the performance requirements headings of earlier chapters (like reliability, logistics, interface, and environmental); so, the reader should consciously work toward avoiding confusion over these two uses of the word *performance*.

A detail specification under MIL-STD-961E is created by simply adding the product-specific requirements in a paragraph that would be a void in the preceding performance specification and adjusting the other content for the design reality. The reader could refer to Office of Secretary of Defense publication SD-15 for guidance in preparing the two kinds of specifications.

5.5.4 Part II Specification Content Development

5.5.4.1 Outline suggestion

Figure 5.5-2 offers a high-level outline for Section 3 of a Part II hardware item specification, using a format the author prefers that groups all the characteristics he refers to together as specialty engineering characteristics.

3	REQUIREMENTS
3.1	Functional and performance requirements
3.1.1	Missions
3.1.2	Threat
3.1.3	Required states and modes
3.1.4	Entity capability requirements
3.2	Interface requirements
3.2.1	External interfaces
3.2.2	Internal interfaces
3.3	Specialty engineering requirements
3.3.1	Reliability
3.3.2	Maintainability
3.3.3	Deployability
3.3.4	Availability
3.3.5	Transportability
3.3.6	Materials and Processes
3.3.7	Electromagnetic radiation
3.3.8	Safety
3.3.9	Human factors engineering
3.3.10	Security and privacy
3.3.11	Computer resource requirements
3.3.12	Logistics
3.3.13	Personnel and training
3.3.14	Design and construction
3.3.14.1	Weight
3.3.14.2	Volume
3.3.14.3	Form factor
3.3.14.4	Center of gravity
3.3.14.5	Nameplates and product markings
3.3.14.6	Producibility
3.3.14.7	Interchangeability
3.4	Environmental conditions
3.5	Requirements relationships
3.5.1	Precedence and criticality requirements
3.5.2	Requirements traceability

Figure 5.5-2   *Part II outline*

5.5.4.2 Content development techniques

The structured analysis techniques covered in this part are not very useful in developing Part II content. Nor is a structured analysis technique necessary. When a Part II specification must be prepared, one has in hand the corresponding Part I specification. So, a good way to build a Part II specification is to build on the Part I specification. The Part I specification should be, and commonly will be, very light on design and construction content; so the first step can be to refine or define the content of this paragraph with design details covering weight, dimensions, volume, center of gravity, and any other factors important to the customer.

Step two should be to review the specialty engineering content for design solution orientation and coordinate with characteristics that are sufficiently important to the customer to require inspection after manufacture and before delivery. Apply this same test to environmental requirements, focusing on those characteristics that will be checked for each production item prior to shipment. Be very careful here to refrain from environmentally testing every unit as you would in qualification. The unit must have a full life potential when shipped. Interface requirements should be phrased in physical terms in the Part II specification that can be measured or inspected using connector part numbers, clocking, pin numbers, and orientation; bolt pattern dimensions; fluid plumbing connection part numbers, size, fluid specification numbers, and seal configuration; and other physical characteristics.

As an example, the aircraft onboard computer has an I/O interface requirement in the performance specification for receiving a radar altitude analog signal from a radar altimeter. It may require that the signal represent a zero- to 5000-foot altitude with a certain maximum altitude change rate based on possible terrain shapes and aircraft performance figures (speed, engine power, drag, and climb and dive rates). A detailed specification requirement for this signal should define the specific signal characteristics in the interface media and coordinate that definition with the real-world physical situation while in flight. Therefore, a detailed specification requirement, given a direct current electrical analog signal interface, could state the requirements in terms of volts DC per foot of altitude change and a degree of linearity between the limits over a range of altitude change rates and specify the pin numbers of the electrical connector on the computer I/O and altimeter where the signal will appear. This content would result in acceptance test plan content calling for a radar altitude simulator being set for 2500 feet and measuring 2.5 VDC across two pins on a particular connector.

6

Requirements Management

Contents

6.1

Process Overview from a Management Perspective

Contents

6.1.1 Introduction

6.1.1.1 Overview

We have discussed, in the earlier chapters, methods for performing requirements analysis and specification generation on a program. In this part we seek to develop the tools to allow us to plan and manage the many tasks needed to make this effort successful. Our approach focuses on using integrated product and process teams (IPPT) in an application of the concurrent engineering approach, with the teams staffed from the functional departments of a matrix organization. It is possible to organize in other ways, but in my view, those other ways are not as effective where the enterprise is large and the problems it solves for its several customers are complex. The functional organization provides the resource base programs need in the form of people who know what they are doing in fields related to the enterprise product line, good tools, and good practices built around a common process for all programs, with the programs providing a customer focus for the current business.

We begin with a brief discussion of the total quality management (TQM) concept that passed through industry consciousness in the early 1990s. TQM offers us a way to work toward perfecting our process, no matter how bad our current process may be. Then, we cover ways that the lead requirements analysis and specifications development engineer assigned to a new program may initially set up the requirements and specifications development infrastructure. This includes adequate provisions for computer hardware and software and to support requirements analysis, specification preparation, editing, publication, online reviewing and viewing, and reliable archiving. The principal engineer must also encourage a conscious selection of one or more requirements analysis strategies and train the team, if necessary, to perform requirements analysis within the context of the selected strategies.

The product program must develop an effective way to manage the development effort and regulate the relative progress of the requirements analysis and concept or design development activities so that requirements analysis leads the design effort. A means is needed to assure that the principal engineers and integrated development team leaders will respect the requirements-first direction. The most effective way for this to happen is to make sure that program managers and team leaders have been motivated to believe this and are effective in managing program activities along these lines. It also helps to give quality assurance the responsibility for auditing programs to discover to what degree they are following the common process approach as adjusted for a particular program.

The principal tool for managing program specifications preparation and maintenance is the specification tree. It is used to identify the architecture subset that will require preparation of specifications and as a placeholder against which responsibility for requirements analysis and specifications is identified. We must also have effective methods for managing a specification baseline on a program. This involves a clear definition of what specifications are required (via the specification tree) and sure knowledge of the current status of each of the documents identified on the tree. Further, the database, electronic word processor files, and paper versions of each current baseline document must remain synchronized.

Whether the program employs a paperless approach or not, it is important that a program library be formed to safely retain the electronic media for all specifications in the interest of minimum cost. If this library is computer networked and provided with the proper infrastructure, it can also greatly reduce the number of paper copies of documents reproduced, use of outdated requirements, document review costs, and document revision costs.

We will complete the process management picture with a replicable pattern of behavior that can be applied as a module of requirements analysis activities for any item. Many of these modules may be fitted together to cover the complete requirements analysis task for a program.

6.1.1.2 Total quality management

Too often we find ourselves driven from one panic to another during proposal and early program execution phases, responding to expressed and implied customer requirements that force us into procedures different from our own preferred internal ones. We tell ourselves that we risk being nonresponsive to a customer's request for proposal (RFP) if we deviate, however slightly, from the customer's requirements. As a result, we let our customers drive us to redesign our organizations and methods for each program. These program differences prevent us from developing an identity of our own and from working toward perfecting our process to better serve all of our customers and our own interests.

Dwight Stones was a world-class high jumper for many years and holder of the world record several times. He did not become a world champion in his field overnight or easily. But by watching Dwight perform just once you could see the secret to his success. You can apply this same secret to your requirements analysis activities beginning today and, over time, become as great as you can be.

As Dwight Stones prepared to make a jump, you could see him mentally going through every step of his jump, reprogramming himself for the jump. During his approach and jump, he tried to repeat a pattern of behavior that he had perfected over years of practice, practice, practice. But he also learned from his mistakes and made small changes to prevent repeating them. Dwight Stones developed a baseline approach that he consciously tried to repeat each time and made small changes based on experience and education. Constant practice against a standard changed only in small increments for sound reasons (as well as a lot of natural athletic ability) elevated Dwight's performance to the peak of human capability in his sport. We can apply this concept in our work as well.

Total Quality Management (TQM) may be defined as a continuous process of improving an organization's methods, resulting in increased customer satisfaction because of lower cost and better performance of products produced by the organization. The essential elements of an effective implementation of the TQM concept are:

1. Understand the current process your organization uses and write it down so that everyone may gain easy access to a description of everyone else's role in the process.
2. Encourage everyone to be consciously alert, while doing his or her job, to the possibility of better ways of doing it that result in less cost and better product quality.
3. Empower someone to play the system engineering role applied to your process and encourage that person to focus on seeking out suboptimization at the lower-tier organization level.
4. Provide an organized method to collect inputs on ways to improve the process and screen and prioritize them for action to make improvements.

5. Develop a minimum but effective set of metrics that measure your performance in critical areas with which you can monitor performance and plot improvements and identify areas most needing improvement.
6. Actively seek out your customers' views on your performance and listen carefully to their inputs.
7. Selectively implement the results of this continuous process study, using your metrics to help identify the most costly parts of the process that will leverage the biggest change and then change the written process description accordingly.
8. Provide an organized way to train the workforce to implement the process and continuous changes to it.

We should also study our competitors and other organizations we respect to learn what constitutes a world-class capability and compare our organization's current capability against these benchmarks. This knowledge can help to steer and set priorities for our improvement efforts. If we repeat this process, continuously taking advantage of lessons learned from each program application, there is a good chance that our organization will improve its performance and achieve increasingly greater success.

But how can we possibly do this and avoid the risk of being branded nonresponsive in our proposals? DoD identified this as a problem that they had to cooperate with industry in solving. In DoD Directive 5000.43, Acquisition Streamlining, they specifically instructed procurement offices to encourage contractors to apply their proven processes on contracts. They based this position on the notion that good contractors will apply TQM principles to their business and products and provide their customers with good value motivated by a self-preservation interest. The DoD Instruction 5000.2 revision issued in 1991 canceled 5000.43, but it includes the same intent. Since that date, this trend has been continued.

It is true that DoD agencies continue to impose many of the same standards, despite this policy change to programs, through system specifications and statements of work. It is also true that this continues to force contractors to comply with external procedures not tuned to their development and production processes nor their facilities and tooling base. But DoD directives do encourage government program offices and contractors to jointly tailor (adjust) these applicable documents to allow the contractor to apply their processes wherever possible. The contractor can offer alternative tailoring in the context of this policy. DoD has also passed through a tremendous conversion from military standards to commercial standards to focus on performance specifications. A performance specification should contain content telling what the product process must do and contain no content telling how to design the product.

The way a contractor can stay on top of this problem is to have sound internal procedures covering how they perform their work and then map these procedures to a set of applicable documents commonly applied by their customer base, whether these standards be military or commercial in nature. Where these customer documents would require the contractor to perform contrary to their preferred process, the contractor should tailor them to make the customer document identical to their internal procedure. When a proposal is prepared against an RFP containing applicable documents, those documents, and tailoring offered by the customer, are compared with the generic tailoring previously prepared and the differences used to generate alternative tailoring initiatives offered with the proposal keyed to the contractor's minimum-cost, best-quality approach.

There will be cases where the customer will insist on their own tailoring (or untailored version), but generally a convincing case can be made for the contractor's version because the result is their least-cost, best-quality approach, satisfying the customer need and giving them best value. In the few cases where the customer cannot deviate from a particular requirement, the contractor makes an adjustment in their internal procedures for a particular program and adjusts the cost estimate accordingly or decides not to bid on the program.

6.1.1.3 Buzzwords forever

TQM followed many other initiatives, such as "Zero Defects," and preceded many others, such as "Six Sigma" and "Lean." Each of these initiatives was enthusiastically encouraged by an author or organization, and each offered good ideas for improvements in one's process. But all too many managers have become totally consumed by one or more of these. The personnel in some organizations have been subjected to so many of these initiatives that when another appears they wonder how long this one will last. Rather than permitting each of these new buzzwords to revolutionize the organization, management should evaluate them for ways to integrate them into their existing common process.

If each program can be performed using the same processes, the contractor will benefit from the "practice, practice, practice" notion through which sports figures become world-class competitors. If a contractor is introspective about their performance on each program and consciously seeks to make small improvements in their process based on lessons learned from programs, they will constantly improve their baseline, and practice will improve their performance. The combination of performance to a standard that is referenced to customer-preferred standards and continuous small improvements of the standard, possibly influenced from time to time by the positive effects of new buzzwords, will move the contractor toward a goal of being as good as they can possibly be.

This book offers a comprehensive process for planning, managing, and accomplishing an effective requirements analysis and specification generation process, but it probably is not perfect for any given organization and program. You are encouraged to pick components from it and mix them with the good parts of your own existing process. Then you should work continuously to improve your process in the interest of better serving your customers' needs. An improved profit picture will follow satisfied customers.

6.1.2 Program Preparation

We should not start work on a new program without a plan nor without preparation. The best time to accomplish this planning and preparation is during the proposal period preceding a contract win, while all of the great ideas that flow into the proposal are still warm. The activities covered in this section should begin during the proposal period, and the product of that work should be retained for use as a basis on which to build during the contract period of performance.

6.1.2.1 Resource overview

Every program should begin with a set of prepared resources that let the program begin requirements work efficiently at program start and pave the way for success while specifications are in development. These resources should be generically available to every program, with a possible need to adjust them somewhat for any program-peculiar factors during a proposal period. These resources include a set of specification templates, a preferred set of models coordinated with the content of the specification templates, a set of tailored applicable documents, and the teaming plans set for the preferred architecture of the product system. Generally, enough work will have been accomplished during the proposal to have clearly defined these entities. It is true that during the proposal effort the proposal team may not yet know enough to have perfectly defined these resources, so further work during the initial contract period may encourage changes in some of them. Figure 6.1-1 illustrates the program preparation process explained in this section.

6.1.2.2 Specification templates

When people from aerospace or commercial firms ask the author to recommend a template for specifications, the author answers MIL-STD-961E for hardware and system specifications, and EIA J STD 016 and associated data item descriptions (DID) inherited from MIL-STD-498 for software. The enterprise should have a template outlined for the following kinds of specifications: (1) system, (2) hardware item performance specification, (3) hardware item detail specification, (4) software item performance specification, (5) software item detail specification, (6) parts specifications (treated as drawings by DoD), (7) material specifications, (8) processes specifications, and (9) hardware and software performance and detail interface requirements specifications. If an enterprise is often called on to develop several kinds of products, the team may conclude that they need a set of templates, one for each family of products, such as operational equipment and support equipment.

These templates can be kept as Microsoft Word documents if the enterprise creates its specifications directly as Word documents. Alternatively, a database system could generate template content of the appropriate type within a database when you identify the template to be used for a specific item. In any case, the templates for a given program should be stored under configuration control in some read-only storage media that everyone can access but that only by those responsible for managing the requirements work on the program can change.

Many are surprised that MIL-STD-961E permits specifications to be captured as models, presumably in computers, though not necessarily so. This might fit in most perfectly for software development efforts, but it could also fit for all specifications a program must develop. One could claim that a RAS-Complete (explained in Chapter 3.10) was a model, and the specification tree could consist of identifica-

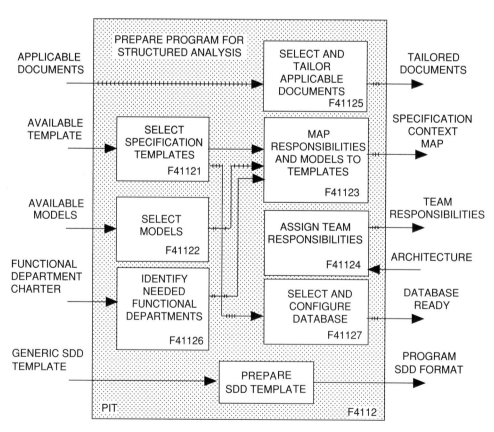

Figure 6.1-1 *Prepare program for structured analysis*

tion of the records related to particular line items in the RAS related to each item in the RAS. This can be done simply by including an architecture model ID or WBS in the RAS. In this case, there would be no formally released paper documents, though one could print them out if desired.

6.1.2.3 Analytical models

The requirements analysis work on every program an enterprise undertakes should be accomplished using structured analysis with a particular model set consciously considered and selected. As this book is being written, a good universal set of models would be: (1) traditional structured analysis (discussed in Part 3) for systems and hardware items and (2) unified modeling language (UML) for software entities (discussed in Part 4). Over the period 2000 through 2010, the system engineering and software engineering communities will have been working toward a common modeling environment within which engineers can develop the requirements and design concepts for systems, hardware, procedures, and software. This model will be a variation of UML that, even in 2005, the author would accept contains traditional structured analysis as a subset. The first effort in this direction is SysML Version 1.0, released in early 2005. See Chapter 4.6 for an expansion of these ideas.

6.1.2.4 Model or template maps

Given that the templates and models have been selected, the developing organization should map the two together so that analysts working on the program can see the clear connections between them. This can take the form of a matrix for each specification template, where for each paragraph title in the template a model or method is listed through which the analyst would derive the corresponding requirements. Once this pattern has become routine in an enterprise, it is unnecessary for the analysts to consult these matrices on a regular basis. It becomes an element of the enterprise's common process with which all long-term employees have become familiar.

6.1.2.5 Planned writing responsibilities

The model or template map can be further developed to include a column for the functional department in a matrix organization from which a person should be drawn to do the analysis using the prescribed model and to write each paragraph defined in the template. The understanding is that the manager of the identified functional department has made certain that the people in that department are skilled in performing that work using the indicated modeling approach. Figure 6.1-2 provides an example of the suggested map for Section 3 of a specification as defined in MIL-STD-961E. The "Preferred Model" column tells what models will be employed to gain insight into necessary requirements, and the "Discipline Specialist" column tells what functional department should be depended on to provide people to accomplish the related analysis on programs.

6.1.2.6 Preparation for structured analysis work product capture

The program should prepare to capture the work products produced by analysts; otherwise this data will be lost. This book encourages the use of a document called a system definition document (SDD), but there is no accepted name or format for such a document because it is so seldom created. The SDD is explained in detail in Chapter 3.11. Where a program applies traditional structured analysis, this document would include seven appendixes to capture the work products:

1. *Functional.* Contains functional flow diagram and function dictionary.
2. *Environmental.* Contains system timeline, system spaces conclusions, and corresponding standards set identification, as well as the list of system environmental parameters and selected variable range. This appendix applies the content of Appendix F (process diagram), Appendix C (architecture), and system environmental parameters from Appendix B in a three-dimensional model that defines end-item environmental requirements. It identifies hostile, noncooperative, and self-induced requirements and integrates them into end-item environmental requirements. Finally, the appendix reports the zones selected for each end item as the basis for component-level environmental analysis.
3. *Architecture.* Contains the architecture block diagram and architecture dictionary.
4. *Interface.* Contains the interface defining diagrams (N-square and/or schematic block diagrams) and the interface dictionary.
5. *Specialty Engineering.* Contains a specialty engineering scoping matrix that allocates specialty disciplines to architecture entities, directing people from these disciplines to develop requirements against the architecture indicated.
6. *Process.* Contains a physical process flow diagram used in end-item environmental requirements analysis and logistics support analysis.
7. *Requirements Analysis Sheet.* Contains the system RAS or references the database system within which it is contained.

Some of these work products can be captured in one or more requirements database applications, but the author knows of no such product that provides a full set of models to fully support traditional structured analysis. Programs requiring development of software, which is nearly all programs today, can capture the models more completely in the software analysis and design tools they employ. In these cases, it may be more effective to accept that the model data is contained within the application. If it were necessary to produce a paper document reporting on the structured analysis work products on a program applying UML for software and traditional structured analysis for systems and hardware, one could apply the system definition document structure noted above with an extended set of appendices as follows:

8. *Use Cases*
9. *Activity Diagrams*
10. *Statecharts*
11. *Sequence Diagrams*
12. *Collaboration Diagrams*
13. *Class/Object Diagrams*
14. *Component Diagrams*
15. *Deployment Diagrams*

6.1.2.7 Applicable document action

The specifications developed for the program will involve some list of applicable documents as a function of the technologies applied and the customer base served. In most program situations, the developer has been in business for

Para Nbr	Paragraph Title	Preferred Model	Discipline Specialist
1	Scope		System Engineer
2	Applicable Documents		System Engineer
3	Requirements	Traditional Structured Analysis	System Engineer
3.1	Functional and Performance Rqmts.	Performance Rqmts Analysis	System Engineer
3.1.1	Missions	Mission Analysis	System Engineer
3.1.2	Threat	Threat Analysis	System Engineer
3.1.3	Required States and Modes	Functional Analysis	System Engineer
3.1.4	Entity Capability Requirements	Functional Analysis	System Engineer
3.1.5	Reliability	Reliability math Model	Reliability Engineer
3.1.6	Maintainability	Maintainability Math Model	Maintainability Engineer
3.1.7	Deployability	Logistics Support Analysis	Logistics Engineer
3.1.8	Availability	Availability Math Model	Logistics Engineer
3.1.9	Environmental Conditions	Three Layer Model	System Engineer
3.1.10	Transportability	Logistics Support Analysis	Logistics Engineer
3.1.11	Materials and Processes	M&P Controls	M&P Engineer
3.1.12	Electromagnetic Radiation	EMI Analysis	EMI Engineer
3.1.13	Nameplates and Product Markings	Boilerplate	Quality Engineer
3.1.14	Producibility	Manufacturing Engineering	Manufacturing Engineer
3.1.15	Interchangeability	Logistics Support Analysis	Logistics Engineer
3.1.16	Safety	System Safety Analysis	Safety Engineer
3.1.17	Human Factors Engineering	Human Factors Analysis	Human Factors Engineer
3.1.18	Security and Privacy	Threat Analysis	Security Engineer
3.1.19	Computer Resource Requirements		Computer Hardware Engineer
3.1.20	Logistics	Logistics Support Analysis	Logistics Engineer
3.1.21	Personnel and Training	Logistics Support Analysis	Logistics Engineer
3.1.22	Requirements Traceability		System Engineer
3.2	Interface Requirements	N-square Diagrams	System Engineer
3.2.1	GFP Interfaces	N-square Diagrams	System Engineer
3.2.2	External Interface Requirements	N-square Diagrams	System Engineer
3.3	Design and Construction	Engineering Domain Techniques	Engineering Domain Engineer
3.3.1	Production Drawings		Engineering Domain Engineer
3.3.2	Software Design		Computer Software Engineer
3.3.3	Workmanship		Quality Engineer
3.3.4	Standards of Manufacture		Quality Engineer
3.3.5	Process Definition	PMP Controls	M&P Engineer
3.3.6	Material Definition	PMP Controls	M&P Engineer
3.4	Precedence and Criticality of Rqmts.		System Engineer/Project Engineer
4	Verification		Verification Engineer (T&E)
4.1	Methods of Verification		Verification Engineer (T&E)
4.2	Classes of Verification		Verification Engineer (T&E)
4.3	Inspections		Verification Engineer (T&E)
5	Packaging		Logistics Engineer
6	Notes		All

Figure 6.1-2 *Coordinated specification responsibility and models*

some time dealing with a particular customer base and product line. A large percentage of the applicable documents that will be called in the program specifications will be well known. Ideally, the enterprise will have built information over time about how those documents should be tailored for the closest coordination with the enterprise processes and product line as discussed in Chapter 5.4. This data should be available to all programs.

During the formative program stages, the proposal or program team must perform an applicable document selection and tailoring exercise. If the customer provides a detailed list with tailoring as a part of a system specification and statement of work, the task is to evaluate the documents on that list against company practices and offer

alternatives that bring the references into alignment with company practices. Customers are interested in the best value for their dollar, and company practices should reflect the best the enterprise are able to perform in developing a good product for a reasonable cost.

6.1.2.8 Teaming planning

In the program preceding study and proposal work, the team will have identified a preferred architecture down to some level of indenture. The program integration team (PIT) will have developed a system specification from available user documentation; and, ideally, the PIT would have also developed the top-level specifications immediately subordinate to the system specification corresponding to the items that the top-level IPPTs will be assigned. Each top-level team, when formed, should begin its work with a top-level item specification and the component of the WBS, SOW, IMP/IMS planning data and associated budget related to the item for which they are assigned responsibility. The teams may be charged with developing the lower-tier specifications, or that can be a continuing responsibility of the PIT as the program or enterprise chooses. As lower-tier teams, if any, are formed on the program, they should also each begin life with a top-level specification prepared by their parent team or PIT. I prefer a federated approach for responsibility, with the parent team responsible for developing the top-level specifications for each of its child teams, passing responsibility for lower-tier specifications within that team on down to the lower-level teams but insisting that they apply the planned models and methods.

The teams should be organized about the system architecture items, physically collocated, and cross-functionally staffed from the functional departments, based on the discipline needs of those teams. The teams should be called on to cooperate across interfaces between the items they are responsible for so as to clearly define those interfaces in a mutually agreed upon fashion, thus forming essentially an interface control working group (ICWG) of the whole program team. Refer to Chapter 3.6 for details.

6.1.2.9 Program specification library

A program has two choices in retaining the master copies of its specifications. It could publish them as paper documents from computer media, in which case it would have to keep the signed paper copies coordinated with the computer data. The alternative is to keep the requirements in a database from which anyone may call up the data on his or her own workstation. In either case, some form of library is required as described in this section. The aggregate of all specification information should fall into two sets: (1) the data that corresponds to specifications not yet released and (2) the data corresponding to specifications that have been released. The PIT system engineering people and any other teams with requirements analysis responsibilities should have control over the former, and configuration management (presumably in the PIT) should have control over the latter.

6.1.2.9.1 Library Initiation A program specifications library (PSL) should initially be set up during the proposal phase and plans laid for full implementation at contract award. The program may have more than one of these libraries, each of which includes specifications under the responsibility of one or more departments or teams; or all of these documents may be merged into a single library for the program. The aggregate of these libraries is likewise referred to as the program specification library in this book. A PSL has the following major components:

1. *PSL Librarian.* A PSL should be operated by a librarian trained in the use of the computer hardware and software used on the PSL, library procedures, and document configuration control procedures. There should be at least one other person qualified to act as the librarian to provide service during periods of regular librarian sickness and vacation.
2. *Library Data.* A central PSL should have control over: (1) the master electronic media for each released specification, (2) all electronic program specification standards

masters for system, configuration items, subsystems, components (in-house or procurement items), parts, materials, and process specifications, (3) program-peculiar specification status database information, and (4) an applicable documents electronic database containing a listing of all customer-applied compliance and guidance documents with tailoring.
3. *PSL Workstation.* One workstation under control of the central library agent should be identified as the PSL workstation. This workstation should be capable of: (a) reading and editing all PSL documents; (b) preparing floppy or CD disk copies of specification standards for principal engineer use or otherwise providing for distribution of the standards; (c) control of the PSL baseline, whether installed on a computer network server, managed on the PSL workstation hard disk, or located on floppy disks under librarian control; and (d) maintaining the status of the electronic baseline.
4. *PSL Workstation Software.* Software must be installed on the PSL workstation to provide the capabilities noted in item 3 above.
4. *Other Workstations.* The program needs one or more other workstations, as a function of the number of documents that might be simultaneously worked, capable of preparing program specifications within the context of the program specification standards.
5. *PSL Library Standard.* The library must be operated in accordance with a company or program procedure. The procedures for all libraries (if more than one) on a program should have some consistency.

6.1.2.9.2 PSL Variations Libraries can be established with a range of characteristics as a function of program needs. Four specific ways they may vary are:

1. *PSL Variations by Responsibility.* A system engineering PSL may be created as a centralized repository for all program specifications. It may be structured to retain only system-level documents, or it may capture something in between. Alternatively the program may elect to have some other department (data management, engineering administration, the vault, or configuration management, for example) manage a central repository for all program specifications. Two or more specialized libraries could be created, each with a librarian (software, parts engineering, materials and processes, and all other specifications, for example).
2. *PSL Variations by Storage Media.* The PSL data may be retained in electronic media in any of several ways. The documents may be stored on floppy disks, removable hard disks, magnetic tape, or a hard disk internal to a stand-alone workstation or network server. In any case, data must be protected against loss of the primary storage media by suitable backup procedures. One could use only paper copies produced on typewriters, but most companies are well beyond this evolutionary stage.
3. *PSL Variations by Connectivity.* The PSL may be organized as a stand-alone computer workstation or located on a network server (the preferred configuration). Wherever possible, the documents should be stored for accessibility by the program population through a computer network for the purposes of reading, reference, or, under suitable controls, for update. Any updating or revising should be done external to the PSL.

4. *Variations by Word-Processing Software.* The program may use any of the many good computer word-processing or desktop publishing software applications as a function of customer preferences, available program computer assets (central Wang center, for example), the preferences of program personnel, or the resources provided by supporting departments. Wherever possible, the program should reach an agreement on minimizing the diversity of applications in the interest of assuring the simplest possible access to documents by the widest population on as many workstation types as possible. Microsoft Word is one word processor that can function on Macintosh and IBM machines equally well. Networking solutions exist that permit a computer network composed of IBM clones and Macintosh machines to jointly share a central repository, with the documents appropriately formatted in a word processor like Microsoft Word. If a company has only one kind of machine in use, the solution only becomes easier.

6.1.2.9.3 Security The data residing in the PSL must be secure from unauthorized change. If located on a network server, it should be protected from change, erasure, or addition in that environment by anyone except the PSL librarian. If located on the hard disk of a stand-alone machine or on a series of floppy disks, the librarian must protect the data using some combination of physical security and file access security. How bad can it get? One company suffered the physical loss of the whole hard disk containing the complete specification library. It was stolen, apparently by a terminated employee.

If the PSL contains classified data (secret or confidential), the data must be handled, stored, and processed in accordance with the customer's classified data regulations. In general, it will be prudent to locate classified data on removable computer data storage media that can be physically marked and safeguarded in an appropriate classified data cabinet. Generally, this data must be processed only on a stand-alone computer workstation specifically approved for use with classified data.

All PSL data must be protected from willful or unintentional loss or damage through an effective data backup procedure. Depending on assets available to the program, this may involve use of a tape drive, floppy disks, CD ROM, or other storage media regularly used to copy master files. The library equipment and procedures must be designed so as to preclude loss of data due to any single point failure of the system, including human users through willful or accidental actions.

6.1.2.9.4 Availability The specification data in the PSL must be readily available for read-only access by anyone on the program. This requirement can best be served by placing the PSL on a network server protected as read-only. Network service to individual engineering workstations and CAD terminals is suggested, but if resources are unavailable to satisfy this capability, a number of well-placed networked view stations can be provided at minimum cost.

6.1.2.9.5 PSL Finances Each specification agent should be held responsible for his or her own budget to maintain his or her PSL. But, if other departments or teams choose to integrate their specifications into a central PSL, or if they are directed to do so by program management, the lead requirements analysis or specifications person must set up

procedures to accommodate these other documents. If the library operating cost is included in the budget for the department responsible for the central library, there will be no problem in allocating cost to users.

Otherwise, a burden-sharing arrangement is required to ensure that all participants are paying their fair share of the cost for library services. Figure 6.1-3 provides an example of a formula for determining cost that a particular department should be expected to pay (Cd) in person-hours per month as a function of the total number of documents (L) in a library and the number of documents that the department (D) has in the library. This formula is based on the illustrated inverse relationship between the total number of documents in a library and the cost per document to maintain them in the library. This relationship reflects an expected economy of scale as the library becomes large.

6.1.2.9.6 Specification Standards Loading The requirements analysis or specifications lead person should be held responsible for acquiring enterprise specification standards and translating them into program specification standards. These standards can be made available to program personnel assigned a specification preparation responsibility either as a floppy disk copy, through network service, or from within the database system employed on the program.

6.1.2.9.7 Requirements Database Interface If the program uses a requirements database system, an interface must be provided between the database system and the library function to couple specifications generated from the database into the library. This could be as simple as some kind of desktop publishing application, such as Ventura or Interleaf, permitting database output, boilerplate text, and graphics to be integrated into a single electronic document. The database is set up to output to a particular file name to be located in a particular directory/folder.

6.1.2.9.8 Data Ownership The specification data that has been reviewed, approved, and released should be under the control of PIT configuration management. Specification data that is planned or in work should be under the control of system engineering in the PIT and subordinate IPPT. If all the data is in a common database system, the records can be assigned access control to accomplish this partitioning of responsibility. If the specifications are being maintained in a word processor, the PIT CM must take ownership of the master copy in electronic and paper formats.

6.1.3 Program Implementation

When a program begins under contract, the team should have available to them the infrastructure described above and be fully ready to run. This is an uncommon condition, but it is a goal that the proposal team should work hard to achieve. Management of the early requirements analysis activity during the subsequent program implementation can then focus on the technical task at hand without being burdened with the administrative details that should already have been disposed of.

6.1.3.1 Program specifications plan

The specifications plan is a subplan to the system engineering management plan (SEMP) or system engineering plan (SEP), when separately prepared. As an alternative, it can be integrated into the SEMP/SEP or both the SEMP/SEP

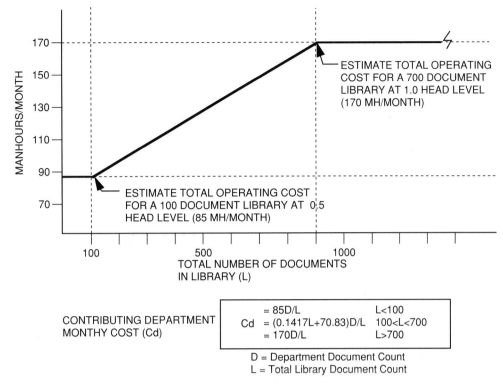

CONTRIBUTING DEPARTMENT
MONTHY COST (Cd)

$$Cd \begin{array}{ll} = 85D/L & L<100 \\ = (0.1417L+70.83)D/L & 100<L<700 \\ = 170D/L & L>700 \end{array}$$

D = Department Document Count
L = Total Library Document Count

Figure 6.1-3 *Cost-sharing formula*

and specification planning, integrated into the program-integrated master plan (IMP). The plan should address the eight specific items explained below as a minimum.

6.1.3.1.1 Program-Unique Document Identification   The plan must contain a clear methodology for identifying what specifications will be required on the program. The methodology offered in Part 3 of this book is to review the items on the architecture block diagram and compare the item characteristics against a specification selection checklist.

6.1.3.1.2 Responsibility Assignment   The plan must define a methodology for assignment of responsibility for development and maintenance of specifications identified, as noted above. This can be as simple as a program specification list with department numbers and names after each document. In some companies, a system engineering organization prepares all specifications, making this list a no-brainer. The plan must also offer an effective means to communicate this list to the assigned integrated product and process teams (IPPT) or principal engineers following a formal review of the list by the chief engineer.

This book uses the term *IPPT* to denote a cross-functional team composed of engineers, analysts, and support personnel assigned under a leader to develop a major system item. The term *principal engineer* is used to identify an individual engineer responsible for development of a lower-tier item, possibly under the cognizance of an IPPT. In the context of this book, the principal engineer will generally be dependent on support from engineers and analysts

not under his or her supervision who may be assigned to a parent IPPT and shared by several lower-tier principal engineers. The term *principal engineer* is most often used in this chapter to denote the leader of the development team, but the reader may also put *IPPT leader* in its place, depending on the item level in system architecture.

At the system level, a program should employ a system team referred to in this series as a "program integration team (PIT)." The PIT should be responsible for top-level specifications down to the level corresponding to the top-level specification for each IPPT. If a system team is not used on a program, a system agent of some other kind is needed. The term *chief engineer* used in this book refers to either the PIT leader or a senior technical person in the PIT.

The preferred method of assigning responsibility for requirements analysis and documentation is to gain the support of the chief engineer and program and functional design supervision in the identification of principal engineers for architecture elements and to identify those principals also responsible for the development of the associated requirements document defined on the program specification tree. This responsibility normally can be broken into four levels and mapped to the program engineering organization as explained in Table 6.1-1. MIL-STD-961E recognizes only system, item, and PMP levels.

Systems engineering could be held responsible for requirements analysis and specifications generation for all items designated by the customer as configuration items. Seldom will any of these items be designated as subsystems, but some may be component-level items, resulting in a potential

Table 6.1-1   *Principal engineer levels*

Level	MIL-STD-490A Scope	MIL-STD-961E Scope	Principal ORG
1	System, Segment, Prime Item, Facilities	System, Item	PIT/System Engineering
2	Subsystem	Item	Teams
3	Component	Item	Principal Engineers
4	Parts, Materials, and Processes	Parts, Materials, and Processes	PIT/Parts Engineering M&F

conflict with the pattern defined in Table 6.1-1. These few ambiguities can be decided on a case-by-case basis.

As an alternative for Level 1 in Table 6.1-1, some complex items in the system that will require cooperative development between several specialized design departments may be assigned to cross-functional concurrent product development teams, as noted above. The team may then assign subordinate items within that architecture branch to lower-tier principal engineers. The same team that is responsible for concurrently developing a product design solution mated to a manufacturing and logistics support concept is tasked to first concurrently develop the requirements for the item.

Ideally, when a team is first formed, the parent team would have already developed and released the specification that corresponds to the team's top-level responsibility, so that the team comes into being with their cost, schedule (in the WBS, SOW, and IMP/IMS), and performance factors (item specification) already defined. The team will then be responsible for developing all lower-tier specifications themselves or through them to their subteam assignments.

6.1.3.1.3 Specification Scheduling and Statusing   The plan must contain a specification schedule in terms of due dates, a means of keeping track of the status of each document in work at any point in time, and how the program team is doing overall in specification development in terms of schedule compliance. You should negotiate due dates with the document principal engineers that are consistent with overall program schedules for development of item concepts and designs and include these dates in the specification list.

Actuals then can be included in this same list and a dual *S* curve generated from these two sets of dates to reflect general program specification development performance for periodic review by the PIT leader or chief engineer. The due dates and actual dates need to be explained in the plan. For example, the due dates could be principal engineer document completion dates, review dates, or release dates. Several dates could be tracked to provide feed-forward information. Where progressive requirements analysis is used in combination with simple concept requirements lists, primitive requirements statement steps can be tracked (attribute list, valuation) followed by one or more formal document preparation milestones.

6.1.3.1.4 Specification Baseline Identification   The plan must identify the program baselines in terms of dates and/or program milestone events that will be recognized. These may coincide with customer-defined baselines (functional, allocated, and product) only or include additional internal baselines as well.

6.1.3.1.5 Baseline Definition Documentation   The plan must define how the program will define a specification or requirements baseline. The suggested approach is to pre-pare a program requirements baseline report (RBR). This report includes a list of all of the program-specific specifications with status, revision, and responsibility information; a list of all applicable documents the contractor is required to comply with; and the contractually binding tailoring associated with each. The formal specification tree can also be included in this report. All of this data could alternatively be appended to a program specifications plan.

6.1.3.1.6 The Physical Baseline   The plan should define what the actual physical specification baseline is. The normal physical baseline is the set of signed, released, paper masters located in a vault augmented by any group-released masters retained in engineering department repositories. In the future, this will probably change so that the masters are actually in electronic media residing in some program computer library. As an interim step toward the future, one could locate the network file server containing the released specifications in the same vault, or under their control, containing the signed paper masters. Somewhere in the release cycle you must ensure identity between paper and electronic baselines. This could be done in the vault or by a preceding release checking function.

6.1.3.1.7 Electronic Specification Library   The program should operate an electronic library of specifications as the working masters for all specifications start becoming available. This may be comprised of a central computer-networked library (under vault control) or a distributed, or federated, set of libraries, each with its own librarian. These libraries can provide the program with online, electronic access to all specifications if networked; but, as a minimum, they must protect the investment in typing labor to support future revisions.

6.1.3.1.8 Specification Change Management   The plan must cover how changes will be handled on the program and how they will be handled differently in different program phases if the process is graduated in some way. This may be as simple as to refer to the change procedures in a company standard.

6.1.3.2 Program specification standards preparation

6.1.3.2.1 Responsibility and Content   The program or proposal requirements lead engineer should be held responsible for preparing of a set of program specification preparation standards (boilerplates or templates). The preferred approach for this task is to convert a previously prepared set of generic specification standards available in electronic media into program standards by including the results of: (1) an environmental use profile and zoning analysis, (2) an identification of the specialty engineering disciplines needed for the program, and (3) the program-applicable documents analysis, as amplified below:

1. *Customer Requirements Compliance.* Company standards can be derived from a set of documents, such as the data item descriptions corresponding to the types of specifications covered by MIL-STD-961E, with possible adjustments in style for electronic publishing compatibility. These changes may not be fully acceptable to the customer, in which case the program standards will have to undo these changes. Other customer requirements must also guide the adjustment of the company standards.

2. *Environmental Requirements Analysis.* It is normally senseless to develop a generic company boilerplate for product environmental requirements, but developing a generic program environmental requirements boilerplate is commonly a money-saving action that makes it unnecessary for each principal engineer to repeat essentially the same analysis. The results from an environmental use profile and zoning analysis should be captured in program specification standards. The component-level principal engineers should have to identify only in which zone of a prime item their component is located and pick off the environmental requirements identified for that zone.

3. *Applicable Documents Analysis.* Another major piece of information to be added to the program standards is the results from the program applicable documents analysis. This should consist of a list of customer compliance and guidance documents with some tailoring. This list should be included in program standards Section 2 as a shopping list. The item principal engineer should exclude any not referred to in the document. Any other documents not on this list that the principal engineer adds as a compliance documents should be carefully considered by the chief engineer for need at the document review.

6.1.3.2.2 Standards Availability

The specifications plan must tell how the program standards will be made available to the principal engineers (on a network, copied to the principal engineer's blank disk by the lead requirements engineer, given out in paper form, or the like).

6.1.3.2.3 Multiple Standards Levels

While the program must convert the company standards to program standards, it may also be useful to convert the program standards into multiple type standards. For example, given that a company standard for in-house requirements documents has been converted into a Program XYZ standard, the support equipment design function may find it advantageous to create a support equipment standard from the more general program standard. Likewise, if 15 valve specifications are required for outside procurement, it may be advantageous to prepare a valve standard from the program procurement specification standard. Alternatively, a general valve specification could be prepared that is referenced in each specific document.

6.1.3.3 Specification tree development

The requirements lead engineer should be held responsible for transforming the evolving system architecture block diagram into a program specifications tree that identifies what requirements documents must be produced, what type they shall be, and who is assigned principal engineer responsibility. The initial, top-level specification tree should be prepared during the proposal period for inclusion in the proposal. This activity must be coordinated with the WBS, make-buy plan (procurement specifications), and configuration item identification activities (where a customer manages the program in this manner).

The program specification tree must cover all requirements documentation on the program. This includes system, segment, configuration item, and procurement specifications, interface control documents, and internal requirements documentation. On a DoD or NASA contract, the customer may require the preparation and maintenance of a contract specification tree that identifies only the system, segment, and configuration item specifications identified by the customer. This contract tree will be a subset of the complete tree.

The specification tree can be prepared in diagrammatic or indentured list format or a combination of these formats. The diagrammatic format illustrates the breakdown of architecture items that must have their requirements formally documented. Figure 6.1-4 shows the top end of such a tree. The alternative is to list the documents in architecture order with some method of indicating indenture. A mixed alternative could include a diagram for top-level documents supplemented by an indentured list for lower-tier requirements documents. If the program maintains an architecture block diagram, this diagram can be annotated for requirements documentation identification to satisfy the specification tree requirement.

There are essentially three kinds of requirements documents that have to be identified in our tree, and these kinds relate to the contractual relationships represented by the documents. A customer will generally define the items for which they require specifications based on the intensity they wish to employ in managing the program. In the days of MIL-STD-490A, these specifications were identified as system, segment, and configuration item specifications, and they were all contractually binding, authenticated (approved) by the customer, and formally delivered to the customer. The term *segment* does not appear in MIL-STD-961E, but there are system and item specifications that the customer may wish delivered.

The second kind of specification is created by the contractor for items that are purchased from their suppliers to ensure that suppliers clearly understand needed product characteristics. Finally, there are some elements that will be designed and manufactured by the contractor for which the contractor will choose to prepare requirements documents. The latter are generally items intermediate between system and configuration item specifications or subordinate to configuration item specifications. They are prepared to capture the flowdown of requirements in an orderly fashion across what otherwise would be gaps in the requirements documentation.

6.1.3.4 Principal engineer selection, assignment, and training

The principal engineer is the person assigned the responsibility for identifying the requirements for an item in the system architecture and capturing them in a document of the appropriate type, as noted on the specification tree.

As these assignments are made, we must make certain that each principal engineer understands his or her role and responsibilities. It may be necessary to offer training for one or more principal engineers in their requirements analysis responsibilities, though that can be very difficult to do amidst the actual program implementation. Working

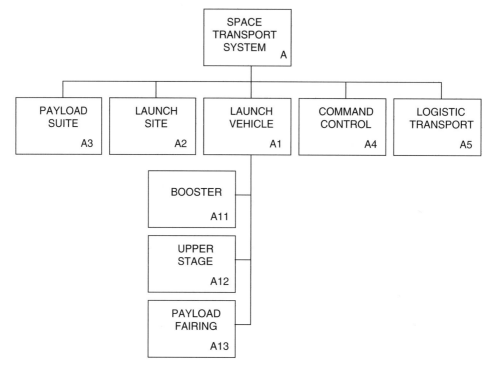

Figure 6.1-4   *Typical specification tree*

requirements analysis luncheons or after-hours classes may provide a means to interlace the work and training needed. Avoid giving the principal engineers training on the whole SRA process. Focus the training on the few elements they need to be effective. Ideally, the responsible functional departments would supply programs with fully qualified engineers requiring only program-specific training, if any.

6.1.3.5 Program specification development methods

The lead requirements engineer must identify any and all specification agents and define and/or negotiate a program specifications development approach with these specification agents, the chief engineer, program management, and configuration management. A specifications agent is responsible for the development of a series of specifications and the configuration management of his or her work electronic media for these specifications.

The recommended specification development environment calls for the use of networked microcomputers containing one or more specification libraries, each operated for the specification agent by a specification librarian in accordance with a published standard. Figure 6.1-5 illustrates such an environment.

On a program where the cloning strategy is used, one or more of these libraries should contain specification standards that are made available to all principal engineers with a specification responsibility in read- or copy-only mode. The principal engineer copies the standard to his or her local workstation and transforms it into a specification for the item based on the results of the requirements analysis activity he or she performs and coordinates. This document

is made available in a buffer on the network for review and marked up or annotated by reviewers using networked computer markup software. After corrections are accounted for, the principal offers the document to the chief engineer for review. The appropriate librarian loads the released, approved document into the expanding specification library, where the document resides in read-only mode.

Where the program is equipped for computer video projection in meeting rooms, the document review can be accomplished with real-time correction by the principal engineer in control of the computer driving the video projector. Alternatively, the principal, chief engineer, and reviewers could hold telephone conference calls on a program properly equipped, with all of these people viewing the specification at their own desks on networked workstations. Once again, the principal engineer should be the only one allowed to change the text during the review.

Where the program uses a requirements database approach, the data should be created, reviewed, and approved while in the database form. Specifications are then created from the approved database content using a specification generator at baseline events. Only one requirements database should be used on a program unless classified data is involved where a separate database for classified data may be appropriate.

During the planning phase, the program should select one or more of four strategies: cloning, structured analysis (preferably using a database approach), question and answer, or freestyle. As a program matures from one phase to another, it may make sense to change strategies. An experienced system engineer might very efficiently create an

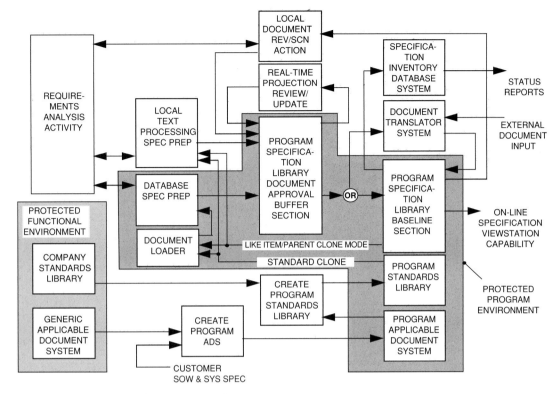

Figure 6.1-5 *Specification development environment*

SRD for a concept phase program using the freestyle strategy. In the demonstration/validation and early engineering and management development phases, the program could profitably use structured analysis. Finally, in the production phase, the large set of baseline documents in the electronic library may make it possible to use like-item cloning very effectively. In general, however, as noted throughout this book, structured analysis is the lowest-risk approach to building good specifications.

6.1.3.6 Modularization of the schedule
The principal engineer should be shown how his or her work fits into the overall program specifications schedule and encouraged to prepare a detailed schedule for that item. Figure 6.1-6 illustrates the expansion of a program schedule into an item schedule. The principal engineers have to be brought into the responsibility network for making the program schedule work by making their own schedules work. The PIT leader or chief engineer must manage the overall engineering development activity and hold the principal engineers accountable for their part in schedule performance.

6.1.3.7 Regulating the plunge
Requirements analysis should precede design or procurement activity. It often happens that the engineering organization panics, stimulated by a pressing schedule, and concludes that they cannot waste time on requirements. The program manager or one or more team leaders conclude that they cannot meet the drawing schedule unless they

begin making drawings right away. This is a tragic conclusion that will almost always lead to a very expensive development activity due to the tremendous number of engineering changes that must be made later.

The development effort should not be driven solely by the drawing release schedule, as important as that is. The overall program phase schedule should be created based on a balanced appreciation for requirements analysis, concept development (predesign activity), and design work with the knowledge that later development activity will be easier if the requirements are carefully thought through before the design drawings are initiated.

Figure 6.1-7 illustrates how the PIT leader or chief engineer can control the movement of the development process downward through the architecture, assuming the program chooses a top-down development direction. This view would have to be modified for a middle-out direction sense. The PIT leader or chief engineer needs a means to communicate his or her authorization for work indicated in Figure 6.1-7 in the several categories noted on this figure. Table 6.1-2 suggests one possible means to do this.

Note on Table 6.1-2 that the left columns identify the item, the responsible contractor (for a case where multiple contractors are involved through teaming, joint venture, or associate arrangements) and principal engineer, document type, and number. Columns are then provided for requirements documentation scheduling and status. The date in the "RQMT DOC DUE DATE" column should be earlier than the date in the "CONCEPT DUE DATE" column. The

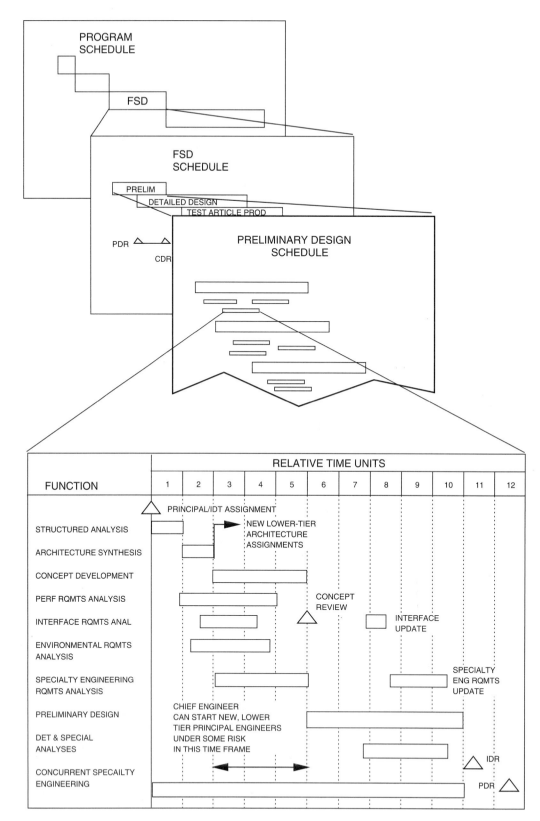

Figure 6.1-6 *Development schedule modularization*

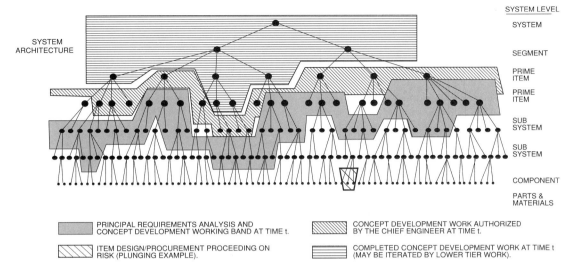

PRINCIPAL REQUIREMENTS ANALYSIS AND CONCEPT DEVELOPMENT WORKING BAND AT TIME t.

ITEM DESIGN/PROCUREMENT PROCEEDING ON RISK (PLUNGING EXAMPLE).

CONCEPT DEVELOPMENT WORK AUTHORIZED BY THE CHIEF ENGINEER AT TIME t.

COMPLETED CONCEPT DEVELOPMENT WORK AT TIME t (MAY BE ITERATED BY LOWER TIER WORK).

Figure 6.1-7 *The advancing wave*

current status carried in the "DOC STAT" (status) and "CONCEPT STAT" columns should reflect a more mature status of the requirements than the concept until they are both noted as complete. Additional columns can be provided for preliminary design, detailed design (coordinated with the drawing release schedule), and test-related events.

The requirements in early program phases can be captured in simple requirements lists, rather than rigid specification formats with reams of confusing boilerplate, using the progressive requirements analysis concept covered in Part 2. The concept data can be captured in an organized set of data generated by the principal engineers and supporting personnel on program microcomputers. This whole set of data provides the principal engineer with briefing data for PIT leader or chief engineer's approval review and formal design reviews and the program with a way to communicate the resultant evolving baseline to everyone.

6.1.3.8 Selective requirements development
Selective development is defined as "the development of requirements and concepts for system elements, the parents for which do not have a set of requirements and concepts developed." It plunges into the architecture ahead of the structured process to develop lower-tier elements without benefit of the information that flows from the structured process. The small crosshatched area on the lower fringe of Figure 6.1-7 identifies such a case.

Selective development offers a third alternative between the extremes of unstructured and top-down structured development. The wholesale application of selective development is probably better termed *unstructured development*. The term *selective development* should be reserved for specific purposes and not abused. There are several appropriate reasons for engaging in selective development. Three are explored below, followed by a discussion about inappropriate applications of the technique:

1. Commonly it is necessary to develop procurement specifications to support a cost proposal bid much earlier

than the structured process has developed the requirements appropriate to these elements. Potential suppliers will generally, and understandably, not provide a firm bid without a clear understanding of the requirements for the element they are asked to bid on. This means that they will require a procurement specification. The requirements for these items may have to be selectively developed.

2. Situations sometimes develop where schedule risks are identified in the development of particular system elements that force early development as the most advantageous abatement technique. This may entail parallel development of two or more approaches, each of which must be supplied with a requirements set before the parent requirements are defined.

3. The customer or the contractor program director may direct that particular system elements be developed with high priority even before the parent requirements are fully defined for reasons other than perceived risk.

Because of our familiarity with our product line, we may conclude, even properly sometimes, that it is necessary to conduct specific concept trade studies far in advance of the requirements development. But selective development should not be allowed to become the norm. The customer schedule not uncommonly panics the engineering community into believing that we must race to drawing development; it's "every man for himself." The system engineering community must counter these attitudes with factual support for the structured approach.

At the time it is decided to plunge, the principal engineer or the IPPT responsible for the selective development activity must be charged with establishing a set of departure requirements. These should be based on the requirements for the lowest-tier parent element in the same architecture branch that these elements appear. During the period of the selective development activity, the structured process may progress further down that architecture branch, providing refinements in the requirements available to the team.

Table 6.1-2 Development control table

Arch ID	Item name	Contractor	Dept	Engineer	Type	Number	Rev	Due date	ERB NBR	Stat	Due Date	ERB NBR	Stat	Remarks
		Principal			Requirements documentation						Concept development			
A14	COSGROVE UPPER STAGE	GSTS	120-1	JONES	CI	72-1934	-	02-25-90	061	G	03-10-90	063	G	
A141	EQUIPMENT MODULE	GSTS	120-1	BURNS	DRD	72-2430	-	03-10-90	062	G	04-20-90	068	G	
A1411	EM STRUCTURE	GSTS	231-3	ADAMS	DRD	72-1122	A	04-05-90	066	G	05-10-90	080	O	
A1412	AVIONICS SUBSYSTEM	GSTS	342-0	PERKINS	DRD	72-3323	-	04-10-90	067	O	05-15-90	081	R	PLATFORM MOUNTING SCHEME
A14121	GUIDANCE & CONTROL SUBSYSTEM	GFG	350-1	SMITH	DRD	72-2579	B	05-20-90	083	R	06-25-90		O	
A141211	ON-BOARD COMPUTER	TELE-BOND	780-2	BROWN	PS	72-2582	-	06-10-90	091	O	07-10-90		O	CHIP GIDEP ALERT
A141212	INERTIAL NAVIGATION SET	REYNOLDS	780-2	WINDHAM	PS	72-2585	A	06-15-90		O	08-10-90		O	
A141213	STAR TRACKER	GFE	100-0	FLETCHER		S-1267-3	A	02-10-90		G	02-10-90		O	
A14122	ELECTRICAL SUBSYSTEM	GSTS	420-0	BLACKMER	DRD	72-9333	-	05-10-90	082	G	06-10-90	093	G	
A141221	BATTERY	RPS	420-1	GOMEZ	PS	72-9354	C	06-10-90	090	R	06-15-90	095	O	STRIKE SCHEDULED
A141222	WIRE HARNESS	GSTS	420-2	CHIN		72-9365	-			G	06-20-90			
A141223	POWER CONTROL UNIT	ESI	420-3	JOHNSON	DRD	72-9202	B	06-10-90	089	G	07-25-90		O	
	STRUCTURE					72-1211	-	05-20-90	084	G	06-10-90		O	

It is important to understand the lowest tier of requirements available in the branch where the selectively developed element resides so that requirements can be made to reflect them as much as possible. It will not be possible to establish traceability connections to these requirements because there will be a gap; that is the nature of selective development. Traceability problems can be minimized by using the available superior element requirements as a guide during selective development events.

When the structured development process does arrive at the previously selectively developed requirements, a traceability audit is required. You may very well find on arrival at the join-up that there are serious problems needing significant changes in the requirements and concept for the plunged item or for the higher-level system elements. When we undertake plunging as a means to mitigate risk, we should understand that we are really only trading one kind of risk for another.

6.1.3.9 Requirements risk management

Chapter 6.2 covers requirements validation and risk management. Chapter 6.3 offers additional risk abatement techniques dealing with value management. Actually, nearly everything that system engineers contribute in support of program and team management is related to risk abatement. This starts with insuring that good specifications are published in a timely way.

6.1.3.10 Process controls

The chief engineer and system engineering community need mechanisms to ensure that the planned activity covered on program and item schedules is happening and producing a cost-effective solution to the customer's need. An effective way to focus management energy is to define a development module that can be applied by each principal engineer. These modules can be connected through multiple layers, if necessary, to provide for development of multileveled architecture items.

6.1.3.10.1 IPPT Meeting Structure Ideally, integrated product and process teams (IPPT) would be physically colocated to encourage communication and concurrent interaction. Each team should capture the results of its requirements work in a networked computer database or public folder system in a read-only form so all program personnel may gain access to the current baseline at any time.

The ideal situation for an IPPT is for all of the team members to be given a space and furnishings arranged to permit easy conversation among team members during their work. A fundamental principal of concurrent development is to encourage practices that minimize the need for communication while maximizing opportunities for communication.

As a minimum means of implementing the concurrent engineering approach during the requirements identification period, the IPPT leader must hold periodic meetings to ensure that team members are kept abreast of the current baseline and suggested changes to it and to solicit input. All team members must be consulted during formal trade studies conducted by the team. Each team member or discipline is responsible for completing their portion of the requirements set for the item and for understanding the complete requirements set for the item.

Figure 6.1-8 illustrates a meeting cycle that can be used to review evolving requirements and concepts developed by the integrated development teams. The meetings are obviously arranged cyclically in this example, with seven-day

centers. This cycle span should be selected based on how many IPPTs the program has and the intensity with which the development process must be managed. The cycle time could be adjusted over time with low span time at the start and less frequent meetings as confidence builds.

The chief engineer can use this series of meetings to monitor and manage program activity as defined in the development control table and item schedules. These data can be updated based on the meeting results. Action items and direction by the chief engineer at the meetings is followed up at future meetings.

The circular symbols indicate IPPT meetings chaired by the chief engineer. A system meeting is included in the cycle to provide a time to evaluate the integration and system optimization of the results of the prior round and focus on actions and review items for the next round. The IPPT leaders may need to hold their own brief meetings (triangular symbols), as indicated on major items within their area of responsibility.

6.1.3.10.2 Requirements Traceability Audit As the requirements fall out of the process defined in this book, they must be hooked up to the process through which they were derived and to item parent requirements. This can be done by requiring all the principal engineers to do their best to include this data in their requirements analysis data and by auditing their work. It may not be possible to audit all of the requirements for traceability. In this case, priorities must be established and a sampling technique instituted. The results of the audit must be to correct flaws in the condition of traceability. Questions about requirements that do not trace to parent requirements must be asked of the offending principal engineers and those questions resolved.

6.1.3.10.3 Status Tracking The scheduling and control measures outlined above provide the framework within which the system engineering community can track the status of the requirements analysis process and compare with planned performance. The resultant error signal is used to identify and correct unsatisfactory performance. The meeting cycle and one-on-one discussions are used to communicate corrective action direction based on observed status problems.

6.1.3.10.4 Integration and Optimization Activity The system engineering community must constantly study IPPT products to assure that the evolving requirements and concepts are optimal and balanced. Conditions of suboptimization are identified, alternative system-optimization approaches evaluated, and a course of action selected and implemented. These influences on the architecture items under IPPT development responsibility may cause recycling of some requirements and concepts.

6.1.3.11 Tailoring the development intensity

This book provides a systematic requirements analysis methodology to supporta DoD weapons system development activity requiring adherence to acquisition process controls and procedures. The book could be used as the requirements analysis process description for such a program referenced in the program system engineering plan.

However, all programs will not require every component part of the process included in this book, nor the level of intensity suggested here. People responsible for engineering management on each program must decide the right intensity for each of these system engineering activities for the particular program phase, based on the customer statement

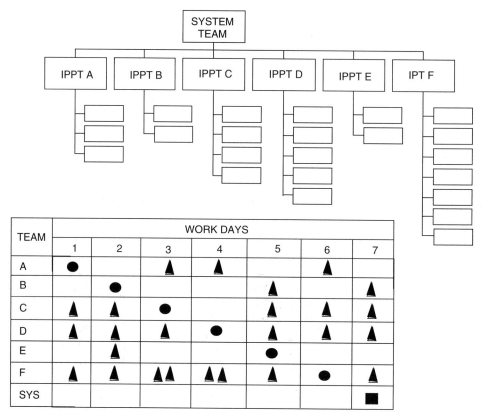

Figure 6.1-8 *Sample IPPT meeting cycle*

of work and company standard practices. A good way to organize this approach is to define system engineering process criteria (including requirements analysis criteria) for your company that are composed of a series of simple statements that define the minimum system engineering process features to be applied on each program. These generic criteria can be included in the company's generic system engineering manual or management plan.

Then, make it the responsibility of each program to define in the program system engineering management plan how they will comply with the company's generic criteria, subject to the approval of company functional engineering management.

A DoD weapons system may have to use the full content of this book to comply with customer requirements. A skunk works operation can get by with much less, but even here there are limits beyond which you court disaster. We close out the discussion of program implementation with sample criteria for early program phases where requirements analysis work is most intense. It may be instructive for you to write down how your program or company normally complies with each of these criteria. Then you might ask yourself how your company should respond. If there is a difference between the two positions, you may have identified a basis for improving your company's capability.

Sample System Engineering Process Criteria (SEPC) for Concept Development, Demonstration-Validation and EMD through Preliminary Design Review

1. A clear statement of system need developed by or agreed to by the customer
2. A set of requirements expanded from the need to define the mission, brief definition of major elements in system, system performance goals or requirements, external interface definition, system environment definition, and a minimum of specialty engineering constraints in the simplest format acceptable to the customer
3. An organized means using structured analysis models to decompose the customer's need into a system architecture consisting of hardware, computer software, facilities, materials, and personnel that can be related to the customer's definition of cost accounts
4. An organized method for assigning responsibility for development of the system elements in the architecture to personnel and/or teams that is mutually consistent with the customer's definition of work breakdown (in support of effective cost collection) and the enterprise development team concept applied on the program
5. A means to define and enforce the definition of requirements for items prior to design work
6. A means to capture all requirements correlated with their architecture item and to make them available for reference by all program personnel
7. Traceability of requirements to sources, other requirements, and the development process

8. An organized way to develop compatible internal and external interfaces among elements depicted on the system architecture
9. An organized way to capture and retain the results of development team product and process concept and design work so as to be easily accessible at any time by all other program personnel and available for iterative improvement by responsible engineers and analysts
10. An organized means to schedule work performed on development teams and to monitor progress to those plans; this is also a C/SCSC, or earned value, criterion
11. An organized method for gaining insight into the results of development teamwork on the part of management, coupled with a means to direct the work of teams based on these insights
12. An organized means to integrate the work of separate product development teams and to optimize at the system level across design, manufacturing, test, operations, and logistics support to overcome problems resulting from suboptimization at the individual team level
13. A means to define what the system and its components must do in the customer's environment to maintain needed readiness and operational capability and to attain mission success. This method must provide for development and evaluation of the system maintenance and logistic support concepts and coordination of those concepts with the evolving system design concept and a mechanism for gaining insight into needed procedural and human task-oriented requirements. This must include a way to feed the results of analyses into these matters back to the design evolution.
14. A concurrent method for evaluation of the system and its components for reliable and safe operation. This must include a way to feed the results of analyses into these matters back to the design evolution.
15. Provision of a means to reach sound decisions where current work plans have come to discontinuous points in the development process requiring direction
16. A means to capture design rationale and traceability of the decision process that led to the preferred system concept
17. A means to clearly define the current system requirements and concept baseline during the period leading up to PDR. After PDR, the baseline is clearly captured in a set of specifications, reports, and engineering drawings listed in an engineering drawing management system and rigorously controlled by configuration management.
18. A means to concurrently perform specialty engineering work and to efficiently integrate the results of that work into the ongoing development process and product definition.

6.1.3.12 Development data package concept

Once engineering drawings start coming off the boards or CAD stations, standard configuration management procedures are very effective in controlling the design baseline. Many organizations find it very difficult, however, to control the evolving requirements and concepts baseline and maintain decision traceability during the sometimes chaotic period leading up to the preliminary design review. Many engineering organizations have found themselves in a sad predicament at a critical design review without the backup data asked for by the customer for a critical decision made months earlier.

The requirements database concepts exposed in Part 7 can provide a means to capture not only program technical requirements but also rationale, sources, and traceability associated with these requirements. So, the database approach can be used to satisfy the need to retain rationale data for requirements but may not totally solve the design concept decision capture problem outside the requirements information component. We seek a solution that embraces both requirements and concept information and ways to manage the evolution of this information resource in early program phases.

Over a period of years, many system engineering organizations have evolved a universal view of all of the information of interest in the early phases of system development and defined a particular organizing structure for that information. One such concept, called a "development data package (DDP)," was advanced by logistics engineers at General Dynamics Space Systems Division in the early 1990s. Figure 6.1-9 illustrates one way this information package can be organized for a project. The horizontal matrix axis lists the sections of the DDP. The vertical matrix axis lists the organizations participating on an integrated product development team focused on developing a particular item that happened to be an avionics box.

The understanding is that there must be a place in the DDP for all members of the team to put all their information work product during the life cycle of the DDP. This includes the kinds of information that many organizations expect engineers to put in their engineering notebooks, journals, or logs. The DDP captures all information of interest to the team and makes it available to all team members. Later we will see how computer technology can satisfy the availability requirement. First let us discuss DDP organization and life cycle.

Under this concept, each IPPT leader and principal engineer must create and maintain a development data package (DDP) for his or her item beginning when the chief engineer authorized the team to start development work. The DDP provides a means to capture development information from all IPPT members in a common format between initiation of team activities and completion of the preliminary design review (PDR). It must provide a place for every concurrent engineering team member in which to put his or her information product. Between PDR and critical design review (CDR), the content of the DDP should flow out to formal documentation destinations such as specifications, engineering drawings, and planning data libraries, and DDP maintenance should be discontinued as each section makes the transition. Table 6.1-3 defines the formal documentation destinations of each DDP section (noted in the vertical axis of Figure 6.1-9) subsequent to PDR. Generally, the team should be allowed the time between PDR and some time prior to CDR to complete the conversion between DDP content and the formal data destinations.

The matrix intersections of Figure 6.1-9 are annotated with responsibility information, as explained at the bottom of the figure. Each DDP section is assigned a principal integration agent, who is responsible either to input information or integrate inputs from others into a coherent story consistent with all other DDP information. For example, the requirements section is owned by a systems development function in Figure 6. 1-9. As you can see, several team members should provide inputs to this section, and others should be held accountable for understanding section content. Still

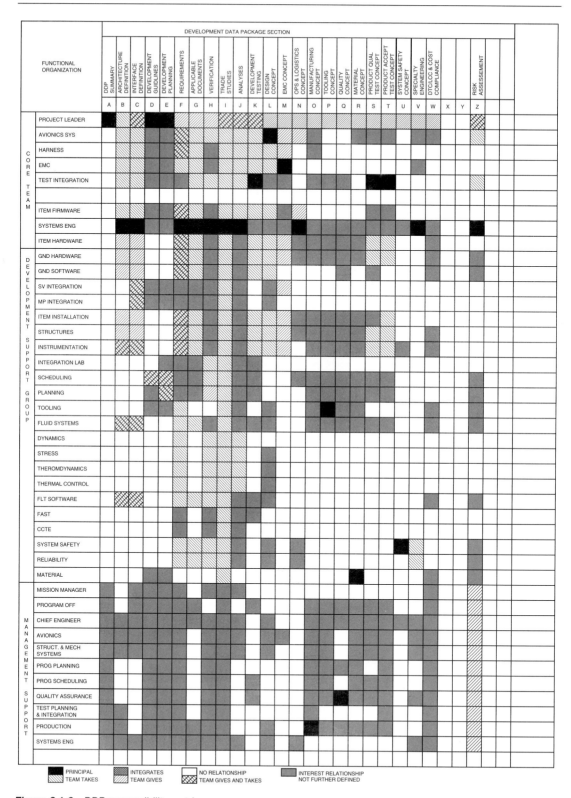

Figure 6.1-9 *DDP responsibility matrix*

Table 6.1-3   *DDP data destinations*

DDP	Section	Formal documentation destination
A	ARCHITECTURE	Specifications, Specification Tree, Specialty Engineering Models
B	INTERFACE	Specifications, Engineering Drawings
C	DEV GUIDELINES	Program Planning, Supplier SOWs
D	DEV PLANNING	Program Planning, Supplier SOWs
E	REQUIREMENTS	Specifications
F	APPLICABLE DOC	Specifications
G	VERIFICATION	Specifications, Program Test Planning
H	TRADE STUDIES	Design Rationale Traceability Documentation
I	ANALYSES	Design Rationale Traceability Documentation
J	DEVELOPMENT TEST	Integrated Test Plan
K	DESIGN CONCEPT	Engineering Drawings
L	OPS/LOGISTIC CONCEPT	Logistics Support Plan
M	MANUFACTURING CONCEPT	Manufacturing Plan, Facilitization
N	TOOLING and STE CONCEPT	Procurement Documents
O	QUALITY CONCEPT	Manufacturing Planning Documents
P	MATERIAL CONCEPT	Procurement Documents
Q	PRODUCT QUAL TESTING	Integrated Test Plan, Test Procedures
R	PRODUCT ACCEPTANCE TESTING	Integrated Test Plan, Test Procedures
S	SYSTEM SAFETY CONCEPT	System Safety Plan, Hazard Reporting
T	COST COMPLIANCE ASSURANCE	Program Planning, Specifications
U	RISK ASSESSMENT	Design Rationale Documentation

others are identified as interested parties with no obligation to interact over section content.

If the program uses computer word processors to prepare specifications rather than a computer database system, the content of the DDP requirements section can be initially provided using the word processor of choice in the primitive style of Part 2. As the item definition matures, these primitive statements may be expanded into full specification text in time for publication as the item specification in the required format (configuration item responsive to the customer data item description [DID], procurement specification, or in-house requirements document).

Where the program uses a computer database to capture item requirements, the DDP requirements section may simply reference the database content or be used as a baseline repository for the most recently approved snapshot of database content while the working database content continues to mature.

Whether word processing or database technology is employed, the team responsible for the particular DDP would apply a sound requirements analysis process, such as that covered in this book, to identify the content of the requirements set for the item.

The DDP could be assembled in paper media using typewriter or stand-alone computer technology, but the most powerful application of the concept requires networked microcomputers tied into the DDP located on a network server. The server is set up with a set of templates for each section that requires a specific application program, a drop box, and a working baseline consisting of all of the work completed to date.

Someone is appointed to manage the database. That person makes sure everyone on the team understands how to gain access to the templates and how to use them to create their product. Each person with a DDP input responsibility, as defined in some variation of Figure 6.1-9, copies the appropriate template to his or her local workstation and proceeds to enter his or her work. At points in time defined by the DDP

data manager, each contributor drops a copy of his or her section in the drop box on the server. Periodically the DDP data manager looks in the drop box for input. New inputs are placed in the working master by the data manager, the only one with the password to change working master content.

At any time, anyone on the program may gain read-only access to anything in the working master. If the program is equipped with meeting room computer network access and video projection capability, periodic concurrent engineering team meetings may be accomplished by projecting directly from DDP resources. Periodically, even on a daily basis, an approved copy of the working master (complete or a subset) could be transferred to customer access. In fact, in early program phases, this would be a much more effective contract data requirement list (CDRL) item than the piles of reports commonly delivered to customer file cabinet resting places.

As useful as the DDP is in solving the information communication and integration problem during early program phases, the DDP concept is probably not the most efficient long-term solution to a company's information needs except on a small project. But, given that a company does not now know what its aggregate system development information needs are or how those needs relate to long-term information needs, the DDP concept can provide a manageable growth path from ignorance to understanding. At the terminal end of this path, after applying the DDP concept on several programs, the company will have an excellent understanding of its needs and be able to phrase requirements for their information system builders on their route to a capability in model-driven development.

6.1.4 Program Closeout

At the end of a program, the program team should archive the specifications so they can be easily returned to an active status in case the program comes alive again. Some programs, frankly, never seem to end. A good example is the

Boeing B-52 program, which will probably run from the 1950s through the 2050s. In 2004, the Department of Defense elected to make the spiral development model its principal approach to system development and, in combination with the policy of cost as an independent variable, promises to stretch programs in time, making it necessary to maintain specifications over a longer term.

Ideally, a program would conduct a postmortem, identifying things that went well and otherwise. Where some activity did not go well, the closing program should recommend to management that changes be made in the resource base supplied to programs by the functional department structure. On a program with a long life, it might be wise to conduct this discussion periodically.

6.2

Requirements
Risk Management

Contents

6.2.1 Validation and Risk

Requirements validation is a subset of program risk management. It is the component of the risk activity that is focused on the quality of the requirements defined for items and assessment of our ability to satisfy them in the design process. Figure 6.2-1 illustrates this relationship. The requirements that flow out of the requirements analysis process should be evaluated in the requirements validation process to determine if they are necessary and, if so, whether or not we have confidence that we can synthesize them, that is, create a compliant design. This is the form of validation that ANSI/EIA 632 refers to as "requirements validation." ANSI/EIA 632 also describes an activity, called "product validation," where users have an opportunity to determine whether the product satisfies their requirements, and I believe this to be an element of verification referred to by DoD as "operational test and evaluation (OT&E)."

The answer to the question of whether we will be successful in synthesizing the requirements is a function of the nature of the requirements, the requirement values, the technology available to us, and our skill as designers. Therefore, the answer to the question cannot be provided strictly as a requirements issue. Rather, it must be answered in context with the skill with which we can synthesize the requirements. We may have to at least define a credible design concept to gain confidence that we can satisfy the requirement. So, validation can be thought of as an early part of the transform between requirements and design development.

There are several techniques that can be used within the validation process to gain confidence that we can satisfy the requirements, including simulation, testing, analysis, demonstration, and examination. Where a key requirement promises to require a considerable period of time for us to gain confidence in our ability to satisfy it, we should apply technical performance measurement (TPM). TPM is a metric approach for tracking in time the required value relative to the current demonstrated capability. A particular engineer is assigned the responsibility for closing the gap. This principal engineer must maintain a control chart and a clear plan of action to drive toward compliance.

Because every program must marshal its scarce resources and selectively apply them to first priorities, it is useful to evaluate all risks, including the potential problems we may have in satisfying requirements, in the context of the two parameters illustrated in Figure 6.2-2. The requirements that pose the problems most likely to occur with the most severe consequences should rank high on our validation list and be pursued with the greatest energy.

There are methods to translate these risk categories into numerical probabilistic values. Some program managers are comfortable with numerical values, but many are not. It is generally true that the consequences and occurrence probabilities are subjectively determined; so, at the heart of all numerical systems is a subjective underpinning, perhaps with a finer granularity than suggested by the Venn diagram offered by Figure 6.2-2.

All risks, including requirements needing validation, can be partitioned into three program parameters: cost, schedule, and performance. While everything can be related to cost, it is useful to relate a risk to a finer granularity than that, and these three are very commonly used. Most requirements pose performance risks, but a broader view of the problem would suggest that most anything can be accomplished, given enough money and time. If a customer absolutely needs a product satisfying a particularly difficult requirement, and it is only a matter of piling up enough time

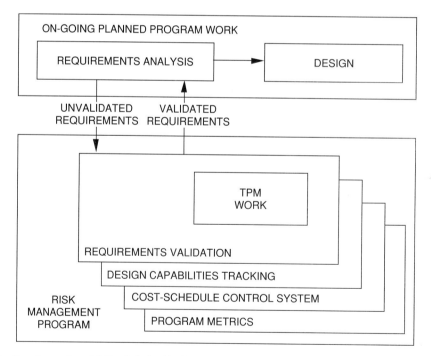

Figure 6.2-1 Requirements validation is imbedded in the risk program

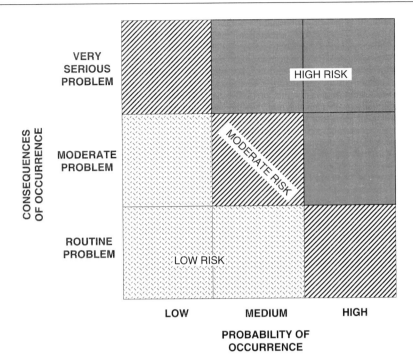

Figure 6.2-2 *Risk level assignment and display*

and money, the risk of failing to satisfy the requirement may be more meaningfully related to a cost or schedule risk associated with success in satisfying the requirement. Some people working on advanced programs that push available technology in several directions at once prefer to include technology as a fourth risk parameter.

Because risk is future oriented, it is very difficult to accurately characterize related data numerically, or sometimes even qualitatively. This problem, however, is insufficient grounds on which to base hostility to risk management. The intent is to identify potential problems so that we can avoid future surprises and to characterize those problems as accurately as possible as a basis for work focused on avoiding the adverse consequences of those problems should they occur. We may make mistakes in initially characterizing the probability of occurrence or consequences of occurrence, but if we continue to concentrate on avoiding particular outcomes, our estimates will become more accurate over time. If we simply disregard future possibilities and let things happen, we will most often be disappointed with the results.

Risks are realities that good managers should avoid by identifying and mitigating them as early as possible. As a cross-check on our decision-making process, we should find no requirements risks that are not embraced by our validation process. The planned validation work for the requirements for an item essentially becomes a major piece of the risk mitigation plan for the item. It is true that some programs must accept considerable risk of failure to press the state of the art, but this can be done with as much safety as possible by consciously trying to foresee the nature of problems that may occur.

Risks come in several varieties, and sometimes one of these risks is more important to avoid than the others,

resulting in increasing one risk to reduce another. For example, a system may have a very pressing delivery date requirement and a very difficult technical requirement for navigation accuracy. In the interest of mitigating the delivery and technical risks, the developer may elect to procure two different navigation units in a parallel procurement, in the hope that one of them will satisfy the demanding accuracy requirement in time to satisfy the delivery requirement. In pursuing this course, the developer may build a cost risk into the program of less concern than the other risks. Both of these sets may have to be carried as alternative installations for some time until the accuracy and delivery date issues become more clearly defined. Throughout this period, the developer will have to pursue validation and verification of the requirements for both of these equipments.

6.2.2 The Validation Time Span

Validation properly takes place over the period beginning with the identification of system requirements and ending with a credible preliminary design solution for all of the elements of the design. For most system elements, the end will correspond with an approved conceptual or preliminary design, and this is a good goal in general. For other items characterized by major concerns and challenges, the validation process may stretch further into the design process. We should understand that the longer it takes to gain confidence in our ability to implement a design compliant with the requirements, the longer we must carry with us the burden of risk that we may not be able to arrive at a sound solution or that we will select a bad design that later must be either changed or accepted with regret by our customer. All of the requirements validation work should absolutely be complete by the time management and/or the customer concludes that the design is complete. In the case of Department of Defense programs, this occurs at the milestone called "critical design review (CDR)," commonly coinciding with 95 percent or more of all planned engineering drawings having been released.

Some readers, especially those with engineering design experience, may feel that the wrong slant has been placed on the validation process in this section. They will feel that validation is a design process and not a requirements-related process. Clearly, it is both. It is through what I call "validation" that we gain confidence that it is possible to synthesize the requirements into a viable design. So, many would say that this is a process of validating the design. For those who feel this way, let us agree that we are both right. Validation is one component of a sound risk-management process that embraces both the requirements definition and design solution processes. Validation applies to the transform between the requirements defined for an item and the design concept created in response.

Between the requirements and the design solution there exists a chasm with the rims separated by the creative design process that takes place within the mind of one or more gifted engineers. The system engineer must not become depressed over the lack of traceability between these two but rather should seek to build the requirements definition as carefully and quickly as possible and, in concert with the designer, designers, or design team, reach a conclusion about their ability to translate these requirements into a compliant design. The validation process spans this chasm between the ordered world of requirements and the chaos (no offense intended) of creative design work.

As requirements for the system and its elements are identified, they should be partitioned into those that will require formal validation and those where, in our good engineering judgment, the cost of formal validation is not warranted by the likely results. So, there is a requirements-oriented component to validation. But, the validation process also extends into the design process because we wish to gain assurance that we are capable of satisfying the requirement. The design concepts conceived by designers may have to be subjected to careful scrutiny through test and analysis to prove that they can be effectively employed. If, for example, the needed system functionality requires a computer speed faster than any machine yet developed, we would be remiss not to build a breadboard or brassboard model based on new technology and test it to show that the technology was sufficiently mature and effective to solve the problem.

There is a natural reluctance to submit requirements or a design concept to validation action in any public way because it implies that they might be wrong. The egos of the designer and management can be caught up in the decision logic leading to a foolish feeling of infallibility. This is very easy to fall prey to because it occurs early in a program, when you have just finished building a winning proposal and everything appears new and possible. You begin to believe the optimistic statements you have been feeding to the potential customer through your proposal and subsequent meetings. The systems engineering community must remain firm at this point in its encouragement of the need for validation work where it is appropriate. The system engineer interested in pursuing validation should recognize that it may be career limiting, because the practitioner appears to be the only obstacle to program progress. So, the need for validation should be sold to the program through the proposal or business planning process and not left under the surface until program implementation. Successful enterprises will make it a fundamental part of their way of doing business.

6.2.3 Avoiding a Null Solution Space

Many system engineers believe that it is impossible to overdo the job of requirements definition because the function of requirements is to "specify" or state in full and explicit terms. This mind-set is encouraged when a system engineer creates the specification and a designer must implement it; that is, two different people do the work. This attitude is also more pronounced in system engineers who have never served as design engineers. We must always keep foremost in our minds that the whole-systems approach to the development of systems to solve complex problems is driven by a need to protect the specialized design engineer or team from knowledge deficit while providing them with the widest possible solution space. A requirements analyst should seek a condition of requirements completeness and sufficiency, not an overpowering number of requirements flowing from excessive zeal. The best specification is not the heaviest possible specification.

All requirements do constrain the design solution space, as they should. Too many requirements constrain the design solution space too much, possibly preventing the designer from selecting an alternative that could be superior to at least some of those allowed by a more restricting set of requirements. When we are doing requirements analysis, there is no simple technique for limiting the requirements we identify to only those that are absolutely essential. There is no way that computers can do this for us, no matter what tools we bring to bear.

The best protection against overspecification is to apply structured requirements analysis processes, such as those described in Chapters 3 and 4, that encourage us to identify only the necessary requirements. These methods also encourage completeness and therefore sufficiency. A second way to prevent requirements excesses is to perform an ongoing traceability analysis during the requirements analysis process. Where it is not possible to establish a traceability link from one item requirement to a parent item requirement or to a structured analysis process, we should ask whether the requirement is really necessary. The requirements in performance specifications should also be devoid of solutions. They should be design independent. We should ask ourselves if each of the requirements defined for an item is necessary. If the requirement were not included, would it be possible to create a design satisfying the remaining requirements that failed to interact synergistically with all other items to satisfy higher-order requirements all the way to the customer need? If the answer is yes, then the requirement should be retained.

Finally, we should supplement a structured approach to requirements analysis, a watchfulness for excluding design content, and an appeal to the discipline of traceability with a selective formal validation activity. The term *validation,* in this book, refers to a process of proving the need for particular requirements, the validity of their values, and that it is possible to synthesize them into a design solution. As noted above, it is possible to define requirements for an item so that they unnecessarily limit the solution possibilities. It is also possible to limit the solution space to a null, so that it is impossible within the laws of science and finance to solve the problem.

Ideally, we would never stray from the norm of capturing only the necessary and sufficient requirements. But requirements are identified by humans, and humans can make mistakes. Therefore, it is possible that we will fail to meet this high standard. Because it is better to discover a null solution space or one unnecessarily restricted by requirements earlier than later, we should seek to uncover these conditions as early as possible. An effective validation activity can be used to detect and correct this condition during the requirements definition process.

It is probably true that the validation work should include at least one person who was not deeply involved in creating the specification content. A fresh mind not contaminated by past wrong thinking will generate many questions that would not occur to the participants, due to the limitations imposed by groupthink. These questions may sting the egos of those who developed the requirements more than a little, so the process should be implemented by secure, mature people. But, even if it results in fistfights, it is a valuable process that will prevent many problems later in the program when there is less flexibility to handle them. Criticism is very inexpensive, and we should take full advantage of it. One who writes foolishness cannot see it, but everyone else can.

6.2.4 Validation Process Description

6.2.4.1 Overview
Figure 6.2-3 illustrates an overall process for validation as it can be applied to a single item for which the requirements are being determined. This process fits within the

"REQUIREMENTS VALIDATION" box of Figure 6.2-1. If the terms are new to you, for the time being simply try to absorb the process flow alternatives. We will discuss each path in some detail in this section. Validation should be applied to requirements as we identify them and early in the period when the designer or design team is working to synthesize them into a concept or preliminary design solution. Figure 6.2-3 focuses on a single requirement for one item, and this process must be applied to hundreds or even thousands of requirements for many different items on a program. Think of this process as a sieve through which we pour the requirements for an item. The decision logic steers the requirement through the chains of activity.

On a large program, we may have a number of sets of requirements in various stages of the validation process at any one time, and the status of these requirements with respect to the validation process will be changing over time as well. So, we need a means to track the status of the validation process throughout the requirements analysis and design synthesis processes. We will introduce matrices for this purpose, but the ideal method is to have this capability built into the tool used to capture the requirements, accomplish traceability on them, and print specifications containing them. While many computer tools supported verification, few supported validation and requirements maturity tracking at the time this book was written. Ideally, that will change.

As shown in Figure 6.2-3, we must first check for completeness and screen identified requirements for correctness and necessity in the "Evaluate Requirements" task.

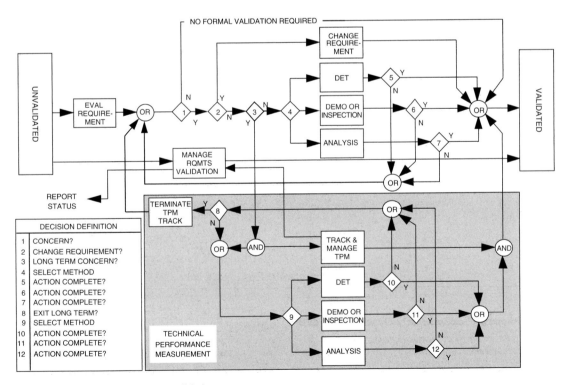

Figure 6.2-3   *Item requirements validation process*

Surviving requirements must then be screened for validation action based on our evaluation. We may conclude that there is no need to formally validate a particular requirement in accordance with a specific criterion, in which case the requirement can be passed on through to validation complete without any formal validation activity. Some requirements may appear very difficult to satisfy because they have never before been satisfied by our organization, and they entail a great deal of risk. Our first act should be to challenge the need for them and the difficult value assigned. In some cases, we may conclude the requirement is unnecessary, in which case it should be deleted. Otherwise, it may be possible to gain customer approval or internal agreement on a change to the value that gives us more confidence.

Requirements with surviving concerns should be passed through one of two remaining channels, both of which are essentially the same and are differentiated by the period of time we expect it to take to complete the validation process. A requirement that will surrender to a near-term validation action, such as a design evaluation test (DET), an analysis, an inspection, or a demonstration can be passed through the upper channel on Figure 6.2-3. If the validation action is going to take a considerable period of time, we should track the work through a technical performance measurement (TPM) program, to be explained in more detail later in this section. If we choose TPM, we have essentially the same situation as far as validating the requirement is concerned, with the difference being that we are tracking the situation more intensely over a longer period of time.

As noted by the recycling possibilities, we may have to apply several techniques in serial over an extended period of time in parallel with preliminary design efforts. The use of decision symbols in Figure 6.2-3 implies a single channel selection process, but we could carry a requirement in several channels simultaneously. Perhaps inclusive "OR" symbols, rather than decision symbols, would have been more appropriate for this reason. For example, we may conclude that the value for a requirement will have to be validated through simulation (a combination of DET and analysis). We may also have a concern for our ability to synthesize the requirements, with our current technology leading to a technology demonstration. The customer may also direct that we track the value as a TPM parameter, regardless of how long we believe it will require to resolve the validation issue. Our challenge is to determine what combination of paths to apply for each requirement that should be validated and to extract ourselves from validation as quickly as good judgment dictates.

Validation work may all be initiated during the requirements analysis process or later when, during the preliminary design process, the design team becomes concerned about their understanding of the requirements or their ability to synthesize them. In either case, the validation process is concerned with requirements for items and should not be disconnected from them. The more general risk-management process may involve risks that are not product-requirements related, so we should think of the requirements validation activity as only a component of a larger activity. One could conceivably run TPM tracking on programmatic or functional management parameters, but these will be referred to as programmatic and functional metrics, respectively. TPM commonly is thought of as related to only product requirements value tracking and reporting, but it is a subset of a larger metric family, as are the techniques of cost and schedule control. Figure 6.2-4 illustrates the rela-

tionship between these several activities in a multilayer Venn diagram. The double crosshatched area represents the overall risk management activity for a program, while the requirements validation activity space relates to the material discussed in this book. You will note that TPM is completely a subset of the validation process while validation is but a part of the overall risk activity.

TPM tracks both the required value and also the current capability of the design solution over time. As the design solution matures, the design people should maintain an understanding about the capabilities of the design solution corresponding not only to those parameters selected for TPM tracking but for all design features. Ideally, the designers should be working to satisfy the requirements in the design capabilities but not exceed them significantly. To the extent that design requirements are exceeded, overdesign is introduced into the design, generally along with unnecessary cost. This is an area that systems people should focus on in doing integration and optimization work on the evolving design and is not related to the validation process.

Figure 6.2-4 illustrates a situation where there are three programs in development by the company at one time. Each must manage its own risk program and the related validation work, as well as the other activities that derive information from the metrics work indicated. All programs may also contribute information to functional metrics that should provide useful management information back to the programs individually as well as comparative between programs, which may help a program gain insight into things that could be done better.

On completion of the planned validation process shown in Figure 6.2-4, the requirements can be considered validated. The possible outcomes for a given requirement as a result of the validation work are: Delete the requirement, change the requirement (its value or context), or retain it as currently written. The decision should be based on facts derived through the validation process. The validation process produces evidence of validity of the requirements or a rationale for changes.

6.2.4.2 Initial screening of the requirements for validation

Figure 6.2-5 expands on the "Evaluate Requirements" task of Figure 6.2-3. The first filter applied in the validation process must be carried through the program until the specification is approved, and it asks whether we have identified all of the appropriate requirements. This is not an easy question to answer because it inquires into things not then known. There is no magic potion that can be applied here, but the structured analysis process offers the best assurance that nothing important has been missed. So, two criteria are advanced as follows:

1. If we have applied structured processes, have we missed anything in those processes? Are our models complete? If we did not use a structured process, there is little hope of finding unidentified necessary requirements.
2. Are there parent item requirements that do not currently flow down to this item? All need not, but those that do not should be consciously reviewed for whether or not they should.

The first filter to be applied to identified requirements is whether the requirement is necessary. If it is not necessary, it should be rejected or deleted. These requirements may

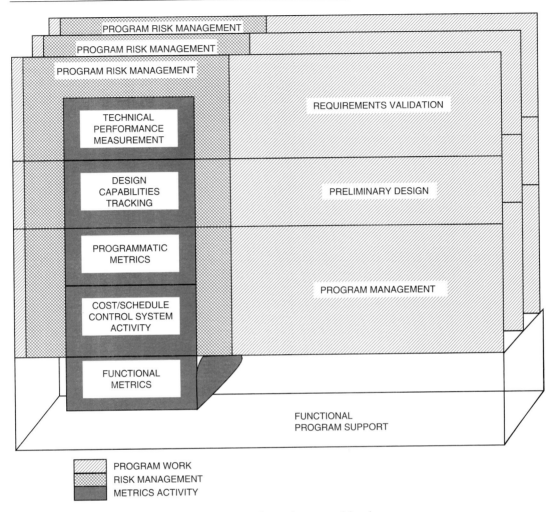

Figure 6.2-4 Correlation of validation with the metrics and program risk universe

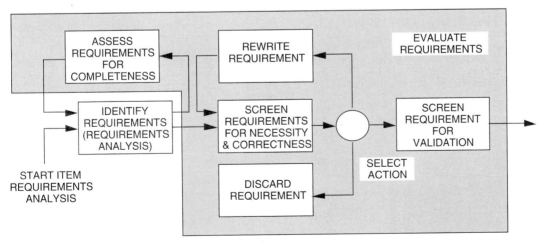

Figure 6.2-5 Evaluate requirements activity

have to be identified in a transitory way as requirements subject to validation, but they should rapidly fade from the scene as they are proven to be unnecessary. An unnecessary requirement invariably adds unnecessary cost and makes it more difficult to derive a compliant design. Criteria for necessity are offered as follows:

1. If the requirement is not recognized, can a design solution be derived for the system that fails to satisfy the need or for the item so that it fails to interact synergistically with other items to the end that the need is not satisfied?
2. The requirement is not contained in another requirement for the item.
3. The requirement is traceable to a parent item requirement or to an entity in an approved structured analysis model from which it was derived.
4. The requirement should be design independent rather than simply narrowing the design solutions directly.

If the requirement is necessary, we must also ask if it has been properly characterized. These questions should include the following:

1. Is the value correct? We can gain confidence in requirements values that were derived from a system math model for the parameter in question because it enforces discipline on value assignment. Simulations properly constructed and operated may also give us confidence. We should be concerned about values for which the person who assigned it cannot give a reasonable rationale.
2. Is the requirement stated correctly? Aside from correct use of the language, the requirement should avoid unnecessary difficult and very specialized words. The requirement should be stated in positive terms rather than what to avoid.
3. Is it possible to obtain more than one meaning for the requirement? The requirement should be perfectly clear so that two or more people will not derive different meaning. One could also ask if the statement makes any sense in a very practical way.

If the requirement is not perfectly stated, it must be rewritten to overcome the critical findings and dumped back into the evaluation pool. This process may require several rounds before the requirement clears the evaluation. To the extent that requirements are commonly found poorly written, the principal engineer, team leader, or other person responsible for the item should find out why and take steps to correct the problem, as it is parasitic in nature and will reduce the resources of time and money available later in the program.

Figure 6.2-6 summarizes the different subsets that a particular requirement may fall into. Up to this point, we have discussed requirements that are not in the current specification that should be and requirements that are in the current specification that should not be. Figure 6.15 breaks the former into two subsets, those previously deleted and those never contained. In both of these cases, it is possible that the specification should not contain them, but it is also possible that a requirement could be deleted in error or not recognized during the analysis. All of the proper content not included and improper content included represent risk, and the validation process should encourage these subsets (identified as a crosshatched space) to go to null.

6.2.4.3 Validation intensity selection

Next, we wish to determine whether the surviving requirements, in the current proper subset of Figure 6.2-6, should be subjected to formal validation. Your organization should have some predefined criteria for selection of the appropriate validation action. If you do not now have such criteria, you may wish to consider the following suggested criteria:

1. Is the value known? If the value is not known, is there a known process in work that will result in a credible value?
2. Does the development organization have confidence, based on data, that it will be possible to satisfy the requirement?
3. Is there a history of success in satisfying this kind of requirement by the development organization?
4. Will the development for the item in response to this requirement entail only technologies with which the development organization has experience and knowledge?

If the answer to all of these questions for a given requirement is "yes," then there is little need to accomplish formal validation action for the requirement. If the answer to all of these questions is "no," it is likely that some kind of formal validation action is needed. In between these two extremes, one must rely on experience and judgment to reach a conclusion about the need for formal validation.

Ideally, these questions should pass through our mind as a matter of habit as we are initially conceiving each requirement; we can thus avoid identifying some requirements in the first place. But we should also pass through a formal requirements evaluation step as the specification content takes on a nearly complete form and before it is initially released. The results should be reviewed at a requirements and design concept review prior to the team being authorized to begin preliminary design. A requirements database should be structured to capture the validation status and to track that status to closure. In the best of all worlds, the analyst would make the binary validation decision when the requirement is first entered into the database. The principal engineer or team leader for the item might review that input and make a final decision and define the validation method and responsibility.

Requirements validation does cost time and money. Therefore, formal validation should be selectively applied. For all of the requirements identified for an item, we should first partition them into two subsets, those for which there is little concern and those that should be subjected to some form of validation. As a result of this binary partitioning, conducted initially early in the program and maintained throughout the requirements identification and preliminary design processes, we should focus one of two principal paths through Figure 6.2-3. The top channel is appropriate where it will be possible in fairly short order to complete the validation action entailing test, demonstration, inspection, or analysis work. The lower channel is appropriate where we will have to track our requirements maturation and design solution process over an extended period of time and provide management with current and historical data on our movement toward completion of the validation work.

The requirements to be formally validated may be included in one or more of four management streams: risk management, TPM, TBD or TBR items tracking, and formal validation. All of the requirements excluded from these

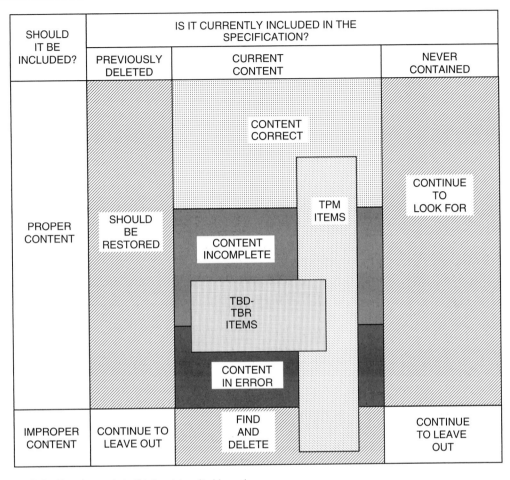

Figure 6.2-6 *Requirements validation intensity hierarchy*

four subsets will have essentially been informally validated by a cursory analytical exclusion. A purist may feel that this will open us to risk because not all requirements are being formally validated. The problem is that we have to obey the fundamental tenets of economics as well as engineering. We have limited resources on every program and must set priorities, recognizing a condition of balance between the different kinds of risk. If we attempt to formally validate every requirement through test and analysis, we may relieve the program of all performance risk except that we may have exceeded the available program funding and realized a cost risk. Program management entails balancing risks and recognizing that it is not possible to study every possibility in complete detail. Wherever possible, we

have to make system development decisions that exclude the need for work based on our experience and history and focus available funding on the most serious issues facing us.

6.2.4.4 Formal requirements validation management
Where the conclusion is drawn that formal validation action is needed, we should create a tracking matrix to clearly identify the responsible parties and needed validation actions, summarize current status, and offer reference to a simple plan for future validation work. Figure 6.2-7 is structured for system-level tracking of all validation actions. If you are using a database system with validation tracking capability, it may be possible to follow this model, but it is uncommon for commercially available tools to provide this function. An

ITEM	REQUIREMENT	METHOD		VALIDATION	PLANNED ACTION
A1342	3.2.5.7	Analysis	Validate value in simulator	Jones	10-30-96
A1342	3.2.5.9	Test	Tech demo our capability	Burns	11-15-96

Figure 6.2-7 *Requirements validation tracking matrix*

alternative is to maintain a separate spreadsheet or database for this data, but this violates the concept of data integrity in that identical data items will be located in more than one place. This will require extreme vigilance to avoid divergence of data entered in different systems. If you are using word processors to create specifications, you could include a validation table in Section 6 while the specification is being prepared. This matrix would not need the item column, and the matrix should disappear when you have completed all of the planned validation actions. This condition should occur before final release of the specification.

A better place to put this table is in an integrated validation and verification document, where it could be used to offer current status of the validation effort. You might also simply refer, in that integrated document, to this matrix, published separately in memorandum report format periodically with revision letters applied to the memo number.

The matrix is adequate to identify and track all of the validation actions, but each validation principal engineer should be required to maintain a current action plan and schedule defining planned work and giving current status. This can be done in a very simple way and does not require extensive documentation, even in a DoD program. A final report should describe results of the validation work. If the work involves several actions, it may be necessary to link up multiple test and analysis reports under the cover of a primary report.

6.2.4.5 Validation through risk management

One or more of the requirements that have been caught by the validation screening activity described above may pose sufficient concern on the part of management to be selected for inclusion in the risk-management program. We are concerned here only with requirements risks, but all risks identified should be included in a risk list for the program and assigned to specific contractors, teams, and persons. It should be clear that the responsible person and organization have the responsibility to fully understand the risk, report those findings, find ways to mitigate the risk (ideally to zero), maintain records of risk mitigation work, prepare and maintain a risk action plan, periodically brief status, and implement approved action plan elements. There should be a definite time when planned mitigation actions must be complete. The action plan may call for development evaluation tests, analyses, inspections, or demonstrations, as noted on Figure 6.2-3.

Generally, the risk associated with requirements will be carried as a technical risk where difficulty is expected in synthesizing the requirement. But, if the problem is stated in terms of the cost or schedule to achieve a result that is known to be possible, then it may be carried in one of those subsets. Likewise, if a program uses technology as a fourth risk category, and the risk is perceived as unavailability of the technology to achieve the requirement, it may be carried in that category. So, the work that will be undertaken to mitigate a requirements risk will have to be colored as a function of the risk category chosen. It may focus on ways to get cost or schedule time out of the design and development process. Alternatively, it may call for one or more technology demonstrations structured to obtain access to new technologies supporting design solutions or manufacturing techniques capable of satisfying the requirement.

One of the most difficult design problems the author ever observed was the design of the wings for the advanced cruise missile in the mid-1980s at General Dynamics Convair (part of Hughes Missile in Tucson at the time this book is being written). The task fell to a terrific mechanical engineer named Bill Doane. Bill had to satisfy all of the normal aerodynamic requirements that resulted in sufficient lift to sustain normal flight within the context of the wings being initially contained within the fuselage prior to launch. That is, the wings had to reliably deploy after launch from a B52, B1, or B2 aircraft. This requirement placed obvious limitations on the size and configuration of the wings. In addition, the missile had to be unobservable by hostile ground or airborne radar, and wings can be excellent reflectors of radar energy. The missile required long range, so low empty-missile weight was very critical to provide for a large fuel load. A successful wing design matured over a long period within the context of numerous technology demonstrations, risk mitigation sequences, trades studies, and design concept developments, balanced against ongoing design work on many other parts of the missile often in conflict.

Because not all program risks are driven by product requirements, even if your requirements database includes validation and risk tracking, it will not satisfy the total program need for risk management reporting. Ideally, one would be using a relational database system with linkage between the requirements database and the risk database. The only way this could be done at the time this book is being written would be to build the database system yourself. Hopefully, as time progresses, the requirements toolmakers will begin to see the benefits to themselves in a standardized and open tool interface architecture that permits one to hook up the internal company relational databases to their commercially available tools.

6.2.4.6 Technical performance measurement

A further subset of the requirements validation activity should entail technical performance measurement (TPM). As in the case of risk items, we must be selective in picking TPM parameters, because each does entail small but real budget and schedule impacts. The program for a very complex system may have only 25 parameters at the system level. The Lockheed Martin/U.S. Air Force F22 program at one point included over 100 parameters at several levels. The number of parameters should be selected for consistency with the way the program is going to be managed. Each organizational structure (program, team, principal engineer) may have assigned TPM parameters chosen from the requirements contained in the corresponding specifications.

The TPM process reports the historical evolution of a requirement value keyed to past and planned program events as a requirement value versus time comparing the required value (which may change over the development period) with the projected or achieved value. Figure 6.2-8 illustrates an example of TPM reporting documentation. The first chart tracks parameter value in time, and the second chart provides the action plan for future work to reach closure. The reason to do this is that management gets insight into how well the program is doing in terms of a limited number of "test points." We should select the TPM parameters with this view in mind, that is, what are the best test points to use in determining the health of the evolving design solution?

As in the case of risks, all TPM parameters should be listed in a management matrix, similar to the validation tracking matrix shown in Figure 6.2-7. This matrix can be extended to include TPM parameter identification and management information; but a better solution, once again, is to link a relational TPM parameter database to the validation

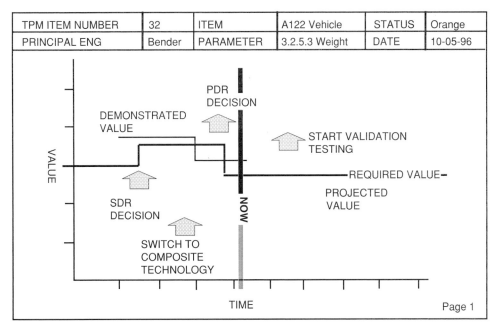

TPM ITEM NUMBER	32	ITEM	A122 Vehicle	STATUS	Orange
PRINCIPAL ENG	Bender	PARAMETER	3.2.5.3 Weight	DATE	10-05-96

a. TPM Parameter Value Chart

TPM ITEM NUMBER	32	ITEM	A122 Vehicle	STATUS	Orange
PRINCIPAL ENG	Bender	PARAMETER	3.2.5.3 Weight	DATE	10-05-96

CONCERN: Current weight exceeds required value by 10% even after a switch from aluminum to composite center section structure.

PLANNED
ACTIONS: 1. Continue to search for ways to reduce weight in other structural sections and on-board equipment. Complete by 11-15-96.

2. During validation testing of the structure, be alert for opportunities for mass removal without adversely affecting strength. Testing will run between 12-01-96 and 04-30-97.

3. Investigate composite outer wing panels and tail surfaces. Complete by 01-15-97.

CRITICAL
POINT: CDR is 08-15-97 and required value must be attained.

Page 2

b. TPM Action Planning Chart

Figure 6.2-8   *TPM parameter documentation*

database (in turn linked to the requirements database). This summary report should identify each TPM parameter, indicate the current status (see Table 6.2-1 for one set of status designations), and name the principal engineer. The database should be structured to print out the TPM documentation charts and status matrix.

6.2.4.7 Requirements maturation control

Throughout the early requirements definition period, we are faced with the gradual maturation of requirements knowledge. Figure 6.2-6 identifies requirements for which an appropriate value has not been determined as TBD-TBR items. TBD means "to be determined" and TBR means "to

Table 6.2-1   *TPM parameter status designations*

STATUS	EXPLANATION
RED	Serious problem exists that must be solved.
ORANGE	Parameter is marginally not under control, but there is a credible plan to get there.
GREEN	Parameter is under control.
BLUE	Parameter is well under control, with possible excess capability, growth potential, or suboptimization.

be resolved." Some people prefer the TBD designation and others TBR. Some organizations use both terms with slightly different meanings. In those cases, TBD is used to mean that no value has been determined, and TBR means that a best guess or challenge value has been included while we work toward resolving the final achievable value.

Values are commonly defined through either an allocation model or an analytical derivation possibly involving modeling and simulation. An appropriate value for most requirements can readily be determined concurrently with identification of the requirement using flowdown based on the parent item value and a logical approach to defining all peer item requirements. Weight and reliability are examples that follow this pattern. Many of the most challenging requirements, however, offer considerable difficulty in value determination through the derivation approach. For example, on a transport aircraft program, we may agonize over optimum gross weight, fuel load, engine power and thrust, and lift and drag figures. These are all interrelated and cannot be easily determined in isolation. Very often, customer needs are not finely determined in early work, leading to some uncertainty about these kinds of requirements.

There is a wide range of opinions on how to proceed with these requirements with value maturity difficulty, including:

1. Identify an initial value and see what happens.
2. Identify goals or targets and see what happens.
3. Initially identify the value as TBD (to be determined) and actively work toward closure.
4. Identify a goal but leave the value TBR and actively work toward closure.
5. Refrain from identifying a requirement until its value has been determined.

The urgency in defining all of the appropriate requirements and their values is at least in part a function of the contractual relationship associated with the specification within which those requirements will appear. If we are dealing with a customer specification (whether written by the contractor or customer), we cannot base a clear contractual relationship on unknown or incorrect values of key requirements. There will be a lot of pressure exerted to release the specification with values, no matter how ill conceived, in order to meet data delivery schedules. This is true of supplier, vendor, or subcontract specifications as well in order to "get the procurement show on the road." This same situation can evolve in an associate relationship, where the two parties and the common customer seek to reach agreement on the contents of an interface control document.

The author would encourage one of the compromise positions, items 3 or 4 on the list above, within an environment that requires timely closure. The other three alternatives are just not very smart. The approaches given in items 1 and 2 are essentially the same. Good luck may be the only salva-

tion. Once the proper value is determined, and it is different from the initial value or goal, there will have to be a cycle of corrective work that simply costs a lot of money. This is a perfect example of faulty system engineering work. The skilled application of the systems approach should lead to a condition of no surprises. The last solution, listed above in item 5, has the same result as the first except that the developer may totally ignore an important aspect of the final product rather than respond to an incorrect value for that parameter.

The alternative in item 4 can be implemented as shown in the following example of an engine thrust requirement in an engine procurement specification:

> 3.2.5.7 *Engine Thrust.* Maximum engine thrust under standard day conditions at sea level shall be equal to 10,000 pounds (preliminary value, final value **TBR-05**) plus or minus 500 pounds.

In the specification, we would have to explain the meaning of preliminary values and how they are identified and that all requirements not so identified are final. We would also have to explain in the contract how changes in value of these preliminary requirements will be covered from a financial perspective. And it is in the contract language that these problems must be addressed. If there are one or more key requirements for which a firm and validated value has not been determined, the contract has to address how the contractor and customer will share in the risk. A fixed-price approach will not protect the contractor where the available technology will not support the value. A cost-plus approach will not protect the customer from a contractor ill prepared to solve the problem. Regardless of the contract type, there should be recognized a class of troublesome requirements, and the contract must tell how requirements may be included in that class or removed from it and how related issues will be resolved. One of the fundamentals in this resolution process is to have a clear list of those requirements. An example of this is shown in the TBD/TBR closure matrix in Figure 6.2-9. This matrix should establish accountability and note a plan of action and time constraints. All of the TBD/TBR items in the specification should be numbered as in the example noted above and coordinated in the matrix. Both notations could be numbered in the same string or in independent strings so long as their notation is unique.

If specifications are prepared directly using a word processor, the matrix could be placed in the preliminary specification notes section under a paragraph titled "Requirements Maturity" and eliminated by the time the specification is initially released. The matrix could also be maintained separately. The ideal method would, once again, entail integration into the requirements tool set, but all commercially available tools do not provide this functionality. The tool should be able to provide a list for a particular specification as well as a

PARAGRAPH NUMBER	PARAGRAPH TITLE	PRINCIPAL	ACTION PLAN	DUE DATE	TBD	TBR
3.2.5.7	Engine Thrust	Logan	Simulation Run	10-30-96		005
3.2.6.8	Fuel Interface	Burnston	Fuel Vol. Study	11-02-96	010	
3.3.5	Lube Oil Pressure	Stone	Max RPM	10-10-96		006

Figure 6.2-9 *TBD/TBR closure matrix*

global program list or count. This is a useful system engineering metric that is very tedious to compute without the automatic computation possible in a computer database.

Figure 6.2-10 gives a database structure subset that supports both TBD/TBR and primitive requirements capture with computer sentence generation. The author has experimented with this structure in a prototype requirements tool and found that all of the normal requirements tool features are supported as well as TBD/TBR. This simply groups all fields together for simplicity of description of the TBD/TBR concept and is not intended to reflect a fully normalized form.

The other TYPE_REQUIREMENT cases envisioned in this database are: (T) text only, (R) reference to an applicable document defined in a companion relational table, and (H) header or title only. The RELATION field values have to account for the following relationships between values and attributes: (L) less than or equal to, (G) greater than or equal to, (S) less than, (M) greater than, and F (equal without tolerance). Requirements of type Q and R would include the key information in the first sentence of the paragraph, and the analyst could thereafter include the content of the TEXT field with additional explanatory information or leave the TEXT void. In text type requirements, the only paragraph content would be the TEXT field content. The user should, of course, be shielded from required knowledge of these codes via the human interface.

With these understandings, case statement code can easily be written to string together the fragments indicated in the SAMPLE DATA column interspersed with simple computer-generated connectives to create the specification paragraph included above in specification format. Code can also be provided to report and manage the TBD/TBR items toward a null condition. An added benefit from this structure is that the requirements analyst need focus only on identifying the attributes to be controlled, appropriate values and units, and the desired relationships between attributes and values. The software code concatenates these data into simple specification language based on TYPE_REQUIREMENT, RELATION, and MATURITY field values. In this case, the computer would generate a boring but clear paragraph as follows:

3.2.5.7 *Engine Thrust. Item* maximum thrust under standard day conditions at sea level *shall be* 10,000 pounds *plus or minus* 500 pounds (TBR-13).

The computer can be easily told to generate the underlined data.

Another benefit from the suggested structure, which carries the value as a separate numerical entity rather than burying it in a string of text, is that the numerical values can be either imported from coordinated specialty engineering models (reliability, mass properties, and so forth) or linked to them such that the desirable information system characteristic is satisfied that any one unique piece of information is stored in only one authoritative place. Margin, budget, and current capability work is simplified as well. Finally, this structure supports automated search for slack and concern cases based on a comparison between required values and current capabilities. In the Figure 6.2-10

FIELD NAME	TYPE	NUMBER OF CHARACTERS	SAMPLE DATA
ARCHITECTURE _ID	CHARACTER	16	A1233
RQT_ID	CHARACTER	3	034
FRAG_ID	CHARACTER	1	1
PARAGRAPH_NBR	CHARACTER	16	3.2.5.7
TITLE	CHARACTER	60	Engine Thrust
ATTRIBUTE	CHARACTER	120	maximum engine thrust
TYPE_REQUIREMENT	CHARACTER	1	Q (Quantified)
RELATION	CHARACTER	1	E (Equals With Tolerance)
VALUE	NUMBER	8, 8	10,000
UNITS	CHARACTER	30	pounds
MATURITY_STAT	CHARACTER	1	R (TBR)
MATURITY_ID	CHARACTER	2	13
TOLERANCE	NUMBER	8, 8	500
CAPABILITY	NUMBER	8, 8	9,200
TEXT	MEMO		

Figure 6.2-10 *Database structure subset supporting TBD/TBR*

example, the engine development is in trouble because capability is 800 pounds short of the required value.

In a commercial situation, there may be no formal customer with whom to interact and no contract through which TBD/TBR action will be controlled. That does not mean that one should dispense with requirements maturation controls. You will have more freedom of action in resolving problems, but you will be constrained by your understanding of customer expectations and time to market forces.

6.2.5 Validation Responsibility and Leadership

Each program of any size must have a systems activity assigned. The leadership in some organizations, impressed by the literature on cross functional teams, has concluded that the function called "systems engineering" was no longer needed, that all of the cross-functional, cross-product, and cross-process work would somehow be accomplished by the several cross-functional product teams, each focused inwardly on its own small part of the overall development problem. Well, that simply will not happen. While cross-functional teams are absolutely necessary, as is their assignment based on product architecture, a system-level team is also necessary, which I refer to as a "product integration team (PIT)."

The overall validation program and process should be under the leadership of the PIT. This leadership should take the form of at least the following provisions and actions:

1. Provide direction that requirements in need of validation be identified for each item undergoing development. Describe for the program the requirements subsets to be respected in the validation program. This may include the classes noted in Figure 6.2-3.
2. Define exactly how these requirements will be identified within databases, word processors, and/or separate media used on the program.
3. Provide the means to identify and capture the requirements validation identification and management information. This may be as simple as a set of matrices for each item, with one or more requirements identified for validation or the logic of use and/or preparation for database use through schema modification.
4. Plan the validation program; review the ongoing work by product teams; provide feedback and direction to those teams on their progress, status, and near-term work; assemble all of the reported information into a program-level status report; brief management on status periodically; and manage the overall program to yield planned results.
5. Identify, plan, and accomplish system specification level validation actions.

We might ask ourselves who should be responsible for identifying the need for validation action. Clearly, the designer or team responsible for the design must identify the need for this testing because it is conducted based on their uncertainty of success and the degree of risk they feel that the program and company are subjected to as a result. Note that this offers the program two failure modes. First, the designer or team may fail to recognize the need for validation action, either through oversight or unwarranted respect for their infallibility and design prowess. Second, validation action may be identified where none is needed through faulty reasoning or excessive caution. In the first

case, the outcome will commonly be surprises late in the program (often during verification testing) that lead to cost and schedule hits and/or a performance shortfall that you are forced to accept because the commitment to a design cannot be undone within the unacceptable cost and schedule constraints that result at the time the problem is uncovered. In the second case, the consequences are added cost driven by unnecessary testing and analysis.

This is a common situation in solving complex problems. You have a choice that is not entirely satisfactory between two bad extremes. But you do have a choice, and it should be consciously made rather than simply evolve as a result of your lack of attention to the details. The use of concrete validation actions also separates the great systems houses from the great integration houses. The latter are very adept at solving horrible problems that arise during production, verification testing, or customer use of the product. These horrible problems simply do not appear in great systems houses because they have thought through all of the possibilities at a time in the program when problems can be solved carefully and inexpensively.

The agency or organization responsible for the overall validation process, ideally the PIT in the context of this book, must recognize these process failure modes and ensure that planned validation actions are motivated by a realistic concern for uncertainty and risk. Validation actions planned by designers and product teams should be reviewed and approved by the PIT as a means of avoiding the two validation failure modes.

6.2.6 Validation Expectations

One way to ensure that a design will satisfy its requirements is to accomplish the design work first and then define the requirements. Yes, this does sound foolish, but it is not all that uncommon that drawings are released before the specification is released in some organizations. While it is necessary to define requirements before creating a design in order for the requirements work to have been at all valuable, it is not done without a certain risk. In our zeal to do a good job in requirements analysis, we may err on the opposite end of the spectrum by defining a problem that has no solution. We can spend considerable time and money working on a design solution, only to find that it is impossible to satisfy the requirements as stated. We should reasonably expect that the requirements will be developed and released in a very timely fashion, that they will be complete and correct, and that they will permit the designer the greatest possible solution space within which to synthesize a solution that will play synergistically with all of the other parts of the system. So, the validation effort can be thought of as a component of the risk-management strategy to uncover any problems with the requirements that would only later be caught after introducing cost and schedule problems.

6.2.6.1 Requirements necessity and completeness
Requirements analysis, when done well, results in a set of organized statements captured in a specification that restricts the design solution space. This set should be characterized as being necessary and sufficient—no more and no less.

If we fail to satisfy the necessary condition, it will be possible for the designer to synthesize the requirements into a design that will not function synergistically with respect to other system elements to the end that the system need may not be satisfied. If we failed to identify a requirement, for example, that a liquid oxygen valve in a cryogenic rocket propulsion system must function at a very low operating temperature, the valve could fail in flight resulting in the loss of the vehicle. You may say, "But that would have been detected in qualification testing of the valve." If the requirement were not in the specification, it is possible that it would be exhaustively tested for that condition. One could argue incorrectly that a test program that examined valve operation at low temperature would be a case of excessive zeal on the part of the test engineer unwarranted by the requirements.

If we fail to satisfy the sufficient condition, we will have overspecified the requirements. The common result is that the cost will be greater than necessary. The increased cost is driven by the greater difficulty of satisfying the requirements. In the worst case, it is possible to achieve a null solution where it is impossible to satisfy the requirements. Thus one of the principal expectations from a validation effort is to encourage the result that we have identified only the requirements that are necessary and that we have sufficiently done so as well. Unfortunately, there is no mechanical way to ensure that the analyst satisfies this condition perfectly. The best ways to encourage this result are to: effectively apply a structured analysis process, maintain requirements traceability, and apply an effective validation effort.

6.2.6.2 Requirement value credibility

All requirements should, ideally, be quantified. Otherwise, how will the designer know how to design the product? How will we test to determine if the product design is adequate for the intended application? Qualitatively stated requirements are an invitation to surprises that are measured in overruns of time and money. But how can we establish a reasonable value for a requirement?

The most common valuation method is budgeting, flowdown, allocation, or apportionment. In this technique, we determine a value for a requirement at the system level and partition that value in accordance with a mathematical rule to determine child values. Many specialty engineers apply this technique, and the following are examples:

1. The mass properties weights model
2. The reliability and maintainability math models
3. Life-cycle cost models

It is very hard to make mistakes when applying this technique. It provides a well-disciplined environment within which to make the numerical value decisions. In the case of weight, the sum of the weights of all of the items at one branch and level in the architecture must be no more than the weight of the parent item for those items, and it is very easy for the human doing this work to determine whether that is or is not the case. The error modes that are possible with this technique are: (1) The system value is too high or too low, leading to either an unachievable or an insufficiently demanding condition; and (2) the values assigned within any one branch at a particular level are not well distributed, leading to local problems that can propagate downward from that point.

We may also include margins in the allocation process to provide management space in the event, that we run into difficulty synthesizing the requirements. A margin is simply an unassigned portion of the value at one or more levels of indenture. In the case of weight, for example, we may allocate only 150 pounds to the next tier for an item that we have previously determined is allowed 160 pounds. We have set aside 10 pounds at the parent item level that we can use during the development process to solve problems the designers encounter. If the child item 2 designer finds that she cannot synthesize the requirements at the weight allocated and also stay within the cost and reliability requirements, she can appeal for an increase in weight allocation from the parent item lead engineer.

This whole process encourages value credibility because it is comprehensive and respects the simple economic principle that the sum of the parts must equal the whole and no more. It is very simple to test the veracity of the models, both by the practitioner who owns the model and those who would audit it.

There are many requirements that will not yield directly to the allocation process, however. And these are commonly the most difficult requirements to quantify. Given, for example, that we have a requirement for aircraft maximum airspeed, how shall we determine the engine thrust needed? It is more complicated than only knowing the airspeed requirement, isn't it? There are many factors involved, including amount of drag entailed in the design, degree of asymmetries tolerated and the surface smoothness planned, and lift versus gross weight considerations. The mathematical relationships among these parameters are complex and cannot accurately be resolved on the back of an envelope. Commonly, we have to apply some form of computer simulation to try various combinations of values working toward identification of a good and achievable mix of values.

In a less grand sense, we can also apply the best mix approach in pairs in what is called "parametric analysis." We may inquire, for example, how reliability changes with cost and determine how much reliability we can afford. Ideally, the cost numbers should include the cost effects of poor reliability on customer loyalty, possibly in the form of two curves. One curve could simply show the effect of the cost of reliability, while the other reveals the cost of unreliability on customer loyalty. Figure 6.2-11 illustrates this relationship.

We could include additional parameters in this parametric decision-making process, but there quickly becomes a point where the human, without special tools, cannot deal with the myriad of possibilities and complexity of the relationships leading to a need for modeling and simulation. Spreadsheet software can be used for low-end models, but specialized, product-specific software will be required for more complex situations. Often this same software or derivatives thereof will be required in the development of the product design so the requirements modeling application is not necessarily money down a rat hole.

Another pathway to value credibility is through historical precedent, either in the form of recognized standards, professional experience on the part of the designer gained from having worked on similar applications in the past, or supervision dictation. These are appeals to authority and can produce valid requirements as well as acceptance of wildly flawed requirements. For all requirements that appeal to authority for their values, it is very important to identify a source and rationale and capture this data in the requirements database or in a tabular form in the notes section of the specification created with word processors. Peers and team or group leadership should subject these source and rationale statements to critical review.

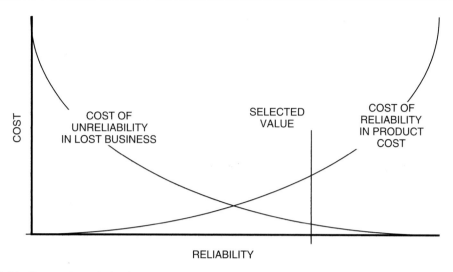

Figure 6.2-11   *Parametric analysis of cost and reliability*

6.2.6.3 Synthesizability

Given that we clearly understand the requirements for an item, have defined what we believe to be appropriate values for them, and have validated those values through the techniques discussed above, there may still be problems in their synthesis. There are two possibilities with any set of requirements: (1) It is possible within the laws of science to synthesize them into a design or (2) it is not. Given that they can be synthesized, based on current science and technology, the synthesis may be otherwise constrained due to cost, available time (schedule), or unavailable technologies. There are, therefore, two sources of nontechnical or programmatic synthesis constraints that we have to guard against as a prerequisite to technical validation of synthesizability—cost and schedule budgets consistent with the nature of the design problem and technology availability.

The designer or design team must review the perceived degree of difficulty in implementing a design based on the approved requirements and programmatic constraints. If the conclusion differs from previously allocated resources in time and money, they must make an appeal to management for a change in the allocations. Those changes may be forthcoming or not, depending on circumstances and the strength of the case made for the change. If the additional resources cannot be acquired, the design approach or technology appeal may have to be changed for consistency with available resources.

The validation process must be accomplished in the context of one or more design concepts. For this reason, many designers think of validation more in terms of design validation than requirements validation. The author has no problem with this conclusion because the validation process covers the transform between requirements and design, but he chose to focus the process on the requirements side of the issue in this book.

Having one or more viable item design concepts that satisfy the item requirements without identification of any associated risks suggests that the item requirements can be synthesized, therefore validating the requirements. To be sure that we have not overlooked anything,

we should have some organized method for keeping track of the relationship between requirements and design features, as suggested in the Venn diagram in Figure 6.2-12. The elements of the requirements set are identified by requirement IDs from the database in use. We could use paragraph numbers appearing in the specification, but at the time the requirements are being validated this can be fairly volatile, whereas the requirement IDs should be stable because they are assigned by the computer in requirements databases and remain stable despite paragraphing and content changes. The design features appearing on engineering drawings or sketches have been arbitrarily identified by codes DF1 through DFn. We could conceivably keep track of the relationship between requirements and design features in a requirements database system; but, as this book is being written, The author knows of no database system on the market that offers this capability.

Clearly, the technique suggested is a traceability technique. The difficulty is in identifying the design features so they can be easily linked to the requirements. The requirements are in text form, commonly in a database within which traceability can be easily maintained with respect to other text information. The design features are graphically expressed on engineering drawings and sketches, and it is difficult to connect these graphical structures in one application to text information in another application. The final answer to this difficulty will have to come through coordination of requirements tools and computer-aided design (CAD) systems. In the meantime, one could identify and maintain synthesis validation traceability by referencing drawing features defined in terms of a feature name and a drawing zone reference. Figure 6.2-13 illustrates one implementation for such a validation traceability record. The understanding here is that a database (ideally an integral part of the CAD system being used) identifies design features using what is called here a "DF ID." The validation traceability database would simply make the connection between the requirements IDs in the requirements tool and the design feature IDs in CAD.

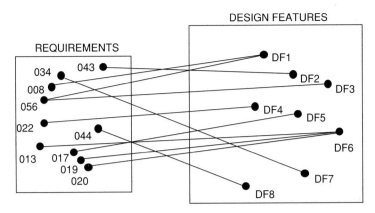

Figure 6.2-12   *Validation traceability*

REQUIREMENT ID	TITLE	DF ID	DESIGN FEATURE	DRAWING REF
043	Power Transfer	DF2	Chain drive	23-3334, ZN 5C
044	Speed-Power Control	DF8	Gear shifting mech	23-3334, ZN 8D

Figure 6.2-13   *Synthesizability validation traceability record example*

These validation traceability techniques provide a comprehensive way of ensuring completeness of the task. The argument can be made quite effectively that on most programs, however, the benefits would not match the cost of implementing. Is there a backup approach that may entail a little more risk but offer some assurance that it will be possible to synthesize the requirements? Yes, one could use the design or peer-review process to achieve credible validation goals. In this approach, two or more reviewers are exposed to the principal design engineer's explanation of the design concept, and they reach a conclusion about the credibility of the design solution relative to the requirements. If the conclusion is that the requirements have not been satisfied, unless it can be shown that it is not possible to synthesize the current requirements, then the designer must try again and submit to

In this book we have referred to this form of traceability as part of the longitudinal traceability case. This path flows from the requirements in the specifications to the design features, to the verification requirements, through the verification plans and procedures, to the verification reports. This whole thread, with the exception of the design features component, can be linked in a requirements database.

6.2.7 Validation Methods

The purpose of validation action is to develop credible evidence of viability. The discussion on values and synthesizability implies that human mental action (which could be called "analysis" in the context or our current discussion) may be solely sufficient to determine if an item's requirements are valid. Neither idle thinking nor analysis is sufficient. The methods that we might apply to accomplish this result are essentially the same as applied in verification. The principal methods involve testing as well as analysis, but

there are cases where inspection and demonstration can be applied effectively as well. The results of this work must be captured in some enduring way, such as in computer media or paper reports, so that they may be referred to over the life of the program.

6.2.7.1 Development evaluation testing (DET)
The testing accomplished in support of the preliminary design process is commonly called "development" or "design evaluation testing (DET)." Given that the designer or team determine that testing is the most cost-effective method for validating a requirement, they must determine how that testing will be accomplished and what resources will be required to accomplish the testing. A set of test requirements should be created and a plan designed to satisfy those requirements.

Testing involves an organized process of stimulating a product (or representation of the product) and measuring responses using special test equipment. The test plan commonly includes a sequence of events, each with a predefined setup and an anticipated response. These test events are structured to prove that a particular design concept offers a viable solution to the item requirements. As a result of the tests, we can conclude with confidence that we either can or cannot satisfy the requirements with current design concepts and technology.

Clearly, testing most often offers the most credible evidence of validation, of the four methods discussed, because it most realistically places the design solution relative to its use and it involves a very controlled stimulus–response exercise that exposes detailed operation characteristics to clear human evaluation. It also most often requires, as might be suspected, more resources in time, money, and supporting resources than other methods. The costs include both labor and material costs. This is the way it should be, of course; the degree of credibility is directly related to cost, and the amount of residual risk is inversely related to cost.

Some examples of validation testing:

1. We have a requirement to provide a particular degree of guidance accuracy that is tighter than anything we have done before. We select a guidance concept, build a breadboard system or buy and integrate off-the-shelf components, and fly the result in a Lear jet with an instrumentation package and autopilot link in accordance with planned mission scenarios. We collect position error data, account for any differences in the control dynamics of the test and planned airframe responses, and compare the results to predictions derived from prediction, analysis, or simulation of system performance. As a result, we gain confidence that a final design can be created that satisfies the requirements. In the process, we also verify our simulation and gain confidence that it can be used to quickly and relatively inexpensively predict performance over a wide range of questions that will evolve during development and subsequently to analyze performance in the context of postproduction engineering changes being considered.

2. We have identified requirements for flight speed, maneuverability, and flight stability for a missile in flight to target. Captive carry space requirements dictate relatively small control surfaces that are unfolded on air launch. We have created a flight simulation that appears to support the current design concept, but we decide to subject the complete airframe to a wind tunnel test, not only for general airframe aerodynamic characteristics (lift, drag, and airframe assembly asymmetry effects) but also for control surface effectiveness. As a second thought, we wonder what would happen if the pilot had to jettison the weapons load in an unpowered state to save the airplane. Further testing reveals that a missile on one station would take one side of the horizontal stabilizer off the launching aircraft. Modifications are introduced that force a missile dropped this way to clear the aircraft in all airspeed–altitude combinations, with the possible exception of takeoff and landing flight conditions.

3. A new sonar system has a requirement to be able to passively detect human-made underwater objects that emit a particular sound spectrum in both fresh- and seawater over a specified depth and distance ranges. To validate these requirements, we construct a test range with salinity control of the water for shorter distances and identify salt- and freshwater test ranges in local water bodies. We intend to determine whether it is possible to detect the required sound sources with laboratory equipment as a precedent to developing design concepts that can also withstand the rigors of naval shipboard use.

4. A new military jet transport must have an electronic flight control system (fly by wire) controls with no mechanical backup. An entirely new airframe, hydraulic system, and control system must be designed. As the control system component and installation designs become clear, a flight control system simulator (iron bird) is developed that duplicates the configuration and operation of the aircraft so that interfaces between the flight control computers (and included software) and the hydraulic system can be tested during development under realistic conditions. Simulated flight control surfaces and hydraulic components must be in the same

relative positions as they would be on the aircraft, so a structural steel frame is built that duplicates the dimensions of the aircraft, with one exception. There is limited space within which to build the iron bird, so the fuselage length must be reduced to keep the simulator components within the available space. Hydraulic lines affected by the shortened simulated fuselage are looped to preserve their true length. All system elements are not available at the time we must start testing, so we begin with laboratory equipment and preliminary models from suppliers that will be modified as the design develops. Flight surfaces driven by the system take the form of structural iron masses adapted to the control system operating interfaces because aerodynamics is not important in this model. As the components are developed, the revised models are installed in place of the currently installed components and tested as part of the entire system. This example illustrates several common characteristics of validation equipment and software. We have to be very careful to duplicate the necessary features of a design concept while economizing on the trivial. Good engineering judgment is needed to make these choices; if it is not present in the team, many bad days will eventually follow.

6.2.7.2 Analysis

Many requirements can be validated very effectively through analysis at less cost than could be done through testing. Analysis requires no apparatus replicating the product or designed to stimulate or monitor product responses. Analysis is conducted by the human mind unaided by external apparatus. The cost in analysis is, therefore, limited to labor cost only (though computing costs may also be involved, depending on how the company financially accounts for computer use).

While the kinds of resources are fewer, it is possible that analysis can cost more than testing in some few cases. Where a human observation is involved, for example, it may be cheaper and quicker to build a simple representation of the product, expose people to it, and evaluate their choices and results than it is to agonize through weeks of analysis over all of the possibilities. The auto industry often will outfit a test car to qualitatively demonstrate suspension or braking system performance rather than conduct a lengthy analysis of performance or computer simulation. The immediate cost is less because of the ready availability of cars and components.

There are cases where testing is impossible or so very expensive as to be prohibitive. For example, it is very difficult to test large items in a zero- or microgravity condition. There are drop towers that can sustain the condition for seconds and aircraft that can fly an arc and sustain this condition for several minutes. But these environments cannot handle very large items nor for very long time intervals. The zero-G performance of a whole space station cannot be tested on Earth because of its size, and it must surrender to analysis as the principle method.

6.2.7.3 Technology demonstration

Demonstration is a process of making something evident through reasoning and exposition. It is a process of showing how some conclusion is supported through actual use of the article or process in question. Demonstration is less useful in validation than verification because you seldom have the physical resources required. During verification work, the product design is complete, and one or more articles have been produced for qualification testing, so the product articles are available for use in demonstrations of their use.

Technology demonstration is most often a special case of demonstration, but it can actually be accomplished through analysis or testing. We may conclude that we do not have access to a technology the designer believes we must have in order to satisfy a particular requirement. Current technology that we have access to simply will not permit a compliant design because the result will be too heavy, insufficiently reliable or durable, or characterized by poor accuracy. A new technology that we have not yet mastered (it could already be in use by our competitors) would solve the design problem and allow us to satisfy the current requirements. In this case, we cannot simply wave our arms and loudly cry during the proposal development period, "We can do it! We're the greatest!" We need some tangible evidence of a new capability we have not previously demonstrated.

By way of example, in the early 1990s there was a lot of interest among space launch vehicle producers, such as General Dynamics, McDonnell Douglas, and Martin Marietta, in finding a way to increase payload weight capability by reducing vehicle structural weight. One promising material change involved the use of a new aluminum alloy with lithium that had evolved from materials research and development work. The problem was how to work it in a manufacturing sense to achieve the needed strength with significant weight savings. The designers needed information on design features that could be manufactured with available or new tooling that would also satisfy demanding structural requirements. Designers were already fully aware of designs and manufacturing scenarios that had proven effective with conventional aluminum alloys and stainless steel, but the knowledge base was not well developed for the aluminum–lithium alloy.

As a result, companies undertook technology demonstrations to gain the knowledge base needed to proceed with confidence in the design of lighter-weight structures providing higher payload performance. These demonstrations involved testing of samples in bending, torsion, and shear under vibration and thermal conditions at levels and frequencies anticipated in the application, as well as evolution of manufacturing methods and tools effective in working the material without degrading its structural properties.

In the interest of simplicity, both demonstration and examination have been eliminated from Figure 6.2-1, but the reader can be simply add them to each string of the methods included.

6.2.7.4 Examination

Examination is a formal or official viewing of an item or process in contrast or comparison with some standard of behavior or configuration. The human senses are the principal instrument of inspection, but simple manually operated measuring instruments may be used, such as rulers, scales, or templates. Specific features of a product or process are observed and studied by one or more humans, generally in a static situation, and features compared with a standard. An inspection of a dynamic condition might better be referred to as a "demonstration."

The most common use of examination is in acceptance of product after manufacturing. Product finish, fit, color, and feature locations and orientations are examples of things that will yield to examination by a quality assurance inspec-

tor. Examination does require a physical reality to be effective, so it is not so commonly employed in validation as it is in requirements verification and product acceptance. On military programs, DoD referred to examination as inspection in MIL-STD-490A but switched to examination with the move to MIL-STD-961. The reason was that the word *inspection* was to be reserved as a general term, and the four detailed methods are all examples of inspection.

6.2.7.5 Combined methods

Often life is more complicated than suggested in the discussion of the four isolated methods. We may have to implement a test or demonstration that produces response data that is very complex and requires analysis to understand. A particular validation action can involve strings of actions, not all of the same method. The four methods offered provide a convenient way to describe all validation actions in a comprehensive way, but it is not all that important what methods are used. It is important that a cost-effective way is determined to validate all requirements.

Simulation offers a very common case in point for combined methods. Early in a program, we create a simulation to help develop a design concept; and, through multiple simulation runs, we uncover a best guess at a set of values for some very troublesome performance requirements. We then develop real hardware and software based on the results of our simulation and update the simulation to replace some of the simulation code with actual product hardware and software. Finally, as the completed product becomes available, we subject it to system-level testing and use the results to verify the simulation. Subsequently, we can use the simulation to predict product performance under new conditions related to modifications or previously untested scenarios. Throughout the development and use of the simulation, we cycle through sequences of test and demonstration, followed by analysis leading to new rounds of testing and demonstration, iteratively tuning the simulation for realism and the preferred product design for optimum performance.

6.2.7.6 Validation by review

You may have reached a tentative conclusion, especially if you have a lot of commercial experience, that this validation process has been overdone and that, as described, it is simply too expensive to implement for the amount of benefit derived. Those with some experience in the development of complex systems will realize that the number of requirements needing something more than a cursory glance in the validation process is very few, perhaps on the order of 5 percent or less. It is common that the requirements offering the greatest benefit from validation are those that most powerfully drive the design solution. It cannot be denied that this thoughtful process of ensuring that the synthesis will be successful does cost money, and the cost can be considerable. However, in the defense and space industry, where some very large and expensive problems have come into view too late to be effectively dealt with, validation is accepted as an example of the famous and expressive Fram oil filter commercial where the mechanic is holding an oil filter and saying, "Pay me now or pay me later."

If cost is a severe program constraint, however, one could implement a partially effective validation effort by insisting that the requirements in each specification be reviewed before release of the document by an engineer with experience on several development programs and who was not involved in the definition of the content of the specification.

This engineer should simply look for problem areas and report them to the specification principal engineer, team leader, or program chief engineer. This simple act, accomplished by someone with no ego investment in the document, will often repay the small cost of three to four hours per specification many times over. This is essentially a peer-review process.

This review process can be expanded by holding in-house design reviews attended by the development team, who present their design concept to a review team, ideally selected from the functional management staff of the company. The presenters should be required to follow a presentation pattern as follows:

1. The team leader or principal design engineer should demonstrate his or her understanding of the item requirements by describing how he or she influenced the design. A subset of the item requirements should be identified that caused the team or designer the most concern, and how the team or designer mitigated these concerns should be discussed.
2. Describe the design solution concept and link the features with the requirements previously discussed.
3. Report on the results of any formal validation efforts undertaken and the status of any actions that have not been completed.
4. Provide time for the person who did the requirements peer review to give a report.
5. Insist that the specialty engineers who reviewed and contributed to the preliminary design work offer a brief statement about the design concept from their specialized perspective. A "no comment" should not be accepted; rather, a clear statement of the viability of the design from their perspective should be expected.
6. Remaining concerns and risks should be cited and the status of the work that remains to be accomplished to mitigate them discussed.

Given that the review team has been populated by people experienced in asking penetrating questions and that they have some knowledge of the product line, this review will pay big dividends. The presenters should not have to prepare special materials to satisfy the needs of this review. It should be possible for them to simply expose the review team to the results of their work using design sketches, results of analyses, and other materials prepared as a normal consequence of doing the design work. If the organization has a computer network on which the product development work has been stored in an organized fashion, it should be possible to project that information directly on a presentation screen in the meeting room where the review is held.

As a result of the review, the review team chairperson should offer a clear response to the development team or engineer supporting the work done to date, if appropriate, or noting specific concerns that the review team still has. Action items should be assigned to accomplish any specific remaining actions to resolve those concerns. These action items should be assigned to a specific person, and a clear completion date should be negotiated with the person responsible for closure. If the review team is generally satisfied with the progress, they may simply require that responses to any action items be reported to the meeting chair as closing action and, so long as the results of those action items do not expose further concerns, accept the design concept and permit the design process to proceed as

scheduled. If serious problems have been uncovered in this process, the review may have to be rescheduled for a specific date agreed upon by the design team or designer.

While reviews are discussed here as an alternative to a formal validation process, the reader should be impressed with the need for a review process regardless what is done about validation. The point is that a sound in-process review can satisfy the *minimal* needs of a validation process.

6.2.8 Product Representations

6.2.8.1 The many views of the product

For many people in engineering, the only view of the product during the development period is of engineering drawings and lists. Just from a documentation perspective, we would also have to accept that the following are representations of the product as well: specifications; test plans, procedures, and reports; analysis reports; and functional analysis data. It could be argued, however, that these documents are also engineering drawings because they can be identified using the same drawing numbering system used for drawings proper.

We would all argue that these documents must refer to a known configuration, and commonly the engineering drawings are used as the master definition of the configuration. After we start preparing drawings on a program, their numbers can be used in combination with dash numbers and revision letters to refer very specifically to items under development. All of the companion documents can be similarly referred to and mapped to their corresponding drawing numbers. Prior to drawings being prepared, architecture IDs can be used as this master configuration identification reference. As drawing numbers are assigned to the architecture items, they may be allowed to take precedence as the primary configuration identification. Figure 6.2-14 illustrates an architecture diagram with architecture ID numbers in the lower right corner of the items represented by blocks. This model uses a base 60 system, making decimal delimiting unnecessary. Alternatively, decimal points can be used to separate levels from one another.

Specific configurations can be defined in terms of sets of documents in particular configurations, forming what are called "baselines." The control of the evolution of these sets of data is the principal activity of a configuration management organization. The methods for accomplishing configuration management are well developed, and most often this job is done very well in engineering organizations. As important as these documentation baselines are, however, they are not the only representations of the product. The configuration of the other representations is seldom managed well. What might these other representations be?

Throughout the early development of a system, special test articles are required to support the validation of certain requirements in development evaluation testing. In the development of an aircraft system, it may be necessary to build a hydraulic control system test bench to evaluate design concepts and their suitability for satisfying required control surface shaft torque under flight conditions. This test bench must comply with the current configuration, as defined by the corresponding product item documentation set. If this test bench configuration includes actuators with different static and dynamic characteristics than those currently identified in the design baseline documentation, the results from any tests involving the test bench are brought into question.

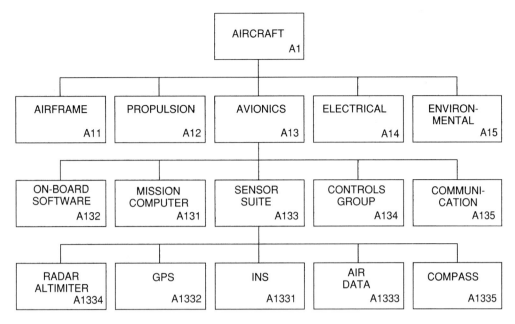

Figure 6.2-14 *Typical architecture block diagram*

If differences between product representations and the current product design go unnoticed and tests or analyses are completed based on those representations, the resultant data may be used to make critical decisions that are flawed, which may lead to other flawed decisions in a pattern that could be very difficult to unravel when we inevitably find that a problem exists.

In addition to special development test articles, other examples of representations include physical models and mockups and mathematical models (finite element analysis, reliability model, weights model, and so forth), as well as simulation software and datasets.

6.2.8.2 Representation identification

The first step in the process of controlling the configuration of product representations is to simply list them all. Table 6.2-2 offers a sample identification matrix showing that we have numbered and named each item. Following that, we have identified the configuration of the representation with a unique drawing number and the corresponding product drawing number. A principal engineer is named who is responsible for knowing the configuration of the item at all times and maintaining it in a specific configuration. Finally, there may be application-specific user data of interest that

one could add to the matrix. We may wish, for example, to have a column for current status.

Table 6.2-5 ensures only that we know, at any one time, the configuration of each representation, provided that we have some way of interpreting the cryptic REP NUMBER. Each line item should be backed up with some viewgraph engineering, one or more drawings, or a more formal set of documentation, all coordinated by the REP NUMBER. This may be as simple as a sketch of the item with features noted with arrows so representations identification and control need not be a costly process.

The process of identifying product representations should begin in the proposal period or commercial equivalent. People from the different functions can be polled or interviewed by a system engineer to uncover representations that people working on the proposal think will be needed during program implementation. Some of these will be identified in proposal write-ups, required material lists prepared in support of the material cost estimate, or be obvious from a description of work required in the statement of work. In some cases, the responsible proposal person may not have overtly identified a representation he or she will need, and it may require some discussion to bring that need to the surface so that it can be listed and included

Table 6.2-2 *Sample representations identification matrix*

Item ID	Item name	Rep number	Product number	Prin eng	User data
001	Avionics Test Bench	34E1212B	34E4242C	E. Jones	
002	Guidance Simulation	34A4511	34E1521	A. Clancy	
003	Reliability Mathematical Model	34A1321	34E1000	P. Gordon	
004	Weights Model	34A5444	34S2400	M. McSwain	

in cost estimates. In other cases, we will find that an item that is listed cannot be justified by the person who added it to the list. It may be possible to achieve the objectives of the representation by other means using resources that are required for other purposes. So, this list or matrix can be established as a part of the proposal work that should be accomplished anyway to coordinate needed product representations with planned work and quantification of the proposal material estimate.

6.2.8.3 Representation management

Given that we have a means to record the identity of each representation and clearly define its current configuration, we should also make provisions for capturing the history of each item. We must know the representation configuration at all times, not just when the program begins. So, we must recognize that the matrix illustrated in Table 6.5 will change over time. As we become better informed about the best product design solution, the design configuration will change. We may also have to modify some representations into different configurations or create multiple versions of them as part of the process of evaluating two or more alternative design solutions.

Ideally, we would immediately think, all active representations should be maintained in a configuration that reflects the current product design status, but that may be both unnecessary and counterproductive as well as costly without corresponding benefit. The first requirement is that we know the representative configuration and be capable of determining its relationship with respect to the current product design. Some representations will have to be applied fairly continuously throughout the development period, and these will have to be maintained in the current configuration. An example of this is the weights model.

There may be long periods of inactivity for some representations following an intense application. During this period of inactivity, the related product design may change several times. The question is: Should we accept the cost burden to upgrade such a representation each time the product design changes or only when and if we have to apply it again? This would be a good factor to include in our identification matrix (as part of the user data). Program management could review the list and place the items in one of several categories as follows: (1) must be maintained current at all times, (2) evaluate the cost benefit for each change and accomplish the change if appropriate, and (3) updates will not be accomplished unless directed by program management.

Audits should be accomplished from time to time of the configuration of active representations, comparing them with the listed configuration. This need not be a costly or elaborate activity. A system engineer can be assigned to audit three items during a two-week period as a supplemental duty. This process can be repeated at other times for other items. Where discrepancies are found, we may wish to modify one or more representations to a specific configuration or only update the matrix to show the current configuration.

Some of the representations will be useful in early program work and others in later program work. Program management should require that the matrix reflect only the active items or that it be annotated with a status field indicating items as active or inactive. When a particular representation is no longer required with some certainty, we may wish to dispose of it or place it in storage in case it may have an application on later product modifications.

Some time after a representation has served its initially conceived purpose and been put in storage, it may be needed again to perform development testing work. The current configuration of the representation and the product configuration of interest must be carefully studied and ways explored to cause equivalence to the extent necessary. It may even be necessary from time to time to deconfigure a representation to explore a product configuration of continuing interest. It is important to recognize that a particular representation need not reflect every feature of a product item it represents. While it would be possible to fabricate a general test article that had all of the features of the planned product, this course of action would probably be more costly and lead to less flexibility than using specialized representations with only the features of interest.

For example, we are developing an air-launched missile and must, at some point in the program, verify that the missile can be safely jettisoned from the aircraft for safety of flight reasons. We could use a complete missile, possibly in a range mode with recovery parachute, or we could use a dead weight with the same size, aerodynamics shape, mass, and center of gravity of the real article but without any of its other features. This problem gets more complicated when the missile must be driven into a particular control fin mode for jettison; in such a case, it might be more effective to use an actual missile with onboard computer to verify safe jettison.

6.2.8.4 Representations documentation

In most companies and programs, there is no recognized way to uniformly document these representations. Let us propose one. During the earliest work on a program, someone should be identified to collect information on these items as discussed above. This information should be captured in a product representations log or report in the form of an identification and status list or matrix and a few descriptive pages for each line item in the list that describes the configuration of the item. The list must also include identification of the engineer responsible for the item and the person to be held accountable for developing and maintaining the descriptive information as well as maintaining the actual configuration of the item.

The person who creates and maintains the overall document could be selected from a systems or configuration management function. The problem with the latter is that there is commonly no budget allocated to this function early in a program when this work must be initiated.

The documentation should be deposited on a computer network server and made available to all in read-only mode. The document need not be assembled into a paper product for publication. It can be assembled from pieces that are assigned to the several principal engineers who have write privileges for their portion of the document. The principal engineer for an item should maintain some form of simple configuration identification documentation, whether this information is collected at program level or not. This whole matter should not add significantly to program cost because the principal engineers should keep track of the configuration of their representations. It will take some systems engineering discipline and force to encourage all of these principal engineers to write down the configuration and share it with the program in general.

Someone in systems or configuration management should be aware of all of the representations the program is depending on and making sure that they are in a configuration appropriate to their use. With the cost distributed as noted above, the cost of the oversight function also will not be significant. What will be significant are the positive results that occur on the program by avoiding failures driven by inappropriate configuration of product representations. This is another one of those situations where systems engineering value goes undetected when it is done well and provides glaring evidence of failure when it is not.

6.2.8.5 Closing the loop on representations

Throughout the early system development process on a program and on into detailed design, we depend on many of these product representations to help define sound requirements values, answer design questions, and expose preferred alternatives in trade studies. These representations permit us to predict how the product system will perform, but these are predictions. When the actual product becomes available and we subject it to tests, we start getting actual values. We should compare the results of actual product use with the predictions derived from our representations. Where they compare favorably under the same conditions, we may conclude that the representations faithfully represent the product. This is a very valuable condition because now we know that the representations can be depended on to predict performance.

When we are evaluating possible modifications to the system in the future, we may depend on our representations in that evaluation to predict performance subsequent to modification. When we are evaluating system performance based on field reports, likewise these representations will produce believable results, encouraging confidence in a particular course of action.

6.2.9 Whole Program Phases

The Department of Defense (DoD) and NASA recognize a multiphased approach for development of systems to solve complex problems. One of the DoD phases is called "demonstration validation," often shortened to "DemVal." The purpose of this phase is ensure that the system requirements are mature and clearly understood, that the technologies are mature, and that a viable concept has been defined. Sometimes this phase can entail designing and building actual full-scale operating product by multiple contractors with flight or ground tests to prove out the design concept. Many aircraft programs in DoD have applied this approach, most recently the F22 program. Two contractor teams each designed and built several full-scale aircraft and completed a flight test program as part of this demonstration. The customer then selected the design that most effectively satisfied their requirements. This is a pretty grand form of validation, not commonly available because of cost limitations and the likelihood that something less costly will be effective in developing an adequate product.

You could look at the spiral development model, involving multiple cycles of analysis, and design, build, and test as a kind of iterative validation process combined with a search for necessary or desirable characteristics. A principal motivation for using the spiral model is a lack of confidence that the problem is understood with sufficient clarity so that a specification can be completed as a prerequisite to design work. This is a valid basis for validation for the whole set of requirements for the item or system involved, and the iterative maturation process of the spiral development model satisfies the validation need.

6.3

Requirements Value Management

Contents

6.3.1 Requirements Value Determination

Every engineer who identifies a need for a requirement through structured analysis has the responsibility to complete an analysis to determine the final specification statement, including the appropriate value. Methods that may prove helpful include: flowdown, partitioning, or allocation; appeal to authority; and synthesis of a value from several related requirements, possibly involving simulation or other form of modeling. The flowdown approach, by whatever name, works well for many specialty engineering requirements where there is a mathematical rule of combination of numerical values. The reliability, maintainability, mass properties, and DTC/LCC math models are examples. Parent values are partitioned into child values following the mathematical rule. A common appeal to authority involves a reference to an applicable document. Values for some requirements, often involving performance issues, cannot easily be determined in isolation. Given an aircraft airspeed requirement, we have to determine needed engine thrust by evaluating the combined effects of thrust, drag, and other factors.

6.3.2 TBD/TBR Management

Specifications do not lead a binary existence, where one minute they do not exist and the next they are complete. It takes time to create a specification. As the specification moves from start to finish, some of its content will be incomplete. We may know that it is necessary to include a requirement for engine thrust but not have yet determined an appropriate value that balances with aircraft payload, life cycle cost, and range. We therefore need a mechanism for identifying an incomplete requirement. Earlier in this volume we introduced a primitive requirement statement in the form attribute-relation-value-units. Of these elements, the only one that might lag in identification is the value. Thus the yet to be determined problem scope isolates on value.

A common term for this is that the value is "to be determined," shortened to the acronym TBD. In any one specification, we may have to deal with more than one TBD at any one time, so it is necessary to distinguish between one TBD and another. The TBDs should be numbered, as in this requirement statement for weight: "Weight less than or equal to **TBD-13** pounds." If you boldface the TBD number, it will be easier to notice in the text.

Some engineers prefer to use the expression "to be resolved" (TBR) when an initial value has been identified but work remains to finally resolve that value. This opens the door to several possible ways of handling these cases. I have encountered three different ways of so doing:

1. *Delay.* Delay identifying the requirement at all in the specification until the value has been finally determined. This encourages ignorance of the potential need to consider one or more important characteristics. As

pressure mounts to release the specification, it may be too easy to simply drop the requirement rather than admit that it is incomplete.
2. *TBD/TBR.* Include the requirement statement with a TBD reference in place of the value until the value is determined or a preliminary value followed by TBR until work has been completed to resolve the value.
3. *Best Guess.* Include a value in the statement that is based on best engineering judgment while continuing to search for a final value. The advantage here is that people can continue related work with some idea of the value. The disadvantage is that they may do just that with more confidence in the value than warranted, leading to major rework later when the value is found to be significantly different than earlier assumed.

Some managers have a very negative attitude toward TBDs and enforce a policy that there shall be none in any specifications. It is a good goal to never release a specification with a TBD included, but there may be cases where that is not possible because of the problem's difficulty. So, a sound policy might be to have no TBDs in a released specification without specific case-by-case approval for a justifiable rationale and then only when they are carried as risks until resolved.

Management should require that every specification principal should maintain a list of current TBDs in the specification and work diligently to reduce this number to zero before release of the specification. One way to do this is to include in Section 6 of the specification a paragraph titled "Specification Maturity" and a table that lists all of the current TBDs such as that shown below:

> 6.1 *Specification Maturity* This specification includes some requirements for which final values have not yet been determined. These are identified in the specification by inclusion of **TBD-N** in boldface in place of the value where *N* is arbitrarily assigned to differentiate one TBD from another. These are listed in Table 6.3-1. When these are eliminated, this paragraph and accompanying table shall be deleted.

While specifications are being developed, team and program management should require periodic briefing of the number of TBD in specifications at regular meetings and take action to encourage the reduction of their numbers. The PIT representative responsible for specifications should accept the responsibility for driving the number of TBDs to zero programwide and depend on a counterpart on each IPPT to do the same within their scope of responsibility. These specification lead people are the principal levers through which management can effectively reduce TBDs.

Table 6.3-1 *Specification maturity*

Para	Title/attribute	Tbd	Principal	Date
3.5.4	Weight	2	Bernson	05-10-06
3.2.12	Maximum Speed	1	Johnson	04-11-06
3.5.15	Reliability	3	Cosgrove	01-05-06

6.3.3 Margin Management

A margin is an amount allowed or available beyond what is necessary. We can apply this concept to the three fundamental program management parameters: cost, schedule, and product performance. In so doing, we inevitably make it more difficult to solve the system problem, so one might ask why we would want to make the manager's job more difficult. The fact is that this technique makes the engineer's job more difficult, but it makes management easier by providing slack that can be used to solve problems when they occur. Margins provide management space. By spending margin judiciously, a program manager can convert risks and problems into nonproblems. If margins are not available, the management space is often nonexistent, and the manager must fail to meet program cost, schedule, or product performance goals.

6.3.3.1 Cost margins

Given that it has been determined how much each program task shall cost to accomplish, it is possible to accumulate or roll up those task costs to determine program cost less the material cost. For each program task identified in the program planning process at some level of indenture, a risk assessment should be done and, depending on the result, margin added to the task estimate proportional to the risk or uncertainty. This cost margin should be accounted for in the planning and accumulated at team and system levels for control by team and program managers.

An alternative approach most often observed in industry involves an honest estimate from the bottom up followed by a program manager skimming off a percentage figure across all the estimates. The skimmed amount is placed in program reserve. The author has never met anyone who liked this approach, but it has the same effect as any other method of margins. Margins make the job more difficult because you are solving a problem within reduced resources or additional constraints. The good news is that most people will be able to respond successfully to this added challenge. This focuses the management problem on those who cannot accomplish planned work within the added constraint. When problems are identified, the program manager has the resources to solve the problem by applying the margin removed from everyone's budget. A cynic would say, of course, that the problem would not have occurred if the program manager had not first skimmed the offender's budget. That may be true, but the overall effect is positive, however painful the details. Margins make good management possi-

ble. The problem with the skimming process preceded by a request for an honest estimate is that we are providing program personnel with a very early statement of approval for dishonesty, in that the experienced people simply include the anticipated 10 to 15 percent anticipated skim magnitude in their "honest" estimate. This encouragement for dishonesty may very well come back to hit you in unanticipated ways.

6.3.3.2 Schedule margins

Schedule margin is more often called "schedule float" or "slack." During the program planning process, the critical path should be determined and, in each of those paths, float established. Float should also be established in other pacing tasks where schedule risk is perceived. Because the customer cannot or will not extend the need date for delivery, the float must come from a demand that tasks be accomplished in less time to establish the float.

6.3.3.3 Characteristics margins

This book is primarily concerned with requirements value margins, but we shall see that cost and schedule margins are related to solving design problems through appeal to margin. Margins may be applied to any numerically stated requirement, but margins do make the design problem more difficult. Therefore, we should be selective in the use of margin for those requirements that we have identified through the validation process as risks. Table 6.3-2 offers a fragment from a margin table for a program. The system (A) is composed of four items (A1, A2, A3, and A4). Four margin accounts are identified (weight, reliability, DTC, and range) at the system level. In this fragment, all values are given for system and item A1 (range does not make sense at this level) plus weight figures for all elements to illustrate how the margin approach works.

Weight and DTC follow the sum rule and follow the simple formula RQT = TARGET + MARGIN, and it is the target values that will challenge the design team to satisfy. The RQT (requirement) value appears in the specification. In both cases, less is better; less weight and lower DTC are better than higher. If you were dealing with a requirement category where more is better, the equation used would be RQT = TARGET – MARGIN. For example, more range is better, so the target for design is greater than the required value. Reliability is a probabilistic quantity, and more is better. So, the formula is different from the simple sum categories, RQMT = (TARGET)(MARGIN). The ACTUAL column is

Table 6.3-2 *Margin accounts*

Arch ID	Category	Values			
		RQT	*Margin*	*Target*	*Actual*
A	Weight	1,340	85	1,255	
A	Reliability	0.98	0.99	0.9900	
A	DTC	135,000	11,000	124,000	
A	Range	1,500	200	1,700	
A1	Weight	342	22	320	319
A1	Reliability	0.96	0.99	0.9697	0.9530
A1	DTC	54,000	1,500	52,500	52,350
A2	Weight	230	10	220	
A3	Weight	348	28	320	
A4	Weight	420	25	395	

included next to the TARGET column so easy comparisons may be made.

Note that the item weight margins add up to the system level margin. One could simply apply a percentage rule for identification of margin, but it would be better to adjust the margin as a function of the risk anticipated.

You may find that some requirements categories keep appearing in program risk and margin lists, and this is a valuable lesson learned from those programs. One should inquire, "Why do we always have to work this requirement category with a margin?" A study of the process related to that requirement is likely to reveal several possible causes for which priorities can then be set as potential ways to reduce the need to manage the category through margin account on future programs.

6.3.3.4 Margin consumption

As the design matures, some of the designers or teams will not be able to comply with the target value challenges. In Table 6.3.1, it appears that the IPPT responsible for item A1 is having trouble satisfying the reliability target. They are doing well with their other requirements, but reliability is a problem. This problem might be solved by giving the A1 team some reliability margin, but that does not look like it would completely solve the problem. The lead designer for the team when asked how he or she could solve this problem might very well say, "Give me another $2000 on my DTC or 10 pounds on my weight, and I can satisfy the reliability target." That is, the manager need not remain within the requirement category where the problem is exposed to solve the problem. Margin can be exchanged often from one category to another.

One of the complaints that one might hear from a customer about the use of margin accounts is that you will end up solving the customer's problem better than needed. Therefore, some of the product cost is unreasonably charged to the customer, encouraging the customer to request some form of consideration. The reply that will generally bring about understanding on your customer's part is that all of the margin will be consumed in the process of managing the program. Few if any programs have been completed with any of the margins unused; certainly this is true for cost margin.

If, in the odd chance, the development process is completed with residual margin, this is good information to know because it may suggest ways that the delivered product capability could be improved either at no added cost or through implementation of an engineering change proposal that might bring in additional program revenue.

6.3.4 Budgets

Budgets are different from margins. The use of margins is a technique to encourage good management while a budget is a technical means for determining child values from the parent value. Managers should encourage the use of budgets by all disciplines where the requirements are mathematically stated and follow a mathematical relationship, parent to child. Weight, reliability, maintainability, LCC/DTC, and signal gain or loss are obvious examples where the budgeting approach can be applied. The benefit is that values are assigned systematically.

Given an item that must weigh 320 pounds, there is a danger that each of the child item principal engineers will feel free to use all of the weight for his or her own purposes. Allocation of the parent value to each of the child items encourages an important discipline.

A guidance accuracy requirement is an easy one to fail to satisfy through a budget failure. Given a particular system guidance accuracy, one could assign the full system figure to the onboard guidance set, only to find that the guidance set satisfies its procurement specification value while the system does not. The problem is, of course, that there are more sources of error in the system than just the guidance set. The guidance set cannot, for example, be installed in the airframe perfectly. Radio and radar path analysis is another example of the application of budgets. Allocation and partitioning are similar concepts but generally thought of as applying to a simpler case like reliability and mass. Budgeting is a more general concept that applies to cases where all of the requirements are not necessarily measured with the same units.

6.4

Requirements Integration

Contents

6.4.1 Who's in Charge?

Many people are involved in the process of developing requirements for each item of a system. Reliability engineers determine appropriate reliability values from reliability models. Mass properties engineers define appropriate weight figures for each item using a weights model. Performance requirements flow into the item specifications from a structured analysis process such as functional flow diagramming. The item principal engineer is at the hub of all of this activity as it relates to the item for which he or she is responsible.

The principal engineer must interact with a host of specialists to ensure that the item receives appropriate requirements from each specialty discipline and must take responsibility for the overall requirements analysis activity for the item. Some kind of system engineering function (a system engineer on the PIT, in my opinion) should be held responsible for the overall requirements analysis activity for the program, but the principal engineers or integrated development team leaders must provide requirements analysis leadership for individual items assigned.

In some companies, the system engineering organization does all of the requirements analysis work. In those cases, the principal engineer for requirements work simply always comes from the same organization, system engineering. Where a distributed system engineering process is employed, engineers from system engineering may be responsible for system, segment, and higher-tier items as well as the aggregate requirements analysis process, while design engineers take over this role in lower system tiers as a preparatory step to developing the design solution.

In any case, some one person must take the responsibility for the requirements analysis process and specification development for each item on the system specification tree. That person is called a "principal engineer" here. The main role for this principal is to integrate the requirements supplied by many contributors into a minimized, coherent set that ensures that the design solution for the item will operate synergistically with other system elements to best satisfy the customer need.

The principal engineer's role in molding the requirements analysis output into appropriate specification content output can be broken down into six fields of play: (1) aggregate requirements integration, (2) specialty engineering requirements integration, (3) interface requirements integration, (4) environmental requirements integration, (5) performance requirements integration, and (6) programmatic requirements integration. We will take up each of these fields in turn starting with number 1 after a discussion of the overall requirements analysis process from the viewpoint of the individual principal engineer.

6.4.2 Item Process View

Figure 6.4-1 offers a view of the generic system requirements analysis process, during the phases when development requirements for Part I, Type B, development, or performance specifications are being identified, from the perspective of a single, individual item principal engineer. The diagram reflects all of these facets of the process previously discussed in terms of the aggregate program requirements analysis process.

All of the engineers and analysts involved in the item requirements analysis process should, as we have discussed earlier, have a means of cooperating easily via physical colocation or excellent communications and computer networking of tools and information products. While not clear from Figure 6.4-1, a concurrent development bond must exist between the several components of the product requirements and the programmatic requirements analysis efforts. This bond calls for close interaction and cooperative development of both the product and process (manufacturing, operational use, and logistic support) requirements.

The diagram covers both the demonstration-validation and preliminary design phases where development requirements are identified. On the first pass through the diagram in demonstration-validation, the system functionality is decomposed through top layers of the expanding architecture and major item (configuration item) requirements developed. During the preliminary design phase, lower-tier development requirements for procurement and in-house build component items are completed as a precursor to actual design work and the beginning of the upward stroke of system development.

6.4.3 Aggregate Requirements Integration

6.4.3.1 Requirements set attributes

The complete set of requirements for an item should satisfy the following five criteria: consistency, completeness, minimized, uniqueness, and balance. The principal engineer must ensure that these criteria are satisfied during the requirements analysis effort, and the management approval process should encourage the principal engineer to give evidence to that effect.

6.4.3.1.1 Consistency   The set of requirements is internally consistent if it does not entail self-contradiction. The set is consistent with all other requirements sets if it does not conflict those other sets.

6.4.3.1.2 Completeness   How can we be sure we have identified all of the appropriate requirements for an item? Unfortunately this is a question that cannot easily be answered; but there is a good chance we will satisfy this criterion if a qualified staff of engineers used a systematic approach to identify the attributes that must be controlled. That is exactly the basis for the structured approach encouraged in this book. If we have conducted a thorough functional analysis of what the item has to do, an effective environmental analysis, a systematic interface analysis, and an integrated specialty engineering requirements analysis and expanded the attributes thus derived into quantified requirements statements using sound quantification methods, there is a good chance that we have satisfied the completeness criterion.

In addition to the set including every needed requirement, all TBDs must be replaced with appropriate values and all figures and tables referenced in requirements text supplied and complete.

If we have used freestyle or cloning strategies, there is less assurance that we have identified all of the appropriate requirements. A boilerplate can be useful in cross-checking specialty engineering constraint categories of a requirements set generated through the structured approach. Remember that cloning approaches can be effective in specialty engineering and environmental constraints but are not effective in assuring completeness for performance requirements and interface constraints.

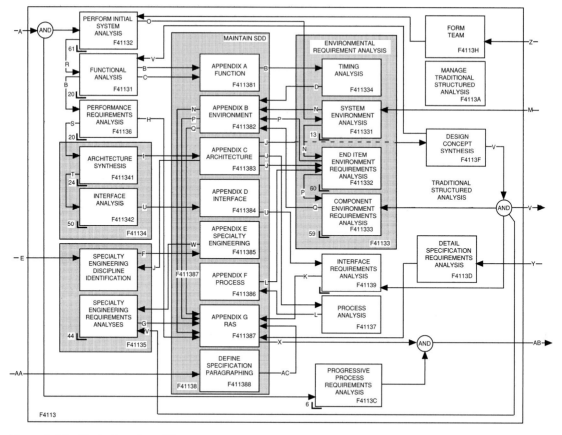

Figure 6.4-1 *Single-item view of the process*

6.4.3.1.3 Minimized But, we just identified one of the criteria as completeness. Now we have a criterion for the opposite situation. Isn't that a contradiction? Strangely enough, it is possible to create too many requirements. Our objective is to decompose the customer's need into a series of smaller problems that will yield to the creative genius of specialized design engineers and their teammates. We need to identify only those requirements that will ensure that the product of the engineer's creativity will work synergistically when integrated into the system.

Requirements have a constraining effect on creativity. We purposely write them for that purpose to ensure that the design solution will have certain important characteristics. Unnecessary requirements constrain unnecessarily and can have the effect of eliminating some potential design solutions that could be better than the remaining options. Requirements do reduce the solution space available to the designer. They should tell the designer the attributes the solution must have in order for the item to function synergistically within the system. They should not define and confine the solution. They should be design-free.

The way to check the need for a requirement is to ask, "What effect would it have if this requirement were deleted? Could the designer, as a result, select a design that would be unacceptable from a system perspective?" If the answer is no, the requirement should be a candidate for deletion.

6.4.3.1.4 Uniqueness Each requirement in the set should be unique in the set with no repetition. Each unique requirement should only appear once in a requirements set.

6.4.3.1.5 Balance Some kinds of requirements are invariably in conflict, and we need to find a reasonable balance point in such cases. Reliability and maintainability requirements, as well as cost and almost everything else, are potential examples. One Atlas rocket had to be destroyed as it diverged from the planned launch path over the Atlantic Ocean. Destruct was commanded by radio, setting off an explosive charge that triggered the propellants causing the vehicle to come apart and its pieces to fall harmlessly into the sea.

There was great interest in what had caused the fault. Luckily the onboard computer was found washed up on the beach and, strangely enough, was intact. When tested, it disclosed that a memory location contained data corresponding to the divergence observed, and the cause was traced to a lightning strike during ascent from the launch pad. This computer was engulfed in a tremendous explosion, fell several thousand feet to impact in the sea, and still worked well enough to determine memory content at the time of the incident. Is it possible that this computer was overdesigned? Is it possible that this overdesign was driven by requirements values way out of step with real-world needs? While it was very helpful to be able to pinpoint the cause in this case, it would not have been unreasonable to expect that the computer would have been totally destroyed.

6.4.3.2 Individual requirements attributes

Each individual requirement in a set should be checked against at least the following six criteria: traceability, style, singleness of purpose, quantification, verifiability, unambiguity, and good sense.

6.4.3.2.1 Traceability Every requirement should be traceable to a driving need in terms of parent requirements, the requirements analysis process through which the need for the requirement was stimulated or a source reference, and the verification process through which it will be proven that the design is compliant. Every requirement should, theoretically, trace up to the customer need. Any requirement that does not is suspect and the rationale for nontraceability should be captured and approved.

6.4.3.2.2 Correctness of Style Every requirement should be written with correct grammar, spelling, and punctuation. It should reflect the customer's style guide and data item description.

6.4.3.2.3 Singleness of Purpose There seems to be a tremendous attraction to writing requirements in long paragraphs. It is not unusual to find many requirements in one specification paragraph. The only valid argument for this, and the argument is marginal, is that some requirements have to be stated as two or more closely coupled thoughts. This should be the exception. In general, we should strive to include only one requirement in a specification paragraph. To do otherwise complicates verification and traceability because of the potential for ambiguity.

6.4.3.2.4 Quantification Requirements should include not only what is required but the needed value or values. Some kinds of requirements must be stated in qualitative terms, but they are exceptions. We should seek to quantify our requirements. Otherwise, it is very difficult to determine if the requirement is satisfied.

6.4.3.2.5 Verifiability One of the most effective ways to ensure good requirements are written is to require that they be verifiable through some practical process. In the process of trying to determine how to verify a requirement, you are forced to write the requirement properly. This is why many people believe that the verification requirements (Section 4 of a specification) should be written by the same person who wrote the Section 3 requirement and that it should be done at the same time.

6.4.3.2.6 Unambiguity Only a single semantic interpretation should be possible. Simple words, simple sentence structure, and avoidance of negatives and passively stated sentences help to achieve this criterion.

6.4.3.2.7 Good Judgment and Good Sense We need to check the requirements set for good sense. Requirements must be written against characteristics of the product and not against things that cannot be controlled with all the money in the world, like the weather. We must not violate the laws of physics or any other discipline. Once again, simple language is a great assist in making this check.

6.4.3.3 Margin check

If margins are used and they apply to the item's requirements, the principal engineer should check to ensure that appropriate margin values have been identified and preserved for the benefit of future program technical risk management purposes.

6.4.3.4 TPM status check

If there are any technical performance measurement (TPM) parameters selected for the item, the principal engineer should check the history of these parameters and ensure that the goals expressed are feasible.

6.4.3.5 Specification format check

The requirements must be fitted into a prescribed format defined by the customer for deliverable specifications and an internally defined format for others. As the requirements become available from the analysts, they should be assigned paragraph numbers from this format.

6.4.4 Engineering Specialty Integration Overview

In Part 3 we discussed each specialty engineering discipline and how it contributes to the requirements analysis effort. In this chapter we are interested in how the product of these analyses can be blended into a coherent story. The facts that engineering specialty engineers are tightly focused on their specialty and that the effects of specialty engineering requirements are frequently in conflict require that the effects of their requirements inputs be integrated or combined to ensure that a condition of balance is realized in the final design and that an unnecessary cost burden is not placed on the customer. Failure to apply these disciplines to a program can easily produce an unfavorable life cycle cost result because of unsatisfied operability needs, poor match between the system and its environment, and extreme support costs. Applying them with uniformly excessive zeal can result in added nonrecurring cost that does not contribute in fair measure to customer benefit. Application of the disciplines with irregular assertiveness can result in unbalanced characteristics of questionable utility. For example, one may end up with an extremely reliable system that cannot, in a reasonable period of time, be tested due to the multiplicity of redundant paths.

One way to integrate the specialty requirements is to hold a meeting with the specialty engineers and ask each engineer to defend his or her requirements under the critical review of the others and the design engineer(s). Figure 6.4-2 offers a more elaborate approach involving integrated checklists.

6.4.5 Interface Requirements Analysis Integration

We need to check that we have identified all of the item interfaces and that we have adequately defined each interface. In addition, we should check the requirements sets for each item interfaced with and verify that their interface requirements are compatible with those for the item in question.

Each principal engineer or integrated development team should cooperate in the formation of a series of internal interface working groups as a means to evolve a joint agreement on each interface his or her item has with the rest of the system and the requirements for each interface. With the exception of interface requirements, all the item requirements are commonly determined within the structure

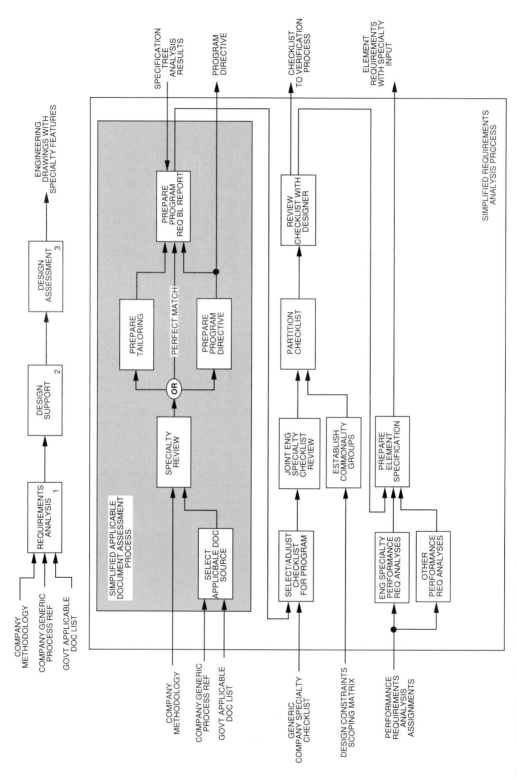

Figure 6.4-2 Specialty engineering integration process

defined by the organization's management hierarchy. We purposely allocate the system's unfolding functionality to items that will associate with how our engineering department or project is organized.

Interfaces commonly run at cross-purposes with this organizational structure, and integration of the interface requirements exposes flaws not only in the product system but in the contractor's program and functional organizations as well. The program team and functional management should be monitoring this process carefully and react to any problems observed to improve the environment for successful teamwork. The symptom to look for is a tendency to withdraw from product interfaces that coincide with organizational interfaces. We need to encourage people to plunge across these gaps, not withdraw from them. Success here may increase intergroup tension on the program, but failure will increase product interface incompatibility problem seriousness.

6.4.6 Environmental Requirements Analysis Integration

Environmental requirements provide opportunities for foolishness. Every environmental requirement should be checked to ensure that the language of the statement focuses on product characteristics and not on an attempt to control the natural forces of the world. Also, we should check carefully for conflicts between requirements that are out of synchronization with reasonable expectations (a snow load on the hot tin roof, for example).

In the process of integrating the environmental requirements using the environmental use profile method encouraged in Part 4, one has to determine exactly how he or she will take the union of the different environmental ranges influencing an item in different process steps. A common approach is to always take the worst case. This solution can have a cumulative effect where each parameter is selected this way, resulting in overspecification and, consequently, overdesign. We should ask ourselves if some other method can be applied, such as the root mean square or a weighted mean.

6.4.7 Process Requirements Integration

The same techniques indicated for specialty engineering integration are generally effective for process requirements analysis. The principal engineer must ensure that the whole team remains focused on the same requirements baseline. Frequent meetings (which can be brief) of all of the team members that encourage synergism among the members are helpful because they expose their understandings about this baseline for the education of their teammates and ventilate their own ideas with teammate criticism. Obviously, the principal engineer must encourage a spirit of togetherness and emotional safety on the team, or the members will not share their views eagerly and openly.

The team must also have an effective means of communicating the baseline among themselves and with other teams. Common computer databases networked throughout the project, good voice communication capability, and physical colocation are helpful to this end. In addition, a team might use a war room concept to post a lot of information for simultaneous view. It is not necessary to have an actual room for this purpose. If the facility has any large expanses of wall, in corridors for example, they can be used very effectively.

The principal integration objective in integration of product and process requirements is to identify inconsistencies between the requirements for each. Some obvious case studies of conflict may be helpful in suggesting many other possibilities:

1. The current product requirements can be satisfied only by a new metallurgical technology used successfully to date only in a laboratory environment. The item manufacturing schedule does not include time to acquire the special tooling required to perform the related operation, and the tooling requirements do not list an appropriate tool needed for the operation. The risk list does not include this potential problem, and no technology work is envisioned.
2. The system will require 132 product items and 25 spares at the level of the item in question. The principal engineer finds that material requirements involve receipt of only 147 of one kind of item component due to an error in addition or failure to update the material requirements from 15 to 25 spares.
3. The wingspan of the aircraft is 53 feet, but the logistics people are planning for a hangar with a door opening of 48.3 feet because that was adequate for a previous aircraft concept version that was compatible with an existing hangar structure at 13 sites where the aircraft is to be operated. The logistics requirements have, apparently, not caught up with the product requirements.
4. The manufacturing tooling requirements call for a master tooling fitting at missile fuselage station 23.5 as a principal support point during missile fuselage buildup. The NASTRAN model does not include this tooling point or the resultant stresses.

There is a fine line between requirements integration and design integration across the product and process valley. Some of the cases included above are very close to this line and perhaps even across it into design integration. There needs to be this concurrent action between the requirements definition for product and process while the requirements are in a state of flux, and this same attitude must carry over into the concurrent design of the product and the process components. In actual practice, it may be difficult to separate this into requirements and design integration, but it is not all that important that we do so. It is much more important in practice to have this activity working well for your project or company than to worry ideologically over whether, at any particular point, we are into requirements or design integration activity.

This work requires a green eyeshade approach to life and can be done very well by people interested in fine details who incidentally also need to be able grasp the grand programmatic view. And, yes, there is a shortage of these people in your company. The best source for these people is among your most experienced employees who have not progressed in their careers above their level of competence. Hopefully, you have some of these people on the payroll who have also not been damaged over the years through poor management.

6.5

Interface Requirements Management

Contents

6.5.1 Internal Interface Control

All system interfaces are more difficult to deal with than the items in the system because of the split responsibility inevitable for interfaces. All interfaces have two terminals; and, unless the same person or team has responsibility for both items that act as terminals for the interface, there will be two different persons, teams, or companies responsible for the interface. It requires a disciplined approach to ensure that every interface has been assigned to a specific person or persons and to follow up that that person or those persons are actually pursuing a clear definition for that interface.

Internal interfaces are those where the two parties on the terminals are both within the developer company. These may be two teams or principal engineers. The PIT should be responsible for identifying all system interfaces down to the level where the highest-level team items are illustrated on the schematic block or N-Square diagram. Below this level, the teams should take over that responsibility. If there is more than one layer of IPPT, each should accept the responsibility for all their internal interface and act as the integrating agent for all interfaces, only one terminal of which touches their item of responsibility.

Internal interface requirements should be included in in-house requirements documents and procurement specifications in a pair-wise fashion. These requirements should be audited in pairs by parent teams and/or system engineers.

6.5.2 Subcontractor Interface Control

Interfaces that suppliers must respect can be controlled through the procurement specification where the supplier must do some development work or in engineering drawings where the supplier is building to print. Suppliers are under contract to your company, and they can be controlled through that contract for interface development as well as everything else.

6.5.3 Associate Contractors

The major interface development problem occurs where the two enterprises responsible for the terminals are not connected by a contract. This can happen when two or more separate contractors are developing a single system for a customer. The customer has a contract with each contractor, but the contractors are connected to each other only through their common customer. This relationship is called an "associate relationship."

6.5.3.1 Formal contractual coverage

The contracts reached with associate contractors should include special coverage for joint work on interfaces across those contractual boundaries. This includes:

1. Joint development and approval of an interface development plan
2. Joint development of an interface control document (ICD) defining the interface between their products
3. Joint staffing of an interface control working group (ICWG) to adjudicate any interface problems that arise in the development of that interface and to provide a technical board overseeing the development of the ICD

The customer should clearly state these requirements in the request for proposal so the contractors can clearly see the need to bid money and time to accomplish these tasks.

6.5.3.2 Principal integrating contractor, SE&I contractor, and IV&V

The common customer may elect to appoint one of the two or more contractors to the position of principal integrating contractor. This contractor may be tasked with development and maintenance of the interface development plan and ICD, but all contractors must have approval authority. Where the integration task is substantial on a large and complex system, the customer may let a contract to one company to act as a system engineering and integration (SE&I) agent for the customer. This contractor normally cannot also have a piece of the contract as a supplier to preserve their independence and encourage fair treatment of the parties to the contract. Some of the companies that often receive these awards are nonprofit enterprises like the Aerospace Corporation.

Another contractual role influencing interface development in this complex situation is called an "independent verification and validation (IV&V) agent." The IV&V agent is intended to give the customer an unbiased perspective from a position of technical strength. Such an agent may audit interfaces contributed to by two contractors to ensure they are describing the same interface.

6.5.3.3 Interface control working group (ICWG)

The principal management forum for developing interfaces across contractual boundaries is the ICWG. Both parties to the interface in question are members, as well as the common customer. In a case where there are three major elements to be developed on a contract, each interfaces the other, and each of these three elements is awarded to a different contractor, there could be three ICWGs formed by contractor pairs and the customer. Each of these forums would develop an ICD and audit design development related to that interface. All should probably function in accordance with a common interface development plan rather than three independent ones. More likely, all three contractors and the common customer would form into a common ICWG while developing three ICDs in accordance with a common interface development plan.

One contractor commonly would be appointed the integrating contractor unless an SE&I contractor is identified. That party would commonly cochair the ICWG with the customer. Each company would send technically qualified people to meetings, with one of them empowered to make decisions for the company on interface issues. This person must then communicate decisions reached internally in his or her own company and coordinate implementation issues with the persons responsible for so doing. This person must also collect issues that require solution and bring them to the ICWG.

6.5.3.4 Interface control document control

The ICD is a specification focused on a relationship between two items rather than on one item in the system, as a specification is. The ICD defines the requirements for that relationship in either a Part I or Part II fashion. A Part I ICD defines development requirements, contains design independent requirements, and is the basis for qualification of the interface. The Part II ICD defines interface acceptance and contains design-dependent requirements that may be expressed in engineering drawings.

One contractor must take general responsibility for the document, but all parties must approve it and any changes to it through ICWG action. Once the ICD has been released and changes are proposed, each contractor must evaluate the impact on their work and product. Engineering change proposals may have to be prepared to provide the customer with a clear picture of aggregate cost, schedule, and performance issues. A trade study may be necessary to determine which of two or more ways to implement the change is most advantageous. Of several alternatives, the balance of responsibility for implementing the change could be significantly different across the contractors. The technical issues will be reviewed by ICWG, but a customer configuration control board (CCB) will probably be the approving authority where it is necessary to commit contract funds to the activity.

6.5.4 Interface Integration Responsibility

Architecture item responsibility is simple in that there is a one-to-one correspondence between things in the system and responsible teams, where a program employs cross-functional teams oriented toward the architecture. Innerface responsibility is also clear in that there is only one party involved who is responsible for both terminals. Outerface will show up somewhere as crossface or innerface in all cases, except where an external interface touches nothing in the system but actually completes a path needed by the system. The problem in interface development is, of course, crossface. I have concluded that the receiver should generally be identified as the principal party for any interface, meaning that they must trigger the conversation and insist that it continue until the interface is fully characterized. But there is one more concern, and that is who should act as the integrator in all cases.

Architecture and interface both require an integrator at all levels, the latter especially because there are two parties to the work and people tend to turn their backs on their external interfaces and focus inwardly on their architecture responsibilities. The architecture integrator is very simple, It is the immediate common parent team. Interface integration is a somewhat more difficult issue in that items in far-flung parts of the system architecture can have interfaces joining them. Therefore, the rule must be altered slightly to read lowest common parent. Figure 6.5-1 illustrates an organizational structure recognizing teams organized relative to the architecture of the product. It also identifies a series of interface control working teams (ICWT) through which interface is defined, integrated, and managed. If, for example, item AX1 interfaces with item AX3, then ICWT AX is the lowest common team and would have the integration responsibility. If item AX1 interfaces with item AY3, then the integrating team would be ICWT A. The parent teams, except for the lowest-tier teams, all operate an ICWT that is active whenever active problems involve it as the lowest common team. The ICWT may also audit all interfaces for which it is the lowest common team.

Figure 6.5-2 and Table 6.5-1 offer a view of interfaces in a complex system involving five architecture levels: (T) total system, (S) segment, (E) element, (C) component, and (I) item. Figure 3.6-16 identifies every kind of interface between the various architectural entities in the system and Table 3.6-1 identifies the lowest common team (LCT) in each case.

6.5.5 Interface Audit

From the examples offered, you can see that the attributes selected to be controlled for a given interface are a function of the interface media: mechanical, fluid, electrical, and so forth. Once it has been determined in what media the interface will be completed, it is very easy to select attributes to characterize the interface. This is another example of the need to do some design concept work in association with constraints analysis work.

The system agent (or lowest common team in the context of Figure 6.5-2) should audit both terminals of all cross-organizational interfaces, interfaces with different organizational responsibilities on the two terminals. This can be accomplished by simply reading the corresponding requirements for the two terminals and reaching a conclusion about whether or not they are describing the interface in a compatible fashion. Generally, mating interfaces will have some form of plug-and-socket relationship, but otherwise they should be the same.

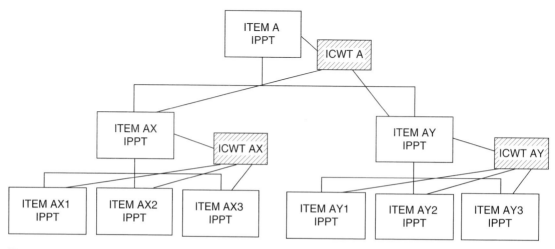

Figure 6.5-1 *Federated interface control working team structure*

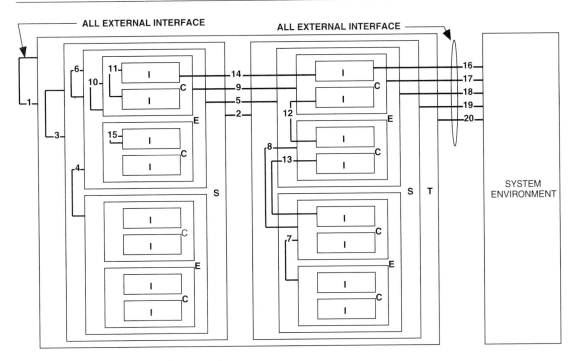

Figure 6.5-2 *Interface integration categories*

Table 6.5-1 *Lowest common team identification*

NBR	Name	LCT
1	All-System Innerface	T
2	Segment-to-Segment Crossface	T
3	Segment Innerface	S
4	Element Innerface, Same Segment	S
5	Element to Element Crossface, Different Segment	T
6	Element Innerface	E
7	Component-to-Component Crossface, Same Element	E
8	Component-to-Component Crossface, Different Element, Same Segment	S
9	Component-to-Component Crossface, Different Element, Different Segment	T
10	Component Innerface	C
11	Item-to-Item Crossface, Same Component	C
12	Item-to-Item Crossface, Different Component, Same Element	E
13	Item-to-Item Crossface, Different Component, Different Element, Same Segment	S
14	Item-to-Item Crossface, Different Component, Different Element, Different Segment	T
15	Item Innerface	I
16	Item External	T
17	Component External	T
18	Element External	T
19	Segment External	T
20	All-System External Interface(Crossface Relative to Environment)	T

Below the system level, interfaces should also be audited. Within the architectural responsibility of any one cross-functional team, the team should audit lower-tier interfaces where both terminals are items under the team's responsibility. Ideally, each team would act as the system integrator for all of its lower-tier teams and principal engineers.

6.5.6 Some Nonstandard Interface Concepts

We have explored the traditional view of interface between two physical objects. There are three other interface possibilities of which the system engineer should be aware. A discussion of these may also help to cement the interface notion more fully in your mind. These include: computer software interfaces, fault-driven interfaces, and interfaces that result from what are called "sneak circuits."

What is computer software? Is it a physical thing? Computer software is an organized sequence of combinations, or patterns, of electrical states within a computer. Some of these patterns relate to information operated on by computer instructions that are a part of a computer program, that is, computer software. It is common practice to design computer software as a series of modules, each of which performs specific functions. In a central computer for an antisubmarine warfare computer system on a U.S. Navy destroyer you might expect to find a computer software module that stabilizes the sonar data acquired from a sonar transducer physically attached to the ship's hull (depending on exactly how the data stabilization problem was solved).

In the software for an aircraft central flight computer, you might find an air data software module to compute true airspeed and barometric altitude from ram and static pneumatic pressures derived from the aircraft's pitot tube. In the computer software for a space transport system launch vehicle, you may find a software module that computes thrust vector control signals to control engine gimballing that steers the vehicle during ascent.

In each of these cases, you would find other computer software modules that require data from these modules and ones that feed information to the ones listed. These would be examples of software-to-software interfaces. While there is no physical situation here that the human can relate to, as in the case of two mating electrical connectors providing signals across an interface, we would have to agree that an interface is present between these computer software modules. We could define what the interfaces are between these modules in terms of information passed between the modules and the timing needs of that transfer. We explored these data interfaces more fully in Part 4

Then we might ask, "Is such a thing as a software-to-hardware (or vice versa) interface possible?" Before answering this question, we need to understand that the computer software has an immediate interface in a somewhat physical way with the memory chips, central processor circuitry, and other parts of the machine within which it operates. The software must honor the physical characteristics of machine construction. So, in every computer application, it would be appropriate to say there was a software-to-hardware interface in this sense. Well, how about beyond this narrow case, though?

A computer software developer would make the case that interfaces exist between the computer software modules and the end instruments controlled by the software. Let us say that we have a computer input/output port that passes an analog command signal generated by the engine gimballing software module through a digital-to-analog converter, two wiring junction boxes, an amplifier unit, and 60 feet of wiring to the elevator (pitch) actuator. The software development engineer would want to know the static and dynamic characteristics of this whole string in order to properly code the controlling software module. The software developer must understand the dynamic characteristics of the actuators and related controls while the aircraft is flying at different speeds.

We may refer to this interface between the software module and the actuator as an extended software-to-hardware interface. We could lump the aggregate effects of the circuitry and actuator together as the actuator and define the requirements for the software across this interface. The engineers involved with the electrical signals between the computer and the actuator, which carry the results of the software module to the actuator, would be more interested in the immediate physical and functional interfaces between the computer I/O port connector, the junction boxes, and the actuator servo controller in terms of connector pin assignments, wire sizes, and voltage levels.

It is important to note that software engineers will commonly lump the effects of an extended interface, such as we have described, in their models. It can develop that computer simulations and test articles created to develop the real flight software have different characteristics than the flight article design. If this is tolerated, you may find that the software installed in the actual flight article exhibits troublesome bugs or even very different and dangerous effects. It is important to control the configuration of not only a flight article in this case, but the test articles and simulations as well and the relationships among them.

Not everyone working on teams developing such a system will be equally positive and helpful in helping the software people understand these extended interfaces. The actuator engineer may foolishly resent a software engineer's interest in his or her completely analog actuator's internal workings. Alternatively, the software developer may not think to press beyond the immediate software or I/O boundary. These are typical cases of specialists looking inward toward their center rather than outward across their interfaces. This is why we need an integration agent at the system level to make sure these kinds of interfaces are properly identified and developed.

Unplanned interfaces can be created as a function of faults developing while the system is in operation. One space shuttle flight resulted in a catastrophic crash as a result of a failed O-ring allowing a solid rocket motor gas jet to impinge on the liquid hydrogen tank. On a Boeing 737, a large portion of the fuselage tore off in flight allowing a direct interface to develop between the natural environment at altitude and the interior of the aircraft as a function of metal fatigue driven by alternating pressure differentials across the fuselage when the aircraft interior was pressurized on each flight and then allowed to relax to ambient pressure at ground level.

Fault-driven interfaces are almost always bad news, and the reliability and/or system safety engineers performing the failure mode effects and criticality analysis (FMECA) should be alert to these possibilities during the evaluation of the evolving design.

An electrical hardware sneak circuit is an unplanned path for a signal to be applied to a load causing undesired consequences. These sneak circuits are not a function of hardware failure; rather, they are a function of a hidden design fault reflected in the engineering drawings from which system elements are manufactured. These sneak circuits can be thought of as another form of unplanned interfaces. Sneak circuits may also occur in computer software and nonelectronic hardware networks, as well as combinations of all of these kinds of system implementation.

A sneak circuit need not always result in undesired consequences, though that is the normal result. When employed as a field engineer for Teledyne Ryan Aeronautical, the author once found that he could program the flight of an unmanned reconnaissance aircraft to achieve a safe airspeed-controlled climb rather than a forced altitude climb rate by issuing a pair of conflicting altitude commands from the onboard programmer, thus avoiding a troublesome flight restriction. When the flight control system designer was told about this, he slyly said it was a planned growth mode. In another case, on an earlier model vehicle, it was found that by issuing a command that would normally never be used with the vehicle at altitude, the vehicle would dive to Earth impact. The latter is a more common result with a sneak circuit than the former. Sneak circuit is, in some companies, a specialty engineering function that evaluates the evolving design to prevent these kinds of unintended interfaces. Some system houses have computer programs that automatically search the design for and identify sneak paths.

6.6

Requirements Verification Management

Contents

6.6.1 The Three-Step Process

Earlier in this book, the author offered the point that system development is best accomplished by applying a three-step process in the right order with all three steps performed within the context of a sound technical management infrastructure. Those three steps are: (1) define the problem, (2) solve the problem, and (3) prove that the resultant product satisfies the definition. The first step entails accomplishing requirements analysis to produce a set of specifications that collectively define the problem to be solved. Step 2 requires three synthesis processes, beginning with crafting an engineering design, followed by procurement of the needed materials defined in the engineering and concluding with manufacturing the product defined in the engineering. The third step consists of inspecting the product to reach a conclusion as to whether the produced product satisfies the requirements that drove the whole process.

There are several ways that a developer can fail in implementing this prescription but only at their peril. The three-step process offers the developer the lowest possible risk approach. Many system engineers have worked in companies where specifications either are not prepared or are done poorly, followed by a design based on the design engineer's knowledge and internal thought process. Others have witnessed verification processes based on the design rather than the requirements. The development process can be even further misapplied when the product is computer software. There are software people who do not waste any of their time trying to understand the requirements appropriate to the design nor by carefully thinking through an appropriate design. No, some software developers start with the manufacturing process—writing code. At least in hardware, design must be accomplished in the form of engineering drawings. In some organizations, the specification is written after the design is complete. This is like writing the plan after the work is complete. One would think that a very good document would result because all of the uncertainty has been removed; but it is wasted effort. Each of the three steps must be applied well and in the right order.

6.6.2 The *V* Words

The third step in this process will be referred to as "verification" in this book. There are many different meanings attached to a pair of words both of which start with the letter *v*. The other *v* word is *validation*. If you look these two words up in a dictionary, you are likely to find that they are very similar in their meaning. And yet, it is somehow not surprising that different people may assign different meanings to the words.

The author applies the word *validation* to gaining confidence that it will be possible to satisfy the requirements in the design process. EIA 632 refers to this as "requirements validation." Validation must, therefore, be accomplished before detailed design. Ideally, this validation work would be accomplished while the requirements are being written. In most cases, requirements that appear in specifications do not offer a great deal of risk because, in a given development organization, the engineers present have a lot of successful experience in crafting designs relative to similar requirements. Validation is a way of controlling risk, as discussed in Chapter 6.2. We should accomplish formal validation on requirements that appear to offer some risk.

In some organizations, management may require that the phrase *to be determined (TBD)* not be used in a specification in place of a numerical value that is not yet known. As discussed in Chapter 6.3, it is desirable to initially release a specification with all of the content complete, but the author would rather see a TBD in a released specification than to have a value of questionable appropriateness inserted to satisfy an arbitrary management decree; because once the specification is released, the requirement takes on the same credibility as other requirements that were more thoughtfully defined. However, a specification should contain no TBDs and should otherwise have all of the related risks resolved by the time of the preliminary design review as a prerequisite to entering detail design.

EIA 632 identifies another form of validation, called "product validation," and defines it as a process by which the user reaches a conclusion about the product relative to his or her user requirements. Note that the user requirements may be different from the requirements covered in a development contract between a contractor and a large customer organization. Often an acquisition manager may have to curtail aggressive user needs in the interest of affordability. In earlier years, when DoD procurement was guided by cost as a dependent value, the cost was what ever it took to satisfy performance requirements. Today, where DoD respects cost as an independent value, it is a more common reality that the user and acquirer agree on a spiral development sequence that satisfies initial needs and, over time, progresses to a product block that completely satisfies user requirements. The author believes that the product validation term in EIA 632 coordinates with the DoD concept of operational test and evaluation (OT&E).

6.6.3 Verification Classes

Verification is a comparison process. We compare one object with a standard of excellence and reach a conclusion about the degree of match between them. In all cases in verification, the standard is the content of a specification. The object we are comparing with the content of a specification is a product entity.

Department of Defense breaks down all verification work into three classes. Item designs are qualified by comparing the engineering expressed in a qualification unit, often manufactured in an engineering laboratory, with the content of an item performance specification, which should be design independent. This work may be accomplished within three different contexts. In a program that will produce many items, there is generally enough funding to support the manufacturing of special dedicated qualification items that are subjected to exhaustive tests, commonly resulting in the life of the unit being used up. Products exist that will be built only as single articles. Factories, space launch facilities, thrill rides, and large bridges are examples. We cannot test these kinds of systems with the same aggressive approach we would apply to dedicated qualification units because the one-of-a-kind article must have a full life ahead of it when the testing is complete. Functional testing is more common in these cases, with less interest in extreme environmental stresses. A middle ground is formed by what could be called high-dollar, low-volume programs. In these cases, only a few articles are produced, which are too costly to support a dedicated qualification unit. In these cases, the first article through the production process is subjected to qualification, applying reduced environmental stresses. Upon completion of the qualification work, the surviving unit is refurbished and passed back into the production process for final acceptance. The first article may end up being the last item shipped.

The second class of verification is item acceptance, applied to the first article through the production process. The first article is inspected to verify that it complies with the content of the detail specification that was crafted based on the design features. The content of a detail specification is design dependent and is intended to drive the acceptance test planning process. After proving that the first article complies with the requirements in the detail specification, every article passes through the same acceptance process as a prerequisite to being delivered to the customer.

The third verification class is system test and evaluation. Assuming that all of the items that comprise the system have been qualified proving that their design is adequate for the application, all of these items are collected and operated as a system to determine if the system satisfies the content of the system specification. It is possible that all items could satisfy their item specifications and yet the whole may not satisfy the system specification. DoD test and evaluation work is partitioned into development test and evaluation (DT&E) and operational test and evaluation (OT&E). In the former case, the contractor accomplishes the work and is trying to establish that system complies with system specification content and thus that the contractor has satisfied contractual requirements. In the second case, the user operates the system, and the frame of reference is not contractual requirements but the original user requirements. The user may very well conclude that the system fails OT&E even though it passed DT&E with flying colors.

6.6.4 Verification Methods

Verification work is accomplished by making comparisons. We compare the characteristics of an article under inspection with a standard. In each case, the standard is a particular kind of specification. In making those comparisons, we apply four commonly accepted methods: (1) test, (2) analysis, (3) demonstration, and (4) examination. The author has elected to use the four methods defined in MIL-STD-961E, which are different in one small way from the four methods applied by the standard it replaced, MIL-STD-490A. MIL-STD-961E changed the name of "inspection" to "examination" because the word *inspection* was reserved for the general case following the quality assurance lead. The four methods are all examples of inspections.

6.6.4.1 Test

A test is an element of verification or inspection that generally denotes the determination, by technical means, of the properties or elements of items, including functional operation, and involves the application of established scientific principles and procedures. The product item is subjected to a systematic series of planned stimulations, often using special test equipment. Performance is quantitatively measured either during or after the controlled application of real or simulated functional or environmental stimuli. The analysis of data derived from a test is an integral part of the test and may involve automated data reduction to produce the necessary results.

6.6.4.2 Analysis

Analysis is an element of verification or inspection that uses established technical or mathematical models or simulations, algorithms, charts, graphs, circuit diagrams, or other scientific principles and procedures to provide evidence that stated requirements were met. Product item features are studied to determine whether they comply with required characteristics.

6.6.4.3 Demonstration

Demonstration is an element of verification or inspection that generally denotes the actual operation, adjustment, or reconfiguration of items to provide evidence that the designed functions were accomplished under specific scenarios. The items may be instrumented and qualitative limits of performance monitored. The product item to be verified is operated in some fashion (operation, adjustment, or reconfiguration) so as to perform its designated functions under specific scenarios. Observations made by engineers are compared with predetermined responses based on item requirements. The intent is to step an item through a predetermined process and observe that it satisfies required operating characteristics. The items may be instrumented and quantitative limits of performance monitored and recorded.

6.6.4.4 Examination

Examination is an element of verification or inspection consisting of investigation, without the use of special laboratory appliances or procedures, of items to determine conformance to those specified requirements that can be determined by investigations. Examination is generally nondestructive and typically includes the use of sight, hearing, smell, touch, and taste; simple physical manipulation; mechanical and electrical gauging and measurement; and other forms of investigation. A test engineer makes observations in a static situation. The observations are normally of a direct visual nature, unsupported by anything other than simple instruments like clocks, rulers, and other devices that are easily monitored by the examining engineer. The examination may include review of descriptive documentation and comparison of item features and characteristics with predetermined standards to determine conformance to requirements without the use of special laboratory equipment or procedures.

6.6.4.5 Other methods

6.6.4.5.1 Similarity On heavily precedented programs involving major modifications, there are many architectural entities that cross over to the new system from the prior one with little or no change. Given that, in its prior development effort, the system qualified its items, then these unmodified items in the new system would have been qualified at an earlier time for the requirements of the earlier version of the system. It is possible that the requirements for this item have not changed, but they are more likely to be somewhat different. To the extent that the developer can correctly claim that the design has been proven against the new system requirements through it prior use, the program can avoid new testing and the associated costs. One way to avoid a new qualification process for this item is to claim that the requirements have been verified by similarity between the current requirements and those of the prior system. This is called "verification by similarity."

To make this a valid activity, an analysis should be crafted to make the comparison between the two sets of requirements accompanied by some logic to show why the prior qualification is adequate for the new application. An example would be if, on the prior program, after the item was qualified, the engineers continued the testing to discover that the margins available and the actual capability of the item proved to be better than the current requirements. This reality could be used as a basis for similarity. This basis is actually a subset of analysis because of the need to coordinate capability with some logic or discussion supporting the comparison.

6.6.4.5.2 Simulation More and more often, requirements are proven satisfied in a design by simulation. This is a valid approach when it is very difficult to set up the conditions in the real world and the simulation adequately reflects real-world conditions. For example, a new radio system for use in amphibious operations will entail 100 radios working in a network, and we have to prove that the system will handle the traffic planned. We could simulate this without any radios being used in a traffic model. Alternatively, a mathematically sufficient number of radios could be arranged in a laboratory with inputs and outputs networked with a test computer.

When the author was involved in system engineering training at a company, an engineer at the firm explained that the company was trying to convert from iron verification to simulation verification. They had a history of building a string of prototypes and shipping a large number of mature prototypes to good customers who would operate them and report back their findings.

6.6.5 Qualification Verification

Qualification establishes whether an item in a system is adequately designed for the application. It does so by proving to what extent the product design satisfies the content of the item performance specification that also drove the design.

6.6.5.1 Verification requirements

Figure 6.6-1 illustrates the relationships between product requirements captured in Section 3 of a six-section specification format like that used by military programs, the verification requirements captured, in the author's view, in Section 4, and the verification tasks through which it will be proven whether or not the product design did, in fact, satisfy its requirements. This diagram is motivated by the RAS Complete notion explored in Chapter 3.10. The product requirements are all derived from the artifacts of some model and allocated to a product entity on the architecture axis. For each product requirement we are obligated to identify the method (Paragraph 6.6.4) through which it will be verified that the product design has that characteristic. In each case, we should go beyond that simple act and identify the corresponding verification requirements (on the verification compliance plane in Figure 6.6-1) that are not requirements for the development of the product but are rather requirements for the development of the verification plan.

The specification should include in Section 4 a traceability matrix that correlates the product requirements in Section 3 with the methods to be applied, and the verification requirements discussed in the section, "Verification Process Design."

The verification compliance matrix is the union of the verification traceability matrices in all of the specifications. It links all the requirements in all the specifications to methods and verification requirements. This is the right document to use in designing the verification process, where the verification engineer assigns or allocates the verification requirements to tasks, thus identifying the tasks to be performed and collecting their requirements. These tasks must be put into a network or Gantt chart schedule format respecting needed task sequence precedence.

6.6.5.2 Verification plans and procedures

Each task is then assigned to a responsible engineer who plans it, determining the needed resources (time, money, people, facilities, and things). These resources are defined in a plan, along with the procedures for accomplishing the work. Some organizations release the plan and procedure separately. Many engineering organizations forego the identification of verification requirements, leaving the verification engineers to plan their assigned tasks without a set of requirements. The danger in doing this is that the tasks may be planned based on the design rather than the requirements the design is supposed to satisfy.

6.6.5.3 Verification implementation, reports, and audits

At the appropriate time, as defined by the schedule, the principal engineer or the test laboratory responsible for performing the task must bring the resources identified in the plan to a state of readiness and accomplish planned work. The results of having accomplished the task are included in a task report, providing the program with evidence of whether the design satisfied the requirements that were intended to drive the design.

All of the reports for a particular item should collectively show the degree to which the design satisfies all the requirements in the corresponding item performance specification. The evidence may show that all of the requirements were satisfied or that some of them were not. The former case offers good reason to proceed with item production. Results that reveal a failure to satisfy the requirements, should trigger either a change in the requirements or a change in the design, followed by a retest.

The units subjected to qualification are commonly manufactured in an engineering laboratory by highly qualified people. The developer should make every effort to counter possible critical customer questions about the differences between the qualification unit and the units that will come off the production line by using the best available manufacturing and quality planning relative to the production of the qualification units.

Large system acquisition agents like DoD, FAA, and NASA often wish to formally audit the evidence produced by the item qualification process. This audit is often called a "functional configuration audit (FCA)." The customer representatives review all of the reports and reach an independent conclusion.

6.6.6 Acceptance Verification

The acceptance process parallels the qualification process, except that it is driven by the content of a different specification. The performance specification, previously referred to as the "development" or "Part I" specification, contains design-independent content intended to drive the design and qualification processes. The detail specification (earlier referred to as the "product" or "Part II" specification) contains design-dependent content and is intended to drive the acceptance process. We cannot prepare the detail specification until we know what the design features are because the detail specification is intended to define the essential characteristics that each article of the item must have to be

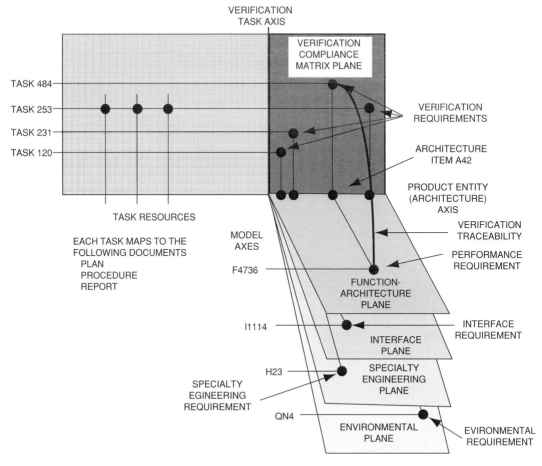

Figure 6.6-1   *The RAS complete view of verification*

acceptable for delivery to the customer. These characteristics will relate to space and mass measurements, operational capabilities measurable from the outside of the item. Any measurements will commonly be input/output related.

The detail specification content, like that of the performance specification, is mapped to the same four methods and corresponding verification requirements, translated to verification tasks, with those tasks then planned and integrated into the production process. Generally, it will be manufacturing people who accomplish the plan content and quality assurance people who attest to the actions having been taken properly. Large and complex items may have to have acceptance actions taken at various points in their assembly process rather than a simple postmanufacturing test.

After the first article has completed its acceptance inspection, some customers will want to audit the results of the acceptance process. This audit is commonly called the "physical configuration audit (PCA)." It may be implemented by inspecting the results of acceptance paperwork or a detailed comparison of the manufactured article with the engineering drawings or code lists. The same matrix relationships exist as discussed under the section on qualification, except that the verification matrix is in the detail specification.

6.6.7 System Test and Evaluation Verification

System test and evaluation, referred to as development test and evaluation (DT&E), are based on the content of the system specification, and the intent is to determine to what extent the system satisfies the content of that specification. The proof is derived in task reports identified using the same machinery discussed above. Only the specification is different. The OT&E process is driven by a different standard than is DT&E. It is based on user requirements contained in documents developed early in the program history and possibly updated over the development period. The OT&E is focused on mission capabilities defined by the user that, ideally, were used by the contractor in developing the content of the contractually valid specifications.

6.6.8 Management Matrices

The verification process is well supported by a series of four matrices, as suggested by Figure 6.6-2. Each of the specifications of the three kinds discussed in this chapter should contain a verification traceability matrix that couples the product requirements to the methods and verification requirements. The union of all of these matrices with a few

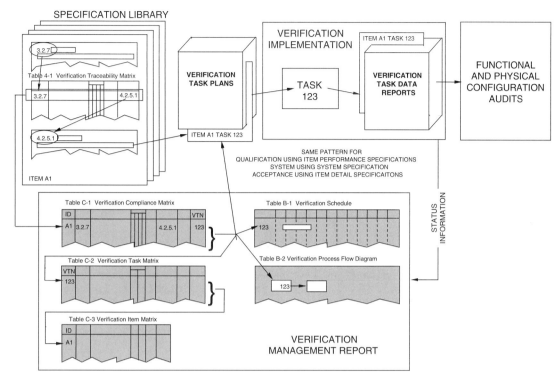

Figure 6.6-2   *Verification matrices*

added fields comprises the verification compliance matrix that includes every requirement in every specification. This matrix can effectively be used as the basis for designing the verification process for item qualification, item acceptance, and system test and evaluation. The matrix includes every requirement identified for the system under development, and the verification design requires identification of all of the tasks, through which it will provide to what extent the design satisfies these requirements. So, verification of design is a matter of mapping verification requirements to tasks of the four kinds (test, analysis, demonstration, and examination), collecting all of the requirements mapped to each task, and preparing a plan for each task. All of the tasks should be illustrated on a verification schedule.

The compliance matrix is the right document to use for designing the process but not the right document to use in managing it. Managers can be effective managing tasks where the cost, schedule, and resource basis are clearly defined and a person has been clearly identified as the responsible party. The task matrix satisfies this need. Each line item of the task matrix represents a task that should map to some number of verification requirements contained in the rows of the compliance matrix. In the case of a very large system, a higher-level matrix is useful listing each item. The fields of this matrix provide status information

on the readiness to enter the related item audit. It can tell the percentage of planned tasks completed, for example.

To make all of the verification work affordable and, therefore, encourage that the work be accomplished, one needs a good requirements database application not only to capture the data but also to allow the linking of that data for traceability purposes. This database application should be the same one that contains all of the requirements data from which specifications are printed. This database can also contain all of the verification plans, procedures, and reports.

As suggested by Figure 6.6-2, there is a collection of documents that should be available to support the verification process. The specifications are needed for other purposes, of course, but they are the drivers of the verification work. We will need a complete set of plans and procedures that have to be configuration managed as well. As the work is completed, we must also capture and retain the reports. Finally, a program should collect the matrices and schedule for verification under the suggested title "Verification Management Report." It is not necessary to actually print this report to paper periodically, but it should be possible to do so if the customer calls for its delivery at particular program milestones. The data for this last document should be contained in the database and more commonly accessed directly from the database with filters set to view the information of interest.

6.7

Specification Development, Review, and Approval

Contents

6.7.1 Specification Development Controls

6.7.1.1 The specification tree

The requirements analysis process is driven by an identification of all of the specifications that must be prepared on a program. The specification tree is the means to define and communicate this information. Figure 6.7-1 illustrates a typical specification tree at the top level for a space transport system during the time when MIL-STD-490A was effective in military procurement. It is an overlay of the system architecture, as evidenced by the architecture IDs in the lower right corner of the blocks. We could place a specification type indicator in the lower left corner of each block. Here, the A means system specification format, B means a hardware item specification format, and C means a software specification format. You might also include type N for no specification required and P for procurement specification. A type I could be established for interface control documents. These letters are not coordinated with the MIL-STD-490A Type letters, which are no longer used formally.

The graphical specification tree is useful at the top level but is not recommended on a large complex system requiring many specifications. It is simply too hard to maintain the graphical document, no matter how it is created and maintained. An indentured list is encouraged in this case, and you really need a specification database anyway just to manage the whole development process. Table 6.7-1 offers an example. They are listed in architecture order with type and specification number identification. With the exception of the interface control document for I23, this table communicates the same information shown in Figure 6.7-1. Obviously, either can be expanded to dizzying depths, but you will easily agree that the tabular list would be easier to maintain, especially if we have included status information with dates in the tabular list.

6.7.1.2 Responsibility assignment

Given that we have identified the need for a particular specification, it is necessary to establish responsibility. Table 6.7-1 offers two columns for this purpose. We have

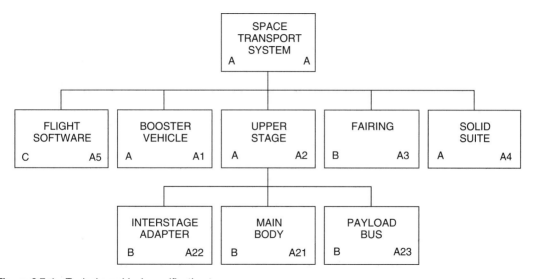

Figure 6.7-1 *Typical graphical specification tree*

Table 6.7-1 *Typical tabular specification tree*

Arch ID	Title	Type	Spec number	Team	Principal
A	Space Launch System	A	A32-33465	PIT	Burnes
A1	Booster Vehicle	A	A32-38434	PIT	Flockensbee
A2	Upper Stage	A	A32-34989	PIT	Alphonse
A21	Main Body	B	B32-44432	IPPT 1	Jones
A22	Interstage Adapter	B	B32-54331	IPPT 1	Adams
A23	Payload Bus	B	B32-55523	IPPT 1	Adkins
A3	Fairing	B	B32-13121	PIT	Tonson
A4	Solid Suite	A	B32-43981	PIT	Smith
A41	Solid Booster	B	785AGF54	IPPT 2	Georgeson
A42	Solid Booster	B	785AGF54	IPPT 2	Baker
A43	Solid Booster	B	785AGF54	IPPT 2	Conovar
A44	Solid Booster	B	785AGF54	IPPT 2	Provance
I23	Vehicle-Launch Complex ICD	I	I32-220981	PIT	Zola

identified, first, the team responsible and then the name of the principal engineer within that team who will write the specification and/or coordinate its development. Let us say that we have established five teams on the space transport system program, as listed in Table 6.7-2. The PIT is a program integration team, the program system agent. The IPPTs are integrated product and process teams assigned item development responsibilities. The PIT has the responsibility for integrating and optimizing across the IPPTs at the system level. Note that we have also made the PIT responsible for the ICD between the complete launch vehicle and the launch complex.

From Table 6.7-2 we can see that we have made the PIT responsible for developing the top-level specifications for each team. Each team would be made responsible for developing any subordinate specifications below that level, as you can see in Table 6.7-2 for the upper stage and solid rocket suite. This is a sound pattern because it permits PIT to give to each top-level team a fully coordinated, complete package at the outset of their work. This package should include:

1. The specification corresponding to the top-level entity for which the team is responsible
2. A contiguous component of the integrated master plan (IMP) and the corresponding component from the integrated master schedule (IMS)
3. The budget corresponding to the work breakdown structure (WBS) item at the team top level, all coordinated with work planned in the IMP/IMS
4. Clear definition of the interfaces they share responsibility for and the identification of the other party with whom they share those responsibilities

6.7.1.3 Process controls

The enterprise should deploy to its programs a generic specification development process, as illustrated in Figure 6.7-2. This process should include some form of peer review and a formal review and approval process. For every specification, a review authority should be identified. One rule might be that the PIT form the review body for all of its specifications and that the program manager or his or her surrogate chair that meeting when the specification reviewed must be approved by the customer and the PIT leader in other cases. Specifications subordinate to the IPPT top-level specifications (which should be a PIT responsibility) might be reviewed by the responsible IPPT. Alternatively, the PIT might require that IPPTs cross-review each other's specifications to introduce a degree of objectivity into the mix, but management will have to be watchful that an excess of interteam competition does not creep into the process.

Table 6.7-2 *Program team responsibilities*

| | Responsibility | |
Team	Arch	Title
PIT	A	Space Launch System
IPPT 1	A1	Booster Vehicle
	A3	Fairing
IPPT 2	A2	Upper Stage Vehicle
IPPT 3	A3	Booster Suite
IPPT 4	B	Launch Complex

6.7.2 Specification Publishing

The specifications move from a planning process in the earliest days of the program through development and on to publishing, managed well throughout. The overall publishing process is illustrated in Figure 6.7-3. The upper portion of the diagram relates to product specifications and the lower portion to processes. The term *generation* refers to placing the specification into a form that can be reviewed. In an organization that uses a database system, this entails simply spinning the database content into a specification on the screen or printed in paper copies. The review and approve process, shown in Figure 6.7-4, is a formal or informal peer review way of comparing the content of a specification with a set of standards that all specifications should meet. Following approval, the specification must be formally released, published, and made available to program personnel either in paper or online form. The released specification must thereafter be formally protected through configuration management. Any changes must pass back through this same process to gain approval.

6.7.2.1 Formal review process

The formal review process should include a conscious evaluation of general specification template faithfulness and overall quality, measured in accordance with a specification checklist. Next, the specification should be checked for adherence with good traceability standards. The program may choose to fully implement traceability standards shown in Figure 6.7-4 or some subset thereof. The final string of checks looks at the specification for residual risk, completeness, and excess content. The reviewers and the review chairperson reach a decision calling for either corrective action or approval of the specification.

6.7.2.2 Peer review

Specifications prepared on small or advanced programs may not have sufficient budget to support a formal review process. In this case, while not as supportive of a low-risk approach, a specification can be reviewed by experienced people in a less-organized fashion, called a "peer review." The team is assembled and asked to review the document either together or online at their desks, followed by a group meeting to discuss the content.

6.7.3 Specification Archiving, Distribution, and Access

The master copy of each specification must be retained by an assigned authority to protect the integrity of the document. Once approved and released, this master must be accurately identified and protected against change. In one organization the author recalls, the master was changed during work on an engineering change proposal (ECP), but the ECP was subsequently canceled. The organization no longer had a master for the specification in effect because it had become corrupted by the change work that did not materialize. It helps to consider each specification build or change a separate campaign that results in the release of a document that will exist forever. If that document is subsequently changed, the change is built anew on the baseline past.

Specifications must be readily available to personnel working on a program. As they are released, the specifications must be distributed to those who need them. As they are revised, the same is true of the revisions. You could

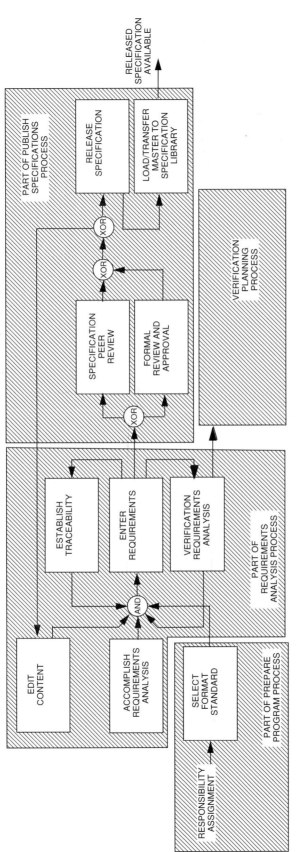

Figure 6.7-2 *Specification development process*

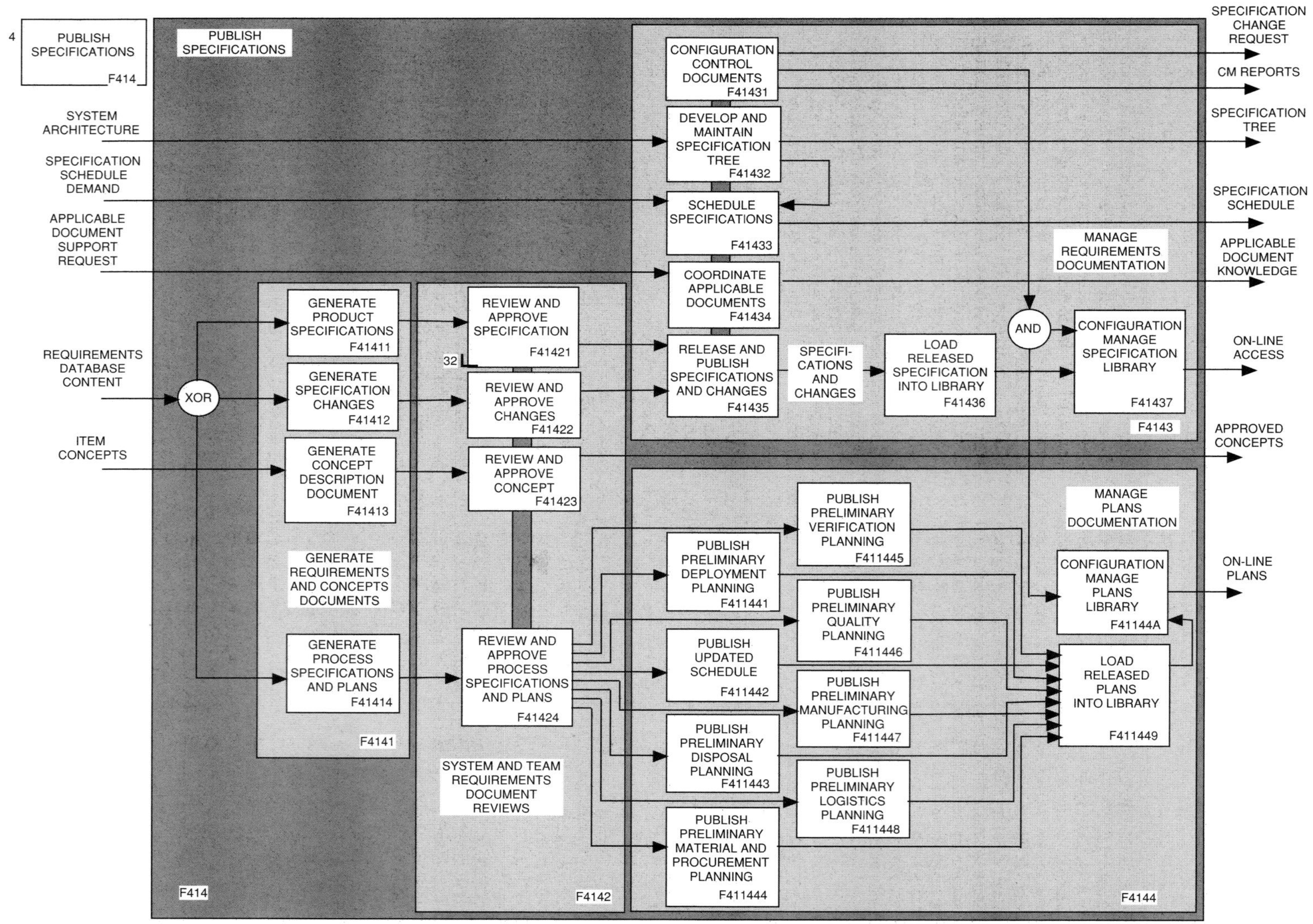

Figure 6.7-3 *Specification publishing*

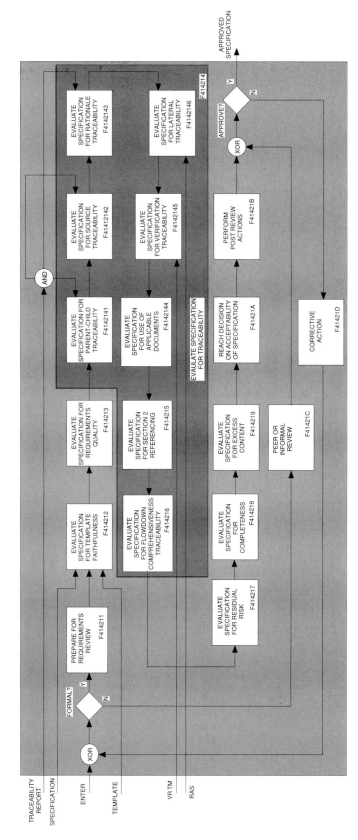

Figure 6.7-4 Specification review and approval

choose to distribute them to everyone you think needs them and expect that others would get a copy from those distributed, place one or more copies in a central library from which interested parties may check them out, or both. There are clear advantages to computer media that should simply scream out of this page at you as you read.

6.7.3.1 Paper methods

Years ago, specifications were crafted with typewriters and typesetting. These were published in paper form and distributed using shoe leather and mail systems. If most of your specifications are in paper media, you may have no near-term option but to place them in a paper document library from which program personnel may check them out physically. But, even if this is the current case, you should be making plans to move to an online specification library for cost, efficiency, and document configuration control purposes.

After a specification is formally released, the master must go to reproduction where sufficient copies are made to cover the needs for distribution and the library. The master should be returned to "the vault," a physically secure facility (not in the classified data sense, unless this is also a valid concern) where all the engineering masters are retained. The copies must then be physically distributed. If the specification in question is also a customer-approved specification, another loop will be required to gain their approval prior to distribution.

6.7.3.2 Networked library

Adios, paper—and good riddance! Even if you are currently using a paper medium for distribution, you probably already have the resources in place to convert to computer media. It requires specifications captured in computer file media, a network with adequate storage capacity and speed, and easy access to terminals or PCs on the part of the people. These features are present in almost everyone's shop today or are not beyond the pale to achieve. It is unimaginable that anyone would use a typewriter today to prepare a specification, so they will always be created on some form of word processor. The results of this work can be passed on to document release function on a disk or via the network connection and thereafter transferred to an online library from which anyone may open it.

6.7.3.3 Web page library

Some organizations have implemented the ultimate specification archiving and distribution system without really publishing specifications. The McDonnell Douglas (now Boeing) FA-18E/F Program, under the leadership of Ken Kepchar, captured all their requirements in RDD-100, an excellent requirements database from Ascent Logic, that, with its complete capability, does require some experience to use effectively. They assembled the software necessary to allow anyone without special skills to access the database from a Web page that listed all of the specifications. A click on the specification of choice brings up the document, automatically formatted into Interleaf. This is the best of all worlds because everyone can have access to any specification from wherever they are in the facility, and that specification will be the current version, while allowing the people responsible for creating and maintaining the specifications to do their work in a rich tool set. This same model can, of course, be applied to every form of documentation on a program including the plans, drawings (CAD library), and reports.

6.7.4 Specification Change Management

Specifications are corrected or updated when necessary, by means of either a change or a revision. A change, containing only the changed portions of the specification, is accomplished by the issue of a specification change notice (SCN) and attached change pages. A revision consists of a complete reissue of the entire specification, all pages being identified by the same applicable revision letter. In general, corrections to only a small portion of a specification should be accomplished by a change, whereas extensive corrections requiring revision occur when: (a) over 50 percent of the pages have been, or will be, involved in the intended correction plus outstanding SCNs; or (b) a revision is economically more practicable than issue of page changes by SCN.

As a general rule, no more than five (5) SCNs should be issued against a particular revision (or original issue); when the sixth modification or correction is required, the outstanding changes should normally be incorporated in a revision of the specification. On the Atlas E space transport upgrade program between Vandenberg AFB and General Dynamics Space Systems Division, the customer allowed the system specification to reach a total of more than 100 change notices because they did not want to spend any more money than necessary on documentation. It was very close to impossible to really know what was included in that specification.

As the practice of viewing specifications via computerized online specification repositories rather than the physical distribution of many paper copies becomes more prevalent, programs should consider more strongly the complete revision of documents to avoid the clutter of change notices.

6.7.4.1 Changes

A DoD or NASA customer will commonly require that anticipated changes to customer specifications be proposed and issued on a specification change notice (SCN). Otherwise changes will be made by complete revision and reissue of the document. Specification sheets are normally changed by revision only. As required by DOD-STD-480, a separate SCN must be submitted as an enclosure with an engineering change proposal (ECP) for each specification to be changed. SCNs so submitted are issued and incorporated only after approval of the ECP and the engineering change ordered. An SCN is also used to issue corrections to a specification unrelated to an ECP.

6.7.4.2 Specification change notice

The SCN is a document used to propose, transmit, and record changes to a specification. A SCN form, illustrated in Figure 6.7-5, can be used as a cover sheet and letter of transmittal; the page changes associated with that SCN are attached and constitute an integral part of the SCN.

6.7.4.3 Proposed SCN

The proposed SCN must identify for the specification approving agency the exact change in specification paragraphs, figures, or other content that will be distributed to users if the SCN is approved. Such modification in content in this proposed form of the SCN may be submitted in final specification change form or as an enclosure on which the proposed changes in sentences, paragraphs, figures, tables, and so on, are described.

SPECIFICATION CHANGE NOTICE

| 1. ORIGINATOR NAME AND ADDRESS | 2. ☐ PROPOSED ☐ APPROVED | 3. CODE IDENT. | 4. SPEC. NO. |
| | | 5. CODE IDENT. | 6. SCN NO. |

| 7. SYSTEM DESIGNATION 8. RELATED ECP NO. | 9. CONTRACT NO. | 10. CONTRACTUAL ACTIVITY |

| 11. CONFIGURATION ITEM NOMENCLATURE | 12. EFFECTIVITY |

THIS NOTICE INFORMS RECIPIENTS THAT THE SPECIFICATION IDENTIFIED BY THE NUMBER (AND REVISION LETTER) SHOWN IN BLOCK 4 HAS BEEN CHANGED. THE PAGES CHANGED BY THIS SCN BEING THOSE FURNISHED HEREWITH AND CARRYING THE SAME DATE AS THIS SCN. THE PAGES OF THE PAGE NUMBERS AND DATES LISTED BELOW IN THE SUMMARY OF CHANGED PAGES, COMBINED WITH NON-LISTED PAGES OF THE ORIGINAL ISSUE OF THE REVISION SHOWN IN BLOCK 4, CONSTITUTE THE CURRENT VERSION OF THIS SPECIFICATION.

| 13. SCN NO. | 14. PAGES CHANGED (INDICATE DELETIONS) | * S | * A | 15. DATE |
| | | | | |

| 16. TECHNICAL CONCURRENCE | | | DATE |

FORM
DD 1 DEC 68 1696

*"S" indicates supersedes earlier page   "A" indicates added page

Figure 6.7-5 *Specification change notice*

6.7.4.4 Approved SCN

An approved SCN is used to transmit the change after approval by the contracting agency. It also provides a summary of pages affected by all approved changes. SCNs are not cumulative insofar as transmittals of previous SCNs remain in effect unless changed or canceled by a SCN of later issue. However, the summary of current changes is a cumulative summary as of the date of approval of the latest SCN.

6.7.4.5 Changed pages

Updated and reissued pages should be complete reprints of pages suitable for incorporation by removal of old pages and insertion of new pages. All portions affected by the change should be indicated by a symbol in the right-hand margin adjacent to and encompassing all changed portions. When change pages are issued for specifications with pages printed on both sides of a sheet, and only the page on one side of a sheet is affected by the change, both sides of the sheet should be reissued. The unaffected page side is reprinted without change and carries a date unaffected by the change. It should not be included in the change summary as being affected by the change.

6.7.4.6 Change numbering

SCN numbers are assigned in sequence, beginning with 1, against the original issue or current revision of a specification. Thus, when a specification is revised, the SCN numbers begin again with 1. The proposed SCN and approved SCN carry the same number. Once an SCN has been submitted to the contracting agency, its SCN sequence number should not thereafter be changed or assigned to another SCN, even if the SCN is never issued. However, SCNs may be approved by the contracting agency out of sequence. Hence, an SCN proposed after a previously proposed but not yet approved SCN may require revision if the later one is approved prior to the earlier one or an earlier SCN is not approved; in which case the numbers assigned will not change, although the contents of the change pages may require a change.

These rules on SCN numbering are a strong encouragement to get the requirements right the first time. If the requirements for several specifications are in a state of dynamic change, keeping track of the current requirements can become very confusing. A program should expend the customer's cash to get the requirements correct rather than changing them through ECP action.

6.7.4.7 Identification and numbering of changed pages

Each changed page is identified by the specification number and the applicable revision letter. The date of issue of the SCN is entered under such a number and must agree with the date entered in the upper-right-hand corner of the SCN form.

For example, assume that the current revision of the specification is A, the date of issue of such revision is 20 June 2005, and two SCNs have been approved. If SCN-2 is issued on 5 June 2006, the pages changed by SCN-2 would carry the following identification on each page:

18D4739A | 5 June 2005

The changed pages furnished with an SCN are numbered with the same page numbers as the pages they replace. If it is necessary to replace one page with more than one, the additional pages may carry the same number as the affected page plus a suffix letter in alphabetical order beginning with *a*. Thus, the numbers of changed pages to change page 5, would be 5, 5a, 5b, and so on. If a page is deleted, that num-

ber is omitted in the current page sequence. As an alternative to lettered pages, decimal delimited page numbering (point pages) may be used. In this case, the page numbering for the previously discussed change would be 5.1, 5.2, 5.3, and so on.

This point page requirement stems from MIL-STD-490A, but the author does not recommed it through revision of the document when an SCN would require point page use. A contract may require delivery of data responsive to Computer-Aided Logistics System (CALS) requirements, in which case the documentation is fully computerized and the physical configuration of the printed document with respect to what page data falls on becomes a function of the particular printout device and font used.

6.7.4.8 Revisions

A revision of a specification is a reissue of the complete specification and is prepared, issued, and identified the same way as the specification that it supersedes, except that the identification number is followed by an appropriate revision letter. Letters are assigned in alphabetical order for each succeeding revision. Revision letter *A* should be assigned to the first revision. Each revision should incorporate all outstanding approved changes against the previous issue, as well as approved changes proposed by the SCN that create the need for revision. Revisions of specifications will include symbols in the right-hand margins of the pages to indicate where changes have been made with respect to the prior issue, including changes. The following note will be included in the Notes, Section 6, of the specifications:

> The margins of this specification are marked with a vertical line to indicate where changes (additions, modifications, corrections, deletions) from the previous issue were made. This was done as a convenience only and the government ("company" in the case of procurement specifications) assumes no liability whatsoever for any inaccuracies in these notations. Bidders and contractors are cautioned to evaluate the requirements of this document based on the entire content irrespective of the marginal notation and relationship to the last previous issue.

One of the following notes may be used instead of the above, if applicable:

> Symbols are not used in this revision to identify changes with respect to the previous issue, due to the extensiveness of the changes.

or,

> Symbols are not used in this revision to identify changes with respect to the previous issue in the interest of cost and the type of word processing or desktop publishing used.

> Specification revisions are issued in the same way as the original issue and do not require an SCN for promulgation.

6.7.5 The Special Case of Interface Requirements Documentation

6.7.5.1 A Profusion of document names

In this Part we have discussed the content, organization, and style of a document called a product-unique specification into which pour the results from all of the product requirements analysis work. There are actually a lot of documents used in the development of systems that contain requirements besides these program-unique specifications, and we wish to be able to write appropriate content for all of them. A statement of work (SOW) is actually a set of process requirements that must be satisfied on a program. Program plans contain requirements that must be complied with in doing the related work. These are all valid outputs from a system requirements analysis process. This section will cover yet another kind of requirements document. The program-unique specification covers requirements for items in a system, whereas interface control documents define requirements for the relationships between the items in a system.

These interface documents have many different names and abide by no single standard, even in the Department of Defense. A very common acronym for these documents is ICD, variously meaning interface control document or interface control drawing. In this section, we will focus on the former, though the document we discuss may include interface control drawings.

Hardware ICDs are often prepared in the same two parts that we find in specifications, what MIL-STD-490A referred to as Part I and Part II, where Part I is for development requirements and Part II is for product requirements. If we were thinking in terms of MIL-STD-961E, these parts would, of course, be named "Performance" and "Detail." In the case of ICDs, the Part I document contains technical design-independent requirements that must be satisfied in the design while Part II contains design-dependent requirements that must be satisfied in manufacture and acceptance testing. The Part II document could be in the form of a collection of book style drawings containing reduced interface control drawings that define the physical and functional interfaces with dimensioned drawings appended to a few pages of descriptive frontmatter. Part I is sometimes called an "interface specification" and Part II an "ICD." Either may also be called an "interface requirements document (IRD)," though this name is often reserved for a very preliminary definition of an interface in the same way that a customer may provide a contractor with a system requirements document (SRD) as a precursor of the system specification that the contractor must create. The whole could be included in a single document that covers both development and product interface requirements with any of these names. In software development, the term *interface requirements specification (IRS)* is very common.

The practitioner should expect some variability in the naming of these documents, depending on the customer, but the content can be discussed generically, as we will do in this section. The outline suggested in this Chapter is based on my experience with several kinds of interface documents and will not necessarily be compliant with any particular standard with which you are familiar. The general term *interface control document (ICD)* will be used throughout the chapter as a generic name for these kinds of documents.

While engineers familiar only with hardware may find it hard to accept, computer software also is characterized by interfaces. The requirements for these interfaces are commonly defined in a document called an "interface requirements specification." These interfaces include software–software interfaces, as in the case of data that must be exchanged between two software modules; for example, software–hardware interfaces are also possible, as

in the simple case of compatibility of the software language selected with the computer processor hardware on which it runs. A more profound example of a software–hardware interface is found in the software that generates the elevator command signal in aircraft flight software and the elevator actuator to which it is applied. While you could claim that the software stops at the computer's edge as in its input/output (I/O) port, the software controlling the elevator must be built with an understanding of not only the static nature of the I/O port but the dynamic nature of the actuator and the elevator interacting with the air mass within the normal speed range and through allowable maneuvers. So, the people who design the software must understand the nature of a complete control loop with which they are interfacing, not just the I/O port address. I call these "extended interfaces," where the parties responsible must extend beyond the immediate item physical boundary condition.

6.7.5.2 Conditions of use

An ICD is used to collect and define the development and/or product requirements for a single interface between two objects or items in a system. Its use is most often driven by the kind of contractual relationship, or lack thereof, between two companies responsible for development of the items on the two terminals of the interface.

An ICD is seldom developed where the same organization is responsible for both terminals of the interface. It would be more likely that the interfaces were defined in the specifications at the terminals of the interface in this case. This is also the case where an item is being procured from a supplier. While one terminal is being developed by a vendor and the other by the prime contractor, the vendor is doing so within the context of a subcontract, so the same prime contractor is in control of both terminals, one of them through the procurement specification placed on contract and the other through internal documentation and management. In any case, where there is a contractual relationship across the interface relationship, that interface will most often be controlled by a pair of specifications, both under the authority of the prime contractor. It should be recognized that the word *prime* is used in a generic sense here. For any one procurement, the supplier is the subcontractor and the buyer is the prime contractor. In a given development project, this relationship could run several layers deep. There is only one prime contractor with respect to the ultimate customer, but there may be many prime-sub layers in the overall program, and the content of this chapter applies to any of these pairs, though it is more often applied in the higher layers.

The area where interface control comes under serious strains is where there is no contractual relationship across the interface. Assume, for example, that a customer contracts separately for an aircraft with Albacore Aviation, Inc., and the engines with Cycle 4 Propulsion, Inc.; there will be no contractual relationship between Albacore and Cycle 4 Propulsion. In the case of a prime-sub relationship, Albacore would have placed an engine procurement specification on contract with Cycle 4, and the specification would have defined the interface requirements with which the engine design would have to comply to ensure compatibility primarily with the aircraft structure design. But, in the case we wish to consider, they are coequal contractors with a common customer, and no contractual relationship between them.

To encourage cooperative work between these two contractors, the common customer should require the two contractors to execute a memorandum of agreement (MOA), letter of understanding, or similar instrument that clearly states the intent to cooperate and the means through which that will occur. This agreement should include acceptance of the following:

1. A joint ICD prepared primarily by one of the two contractors with content negotiated with the other
2. Implementation of an interface control working group (ICWG) made up of representatives from both contractors and the common customer through which interface concerns and changes are negotiated
3. An interface control or development plan giving procedures that they will jointly use to develop the interface in question in the context of the ICWG and ICD

6.7.5.3 Living and dying documents
There are generally two approaches to publishing interface requirements. Some customers prefer one approach and others the other. Recall that interface control documents define requirements for relationships among things in a system. Therefore, they define requirements for interface media within the system that may be composed of wiring harnesses, plumbing fittings and runs, signal formats, and so forth. The specifications define requirements for the things in a system, and part of the requirements for the things is their interface relationships. If we are not very careful, we could define the interface requirements for item XYZ with item UVW in the XYZ specification, the UVW specification, and the XYZ–UVW interface control document. You may think this is a good situation, but the reality is that we would have the same requirements in as many as three different places. As a result, we will violate an important principal in information management that encourages that any unique piece of information be located in only one place. The reason for this rule is that information intended to be the same in more than one place will invariably diverge over time because of the difficulty of maintaining that commonality.

Assuming the interface under discussion is one that would be managed through an ICD, there are two ways to avoid this problem. One of them entails a living ICD that remains effective on the program through the life of the program. In this case, the terminal specification references the ICD rather than repeat the interface requirements. We should recognize that the interfaces on the two items will often be different by virtue of male–female mating connectors, data transmission versus reception, or other unique unidirectional aspects; and, under the living ICD method, that single document will have to address these differences as well as the common requirements. As a result of these directionality differences, it is not a violation of the principal discussed above regarding double-booking the requirements.

The other method uses an ICD as a negotiating or management instrument to get to a condition of agreement between the parties as expressed in their independently maintained item specifications. As the name implies, the dying ICD passes from the scene when both contractors agree that they have defined the interface correctly in their terminal specifications. En route to this condition, the agreed-upon interface requirements may be removed from the ICD as they are moved to the pair of item specifications or left in the ICD and marked in some way to indicate that the terminal specifications are now the first authority on them. Theoretically, when the ICD content is reduced to a

void, it is no longer needed and passes from the scene. Thereafter, the requirements must be maintained in the specification pair.

For purposes of this section, let us assume that we will create and maintain a two-part ICD in the living mode unless otherwise noted.

6.7.5.4 Interface definition and document organization
The interface covered by the ICD must be very clearly defined. In a specification, we are interested in all of the interfaces that touch the item, and there may be several items involved. Because an ICD is focused on a relationship between two items, we can define the interface as a function of the two items. This interface could conceivably be a null, that is, no interface; a very rich relationship involving electrical, mechanical, fluid, and software relationships; or anything in between. If an ICD applies at a very high level in a system, the interface may entail hundreds of individual relationships in several different media between several lower-tier items on the two sides of the interface. If the ICD applies at a lower level of aggregation, there will generally be a simpler relationship in terms of the number of different media types and numbers of individual elements.

In the process of developing the system functionality and allocating that functionality to physical things populating the product architecture, the developer should achieve an optimum set of interfaces between these physical things characterized by minimized interfaces, especially where there are different organizations on the opposite terminals of the interface (minimized cross-organizational interface). This is achieved by the way we allocate functionality and the way in which we assign responsibility for development of the things. We should also find that there is no operating functionality within the interface media; rather, all of it should be in the architecture items.

As a prerequisite to developing the ICD, we should apply an organized method for identifying all of the interfaces in the system being developed. Chapter 3.6 offered two good tools for this purpose, schematic block diagramming and N-Square diagramming. In the former, we draw a block diagram where the boxes are the items in the system from its architecture diagram and the interconnecting lines indicate a needed interface relationship. In the latter case, we build a square matrix with n boxes shown on the diagonal and mark the intersections to indicate which ones are characterized by a needed interface relationship. Figure 6.7-6 illustrates both of these techniques (a for an N-Square diagram and b for a schematic block diagram) for the same interface. More people are probably able to use the schematic block diagram more directly than the N-Square diagram, but the latter is much less trouble to create and maintain, and therefore you would expect it to cost less to use. For the purposes of supporting a discussion of ICD organization, let us assume we are going to develop an ICD for the ICD plane identified on Figure 6.7-6 between architecture items A11 and A13. Item A11 might, for example, be an aircraft engine separately procured by the customer and item A13 the onboard computer that will control the engine and many other system items.

The author recommends that an ICD include a schematic block diagram illustrating the interfaces controlled by the ICD and that each interface element be numbered in some way so it can be coordinated with the corresponding paragraph numbers. One alternative for identification of interfaces in ICDs and specifications is to use the engineering

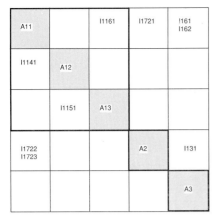

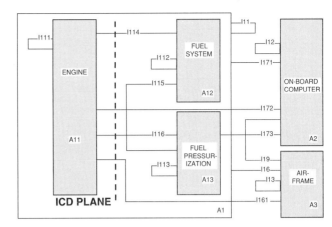

a. N-Square Diagram   b. Schematic Block Diagram

Figure 6.7-6 *Interface definition*

system interface IDs, shown in Figure 6.7-6 directly in the document interface diagrams, but this can result in confusion. Figure 6.7-7 offers a compromise where the lines are annotated with the paragraph numbers and also parenthetically identified with the engineering notations and interface names.

It is much better to use a simpler numbering system where the numbers on the diagram can be coordinated with the paragraph numbers containing the corresponding requirements. In a very simple case entailing ten interface elements, the lines between the items on the interface figure could simply be numbered from 1 to 10 and the paragraphs could include these same numbers in the lowest-order paragraph numbering. If, for example, the interface requirements paragraph is 3.2, the interface requirements paragraphs would be numbered 3.2.1 through 3.2.10.

In complex situations involving many interfaces within several kinds of interface media, we could use two or more layers of paragraph numbering. In more complex situations it may be necessary or desirable to group the interfaces into collections and number the individual interface paragraph numbers within paragraph numbers for the groups. For example, we could establish partitions for item partitions as we did in Figure 6.7-7 and media partitions for mechanical, electrical, and fluid interfaces. Figure 6.7-8 offers a tabular linking of interfaces, their engineering identifications, and the paragraph numbers, recognizing a media grouping scheme (not coordinated with Figures 6.7-6 and 6.7-7). As a further organizing aspect, we might, in this case, also have partitioned the interfaces into physical and environmental classes. Note the coordination between the engineering identifications and the paragraph numbers. However we group interfaces in the ICD or specification, a particular interface may require several subparagraphs to define all of the relevant characteristics.

6.7.5.5 Interface terminals and media

Interfaces have three characteristics, two terminals and a medium, between them. Some examples of these characteristics are:

1. A radio signal is transmitted from a directional antenna on the ground (terminal 1), passes through the atmos-

phere and space (the medium being radio waves in space), and is captured by a receiving antenna on a spacecraft (terminal 2).

2. A 28-volt DC level is applied to a wire by the I/O port of the onboard computer (terminal 1) in accordance with stored program action and causes current flow through a wire in a wire harness (medium) interconnecting the onboard computer and the fuel tank pressurization control valve, resulting in current flow through the valve solenoid, valve activation (terminal 2), and pressurant flow into the fuel tank.

3. The rotor of the electronic ignition system in a Lincoln Mark VII is driven by a gear on the camshaft. Terminal 1 is the gear on the camshaft, the medium is the direct contact between the gears, and terminal 2 is the rotor gear and shaft. As a result of this action and its effect on an electronic circuit and the coil primary, a spark is generated at the corresponding spark plug by the coil secondary via a high-tension cable (medium) running through a distributor cap conductor (terminal 1) to the spark plug terminal (terminal 2).

4. As a liquid-fueled space launch system lifts off from Cape Canaveral, tons of liquid oxygen rush under pressure from the liquid oxygen tank (terminal 1) to a turbine-driven turbo pump (terminal 2) via a large diameter LOX duct (medium) on its way to the engine oxidizer inlet.

In the first example, the medium was in the system environment, but in the other three cases the medium was within the system. Note that an ICD can control the medium where it is part of the system but can only recognize the nature of the medium where it is part of the environment. While working as a field engineer for Teledyne Ryan Aeronautical at Bien Hoa Air Base, South Vietnam, in 1968, the Air Force detachment commander asked the author to help him write a message to Strategic Air Command (SAC) Headquarters about a material problem. Depending on the season, the unit's unmanned reconnaissance aircraft were returning from overflight of North Vietnam with fiberglass leading edge damage on wing tips, horizontal tips, and vertical stabilizer tip. The conclusion was that it was hail damage, and

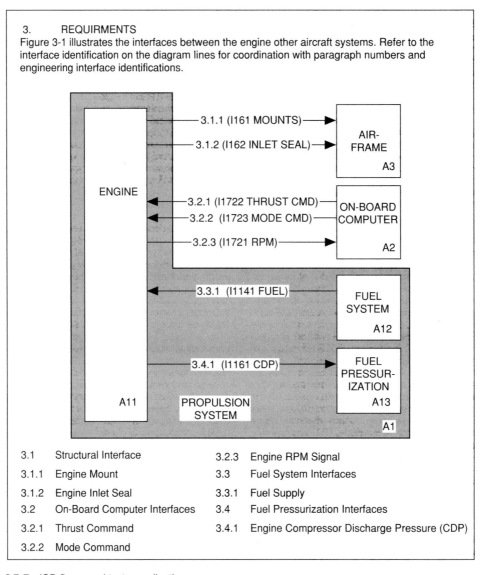

Figure 6.7-7   *ICD figure and text coordination*

a simple message was crafted asking for correction of the material deficiency. Shortly thereafter, a message was received from SAC Headquarters, which you may recall was located quite close to Heaven, reading as follows: "There shall be no hailstorms in the path of the vehicle."

6.7.5.6 System environmental interfaces

We speak commonly of interface influences when considering the effects of the interface on the system and of interface impact when considering the adverse effects of the system on the environment. These are simply the possible directions of the environmental interface elements and part of the complete environmental interface. There are five envi-

ronmental subsets that should be covered at the system level: natural, hostile, cooperative, noncooperative, and induced. Of these, the natural, hostile, induced, and noncooperative elements should be covered under environmental requirements. The cooperative environmental elements should be included in the external interface requirements.

All of these classifications relate to the kinds of relationships among parties responsible for the terminal pairs. A natural element has the natural environment on one terminal and our system on the other. A hostile element makes a connection between a source of hostility (enemy aircraft or computer security cracker) and our system. A noncooperative element joins a source of influence from outside the sys-

ENGINEERING IID NUMBERS	ICD FIGURE NUMBERS	ICD PARA NUMBERS	PARAGRAPH TITLE
I15		3.1	Physical Interface Definition
		3.1.1	Mechanical Interfaces
I1531	1.1.1	3.1.1.1	Mechanical Interface 1
I1532	1.1.2	3.1.1.2	Mechanical Interface 2
		3.1.2	Electrical Interface
I1541	1.2.1	3.1.2.1	Electrical Interface 1
I1542	1.2.2	3.1.2.2	Electrical Interface 2
I1543	1.2.3	3.1.2.3	Electrical Interface 1
I1544	1.2.4	3.1.2.4	Electrical Interface 2
		3.1.3	Fluid Interface
I1551	1.3.1	3.1.3.1	Fluid Interface 1
I1552	1.3.2	3.1.3.2	Fluid Interface 2
I1553	1.3.3	3.1.3.3	Fluid Interface 3
I1554	1.3.4	3.1.3.4	Fluid Interface 4
I16		3.2	Environmental Interface Definition

Figure 6.7-8   *Two-layer media-partitioned interface definition*

tem that is not purposefully interfering with our system but simply is sharing the same operational space. An example of such an element is the electromagnetic interference generated by other systems as a function of their normal operation.

An induced environmental element begins life as a system environmental effect, commonly an energy source, that excites some aspect of the natural environment that, in turn, has an impact on some item in our system. An example would be the tremendous acoustic noise generated by the rocket engines of a Titan IV space launch vehicle while rising off the launchpad. This acoustic energy then bombards the vehicle and its parts during rise off. This noise would not be applied if the system were not present to induce it.

All of these environmental effects have one thing in common, and that is that there is no one with whom we can negotiate to characterize the related requirements. We can only do our best to name the influences and bracket the extremes of their influences. It is appropriate to include these as lines on our system schematic block diagram, but they should be included in a specification or interface control document under environmental influences.

The cooperative system environmental subset involves the relationships between our system and other systems with which it must cooperate. In these cases, there will be another party with whom we can negotiate those relationships as interfaces and arrive at agreements on their nature and appropriate values. These environmental influences should be treated as interface requirements and so captured in our specifications and ICDs as appropriate to the items and interfaces.

6.7.5.7 Transforming lines into requirements

The first step in developing ICD content is to identify the interfaces that must exist. Schematic block and N-Square diagramming are useful tools toward this end; however, these interfaces thus identified are really driven by the way we have chosen to associate system functionality with the things in the system, and it is the functionality of these items that we must consider when constructing the schematic block diagram of the system. We must select particular interfaces defined on our system schematic block or

n-square diagram for coverage in an ICD based on the nature of the contractual relationship, anticipated development difficulty, and interface complexity. The piece of the system interface view should then be transposed to the ICD in question. Finally, we need to refine the interface definition at this plane to show every individual interface at the lowest possible level. This should include every plumbing line, every separate wiring function, and every mechanical connection. Each of these most elementary interfaces should be represented on our schematic block diagram as a directed line segment.

For each line on the ICD interface figure (or marked n-square intersection), we must define the necessary characteristics based on the selected technology. If electrical wiring is the medium, we can describe the requirements in terms of voltage, power, frequency, pulse rate, or network protocol. If fluids are involved, appropriate units could include pressure, flow rate, viscosity, or pressure stability. A mechanical attachment interface can be defined in terms of forces acting at those points in three axes, torque to be applied to attaching bolts, or polarizing and alignment methods.

It is very difficult to define the requirements for interfaces until you define the medium technology, and to do this some preliminary design work is necessary, as is true for all design constraints (interface, environmental, and specialty engineering requirements). Some system engineers feel that all development requirements should be defined before any design work is accomplished. While this is possible in the case of performance requirements, it is just very hard to do in the case of design constraints. For example, it would be possible to complete a pressure-sensing interface between a fuel tank and a pressure control unit in a number of ways. The pressure transducer could be physically located within the tank and electrically connected across the interface by wiring. Alternatively, the pressure sensor could be located in the control unit and connected by plumbing to the tank. It is true that this interface could be characterized as simply a pressure range to be measured without having yet defined the intended design partitioning of the function, but it will be necessary to refine this interface definition more precisely if we are to procure the control unit and possible separate pressure sensor from vendors. The procurement specifications or source control drawings will have to identify the precise interface implementation in terms of line pressures or electrical characteristics, based on the media completing the interface.

6.7.5.8 Document organization

There are many sources of recommended outlines for ICDs, which is to say that there is no real universal standard for them. There are some fundamentals suggested by all of these standards that should be covered, however.

6.7.5.8.1 Hardware ICD   The author taken the liberty of fashioning a hardware ICD structure, shown in Figure 6.7-9, based on many ICDs observed over the years. Section 1 is introductory in nature. Section 2 provides references to documents called out in the text of the document. Section 4 contains requirements for proving that the requirements in Section 3 have been satisfied in the design created in response to the requirements. The content of Section 4 should be used as the basis for test and analysis planning work to be accomplished on the program with respect to the interface covered by the ICD. The author tried to stay with a six-section document because of his own background. Section 5 in this case has no utility and would be a void, or one could make it a traceability section, as does the interface requirements specification structure recommended in MIL-STD-984/EIA-J-STD-016 for software. Finally, Section 6 can be used to capture any kind of information about the interface thought useful that is not covered elsewhere.

PARAGRAPH NUMBER	PARAGRAPH TITLE
1	Scope
1.1	System overview
1.3	Document overview
2	Applicable documents
3	Requirements
3.1	Interface identification and diagrams
3.2	Physical interface requirements
3.3	Functional interface requirements
3.4	Other Interface requirements
3.5	Requirements traceability
3.6	Precedence and criticality of requirements
4	Verification
5	Not used
6	Notes

Figure 6.7-9   *Hardware ICD outline*

Obviously, the outline has been influenced by the six-section style encouraged by military specifications. There is nothing immoral or illegal about using this structure, even for a person interested in systems with a commercial flavor. The part that we are primarily interested in is Section 3, the requirements, but any formal document needs additional content captured here in five other sections. The author has seen three-section and seven-section ICDs, and there is nothing wrong with any of these arrangements. The important point is that the interfaces should be well defined and their requirements carefully identified and controlled.

In hardware interfaces, it is not uncommon to collect them into physical and functional classes, as suggested in the outline. Depending on the interface, there may be other functional categories of interest as well that could be

either captured in paragraph 3.4 or singled out in their own paragraph number, depending on their importance or complexity.

6.7.5.8.2 Software Interface Requirements Specification   MIL-STD-498 (which became EIA J STD 016) and related data item descriptions define a good software interface specification structure. While this is a military standard, and people in commercial practice may take exception to this advice, you should know that DoD approved this standard for only a limited period of time until the IEEE and others could fashion it into a commercial standard. Thereafter, DoD planned to discontinue the military standard and apply the commercial standard in their own procurements. When this book was being written, this military standard had transitioned to EIA-J-STD-016 and is a commercial standard.

PARAGRAPH NUMBER	PARAGRAPH TITLE
1	Scope
1.1	System overview
1.3	Document overview
2	Reference documents
3	Requirements
3.1	Interface identification and diagrams
3.X	Project-unique identifier of interface
3.Y	Precedence and criticality of requirements
4	Qualification provisions
5	Requirements traceability
6	Notes

Figure 6.7-10   *Software ICD outline*

In this structure, paragraph 3.1 would include the schematic block diagram (or reference a more extensive coverage in an appendix) with interfaces numbered as discussed above for coordination with the paragraph numbers in Section 3. Each of these N-1 interfaces is then treated in a different paragraph numbered 3.2 through 3.N in place of the 3.X in the outline. Paragraph 3.Y (where $Y = N + 1$) gives information on the relative criticality of these interfaces if necessary. Please note that the author has capitalized only the leading word in the outline, as is done in MIL-STD-498, even though the author's personal preference would be to capitalize each word in the paragraph title.

You should be very careful in Section 4 to tune in to the customer's accepted meaning of the words validation and verification. In software, the word *validation* is often used to mean the process of developing the evidence of compliance with the content of Section 3. A hardware person would refer to this as "verification," and in MIL-STD-961E, covering program-peculiar specifications for the military, Section 4 is titled "Verification."

6.7.5.8.3 Mixed ICD   Some system interfaces at some level will entail both hardware and software elements, and some of these interfaces may have to be covered in an ICD at a sufficiently high level to include both hardware and software elements. The paragraphing structure should recognize either the functional or the structural organization arrangement. Software may be one of these high-level functional partitions or be a functional subset of one or more structurally arranged entities. Within each software subset, the specific interfaces should be covered as discussed above.

In a very complex interface like this, it may be useful to use a very-high-level schematic block diagram in paragraph 3.1 and include a more detailed expansion of each interface in the subordinate paragraphs, which number the interfaces for coordination with paragraph numbers.

The author has, perhaps, made too much of paragraph numbering in this chapter, but interfaces are by their nature complicated to define and discuss. Anything we can do to make it perfectly clear what we are talking about in interfaces is worth the effort. Use of an illustration coordinated in some way with the paragraphs where the details are found will be very helpful.

6.7.6 Electronic Style Guide

6.7.6.1 Documents of the past

MIL-STD-490A does not address preparation of specifications for electronic delivery via computer networks. Some customers had already begun to show an interest in this mode of delivery at the time this book was published. At some point in time, DoD and NASA will evolve a mature standard to cover such delivery. You or your company may wish to begin to implement this capability either with your government customers or your suppliers before the standards reach maturity. If so, you should establish liaison with other companies, trade associations, and federal government agencies working toward mature standards so that you can stay current in a very dynamic field.

Prior to government or commercial entity issue of mature standards, you may also tailor the data item descriptions for contract data deliverable specifications to permit some changes to MIL-STD-490A or MIL-STD-961E preparation instructions that are particularly difficult with low-end word processors.

The advantage of using a common word processor, such as Microsoft Word or Word Perfect, is that a large population will have the skills to use the capability directly, and you will have more options in the preparation and change of your documents. If you use a high-end word processor or desktop publisher, you will probably have to provide your team with a centralized typing pool, which can become a bottleneck. A common word processor can allow your engineers the freedom, if they choose to use it, to make small changes themselves. It can also enable real-time specification review methods using computer projection.

Electronic specification viewing and delivery capability requires a compromise between a good style for paper publication and computer screen viewing. Four specific changes are listed below that encourage electronic viewing and delivery. There are many others that a company standard could

elect as well. These changes then can become a source for tailoring for MIL-STD-490A or MIL-STD-961E and associated data item descriptions called out in contracts:

1. *Specification Number Headers.* The header containing the specification number shall not be required to include the revision letter appropriate for the page.
2. *Point Pages.* Point pages, where decimal-delimited pages are introduced between normally numbered pages, are to be discouraged because of the difficulty in handling these structures in many low-end word processors. The whole interest in page numbers should be deemphasized to allow object-oriented treatment of content and automatic tables of contents.
3. *Title or Paragraph Flow-in.* The first line of the paragraph text need not necessarily begin immediately after the paragraph title on the same line, in the interest of taking full advantage of outlining and table of contents features of some word processing programs at minimum cost.
4. *Change History Data.* The change history, where required, is not required to refer to the numbers of the pages on which listed paragraphs are located.

6.7.6.2 Database-generated specifications

A company is well advised to move into database use in capturing requirements, and the tools used should be capable of generating the final specification without recourse to subsequent formatting. Computer database systems can be programmed to output the document in any form desired, including Microsoft Word, for example, by inserting rich text format strings along with the data. Even an old DOS system can be made to generate Windows-compatible documents. The advantage to this capability is that the information can be efficiently retained in database form where traceability, verification, and other relationships can be easily maintained, but the data can be viewed in any organization desired anywhere in the company through network connections. This topic is more fully covered in Part 7.

6.7.6.3 The end of the paper specification

MIL-STD-961E permits the use of specifications that exist in models. In this book we have covered many models of use in gaining insight into the identification of appropriate requirements. Rather than translating these requirements into specification formats, we may use the models directly as specifications. It may be some time before DoD updates its procurement and program management practices to enable this alternative, however.

7

Computer
Applications

Contents

7.1

The Computer Tool Infrastructure

Contents

7.1.1 Why Have We Waited So Long?

Throughout the previous chapters, the notion of computer applications kept arising; but an attempt has been made to keep the computer out of the earlier material as much as possible to focus on the human process, because it is people using their brains who develop requirements. Also, because everyone who uses this book will not be so fortunate as to have easy access to a computerized requirements analysis and capture tool, the author believes that those without computers should be able to get some benefit from the book. If, by the way, you are in this condition, without a computer tool, you should recognize that you are probably falling behind your competition. Figure 7.1-1 illustrates an evolution going on in system development that is accelerating over time. Dates offered are notional and not based on serious historical research or thoughtful contemplation of the future.

7.1.2 Evolution of Methods

Throughout the 20th century, specifications were generally published using typewriters or word processors and reproduction or typesetting and printing. The content of these specifications was often determined through ad hoc approaches using specification standards templates. In the 1960s, people in large customer and development organizations began experimenting with the use of computer databases to capture the requirements. Commonly, the database was used as the source for printing paper specifications as well as making it possible to capture good traceability data and to implement other useful management activities. That is, one can implement the document-driven approach using typewriters, word processors, or computer databases. The specification content can be determined through an ad hoc or a structured analysis process.

As databases became more common, it became obvious that it was not really necessary to publish paper documents, in that you could observe different views of the require-ments directly on the computer screen and always know that you were looking at the latest version of those requirements. In the isolated database-driven approach illustrated in Figure 7.1-1 the specifications are not published in paper form. Some of the commercial tools available include analysis support machinery (like functional analysis or behavioral diagramming), permitting structured analysis within the tool; but, in general, the content of these databases can be created using structured analysis or an ad hoc approach, as in the document-driven approach.

The database-driven approach is actually serving as a transition between the document-driven approach and a new approach called "model driven." The model-driven approach commonly includes a requirements database; but, instead of its content having to be entered separately from the source tools for many of the requirements, the database is linked to source tools so that the whole system of tools implements the fundamental principle of a good information system that any one unique piece of data should be located in only one place under the responsibility of one authority. In a database-driven environment, the reliability requirement values for each item must be copied from a reliability math model, where they are determined by a reliability engineer, into the text structure of the requirements database in a form such as, "3.X *Reliability*. Item failure rate shall be less than or equal to 0.0001." In a model-driven environment, the failure rate number is linked from the reliability database, and whatever it happens to be at the time is what appears in the specification exposed to print on paper or the workstation screen. There are some obvious advantages in this direct linking of tools, one of them being the elimination of double booking, which cannot be maintained effectively.

Additionally, in the model-driven approach, all of the models and simulations through which appropriate values of requirements and useful design features are discovered and verified are linked together interactively. As this is

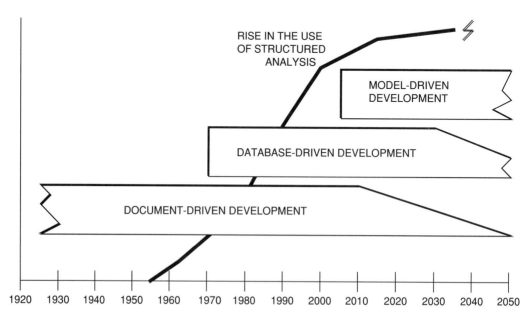

Figure 7.1-1 *Evolution of system development*

being written, there are no complete tool sets one could purchase and implement on a program painlessly. Rather, they had to be assembled from off-the-shelf computer tools by the using organization with added scripts and code to accomplish the integration of tool data.

The development organization that is in some peril today is not the one that has not yet perfected model-driven development; it is the one that, regardless of the driving approach, has not yet mastered the application of structured analysis as the basis for capturing the requirements that populate their specifications and/or databases. It is unlikely that an organization in this condition can leapfrog all the way to model-driven development as some developing societies have leapfrogged the evolution of wired phone systems by installing a cell phone system providing good communication service to an otherwise impoverished land. As soon as possible, such an organization should begin its growth toward structured analysis and database support of that process.

7.1.3 The Computer Tool Environment

If you are blessed with computer tools at your place of work, you or your company management may have been disappointed that there was not an immediate improvement, commensurate with the cost of the tool set, in the organization's ability to do requirements analysis work. To gain full benefit from effective tools, you must have an effective requirements analysis process to begin with, one that the engineers understand how to perform. Tools also have to be well matched to your requirements analysis and system engineering processes and practices.

The promise of the computer as an aid in requirements analysis and specification development was grasped early by visionary system engineers in many organizations, at a time when the only machines available were mainframe computers with 80-column Hollerith punch card input/output. The analysts would enter the results of their work on an 11″ by 17″ paper form that, after approval, would then pass on to the keypunch operator and into the computer through a card reader. Notice the distance between the analyst and the computer content. The owners of the data, in this case, were the computer people, locked away from normal humans in an elevated room—apparently to reflect their exalted status, unapproachable by mere mortals. Mere system engineers could only press their faces to the window through which one could view the wonderful computer.

Early computerized requirements database tools were created for mainframe and minicomputers before networked desktop computers and workstations were available in the numbers we now find in industry. Some of these also predate the desktop graphical user interface (GUI) concept popularized by the Macintosh and Microsoft Windows. Many early tools were derived from tools built for the development of large-scale weapons systems like intercontinental ballistic missiles. Some of these were capable of supporting functional analysis. As computer software evolved into an increasingly significant part of systems, system engineers attempted to apply computer-aided software engineering (CASE) applications that provided overkill on one version of structured analysis and underachievement in applicable documents analysis, constraints analysis (interface, environmental, and specialty engineering), verification, traceability, and specification publishing.

Tom Demarco, an early pioneer in software structured analysis, wrote in *Structured Analysis and System Specification* that he would prefer not to use a computer tool to do the analysis because it removed him too far from the data. The generation Demarco and the author are from might find it difficult to do structured analysis on the computer screen rather than with pencil and paper because there were no computers when we were growing up and entering the workforce. Any structured analysis was performed with pencil, paper, and eraser. The author admits to a preference for doing structured analysis this way today, no matter what structured analysis method is being used. It is likely that people entering the workforce in the early 2000s have a very different experience, having come of age immersed in computer interaction from grade school onward. They have a very different hand–eye coordination relationship than do people from earlier generations and can probably perform very well using the screen, mouse, and keyboard.

Some people in aerospace are concerned about the people joining engineering organizations in aerospace companies in the early years of the first decade of the current millennium, as reported in *Aviation Week* and *Space Technology*. The concern is that the human mind is wired most intensely in the vicinity of age 5, and if the mind is overloaded with visual stimulus at that time the imagination is not fully developed. Contrast the child growing up in the 1930s, listening to the radio for the latest "Lone Ranger" story, with the child of today, bombarded with visual stimulus from television and video games. These concerns may or may not be valid; but if true, it is not going to be easy to correct the problem. Just how does one teach creativity? At the same time, the visually stimulated child may turn out to be exactly the right person to perform in the computer-centric program environment of today.

Throughout this part, we will use dBase-III+ structures in examples because the author is familiar with the product, there is a very large population of engineers also familiar with dBase, and the programming code exposed in a few examples is fairly intuitive, requiring minimum additional explanatory text. Almost any good database product can be used in much the same way.

The material from which this book was drawn was developed concurrently with an effort to develop an effective requirements analysis tool set that began with dBase-II on an Apple IIe and passed through dBase-III+ on an IBM-PC and on to other application programs in a networked environment.

7.1.4 Requirements and Specifications Electronic Environment

Let us see if we can sketch an overall environment within which several computerized tools and computer-aided processes may be used by requirements analysts, system engineers, and a specification publishing agent in an efficient workgroup computing arrangement. Figure 7.1-2 illustrates the suggested environment, and the blocks are explained below.

The suggested environment supports all four requirements analysis strategies introduced in earlier chapters (structured analysis, cloning, question and answer, and freestyle). In the structured strategy, requirements analysts input requirements against architecture elements into the requirements database via a series of structured analysis tools. Within the database they are linked by traceability

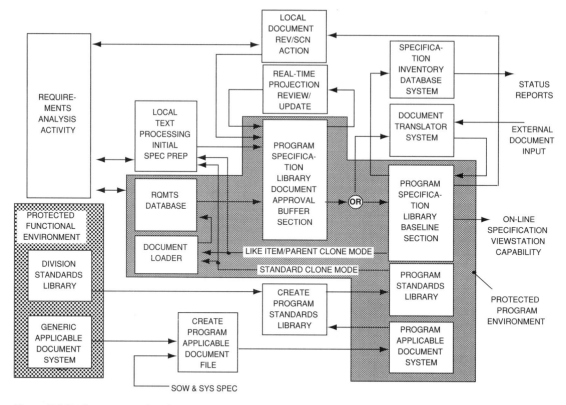

Figure 7.1-2 *Computer tool environment*

codes and related to verification events. Requirements may also be entered directly into the database using the freestyle strategy.

The three cloning strategies of boilerplate, parent-item, and like-item are supported by loading any one of these source documents into the database from the electronic specification library against a particular architecture ID. The database content is then edited, based on the requirements analysis activity, and linked up for traceability purposes.

A specification publisher converts database content for any architecture item into an appropriate specification format. The document in electronic media is retained in a review buffer, where it is exposed to peer review by engineers using markup software followed by real-time markup in a specification review, during which the document is video projected directly from the computer in a meeting room. Changes directed by the decision maker can be made immediately in the database and regenerated in specification format subsequent to the review. While database content can be reviewed directly in the same way with some savings, some people cannot relate to a set of requirements unless they are in the familiar specification format.

Approved documents are passed by a release group function to the electronic library, where they may replace a prior revision or add to the documents stored. Documents may be printed from the library, viewed on workstations (including a window of a CAD terminal), or datalinked to the customer or supplier.

If a program team concludes, after a condition of maturity is achieved, that they need no longer maintain the database with its overhead of traceability and verification data, they could choose to switch to a totally document-oriented environment using word processors to maintain documents in the library and cloning to generate any new ones.

It is not uncommon to have to begin the requirements analysis activity on a program with a system requirements document or system specification provided by the customer. So, our environment may also use the database specification loader to read these documents into the database.

7.1.5 Networking and Workgroup Computing

The computer application examples covered in this book are drawn from a single-user application that will work on a single program where a single engineer does all of the requirements work. This is a very uncommon situation, except possibly on the concept-development phase of a project. On large programs in the engineering and manufacturing development phase, involving distributed requirements analysis responsibilities, we need a much grander solution. Such a program needs a networked system capable of multiuser operation that encourages concurrent development of the requirements among the several specialists contributing.

Some engineering organizations have a central-system engineering group to do all requirements analysis and specification development. Others use a distributed approach, where each principal engineer or integrated development team is responsible for their own requirements analysis work, with a system engineer being named the principal for the system level and radiating downward to the level where the architecture breaks up neatly into design group or integrated development team responsibilities.

The author believes that the distributed approach is the wave of the future and that the networked computer will be the tool that assembles the gaggle of good specialized engineers back together into the equivalent of one great, all-knowing engineer. The networked computer will permit the many specialists to concurrently and rapidly blend their specialized knowledge while evolving an appropriate set of requirements for an item. Then this team will press on to concept development and a design for the item that is responsive to the requirements.

Computer-aided specialty engineering tools interlocked with computer-aided design tools will advise the design engineer immediately when he or she has broken a specialty design rule, such as locating an electronic component junction too close to a heat source on a circuit board layout. It will not be necessary to await a human specialty engineer to discover the design fault during a drawing sign-off review. The principal application of specialty engineering will be to support the design of the specialty tools, but these engineers will also have to interact with the designers on things that no one has been smart enough to introduce into the tools.

7.1.6 A Basic Requirements Database

We have said at a number of points in this book that we should be using a computer database to capture our requirements rather than a word processor file. The logic behind using a database is that many of the interrelationships we have found useful (such as traceability, verification, sourcing, and rationale data) can be linked directly to a single source of requirements in the database.

When the requirements are contained in word processor documents, it is very difficult to interrelate this data without keeping two sets of books that can easily become divergent. When the very tool we are using to maintain traceability, a database for example, becomes divergent from the requirements themselves (in word processor documents), it creates great concern for the integrity of the complete data set. A well-designed database requires us to hit the keys only once for each unique piece of information, so our data cannot diverge.

This is all very well; but, in order to pull it off, it is necessary to be able to print the specification from the database.

Most readers will accept that it would be possible to hook up much of the data we have discussed in the previous chapters through a database, but some may not be ready to accept how easy it is to make a database print the specification. We know that the requirements in a database are stored in fields and records, not in integrated documents. How then can we bring all of the right fragments together to form a recognizable specification?

In the future, requirements documents may be unnecessary. Rather, we may use database content directly. But, for the time being, we cannot gain maximum benefit from a requirements database solution unless we can print specifications directly from the database, simply because our customers expect them.

So, let us set up a simple database structure using quantified requirements types to illustrate how easy it is to do this. This should set us free to continue to build on the database concept without fear of failure on this point. Table 7.1-1 provides a list of fields that defines a simple dBase-III+ requirements database. For those not familiar with computer databases, imagine the fields as the column headings in a table and the records as horizontal lines of data in the table. Unfortunately, it is convenient to switch this relationship in Table 7.1-1 in order to get all of the data on the page easily. You may wish to reread the last two sentences to make sure you understand the record and field relationships expressed in Table 7.1-1.

The ARCH_ID identifies the architecture item the requirement relates to and, in combination with the REQ_ID, defines a unique record. The REQ_ID (requirement ID) is used to establish traceability between requirements. For a specification, we are interested in all of the requirements with a given ARCH_ID. For each requirement we have identified a PARAGRAPH number, TITLE, ATTRIBUTE to be controlled, a numerical VALUE, and the UNITS of measure. The SIZE column tells how many characters there are in the field, and the TYPE column tells whether the field is a character (C), numerical (N), memo (M), or date (D) type. The RECORD N through RECORD N+2 columns each list a typical set of data. The table includes three requirements (keep in mind that they would be horizontal records in our picture of the structure rather than in the columns of data in our sample table). Two of these requirements are for item A136 and one for parent item A13.

The following dBase-III+ command structure will string these together, using the data listed under RECORD N, into this specification paragraph:

> 3.5.5 *Reliability.* Item reliability shall be greater than or equal to 0.998.

Table 7.1-1 *Sample database structure and data*

Field name	Size	Type	Record N	Record N+1	Record N+2
ARCH_ID	10	C	A136	A136	A13
REQ_ID	4	C	13	131	23
PARAGRAPH	10	C	3.5.5	3.2.6	3.2.1
TITLE	20	C	Reliability	Weight	Range
ATTRIBUTE	20	C	Reliability	Weight	Flight range
VALUE	8	N	0.9888	1200	2000
RELATION	1	C	$\geq$	$\leq$	$\geq$
UNITS	20	C	None	Pounds	Nautical Miles

DO CASE

    'CASE RELATION= '>'

        STORE 'greater than or equal to' TO prel

    CASE RELATION='<'

        STORE 'less than or equal to' TO prel

    CASE RELATION='='

        STORE 'equal to' TO prel

    OTHERWISE

        STORE '_____' TO prel

ENDCASE

? 'TRIM(PARAGRAPH)+' '+TRIM(TITLE)+'. Item ';

+TRIM(ATTRIBUTE)+' shall be '+prel+' '+TRIM (STR(VALUE,8))+'.'

By putting this command structure in a loop that looks only at requirements keyed to architecture ID A136 (SET FILTER TO ARCH_ID='A136') and indexing the database on paragraph number, it is possible to print a whole specification for the item from the database. The requirement for A13 and all others keyed to architecture other than A136 will be ignored. The program can be set up to print to a printer, to go to a computer screen or be projected on a meeting room screen for review, or to download to a computer text file that can subsequently be electronically integrated with graphics prior to printing.

This command string works for quantified requirements only, so other cases (DO CASE) would be required to handle requirements statements that contain applicable document references and text- and header-only types. This requires some additional fields in the database structure as well. The following fields, shown in Tables 7.1-2 and 7.1-3, will allow us to handle the other two types of requirements statements.

Table 7.1-2

Field name	Size	Type	Typical data
TEXT		M	text material
DOCUMENT	20	C	MIL-STD-456A
TYPE	1	C	See below

Table 7.1-3

Type code	Meaning
N	Numerical
T	Text Only
H	Header Only
R	Document Reference

The database command files then are arranged to detect the TYPE field content and run cases unique to each type of requirement statement. The program in each case picks out the requirements data from the database and mixes it with some standard specification language to produce a normal sentence. The output of such a specification generator will not win any prizes for its literary qualities, but we want just the facts in a specification anyway. Dull, technically correct sentences are much preferred to poetic stupidity here.

The print code listed above does not include special rich text format code, so the text would be displayed as ASCII. Introducing the RTF codes appropriately in the strings will result in data that can be opened directly in Microsoft Word or any other preferred word processor.

7.1.7 Traceability Hooks

We can create a traceability capability by building a second simple database with two pairs of fields identical to the ARCH_ID and REQ_ID fields of the requirements database. One of these pairs is for the child requirement and the other pair for the parent requirement. Table 7.1-4 offers a sample printout from such a database for two A136 requirements listed in Table 7.1-1. The weight and reliability requirements are traced to A13 requirements for weight and reliability. The REQ_ID numbers (codes) do not have to follow any particular order so long as they are unique within each architecture item requirements. The traceability database simply hooks up unique pairs of ARCH_ID and REQ_ID codes.

A particular requirement may be traceable to one or more parent requirements and from one or more child requirements. Our traceability database must, as a result, handle a many-to-many relationship. The simple flat file system described should handle this. The system should, of course, provide a simple, intuitive screen to allow the engineer to easily build and edit these traceability structures. The traceability database could also include other data, such as traceability analysis notes, the name of the engineer who established the traceability entry, the data, a status field, and a reference to an analysis report.

You may recall that we said earlier in this book that every requirement should theoretically be traceable to the original customer need statement. The traceability database system should be provided with a utility to trace from any given requirement until it finds an open link, arrives at the customer need statement, or dead ends at an associate interface requirement. Any trace that ends in a dead end prior to one of these terminal conditions should be reviewed in depth to find out why the traceability terminated. This is a wonderful audit tool for use where the requirements analysis process is distributed.

7.1.8 Verification Tracking Tool

A single paragraph containing multiple requirements can generate a dozen verification events. Figure 7.1-3 illustrates one method of organizing verification data for database design. We assume that the requirements are in a database,

Table 7.1-4   Sample traceability data

Parent arch_ID	Parent req_ID	Child arch_ID	Child req_ID	Other data
A13	18	A136	131	
A13	141	A136	13	

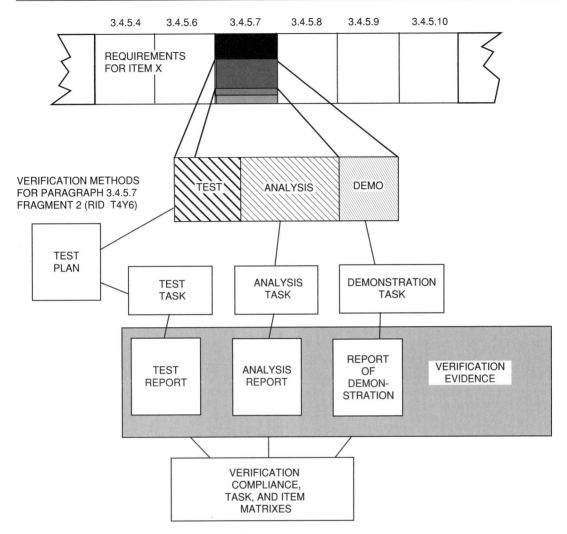

Figure 7.1-3   *Verification tracking links*

as discussed above, to which we can now link the verification data. Because one paragraph may contain more than one requirement, we need to be able to fragment, or parse, paragraph text into single requirements. This is commonly done by assigning REQ_IDs to each fragment. Let us use a VER_ID field to focus on each unique requirement. When we write requirements properly in our own excellent company, there will be only one VER_ID for each paragraph, but we may have to verify the requirements in someone else's specification (our customer's).

One VER_ID may have to link to more than one verification event, as defined by the verification method. When the verification method is assigned, it creates a verification record that maps to either a test or an analysis event (disregard other types for the present). All test verification events are picked up by a test engineering function. Test engineering must integrate all the test events into an inte-grated test plan that minimizes test program cost and schedule demands.

The requirement author or a test engineer creates a verification test requirement that flows into the integrated test planning activity and into Section 4 of the corresponding specification. Test planning and report data and analysis report data can be linked, using our VER_ID, to the corresponding requirement ID.

7.1.9 Requirements Management Data Fields

You can probably already see how to expand our simple database to manage the requirements analysis process. There follows in Table 7.1-5 a list of some fields that will prove useful. These have to be woven into the infrastructure of the application programs, in some cases to control operation of the database.

Table 7.1-5 *Management data fields*

Field	Sample data	Sample field data meaning
STATUS	A	Requirement approved
INITIAL DATE	01-01-05	Date requirement created
REV DATE	10-08-05	Date last revised
REV	B	Revision letter
ORIGINATOR	JONES	Engineer who created the requirement
LAST REVISER	BURNES	Engineer who changed it to Rev B

7.1.10 External Model Hooks

A simple demand-driven specialty engineering mechanism can be created using a combination of a database and a spreadsheet application program (such as 123 or Excel). The database contains the fields listed previously, plus a field CAT that identifies requirements categories. Let us say that we use the character W in the CAT field to denote a weight requirement.

In the requirements database we create a record for each architecture item requiring a weights figure. Then we need a spreadsheet overlay for weights with columns equivalent to the quantitative database fields. The weights engineer is able to load his or her weights spreadsheet from the database using a utility. Within the spreadsheet application, the weights engineer can assign weights figures and play "what if" games, observing the effects driven by spreadsheet formulas.

When these manipulations are complete, the weights engineer refreshes the database from the spreadsheet, updating the requirements marked W in the CAT field. The weights engineer must respect the discipline not to add or delete rows in the spreadsheet; rather these actions must take place on the database side. This process may be repeated indefinitely until engineering management baselines the data. Database baselining can be done across the whole database or by architecture, with additional controls applied to this refresh capability.

This same pattern can be repeated for other quantifiable specialty engineering disciplines, such as maintainability, reliability, and cost with other spreadsheet models. The result is a database concept like that shown in Figure 7.1-4. A simple, central database with user-friendly features is surrounded by specialized spreadsheet tools supporting, and under the control of, specialized analysts. Commonly, companies have more complex specialty engineering tools than simple spreadsheets, but they too can be linked to a requirements database to provide the same effect discussed.

External model hooks are facilitated by the separate numerical field in our database. If you capture the quantitative data mixed in with the text of your requirements, you must include a separate numerical field to enable this mechanism. A separate field can diverge from the same data contained within the text. The computer-generated text concept in the database structure we have studied was inspired by a need to link up to specialty databases and preserve the concept of "one place for each unique piece of information."

This set of tools can be used in a demand-driven fashion by respecting the discipline that requirements can be added and deleted only in the database and edited only in the spreadsheet or other specialized tools. The principal engineer for an item or a system engineer for the whole system can place a demand for specialty engineering work by sim-ply adding a requirement with the right CAT code. The database can be provided with a status utility that allows the specialty engineer to check the database for requirements of his or her category that are currently voids.

7.1.11 Traceability to Process

We have shown how to include requirements traceability in the database between requirements and between requirements and verification events. It may also be required by a customer or desired by a world-class engineering organization to be able to capture the traceability of the requirements to the process used to gain insight into them.

It is a simple matter to capture the traceability to source reference documentation by including a REFERENCE field. A RATIONALE field can be used to allow the analyst to annotate the requirement record with background on why the value was selected or why that requirement is needed, and a SOURCE field can be used to reference a document that covers the requirement source. A NOTES field could even be included to capture a running log for each requirement.

It may also be necessary, however, to link up to the structured analysis processes used to identify performance requirements and constraints. This is not so easy because it requires that you also have a database system for each one of these structured processes. Given that you have them, you can establish links between these databases and the requirements database.

For example, let us accept that we have a functional analysis database that includes a record for each function illustrated on the functional flow diagrams. This database will be used to contain the function dictionary and timeline data as well as function title, description, and initiating and terminal events. Let us also have an architecture database that includes a record for each architecture item on our architecture block diagram. This database will capture the architecture dictionary and provide for other useful data. Finally, we need a database that allows us to allocate the functions to architecture.

This third database includes ARCH_ID and FUNCTION_ID fields. When we allocate a function to an architecture element, we create a record in this database, and each of these allocations should place a demand on our requirements analysis database for someone to write one or more performance requirements corresponding to this allocation. In the process of forcing this performance requirements analysis work, we can include a HOOK field in the requirements analysis database with the FUNCTION_ID. When a function is allocated, it can create a record in the requirements database with the ARCH_ID of the item to which it is allocated, the HOOK corresponding to the function from

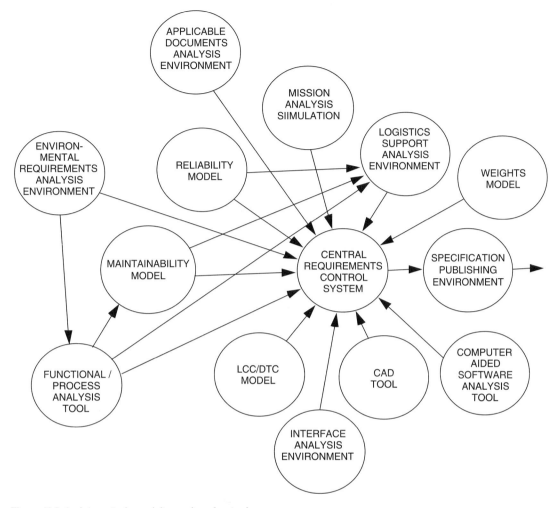

Figure 7.1-4 *Integrated specialty engineering tools*

which it was allocated, and a status field entry signifying that the requirement has been allocated but not converted into a performance requirement.

Note that we have discussed here the allocation of functions to architecture. Back in Part 3, we said there were two alternatives to functional allocation. First, we said we could allocate raw functions. Alternatively, we said that it would be desirable on early studies for unprecedented systems to be able to allocate complete performance requirements instead of functions. The latter is more difficult to implement in a database, but possible. It is left for you to work out a data concept to accomplish this using the examples provided up to this point.

Given that we have an interface dictionary database whose records mirror the lines on all of the system schematic block diagrams, we can hook interface requirements for a specification to this database because we have all of the requirements coded by ARCH_ID, and each interface in the interface database is coded by the ARCH_ID of its two architecture terminals.

There is an added complication for interface requirements. We may have to print interface requirements for either an ICD or a pair of specifications. Within the interface dictionary we should identify, for a given interface, which of these two methods will be used. For an interface using ICD capture, the requirements should flow into our requirements database, keyed to in interface ID rather than an architecture ID. Otherwise, they should flow into pairs of requirements linked to the terminal element architecture IDs.

Environmental requirements can also be driven by a structured database concept. This requires considerable structure to fully implement, however. We will need databases that capture the process–architecture and process–environment relationships introduced in Part 3. The data contained in the combination of these databases provides us with environmental information about each architecture item across its use profile, as defined in the process and environmental databases. Our system should have a utility that allows us to view the aggregate environmental definitions associated with an item based on the processes within which it must function and the environments associated with those processes. The utility should allow us to identify the selected range of the variable based on worst-case or other rationale.

This will be adequate for defining environmental requirements for prime items in the system. We may wish to push further to provide a prime item zoning capability. We cut up the item, as discussed in Part 4, into zones of like environment and assign environmental parameters to those zones based on the item environment previously defined. We then need a database for assigning component-level items to zones of their parent item. The corresponding zone environment becomes a part of the requirements for the item mapped to the zone. You may want to try to apply the dataflow diagram (DFD) concept covered in Part 4 to defining data relationships for such a database concept.

Similarly, we can hook our requirements database to specialty engineering databases where engineers can find all of the backup data the specialty engineering requirements are based on. As discussed earlier, we can also link up the numerical content of models used by specialty engineering disciplines so that we have a unique piece of information in only one source.

Chapter 3.10 offers a concept where every requirement in the database is linked to the structured analysis model from which it was derived. Whether these links are automated or simply refer to graphical data captured in a system definition document as described in Chapter 3.11, we would have traceability to the structured analysis process.

7.1.12 Data Integrity

The combination of interrelating requirements, verification, and process requirements traceability is very powerful, so long as the creation and maintenance of the data is fairly simple to do and can be accomplished during the development process as the development team works its way down through the expanding architecture. To capture this data without an integrated database, where the requirements are in word processor files, would require several independent databases.

You could never be sure, with multiple entries for the same data in different databases, that needed changes would ripple through all of the databases with different people in control of each database. With a central or federated database structure, where a particular piece of data resides in only one place, you have much better assurance of data integrity. The watchword for computer databases is, "Only hit the computer keys once for any one unique piece of information and only store that unique piece of information in one place."

Every piece of information in a database should have associated with it a proper source organization or engineer, and only that organization or engineer should be allowed to change it. In the case of requirements, we know that we have specialized engineers who look across the whole of the architecture from one narrow perspective, such as weight, reliability, or maintainability. By the same token, principal engineers focused on architecture items need to be able to integrate all of the requirements for that piece of architecture provided by a host of specialists. The database concept must provide an environment within which people working these two axes (requirements specialty and architecture entity) can interact in an organized way to identify and refine the requirements for system elements.

7.2 Computer Tools for Requirements Analysis

Contents

7.2.1   A Little History

Early attempts to build requirements database tools on DoD programs involved the use of mainframe computers using 80-column Hollerith cards for input that were punched by keypunch operators from paper forms prepared by analysts. On USAF ballistic missile programs, these forms and their content were defined in painful detail in AFSCM 275 series and specific program SRA standards. These systems left much to be desired, especially because they placed such distance between the analyst and the data. At that time, the mainframe computer space was treated as a separate country, complete with border guards at a hole in the wall through which one would pass their punch cards or forms.

Distributed computing using networked microprocessors in desktop computers and workstations opened the door to development of new requirements database tools that operated directly on the analysts' machine. This resulted in analyst ownership of the data for which they were responsible, a tremendous improvement in the process of specification development. The reader of Tom DeMarco's excellent book *Structured Analysis and System Specification* finds dataflow diagrams hand or template drawn rather than generated by a computer modeling tool. DeMarco felt, at least at the time the book was written, that the use of computer tools to do the modeling separated him too far from the data. The author identifies with this attitude because he grew up with no experience with computers and his hand–eye coordination experiences were derived from pencil and paper work. Thus the author prefers to do the modeling work with a pencil and paper and enter the resulting requirements in a big dumb database where they can be linked and used to generate specifications. Engineers coming into the workforce now had hand–eye coordination experiences in association with computers and video devices from a very young age. It is entirely possible that these engineers will be able to accomplish effective modeling work using hand–eye coordination linked with mental activity using a keyboard, mouse, and screen.

7.2.2   Buy or Build

At one time there were few good requirements database systems available on the market, encouraging some companies to build their own. Indeed, many of the tools now commercially available came into being through this route. Texas Instruments marketed their in-house tool, leading eventually to the formation of a separate company, TD Technologies, marketing the tool under the name SLATE. Ivy Hooks and David Hoffman, while employed by a consulting company, developed a tool for use in their work with NASA, leading to creation of a separate company named Compliance Automation, with Hooks as president, that markets Vital Link.

General Dynamics Space Systems Division (GDSS) created a requirements tool over a period of several years in the 1980s. The systems engineering organization started with a dBase III+ system because they had many engineers familiar with the code and could understand and repair problems fairly quickly. The first generation was built with great support for configuration control and approval of changes, only to find that in practice it was not possible to get any work done. The CM features were detuned, and the tool provided some useful service. The next generation was built in Omnis 5, both for the human interface and for the backend database because that application would allow precisely the same human interface on both Macintosh and IBM machines. A third generation used Omnis 7 for the human interface on both Macintosh and IBM machines but switched to Sequal Server for the back end. In 1992, GDSS went so far as to make a marketing videotape touting the tool's benefits in preparation for a potential sales campaign. This opportunity passed from the scene when the vice president for research and engineering viewed the tape and decided that the division made rockets, not databases. Perhaps this was just as well, in that some people said that in this case the movie was better than the book.

The motivation for moving toward outside sales was to fund further development of the tool for internal use in an environment dominated by great difficulty in gaining support for continued development funding. This brings us to the principal downside of building your own tool. The person responsible for development becomes excessively focused on continued development at the risk of not paying sufficient attention to other matters. There is also no one to blame when the tool fails to function and no outside agent who can be brought in to fix the problems. Many companies start down the road to building their own database tool internally, motivated by cost, thinking that they cannot afford the $10,000 to $40,000 per-seat cost of a first-class tool. Over time, if they were honest with themselves and management, however, they would find that they could easily have invested over $1,000,000 in their tool and still not have improved on the tools available commercially.

The upside of building your own system includes:

1. The tool should be perfectly matched to the preferred process and specification formats. This advantage is partly mitigated because good commercially available tools have a variable schema that permits identification of additional fields and relationships.
2. You would have intimate knowledge of the workings of the system and presumably a resultant ability to train your personnel in its use.
3. Building a tool is an excellent first step with the intent to move to a commercial tool as soon as possible. In the process, you will gain a clear understanding of features you want in a commercial tool.

7.2.3   Available Tools and Their Features

Today there are several excellent requirements tools on the market, and they are improving every year. Their cost is initially a turnoff to some engineers and managers, but one should make a decision on total cost. Will the tool reduce the cost of creating specifications? And when might it be expected to break even compared to continuing with current practices and tools? If your organization were using word processors to create specifications, conversion to a database tool would cause an initial slowdown due to the learning curve. However, a good tool would soon allow you to overcome this temporary problem and move on to significant productivity improvements, coupled with many other desirable features that are just very difficult to do without a tool, such as traceability and verification program linkage. Alternatively, these tools will permit you to do a much better job of requirements analysis and management for no increase in cost. Traceability is very tedious to implement when using word processors but relatively simple to implement using a good database.

7.2.3.1 CORE

CORE is a product of Vitech, based in Vienna, Virginia. It was developed by James Long and his sons and employees, based on work he had accomplished while employed at TRW in Huntsville, Alabama, as well as a long career as a system engineer in industry. The history of the early system engineering work done by the U.S. Army at the Huntsville center and TRW in support of it, when finally written, will provide a rich story of original thinking. CORE uses enhanced functional flow block diagramming as the principal model but can be used to support computer software development as well. CORE includes a graphical modeling capability linked to database content and is one of the few that does so among system engineering tools.

7.2.3.2 DOORS

The dynamic object-oriented requirements system (DOORS) was developed by Richard Stevens, who had worked in aerospace industry in England for many years. QSS marketed DOORS in the United States until the company was acquired by Telelogic in 2000. DOORS is a repository for requirements that have been determined through external methods, human thought, and analysis. That is, the tool does not include a built-in problem space modeling capability but is a fine repository for requirements that are derived from all models implemented in other media.

7.2.3.3 RDD-100

For years, RDD-100 was the top-of-the-line tool for requirements analysis. It included a sophisticated front-end modeling capability with system simulation capability. It was a powerful tool in the hands of a qualified person but somewhat difficult to maintain currency with unless used on a regular basis. The tool was the creation of Mack Alford. He applied a variation on the software model Input/Process/Output (IPO) referred to as behavioral diagramming combining process flow and data flow. The producer ran into financial troubles in 2000 and ceased operation. The residual of the company was acquired by a company with the intention to return it to the market at some point.

7.2.3.4 SLATE

This tool was originally developed by Texas Instruments (TI) for use internally. It matured into an excellent tool that was marketed externally. TI spun off the tool component into a new company called TD Technologies that was subsequently purchased in 2000 by SDRC.

7.2.3.5 RTM

Marconi originally developed RTM for use internally on its development projects. Integrated Chipware purchased the tool and continues to market it. Note this trend for companies, often spun off from a company with a nontool product line, to be bought up by relatively larger tool companies so that the requirements tool is but one tool in the overall product scheme. This is probably a positive sign that will support the move to model-driven development.

7.2.3.6 Other requirements tools

In addition to those briefly described above, other tools have been offered in the past with good features; but some of these departed the scene before becoming competitive. Space Systems Division of Rockwell Aerospace developed an excellent tool named Cassets that operated over Microsoft Office. Anyone familiar with Office could very quickly become proficient in this tool and maintain currency easily over time. When this component of Rockwell was purchased by Boeing, the tool was withdrawn from the market.

PRC developed a pair of tools reflecting the way the USAF preferred to perform system requirements analysis on intercontinental ballistic missile systems along the lines covered in the chapters on traditional structured analysis of this book. One of these components was called Spec Writer. PRC awarded the principal developer, Archie Vickers, rights to the system in 1998, and he continued to support the system through his consulting enterprise.

7.2.3.7 Software modeling tools

There are many computer software modeling and analysis tools that maintain discipline in the modeling environment and even generate code based on the model created and offer some testing capabilities. Software through Pictures is one, and Rational has tools supporting OOA and UML. These tools are generally not well-suited as a general program requirements management application.

7.2.4 Features Not Generally Supported

Commercially available tools continue to improve rapidly, but there are some features that the tool companies have been slow to implement. Some of the tools covered above that are still available include a modeling environment where the problem can be modeled in a functional flow, IDEF-0, state transition, behavioral, or dataflow diagram that encourages identification of performance requirements. Functionality is identified and allocated to architecture items, placing a demand for a performance requirement based on that allocated function against the item to which it was allocated. Most of the tools listed do not provide this capability but provide a sound database system into which requirements derived from various manually implemented models can be entered, linked through traceability, and generated into specifications.

7.2.4.1 Design constraints identification

Available tools generally do not provide models adequate to encourage identification of appropriate design constraints: interface, specialty engineering, and environmental requirements. Physical system schematic block diagramming or N-Square diagramming to which block diagram lines or marked N-Square intersections required interface requirements identification for the pair of items thus identified would be helpful. Some of the ideas in the environmental requirements analysis chapter involving service use profile and end item zoning could be employed to support environmental requirements identification.

The specialty area is very extensive, and different disciplines need different kinds of support that lead to another area of commercial tool shortcomings addressed in the next paragraph. One could appeal to the old USAF notion of a design constraints scoping matrix, covered in Chapter 3.8, but all that would do by itself is to encourage the right disciplines to write requirements about specific items in the system architecture.

7.2.4.2 Tool linkage

Many tools are structured such that a builder consultant can provide interfaces between their tool and others that the client wishes to use in combination. This may be especially helpful between separate hardware and software requirements tools. But, ideally, tools would provide links to the specialty engineering tools used for mass properties, reliability, maintainability, and life cycle cost math models, to name a few. These requirements are in the form of numbers in the source models, and these values should communicate into the requirements database without rekeystroking. We should be able to easily update the specification model from selected specialty engineering databases.

7.2.4.3   Primitive capture and numerical content

One of the things that stands in the way of specialty tool coupling noted above is that the requirements values in specification statements are generally retained mixed with text in an ASCII string. While it might be possible to seek out these ASCII values for the numbers, it would be difficult to differentiate between numbers used in requirement quantification and for other purposes in the text. An alternative approach would be to capture requirements in primitive form where we capture paragraph number, paragraph title, characteristic name, numerical value, and units plus a trailing text field. It is possible, as shown in the prior section, to string these components together to automatically form a complete sentence.

Once you have the numerical value in a separate numerical field, it could be interacted with other numerical fields for margin and demonstrated design values, permitting the tool to search for management space when problems develop in satisfying particular requirements. Margin accounts could be maintained well in context with the actual requirements values rather than in some double-booking scheme. But the additional powerful capability availability would be to permit refreshing the requirements database numerical fields from corresponding specialty engineering models for baseline updates—an end specialty engineering requirements double booking.

7.2.5   Implementation Suggestions

7.2.5.1   Overcoming use difficulties

One of the most serious problems with the use of requirements tools is the difficulty of acquiring skill in their use and maintaining that skill with less than regular use. Some companies have tried to move to a particular tool on a grand scale but have failed to reach their goal for this reason. Some companies have perfected a good compromise in the use of the very best tools, which tend to be difficult to master. They train a relatively small population to use the tool in its full capability, and these people accomplish the requirements analysis work on programs, entering data into the tool or acting as data-entry persons. The rest of the population is provided read-only access to the data through a Web page. A click on a requirements icon brings up a list of all of the documents in the system. A click on one document in a list brings it up on the screen in a word processor document format rather than a multiscreen database system.

This same approach can be implemented for any other application, of course. The beauty of this solution is that the operators of the application can use the very best applications available, updating to take advantage of new capabilities, while users may gain access with a generic easy-to-use application. Everyone does not have to learn and maintain skills on everyone else's applications.

7.2.5.2   Networking

Your tool should be located on a server and operable from multiple workstations. You should wire not only your workspaces but also your meeting rooms for network connection, so that a computer can be located in the meeting rooms, supporting computer projection of database content for use by cross-functional teams doing real-time concurrent engineering on the requirements as they are developed and subjected to changes. This capability also enables projection review of specifications.

8 Closing

Content

8.1

Where Have We Been?

Contents

8.1.1 What Is the Essence of Our Story?

System engineering is an organized method for decomposing a large problem into a series of smaller, hierarchically arranged problems and the integration of the solutions to these smaller problems into a solution for the large problem. The need for this seemingly complex approach has been spawned by the tremendous amount of knowledge available to humanity, the complexity of our problems, and the knowledge limitations of the normal human beings working to solve those problems within an environment of competition and customer cost and schedule constraints.

8.1.1.1 Teamwork and Concurrency

The need for specialization has exploded in step with the expanding scope of humanity's knowledge, leading, in some companies, to serial work patterns to the detriment of our efficiency. We need to find ways to glue the all-knowing design engineer back together. Figure 8.1-1 illustrates this problem. Many years ago a design engineer could master the complete field of design, at least within one domain. In simpler times, the problems design engineers drew would often surrender to a single engineering discipline (mechanical, electrical, and so on). As specialties grew, as a result of knowledge growth, and each expanded its knowledge base, it became impossible for designers to master their own specialized trade and these others as well. The result has been to create a tremendous communications problem for the design engineer.

We have suggested ways in this book to bring about a condition of teamwork and concurrent development of the product design, production, verification, and sustainment processes. These methods have the effect of bonding the several specialists together into a team that is once again capable of an enlightened singleness of purpose that only a few older design engineers may still remember tales of.

We have emphasized that the development organization should seek to first understand the requirements, evolve alternative concepts supportive of those requirements, select the most effective concept, refine the requirements, and proceed to design. The identification and definition process should, ideally, work in a downward direction within a moving band of architecture under the control of

a PIT Leader or chief engineer by whatever name. Once the requirements are known, the design process will generally work best in a bottom-up direction for hardware. Software can be defined and implemented in about any direction.

8.1.1.2 Developmental Directionality

A development organization should consciously choose their preferred developmental directionality. In this book, we have emphasized flexibility in that selection with a preference for the top-down direction. This was based on the notion that we should work from the simple to the complex, from the general to the specific. The alternatives of middle-out and bottom-up offer advantages in narrow circumstances, but our structured requirements analysis approach was focused on the top-down approach. It is not entirely clear to me how to organize an equivalent comprehensive structured requirements analysis process for the bottom-up case when dealing with the hardware aspects of a system.

8.1.1.3 Multiple Requirements Analysis Strategies

We have discussed four requirements identification strategies and indicated the utility of each. While a program may choose to enforce one of these strategies, they may also elect to permit local selection within some constraining rules. My opinion has been clearly stated as favoring structured analysis in some form. In particular, as this is being written, the author would encourage traditional structured analysis for systems and hardware entities, as explained in Part 3, and UML for software, as explained in Part 4. In the not-too-distant future, these two modeling approaches will merge into a truly universal modeling approach with tremendous HW–SW integration benefits for those organizations that can master it. The work has been started by International Council on Systems Engineering and Object Modeling Group with their release of SysML Version 1.0.

8.1.1.4 Demand-Driven Requirements Analysis

We have tried to find ways to encourage analysts to write requirements in response to a demand characterized by primitive attributes. Each of these demands can be keyed to a specific structured modeling artifact and a specific person, team, or department, who must then respond with a value

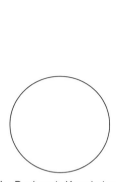

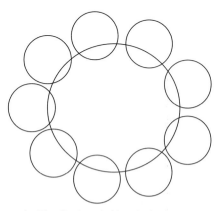

a. The Designer's Knowledge
Base In Earlier Years

b. The Designer's Knowledge
Base Today

Figure 8.1-1 *Putting Humpty Dumpty back together again*

or qualitative phrasing to complete the requirement. This helps to remove an impediment to getting started.

8.1.1.5 Progressive Requirements Writing

Many engineers are confused by the language they see in specifications. In an attempt to strip away all of the trivia, we have emphasized a progressive writing scenario. First, we use structured analysis techniques to identify attributes. Then we apply domain-specific analysis, allocation, models, or good engineering judgment to determine a sound value for the attribute. Finally, we complete the sentence in a format that is consistent with a customer's expectations defined in a specification standard.

8.1.1.6 The Computers Are Coming!

No, they are here! Engineering organizations must find out how to employ them to rebuild the engineering team and glue the designer back together. The computer offers us a communication tool to blend our minds together into a powerful engine of creativity in a concurrent development environment. We have also tried to show how computers may be used to provide broad access to system requirements and a ready repository for the work of many analysts. As industry moves toward model-driven development, our computers will become linked together through networks, forming a nervous system for program teams working on the parts of the system development effort that extends our human communications system, consisting of speech augmented by reading and writing.

8.1.2 Overcoming Impediments to SRA Success

The two greatest impediments to successfully implementing an effective system requirements analysis capability in an engineering organization are: (1) a top-management-supported notion that design groups should have autonomy to protect their creativity and (2) a narrow, religious interpretation of the system-engineering process as a centralized activity by the members of a system engineering department. These two attitudes are poles apart, and closure on a solution cannot be attained until these two buttresses are bridged or knocked down.

The first step must be taken by the system engineering community, and that is to apply self-discipline and educate its own members in a flexible, distributed approach to system engineering. The next step is to gain understanding on the part of design groups and engineering management that SRA is a systematic method for doing what they probably call "a process of deriving requirements."

Ask some design managers, particularly those most hostile to the notion of a systems approach, to sketch out the process their engineers use to derive requirements. You will get a lot of different ideas but seldom a systematic, closed-loop process description. It will be orally presented by the manager because it is not written down anywhere.

Explain that the requirements derivation work that his or her engineers do can be done in an organized way or an ad hoc way at about the same overall cost. If you can then get the point across that the 3000 avoidable engineering changes subsequent to CDR on a recent program were a symptom of lack of up-front requirements analysis within a systematic environment, you are on the road to recovery. The final step is to help the design groups through a requirements analysis training phase tailored to their needs (do not force them to sit through hours of the stuff system engineering cultists adore). In Chapter 1.3, this book offered a simple eight-step prescription that could be useful in educating the staff.

SRA has been applied to some programs very well in the past, but it has also been applied in ways that consume forests, recycle money, and drive engineering managers to early retirement. It should not be so. Applied with skill and sensitivity, it can liberate the design engineers to soar to the creative heights they deserve within a protective framework that prevents failure. The rigid application of an SRA process has been likened to placing the participants in an old-fashioned zoo with cages. This book proposes the equivalent of the San Diego Wild Animal Park, where the animals roam free of cages, as an alternative. You may say that this is a distinction of little substance because we are still talking about a zoo. But perhaps we have already carried that analogy too far.

We have to understand that the development of complex systems is the most difficult technical undertaking humanity has conceived. It may never surrender to the beautifully simple situation described in the management literature. The chief executive officer tells his or her staff to bring their department manuals to the next board meeting. As they arrive for the meeting, the CEO's assistant takes the manuals, throws them into a 55-gallon drum, and burns them to a crisp. As the staff sits down at the table wringing their hands in anguish, the CEO passes out a two-page summary of the new company policy. There are companies that can and actually do operate this way. The development program for putting humans on Mars will not be run like this.

System development organizations can choose to operate in an autonomous, internally competitive fashion until their customer base has escaped; or they can, as they see fit, convert to a fully concurrent approach emphasizing teamwork. But it is unlikely, as the competitive economic squeeze becomes more binding, that the old style ad hoc organizations will be among the survivors.

When you cast an eye toward some countries where capitalism has been economically embraced without benefit of the rule of law and the protections contained in the U.S. Constitution, you might wonder whether unrestrained competition might produce a human condition as bad as the forms of totalitarianism humanity defeated in the 20th century. The author believes that we will be sufficiently creative in the structuring of development methods and work that humans will not become slaves to system development but rather will come more to enjoy the experience because we will unload much of the tedious busywork into computers and expose the system engineer to the interesting and challenging problems humans are most efficient in dealing with, those that require human thought and creativity.

Acronyms

ABD	architecture block diagram
ASIC	application specific integrated circuit
ASCII	American standard code for information interchange
C4ISR	command, control, communications, computers, intelligence, surveillance, and reconnaissance
CAD	computer aided design
CRD	critical design review
CEP	circular error of probability
CMM	capability maturity model
CONOPS	concept of operations [document]
COTS	commercial off the shelf
CPM	critical path method
CRL	concept requirements list
DBS	drawing breakdown structure
DDP	development data package
DET	design evaluation test
DFD	dataflow diagram
DoD	Department of Defense
DoDAF	Department of Defense architecture framework
DTC	Design To Cost
DT&E	development test and evaluation
ECP	engineering change proposal
EDD	enterprise definition Document
EFFBD	enhanced functional flow block diagram
EID	end item description
EMC	electromagnetic compatibility engineering
EMD	engineering and manufacturing development
EMI	electromagnetic interference
ERD	entity relationship diagram
EW	electromagnetic warfare
FAA	Federal Aviation Administration
FCA	functional configuration audit
FDA	Federal Drug administration
FFD	functional flow diagram
FMECA	failure modes effects and criticality analysis
FRACAS	failure reporting and corrective action system
FRAT	function requirements answers test
GIDEP	government industry data exchange program
GFP	government furnished property
GPS	global positioning system
HW	hardware
ICBM	intercontinental ballistic missile
ICD	interface control document
ICWG	interface control working group
IDEF	integrated definition [language]
IMP	integrated master plan
IMS	integrated master schedule
INCOSE	international council on systems engineering
IOC	initial operating capability
IOR	Inclusive OR
IPO	input process output
IPPT	integrated product and process team

IRFNA	inhibited fuming nitric acid
ITER	international thermonuclear experimental reactor
IV&V	independent verification and validation
JCS	joint chiefs of staff
LCC	life cycle cost
LSA	logistics support analysis
MID	model identifier
MOE	measures of effectiveness
MRA	manufacturing requirements analysis
MTBM	mean time between maintennce
MTRR	mean time to remove and replace
MTTR	mean time to repair
NASA	National Aeronautics and Space Administration
OOA	object oriented analysis
ORD	operational requirements document
OT&E	operational test and evaluation
PERT	program evaluation review technique
PFD	process flow diagram
PID	product identifier
PID	prime item development [specification]
PIT	program integration team
PMP	parts materials and processes
PSL	program specification library
QFD	quality function deployment
RAS	requirements analysis sheet
RDD	requirements driven development
RFP	request for proposal
SADT	structured analysis design technique
SBD	schematic block diagram
SCN	specification change notice
SDD	system definition document
SDR	System Design Review
SE&I	system engineering and integration
SOW	statement of work
SRA	system requirements analysis
SRD	system requirements document
SRR	system requirements review
SRS	software requirements specification
SW	software
TAF	test analyze fix
TBD	to be determined
TBR	to be resolved
TLF	time line diagram
TPM	technical performance measurement
TQM	total quality management
TSA	traditional structured analysis
UML	unified modeling language
USAF	United States Air Force
VPM	verification planning analysis
WBS	work breakdown structure
XOR	exclusive OR

Index